America's Energy Gamble

How can America get back to an energy transition that's good for the economy and the environment? That's the question at the heart of this eye-opening and richly informative dissection of the Trump administration's energy policy. The policy was ardently pro-fossil fuel and ferociously anti-regulation, implemented by manipulating science and economic analysis, putting oil and gas insiders at the helm of environmental agencies, and hacking away at democratic norms that once enjoyed bipartisan support. The impacts on the nation's health, economy, and environment were – as this book carefully demonstrates – dire. But the damage *can* be reversed. Ordinary Americans, civil society groups, environmental professionals, and politicians at every level all have parts to play in making sure the needed energy transition leaves no one behind. This compelling book will appeal to course instructors and students, government and industry officials, activists and journalists, and everyone concerned about the nation's future.

SHANTI GAMPER-RABINDRAN is an associate professor at the University of Pittsburgh, with a Ph.D. in Economics from the Massachusetts Institute of Technology and an M.Sc. in Environmental Management and BA in Jurisprudence, both from Oxford University where she was a Rhodes scholar. She served as the August-Wilhelm Scheer Visiting Professor at the Technical University of Munich, Germany, at the Department of Environment and Climate Policy. She is the contributing editor of *The Shale Dilemma: A Global Perspective on Fracking and Shale Development* (University of Pittsburgh Press, 2018), which received critical acclaim.

"With *America's Energy Gamble*, public policy expert Shanti Gamper-Rabindran lays out a stark case that powerful oil and gas interests have, with considerable help from the outgoing Trump administration, gained control of the lever arms of our energy and environmental policy apparatus. Our economic competitiveness, the health of our environment, and the livability of our planet are all now threatened. Read this book to be informed about the threat and armed with the knowledge of what can be done in the Biden era to undo the damage and right the course."

– Michael E. Mann, Distinguished Professor of Atmospheric Science at Penn State University, and author of *The New Climate War*

"*America's Energy Gamble* deserves a wide audience. It makes two important contributions to our understanding of the Trump era. As the first book-length treatment of Trump's aggressive environmental deregulation, it thoroughly exposes the tenuous moorings of that campaign, including its shaky connection to its economic goals. Equally importantly, it reveals the affirmative side of Trump's agenda: not just opposition to business regulation of every kind (though that was a factor), but the vision of fossil fuels as a route to national prosperity."

– Daniel Farber, Sho Sato Professor of Law at the University of California, Berkeley, and author of *Contested Ground: How to Understand the Limits of Presidential Power*

"The book traces how US decades-long policies to favor oil and gas extraction, while running rough shod over communities in 'energy sacrifice zones,' paved the way for the Trump administration's destructive policies. It highlights how the administration's anti-science and anti-democratic decision-making perpetuated the grip of the oil and gas industry when economic prudence and human survival demand a transition to renewable energy."

– Daniel Kammen, Professor at the University of California, Berkeley, and Senior Advisor for Energy, Climate & Innovation, US Agency for International Development

"In detailing the actions and consequences of four years of the Trump administration, Professor Gamper-Rabindran has wisely focused in on one important sector: US oil and gas. Her target audience is both scholars and the informed public, a target she has reached admirably, despite the difficulties in avoiding appearing partisan. Her writing is very readable, well-documented, comprehensive and enlightening."

– Charles Kolstad, Professor at Stanford Institute for Economic Policy Research, and author of *Environmental Economics*

America's Energy Gamble

People, Economy and Planet

SHANTI GAMPER-RABINDRAN
University of Pittsburgh

CAMBRIDGE
UNIVERSITY PRESS

University Printing House, Cambridge CB2 8BS, United Kingdom

One Liberty Plaza, 20th Floor, New York, NY 10006, USA

477 Williamstown Road, Port Melbourne, VIC 3207, Australia

314–321, 3rd Floor, Plot 3, Splendor Forum, Jasola District Centre, New Delhi – 110025, India

103 Penang Road, #05–06/07, Visioncrest Commercial, Singapore 238467

Cambridge University Press is part of the University of Cambridge.

It furthers the University's mission by disseminating knowledge in the pursuit of education, learning, and research at the highest international levels of excellence.

www.cambridge.org
Information on this title: www.cambridge.org/9781316510742
DOI: 10.1017/9781009039567

© Shanti Gamper-Rabindran 2022

This publication is in copyright. Subject to statutory exception and to the provisions of relevant collective licensing agreements, no reproduction of any part may take place without the written permission of Cambridge University Press.

First published 2022

A catalogue record for this publication is available from the British Library.

Library of Congress Cataloging-in-Publication Data
Names: Gamper-Rabindran, Shanti, author.
Title: America's energy gamble : people, economy and planet / Shanti Gamper-Rabindran, University of Pittsburgh.
Description: 1 Edition. | New York, NY : Cambridge University Press, 2022. | Includes bibliographical references and index.
Identifiers: LCCN 2021034847 (print) | LCCN 2021034848 (ebook) | ISBN 9781316510742 (hardback) | ISBN 9781009039567 (ebook)
Subjects: LCSH: Oil-shale industry – United States. | Petroleum industry and trade – United States. | Power resources – United States. | Environmental policy – United States. | Environmental protection – United States. | Natural resources – United States – Management. | Wealth – United States. | BISAC: POLITICAL SCIENCE / Public Policy / Environmental Policy
Classification: LCC HD9566 .G43 2022 (print) | LCC HD9566 (ebook) | DDC 338.2/72820973–dc23
LC record available at https://lccn.loc.gov/2021034847
LC ebook record available at https://lccn.loc.gov/2021034848

ISBN 978-1-316-51074-2 Hardback
ISBN 978-1-009-01801-2 Paperback

Cambridge University Press has no responsibility for the persistence or accuracy of URLs for external or third-party internet websites referred to in this publication and does not guarantee that any content on such websites is, or will remain, accurate or appropriate.

For Felix

Contents

Acknowledgments		*page* ix
1	**Introduction**	1

Part I   America's Energy

2	**Oil and Gas: The Quest for Energy Dominance**	17
3	**Renewable Energy: Setbacks, Successes and Strategies for the Energy Transition**	58

Part II   America's Lands

4	**Public and Private Lands: Extraction and Infrastructure versus Competing Economic Pursuits**	109
5	**Native American Lands: Respect for Tribes' Rights versus Encroachment**	156

Part III   America's Seas

6	**Oceans: Drilling versus Competing Use of Coasts and Seas**	217
7	**Backtracking on Safety: Risking Another BP Oil Spill**	253

Part IV   America's Regulatory Process

8	**Science: Undermining Facts to Understate Regulatory Benefits**	285
9	**Economics: Skewing Analyses to Justify Weaker Regulations**	326
10	**Law: Anti-regulatory Statutory Interpretations and Reshaping the Judiciary**	369

Part V The Global Climate

11 Endangering the Climate: Attacking Global Cooperation, State Governments' Leadership and the Private Sector's Economic Restructuring   415

12 America at a Crossroads   463

Index   503

Acknowledgments

My thanks to many in southwestern Pennsylvania who have worked to chart the region's economic, energy and environmental path. I am especially grateful to community leaders and to colleagues from the renewable energy and oil and gas industries, from farming, labor, environmental and grassroots organizations, and from government agencies, who generously shared their knowledge, views and experiences.

I am immensely grateful to Joshua Ash and Rita Flanagan for their research assistance. Josh critiqued every single chapter of the manuscript over multiple iterations and shared his insights from Appalachia. Rita's completion of the bulk of the references is no small feat. I am excited for Josh's research and work as an attorney on environmental justice issues and Rita's studies at Pace Law School. My thanks to librarian Chris Lemery for his detective skills and to Chet Litteral and Gabrielle Sampson for their help with the references.

I am indebted to readers who provided detailed comments and insights on various chapters, including Joshua Galperin, Michael Gross, Eric Hittinger, Hilary Hoffman, Justin Pidot, David Popp and Aurora Sharrard. Five anonymous reviewers also provided detailed comments and insights. David Owen, Lucy Newman, Lainey Newman, Larry Newman and Silvija Singh shared their helpful comments. Errors are mine alone. Enormous thanks to colleagues at the Bureau of Indian Affairs, the Bureau of Ocean and Energy Management, the Bureau of Safety and Environmental Enforcement, the Energy Information Administration, the Environmental Protection Agency and the National Renewable Energy Lab for answering my inquiries. Jeffrey Baron at the Energy Information Administration and Sandra Begay at Sandia National Laboratories are incredibly generous with their knowledge.

I am indebted to many colleagues in the academic, business, environmental, energy and policy communities for their input. They include Peter Adams, Vic Adamowicz, Ramon Alvarez, Ariel Armony, Josh Axelrod, Ed Barbier, Dan

Bilello, Robert Brulle, Hilary Bright, Marilyn Brown, Sharon Buccino, Heather Campbell, Laura Castellucci, Louise Comfort, Alison Cullen, Alessio D'Amato, John Dernbach, Dan Farber, Christopher Frey, Patrice Geoffron, Alexis Goldstein, Keith Hay, Chris Hendrickson, Dan Kammen, Fiona Kinniburgh, Florentine Kloppenberg, Sarah Krakoff, Alan Krupnick, David Lyon, Thomas Lyon, Michael Mann, Charles Mason, John Maxwell, Gina McCarthy, Al McGartland, Grant McIntyre, John Mendeloff, Grainger Morgan, Len Necefer, Greg Nemet, Destinie Nock, Dörte Ohlhorst, Naomi Oreskes, Pat Parenteau, Karen Pittel, Jesse Prentice-Dunn, Barry Rabe, Sarah Bloom Raskin, Genna Reed, Steven Rosenthal, Costas Samaras, Miranda Schreurs, Jim Skea, Jeannie Sowers, Beia Spiller, Thomas Steiner, Jennie Stephens, Larry Susskind, Laura Taylor, Avner Vengosh, John Walliser, Erika Weinthal, Amy Wildermuth and Mariangela Zoli. Thanks also to students, including Tara Devezin, Mia DiFelice, Hope Finch, Jacob Garcia, Justin Giannantonio, Divya Nawale, Paige Neid, Eric Raabe and Joseph Skibbens.

I benefitted from feedback from participants at the Society for Environmental Journalists panel on science reporting, the Association for Public Policy Analysis and Management panels on energy and the environment, the University of Florida's law and policy workshop, the Potsdam Institute for Advanced Study in Sustainability workshop on law and sustainability, the European University Institute workshop on democratic backsliding, Rencontres Économiques d'Aix-en-Provence future of energy panel and the World Congress for Environmental and Resource Economists energy transition panels. I also thank seminar participants at the Massachusetts Institute of Technology, the Mercator Research Institute on Global Commons and Climate Change, the Pennsylvania State University, the University of Pittsburgh, the Technical University of Berlin, the Technical University of Munich, the University of Innsbruck, the University of Munich and the University of Rome Tor Vergata.

Thanks to colleagues who shared my work with the general public: Reid Frazier of National Public Radio and the Allegheny Front Radio Show, Dominic Boyer and Cymene Howe of the Cultures of Energy podcast and Tara Sheehan at the Policy Wonk podcast.

Miranda Schreurs generously hosted me at the chair for Climate and Environmental Policy at the Technical University of Munich (TUM) in summer 2018 and as the TUM August-Wilhelm Scheer visiting professor in summer 2019. Funding from several units at the University of Pittsburgh are acknowledged: the Mascaro Faculty Fellowship for Sustainability, the Center for Climate and Global Change, the Central Research Development Fund and the University Center for International Studies Hewlett grant. Many thanks to

Allyson Delnore, Gena Kovalcik, Aurora Sharrard and Jae-Jae Spoon for their support; Peter Timmer, Don Davis, James Mendelsohn and the late James McCarthy for their lessons; and Becca Grainger, Sarah Lambert and Matt Lloyd for shepherding this book to publication.

I am so very fortunate to have the wise counsel and incredible support of family and friends: Armin Gamper, Felix Gamper, Maureen O'Sullivan, Lawrence Pierce, Corinne Branquet, Anna Coppa, David Guillou, Daniele Nardi, Lucy Newman, David Owen, Valérie Sene, Bettina Seri and Silvija Singh.

1

Introduction

Pittsburgh, my hometown, sits in the heart of the Marcellus Shale, a geological formation rich in natural gas. The arrival of fracking in the mid-2000s unleashed the shale boom, bringing tremendous benefits: royalties to owners of mineral rights, lower gas bills for consumers and jobs and revenue to communities blighted by the decline in mining, manufacturing and agriculture. At the height of the boom, the shale industry provided 74,000 jobs in the Marcellus region of Pennsylvania, Ohio and West Virginia.[1]

However, the bust that followed a decade later dried up royalty checks, eliminated jobs and saddled large numbers of residents with damaged roads, abandoned wells and contaminated air, water and land.[2,3] Pennsylvania taxpayers are on the hook for $300 billion in costs to remediate an estimated 200,000 abandoned wells in the state from the recent boom and from past oil and gas extraction.[4–6] These boom–bust cycles have prompted debates within communities in this region on whether to continue down the path of greater US oil and gas extraction or to commit to diversifying the energy economy away from fossil fuels.

This dilemma, faced by oil- and gas-reliant regions across the US heartland and coasts, exemplifies Americans' predicament in charting our energy pathway. America's oil and gas extraction grew from 2005 through 2019, thanks largely to fracking and deepwater drilling technologies.[7–9] That development transformed global energy markets and reshaped Americans' living space. The United States became a net exporter of natural gas by 2017[10] and the largest crude oil producer in the world by 2018.[11] However, at least 17.8 million Americans who live close to oil and gas wells have good reason to worry about their health and livelihoods.[12] Many fear greater reliance on oil and gas extraction can seriously damage both the US economy and the global environment.[13,14]

The oil industry (and the related gas industry)[i] has been seen as a critical source of energy and a powerful engine for economic growth since its birth in the 1860s. In the century and a half since then, oil companies have won (or, depending on one's point of view, extracted) a remarkable array of rights and privileges from grateful politicians at all levels of government. But in the past 50 years, American leaders from Richard Nixon[ii] to Jimmy Carter, George W. Bush and Barack Obama have come to recognize what oil dependency costs the nation and have pledged to end America's oil addiction.[15–18] Yet, despite those presidential promises, the United States has become ever more reliant on oil consumption and on domestic oil extraction.[19] America, and especially its political system, seems unable to kick its addiction.

Even against the historical backdrop of unflagging political support for the industry, the Trump administration's partiality for the oil and gas industry stands out. President Trump's America First Energy Plan was simply stunning in the breadth and depth of its favoritism. As detailed in this book, from January 2017 through January 2021, the Trump administration, with the backing of key members of Congress, swung policies dramatically and unprecedently toward the industry, opening more public lands and seas to drilling and rolling back public health and environmental regulations. The administration offered the oil and gas industry all the help it could, working on multiple fronts to give oil and gas producers a slew of additional institutional advantages.

The administration asserted presidential powers to shrink national monuments and marine monuments and thereby allow more drilling.[iii] It flexed executive powers to curb states' powers to protect their waters from oil and gas operations. It reshaped institutions that oversee energy and environmental policies; reframed science, legal and economic analyses; and jettisoned the norms of evidence-based rulemaking – all to further favor extraction and deregulation. It attacked air pollution regulations (which are among the most cost-effective and successful at imposing compliance costs on industry rather than on society) and sowed doubt about the science underpinning those

---

[i] The oil industry and the gas industry in the United States share some characteristics but diverge in other respects. Players in the oil and gas industries include those that are top producers in both oil and gas (e.g., Exxon, Devon, BP, ConocoPhillips and Chevron) and some that are top producers only in gas (e.g., Chesapeake, Encana, Southwest and Williams). I discuss these industries together or separately, depending on the divergence or overlap of their interests or their impacts on the economy, health and the environment.

[ii] President Nixon said, "if we permit ourselves to slacken our efforts and slide back into the wasteful consumption of energy, then the full force of the energy crisis will be brought home to America in a most devastating fashion" [15].

[iii] National monuments and marine monuments, proclaimed by presidents under the Antiquities Act or created by Congress, protect public land and seas from activities that can diminish their values such as oil and gas extraction.

regulations. It eliminated regulatory efforts to curb methane leaks from oil and gas operations. It abandoned US commitments to the Paris Climate Agreement.

Despite the many advantages bestowed by the Trump administration, by mid-2019 the US shale sector, which had driven US oil and gas production growth since 2005, faced headwinds. Banks and investors widely acknowledged their large-scale losses from financing that industry. The shale industry spent $190 billion more than its earnings from selling oil and gas over the course of a decade[20] and faced $200 billion in debt, coming due by 2023.[21,22] In March 2020, as the COVID-19 pandemic engulfed the United States and world oil and regional gas prices crashed, Trump vowed to help the oil and gas industry.[23] Thanks to Congress's economic rescue package and the Trump administration's policies, oil and gas companies indirectly sold millions in junk-rated bonds to the Federal Reserve. They also secured millions in taxpayer-funded loans and many millions more in tax refunds without strong commitments to keep their workers on the payroll.[24-26] Trump issued an executive order that instructed federal agencies to expedite approvals for oil and gas and other energy projects. The order asserted that it was legal to suspend procedures set out under federal environmental laws during a national emergency.[27]

While the Trump administration was doing more than ever before to favor the oil and gas industry, Congress was doing less than ever before to check this one-sidedness. Congress exercised its powers, used only once before, to rescind numerous regulations. It chose to end its 40-year-old moratorium on drilling in the Arctic National Wildlife Refuge. It did nothing to prevent an offshore-drilling moratorium from lapsing, thereby allowing the administration to expand drilling in federal seas in the Atlantic and the Pacific. Faced with an administration determined to reverse measures introduced by previous presidents to protect health, the environment and the climate, Congress largely sat paralyzed. Senate majority leader Mitch McConnell fast-tracked the confirmation of lifetime-tenured federal judges, many of whom are skeptical of the role of the administrative state in regulating the economy, even as a means of protecting public health and the environment.

These actions and inactions were not merely harmful in the short term. If left unchallenged, these blatant moves to help the oil and gas industry will – as the Trump administration and its congressional allies clearly wanted – entrench expanded extraction and deregulation for generations to come. The implications of these long-term impacts are only exacerbated by the urgency with which the United States needs to transition away from fossil fuels to mitigate the climate crisis.

How This Book Is Organized

To assess how America governs – or doesn't – its domestic oil and gas industry, this book looks at six different but interrelated subjects: the current US energy balance; governance of oil and gas extraction on private and public lands; governance of that extraction offshore; the US regulatory process that is meant to oversee energy resources and protect public health and the environment; the interlinkages between US energy and climate policies; and how Americans can and have participated in the governance of the oil and gas industry.

In Part I, I examine the two competing energy visions in the United States: one that foresees a deepening reliance on domestic oil and gas extraction and consumption and one that supports increasing diversification of energy sources and a much larger role for renewable energy.

Chapter 2 begins by briefly sketching the historical context behind US energy policy and spotlights the political influence of the oil and gas sector and its think tanks that was already at play when the Trump administration took office. The chapter then dissects the Trump administration's America First Energy Plan, implemented by its appointees with strong financial ties to the oil and gas sector. I catalog the monetary gains of the US shale boom and offshore oil production but also tally the associated costs from these operations. And I warn of the adverse effects of expanded drilling and deregulation to the US economy, including the increased US exposure to volatile global oil and gas markets, to the industry's inherent boom–bust cycles, and to negative health and environmental impacts.

A less ominous, indeed altogether more encouraging note is sounded in Chapter 3, which looks at the remarkable progress being made not only in blue states but also in red and purple states in diversifying to renewable-energy generation. The chapter describes how renewables have won the support of state governments by offering cost-competitive electricity, revenue to rural landowners and job opportunities. Cities, utilities and private companies are shifting to renewable energy because of declining costs, greater reliability and greater resilience, although there are challenges to be overcome in this transition. Sadly, the Trump administration, in its devotion to oil and gas, seemed determined to ignore the benefits that renewable energy can bring to America.

Part II, spanning two chapters, lays out how the oil and gas industry competes with rival economic activities on private and public lands. I show how the Trump administration sought to secure further advantages for the industry and the pushback this provoked from a number of state governments and environmental groups.

Chapter 4 probes how oil and gas extraction competes against rival economic activities thanks to the rights and privileges it enjoys. Even before the Trump administration took office, existing state laws prioritized the rights of mineral owners over those of surface owners and prevented local governments from restricting oil and gas operations. The Trump administration facilitated the building of oil and gas infrastructure on private lands. It also opened more public lands to drilling, despite companies not drilling on half of the 23 million acres already leased and despite the sizable benefit from conservation. It rolled back environmental regulations, further weakening legal oversight of oil and gas operations that already enjoy exemptions from a number of provisions under federal environmental laws that apply to other heavy industries.

Chapter 5 explores Native Americans' battles to protect their land and water resources from oil and gas extraction and pipelines. For decades, companies have sought to extract oil and gas on Native Americans' reservations and on lands covered under treaties. Companies have also sought to build oil and gas infrastructure on these lands, even when affected communities are opposed to these activities. Trump asserted disputed presidential powers to permit the construction of the Keystone XL pipeline and to shrink national monuments, including the Bear Ears National Monument, that protect Native Americans' cultural sites. I discuss legal challenges mounted by Native American tribes against Trump and his administration and the tribes' early successes.

Part III, consisting of two chapters, describes how offshore drilling vies with competing economic sectors. As in the case with public lands, the administration's actions have been met with significant pushback from a number of state governments and environmental groups.

Chapter 6 examines the Trump administration's proposal to open drilling in the Atlantic, the Pacific and parts of the Arctic where little to no drilling has occurred. The administration proposed to open more offshore areas to drilling, notwithstanding the dangers posed to coastal economies and the fact that companies have drilled on only one-fifth of the 28 million acres already leased offshore. The fire sale of leases cost the American public a potential fortune in terms of revenue while seeking to foreclose other options for the use of public seas. Not surprisingly, a number of state governments, some in the hands of Republicans, challenged drilling in federal waters off their coasts.

The heightened risks as offshore drilling expands to deeper waters in the Gulf of Mexico and to more remote waters in the Arctic are spelled out in Chapter 7. The Trump administration worsened these risks by rolling back regulations that had been enacted in response to the 2010 BP oil spill and to near misses during the 2013 Shell Artic expedition. These rollbacks, plus poor government oversight of companies and scant attention to companies' safety culture, risk a repeat of past oil spill disasters that devastated coastal communities.

Part IV, spanning three chapters, focuses on the US rulemaking process and threats to its functionality. The administration blocked scientific evidence, reframed economic analysis to understate the benefits of regulation and interpreted laws narrowly to blunt federal agencies' regulatory actions.

Chapter 8 tracks how the Trump administration attacked science to weaken regulations that affect the oil and gas industry. At the same time as top scientific advisory jobs were given to individuals who were prepared to cast doubt on perfectly sound scientific studies, the administration sought to block federal agencies' consideration of studies on the health impacts of pollution. Checks on fact-based regulations at federal agencies were curbed by blocking input from independent scientists and overriding recommendations of career scientists.

Chapter 9 delves into the Trump administration's strategies to dismantle regulations by understating their economic benefits. I detail how the administration systematically disregarded major health benefits from reducing pollutants. The interests of future generations were sacrificed in the name of deregulation, and climate damages that impact the entire world were simply disregarded.

What role, if any, courts have had in challenging the expansion of drilling on public lands and seas and the weakening of regulations is the subject of Chapter 10. I detail how a docket of cases raises questions about executive versus legislative powers, federal versus state powers and federal agencies' statutory interpretations. Meanwhile, the judicial philosophy of the courts has shifted further toward deregulation as a result of President Trump's appointment of federal judges, including three Supreme Court judges and one-quarter of the active judges who serve on the Court of Appeals.

In Part V, Chapter 11 describes how the Trump administration blocked climate action. It attempted to derail financially prudent responses by state governments and the private sector to climate change-induced shifts in product and investment markets. The chapter argues that Americans can still take decisive actions to mitigate the worst impacts of climate change. Congress can reset the economic rules, within which private actors operate, to help investments shift toward low-carbon energy and economic sectors and assist workers and communities to make the possible, but challenging, transition. Congress's "climate legislation" at the end of 2020 demonstrates that climate actions that create jobs and economic opportunities can win bipartisan support.

In Chapter 12, I draw lessons from the preceding chapters. The paramount lesson is that Congress's failures to improve oil and gas governance and public health, environment and climate protections paved the way for the Trump administration to inflict long-term damages. Unless those failures are rectified, Americans' economy and health and the environment remain vulnerable to a future, savvier pro-oil anti-regulatory administration. I also discuss lessons from the preceding chapters on how the Biden Plan for Clean Energy Transition and Environmental Justice can support America's energy transition, rebuild integrity into the regulatory process and protect Americans' health, land, seas and global climate.

This Book's Inspiration, Sources and Approach

This book project emerged from discussions with colleagues and students at the University of Pittsburgh and with communities in the tristate region of Pennsylvania, Ohio and West Virginia located on the Marcellus Shale. This region has served as a microcosm of the national and global debate over energy. Which of the competing visions of America's future economic, energy and environmental pathways should the region embrace? Should it drill down on the path of greater shale gas extraction and build its economy and energy pathways around shale? Or should this region, which is still reeling from the collapse of coal extraction and the boom–bust cycles of shale gas, persist in pursuing the goal of economic diversification, even though it poses its own challenges?

As I worked with colleagues and students to comprehend the direction and impact of US energy and environmental policies, the need for a book like this became clear. Few books have focused on the Trump administration's energy and environmental policies. The outlets that report on the Trump administration's energy policies focus on the public's need for immediate updates on the administration's rapid-fire policy pronouncements and deregulatory actions. These include the electronic publication *Environmental Protection in the Trump Era* (2018) by the Environmental Law Institute and the American Bar Association; trackers and commentary on regulatory rollbacks by NGOs such as the Environmental Integrity Project and by law schools such Harvard Law School; and podcasts such as the "Trump on Earth" podcast by National Public Radio.[28] However, these outlets that are focused on keeping the public up-to-date with policy changes, understandably, have not provided in-depth analyses of the broader historical, economic and political context of the administration's actions and detailed assessments of their long-term implications.

I decided to write a book that is aimed at three audiences – professionals working on US energy and environmental policies, scholars and students of these policies, and members of the general public interested in these debates – and that lays out the administration's policies, scrutinizes the logic for and against these policies and their likely consequences, and explores the mechanisms to support or to reverse these policies.

To evaluate these policies, I searched for uniform yardsticks. There are at least three criteria that seem to rise above partisanship. First, advocates of expanding extraction and deregulation argue that these strategies will achieve energy independence and make Americans more prosperous. Thus, one criterion is to determine whether or not these strategies will actually achieve these stated goals. Second, a significant proportion of oil and gas extraction takes place offshore and a substantial amount takes place on public lands.[iv] The American people, present and future, own these resources and, according to laws enacted by Congress, the federal government is supposed to manage them for the benefit of the public. Thus, a second criterion is the extent to which these strategies share benefits broadly or instead concentrate wealth in the hands of a few. Third, the United States has historically respected factual evidence, not least scientific evidence, in its rulemaking processes. Thus, a third criterion is whether the rulemaking process gives a fair hearing to the facts or rejects evidence that does not accord with ideological prejudices and special interests.

In my research, I have relied on various sources. To track the administration's policies to expand domestic oil and gas extraction and roll back regulations, I refer to government documents, including federal agencies' publications in the federal register. To understand the likely economic impacts of these policies, I delve into peer-reviewed studies and into reports by the Congressional Budget Office and the Government Accountability Office, which are tasked to provide factual analysis to Congress. To probe the public health, environmental and climate impact of these policies, I rely on peer-reviewed literature and reports by scientific panels that have been scrutinized by other scientists. To investigate how the administration altered scientific and economic analyses, I compare the administration's regulatory impact assessments to those of previous administrations. I also dig into reports by environmental organizations that are well-regarded for their expertise on science, law, political science and economics. An equally important source is investigative journalists' reports on the actions of Trump-appointed leaders of federal agencies. Finally, to track how the administration's actions have fared amid legal challenges, I examine briefings filed on these lawsuits and on the initial legal decisions.

I also draw on a number of previous studies of related topics. Although few books have, as yet, chronicled the Trump administration's energy policy, a rich

---

[iv] A larger share of the extraction takes place on private lands.

literature exists that examines the centrality of the US oil and gas extraction industry to US energy policies and the industry's far-reaching impacts on the economy and the environment. A number of academic books, while acknowledging the contribution that US domestic oil and gas production makes to the economy, condemn the failure of both Republicans and Democrats to address America's decades-long oil dependency and the adverse economic and environmental consequences. Yale Law School professor Michael Graetz's *The End of Energy: The Unmaking of America's Environment, Security, and Independence* (MIT Press, 2011) and Butler University economist Peter Grossman's *US Energy Policy and the Pursuit of Failure* (Cambridge University Press, 2013) lambaste both presidents and Congress, and both parties, for pursuing ad hoc policies that prioritize domestic oil and gas extraction. These policies, they argue, yield benefits to narrow interest groups but hurt Americans as a whole.

My edited volume *The Shale Dilemma: A Global Perspective in Fracking and Shale Development* (University of Pittsburgh Press, 2018) explores the surge in US oil and gas extraction thanks to the shale boom that lasted from the mid-2000s through the mid-2010s. Americans' thirst for oil and gas resources to fuel the economy often eclipsed concerns about the costs of shale extraction to local communities, public health and the environment. In some other countries – such as France, Germany and the United Kingdom – sections of the public have been more circumspect about the promised benefits and more wary of the potential costs from shale and have chosen to support stronger commitments to transitioning away from fossil fuels. As documented in *Fractured Communities* (Rutgers University Press, 2018) by sociologist Anthony Ladd and his colleagues, communities in the United States have faced uphill challenges in limiting the industry's operations; state governments have typically sided with the shale gas industry in legal actions.

The dearth of books on the Trump administration's energy and environmental actions contrasts with the plethora of bestsellers by academics, journalists and political commentators on the administration's flouting of democratic norms. As I document in the following chapters, these two threads are closely intertwined. The administration's strategy to expand oil and gas drilling while rolling back regulations has chipped away – perhaps "hacked away" would be more accurate – at democratic norms.

Steven Levitsky and Daniel Ziblatt, both of whom are political scientists at Harvard University, dissect how democracies past and present falter when politicians pursue their goals at the cost of institutions that serve the public interest. In *How Democracies Die* (Crown, 2018), Levitsky and Ziblatt detail how the Trump administration inflamed public mistrust by denouncing the role of the media and the judicial branch in checking executive power. Michael Lewis, a financial journalist, delves into the Trump administration's intentional mismanagement of government agencies in his book *The Fifth Risk* (W.W. Norton, 2018). Lewis recounts the administration's contempt for government agencies' responsibility to protect the public and to serve the vulnerable.

Historians James Morton Turner at Wellesley College and Andrew Isenberg at the University of Kansas trace the administration's ability to dismantle health and environmental protections to the political shift in the Republican Party that began in the late 1970s. In *The Republican Reversal: Conservatives and the Environment from Nixon to Trump* (Harvard University Press, 2018), Turner and Isenberg chronicle how the Republican Party deviated from its conservation credentials and became instead an unalloyed champion of cheap fossil fuel and a fierce critic of climate protection. Their narrative comports with analysis by the late Judith Layzer, a political scientist at MIT, whose *Open for Business* (MIT Press, 2014) examines how special interest groups worked with their political supporters to chip away at environmental protections.

Another important focus of books on American energy and environmental policies is the important, albeit so far limited, steps taken to shift away from fossil fuels. In *Political Opportunities for Climate Policy: California, New York, and the Federal Government* (Cambridge University Press, 2016), political scientist Roger Karapin at City University of New York identifies windows of opportunity to advance renewable energy and highlights a number of effective policies at the state and federal level. Energy experts Jennie Stephens, Tara Rai and Elizabeth Wilson, from Northeastern, Texas A&M and Dartmouth respectively, delve into the challenges of the energy transition from both technical and socioeconomic perspectives. Their book *Smart Grid R(Evolution): Electric Power Struggles* (Cambridge University Press, 2015) offers insights on federal and state policies that helped or hindered the energy transition. Political scientist Leah Stokes at University of California, Santa Barbara details in *Short-Circuiting Policy* (Oxford University Press, 2020) how utility companies that operated fossil fuel power plants sought to delay the transition to renewable energy.

The focus of *America's Energy Gamble* is different from all of these studies, but I am indebted to them and their authors for helping to shape and support my arguments and to situate them in a broader context.

A Note about Coal

This book does not discuss coal extraction, even though coal is an important feature in the landscape of US fossil fuel extraction and related pollution, with coal-fired power plants contributing of one-fifth of US greenhouse gas emissions in 2018. Coal deserves a study of its own because its tale diverges from that of oil and gas in at least two ways. First, in contrast to rising US oil and gas extraction, coal extraction in the United States has been on a downward trajectory, despite a temporary resurgence in production due to exports.[29] Coal extraction is expected to decline further due to weakening demand for coal in US electricity generation and diminishing export opportunities. Second, coal's fortunes dwindled because of competition from oil in transportation and gas in electricity generation. A century ago, oil outcompeted coal in powering ships; in the past decade, natural gas outcompeted coal to fuel US power plants. Coal shares one feature with oil and gas under the Trump administration: the administration rescinded a number of regulations on coal extraction and on coal-fired power plants. Although likely to make pollution worse, these proposals are not expected to revive the coal industry.[30]

In contrast to declining coal production, oil and gas extraction vastly expanded in the United States in the early 2000s. The next chapter delves into factors that spurred this expansion and the consequences – both positive and negative – of this expansion.

References

1. E. N. Mayfield, J. L. Cohon, N. Z. Muller, I. M. L. Azevedo and A. L. Robinson. "Cumulative Environmental and Employment Impacts of the Shale Gas Boom." *Nature Sustainability* 2, no. 12 (2019): 1122–1131.
2. S. Gamper-Rabindran, ed. *The Shale Dilemma: A Global Perspective on Fracking and Shale Development*. Pittsburgh, PA: University of Pittsburgh Press, 2018.
3. M. L. Finkel, ed. *The Human and Environmental Impact of Fracking: How Fracturing Shale for Gas Affects Us and Our World*. Santa Barbara, CA: Praeger, 2019.
4. Pennsylvania Department of Environmental Protection. *Pennsylvania Oil and Gas Report* (Harrisburg, Pennsylvania, 2017).
5. L. Legere and A. Litvak. "PA Faces New Wave of Abandoned Conventional Oil & Gas Wells." *Pittsburgh Post-Gazette*, April 5, 2020.
6. Fractracker Alliance. "Abandoned Wells in Pennsylvania." *Fractracker.org*, August 8, 2019.
7. Energy Information Administration. *EIA Adds New Play Production Data to Shale Gas and Tight Oil Reports*. Principal contributors: Jack Perrin and Emily Geary (February 15, 2019). www.eia.gov/todayinenergy/detail.php?id=38372.
8. Energy Information Administration. *US Crude Oil Production Grew 11% in 2019, Surpassing 12 Million Barrels Per Day*. Principal contributor: E. Geary (March 2, 2020). www.eia.gov/todayinenergy/detail.php?id=43015.

9. S. P. A. Brown and M. K. Yucel. *The Shale Gas and Tight Oil Boom: US States' Economic Gains and Vulnerabilities*. Council on Foreign Relations Report (Washington, DC: October 15, 2013). www.cfr.org/report/shale-gas-and-tight-oil-boom+&cd=1&hl=en&ct=clnk&gl=ca.
10. Energy Information Administration. *Natural Gas Explained: Natural Gas Imports and Exports* (July 21, 2020). www.eia.gov/energyexplained/natural-gas/imports-and-exports.php.
11. Department of Environment. *US Becomes World's Largest Crude Oil Producer and Department of Energy Authorizes Short Term Natural Gas Exports* (September 13, 2018). www.energy.gov/articles/us-becomes-world-s-largest-crude-oil-producer-and-department-energy-authorizes-short-term.
12. E. D. Czolowski, R. L. Santoro, T. Srebotnjak and S. B. C. Shonkoff. "Toward Consistent Methodology to Quantify Populations in Proximity to Oil and Gas Development: A National Spatial Analysis and Review." *Environmental Health Perspective* 125, no. 8 (2017): 86004–86015.
13. A. Tsvetkova and M. D. Partridge. "Economics of Modern Energy Boomtowns: Do Oil and Gas Shocks Differ from Shocks in the Rest of the Economy?" *Energy Economics* 59 (2016): 81–95.
14. H. McJeon, J. Edmonds, N. Bauer, L. Clarke, B. Fisher, B. P. Flannery, J. Hilaire et al. "Limited Impact on Decadal-scale Climate Change from Increased Use of Natural Gas." *Nature* 514, no. 7523 (2014): 482–485.
15. R. Nixon. "President Nixon's Nationwide Radio Address on the National Energy Situation." *New York Times*, January 20, 1974. www.nytimes.com/1974/01/20/archives/transcript-of-nixons-speech-on-energy-situation-a-call-for.html.
16. J. Carter. "President Carter's Address to the Nation about Energy Problems." *New York Times*, April 19, 1977. www.nytimes.com/1977/04/19/archives/transcript-of-carters-address-to-the-nation-about-energy-problems.html.
17. G. W. Bush. "State of the Union Address." George W. Bush White House. January 31, 2006. https://georgewbush-whitehouse.archives.gov/stateoftheunion/2006.
18. B. Obama "Address to the Nation on the BP Oil Spill." Barack Obama White House. June 15, 2010. https://obamawhitehouse.archives.gov/blog/2010/06/16/president-obamas-oval-office-address-bp-oil-spill-a-faith-future-sustains-us-a-peopl.
19. Energy Information Administration. *Short-Term Energy Outlook* (February 11, 2020). www.eia.gov/outlooks/steo/pdf/steo_full.pdf.
20. D. Wethe and K. Crowley. "Shale's Bust Shows Basis of Boom: Debt, Debt and Debt." *Bloomberg News*, July 23, 2020.
21. R. Dezember. "Energy Producers Face Big Tab after Shale Bonanza." *Wall Street Journal*, January 2, 2020.
22. C. Williams-Derry, K. Hipple and T. Sanzillo. *Shale Producers Spilled $2.1 Billion in Red Ink Last Year* (Institute for Energy Economics and Financial Analysis, Cleveland, Ohio: March 2020). https://ieefa.org/wp-content/uploads/2020/03/Shale-Producers-Spilled-2.1-Billion-in-Red-Ink-Last-Year_March-2020.pdf.
23. R. Frazin "Trump Floats Funding for Oil after Historic Market Loss." *The Hill*, April 2, 2020.

24. G. Gelzinis, M. Madowitz and D. Vijay. *The Fed's Oil and Gas Bailout Is a Mistake: Financial Stability, Public Funds, and the Planet Are at Risk* (Center for American Progress, Washington, DC: July 31, 2020).
25. J. A. Dlouhy. "'Stealth Bailout' Shovels Millions of Dollars to Oil Companies." *Bloomberg News*, May 15, 2020.
26. S. Gamper-Rabindran. "Fracked Communities and Taxpayers: Shale Economics in the US and Argentina." In *Oxford Handbook of Comparative Environmental Politics*, edited by J. Sowers, S. VanDeveer and E. Weinthal. Oxford University Press, 2021. Online.
27. D. J. Trump. *Executive Order on Accelerating the Nation's Economic Recovery from the COVID-19 Emergency by Expediting Infrastructure Investments and Other Activities.* D. J. Trump White House. June 4, 2020. www.whitehouse.gov/presidential-actions/eo-accelerating-nations-economic-recovery-covid-19-emergency-expediting-infrastructure-investments-activities.
28. Environmental Integrity blog and tracker, https://environmentalintegrity.org; Harvard Environmental and Energy Law tracker, https://eelp.law.harvard.edu/regulatory-rollback-tracker/; Trump on Earth podcast, www.npr.org/podcasts/512656404/trump-on-earth.
29. Congressional Research Service. *The US Coal Industry: Historical Trends and Recent Developments.* Report by M. Humphries. Report R44922 (August 18, 2017).
30. R. Elliott. "Trump's Promise to Revive Coal Thwarted by Falling Demand, Cheaper Alternatives." *Wall Street Journal*, September 16, 2020.

PART I
America's Energy

2

Oil and Gas
The Quest for Energy Dominance

At the start of the twenty-first century, the United States relied on oil and gas for 63 percent of its energy consumption. That reliance had grown throughout the preceding century.[1] Oil had become a critical asset to the United States, particularly for its role in powering America's modernizing transportation sector. Natural gas had grown in importance as a source of heating for buildings and subsequently for electricity generation. Oil and gas had become key ingredients in the burgeoning petrochemical sector. Other energy sources such as coal,[i] nuclear, hydropower, wind and solar, while used to generate electricity, had not been able to serve as direct substitutes to oil or gas as the bulk of US transportation systems and heating systems were not electrified.

In 2001, President George W. Bush's energy advisory group sounded the warning bells about America's growing imports of oil and gas. US consumption had continued to rise unabated while US domestic production was plateauing. In 2005, the shale boom took off, postponing America's reckoning with its oil and gas reliance. The unanticipated growth in US oil and gas production from shale was made possible with the successful application of high-volume hydraulic fracturing and horizontal drilling to extract shale resources.[ii] That production[iii] helped transform the United States from a net importer of energy (having imported as much as 30 percent of the energy it consumed in 2005) to a net energy exporter by 2019.[2] The shale boom from 2005 through 2014

---

[i] The share of coal in US energy consumption began to decline in 2009, falling from 20 percent in 2009 to 11 percent by 2019 (Figure 2.1).

[ii] In conventional oil and gas reservoirs, vertical drilling operations can be used to tap into oil and gas that are within a permeable layer and trapped between two impermeable layers. Conversely, in shale reservoirs, oil and gas are within relatively impermeable shale rock. High-volume hydraulic fracturing is used to break up the rock to release the oil and gas. Horizontal drilling increases the area of a well exposed to the fractured shale rock and thus enables more oil and gas to seep into the well and increases oil and gas extraction from that well.

[iii] The rise in renewable energy generation also contributed to closing the gap between energy consumption and production.

reduced consumers' gas prices, created jobs, provided mineral owners' royalty income and generated tax revenue for local, state and federal governments. However, the decline in regional natural gas prices as early as 2012 in some shale basins caused the shale gas industry to contract, while the decline in global oil prices from mid-2014 through 2016 caused the shale oil industry to sputter. These painful busts, with the losses of jobs, income and tax revenue, plus growing evidence of health and environmental impacts from shale extraction, revived debates about the prudence of America's dependency on oil and gas extraction as its economic and energy strategy.

In 2017, newly elected President Trump announced the America First Energy Plan as the cornerstone of his administration's energy and economic policy agenda. The plan proposed the expansion of America's already sizable oil and gas production as the best way forward to promote economic progress for all Americans. That strategy aimed to "take advantage of the estimated $50 trillion[iv] in untapped shale, oil, and natural gas reserves, especially those on federal lands that the American people own" and offshore in federal waters including at the Atlantic and Pacific coasts that had not seen recent drilling.[2–4] That extraction, the plan claimed, would "bring jobs and prosperity to millions of Americans" and achieve American "energy dominance," that is, the United States would become "independent of the OPEC[v] cartel or any nations hostile to [its] interests." Absent from Trump's energy and economic agenda was support for renewable energy, which had been growing across the United States and contributing to cost-competitive energy, jobs, tax revenue and the economic revitalization of rural communities (Chapter 3).

The plan also set out contradictory claims on how oil and gas extraction would be pursued in relation to environmental protection. The plan blamed regulations for imposing burdens on the energy sector and pledged to repeal major air and water regulations. At the same time, the plan stated that "energy must go hand-in-hand with responsible stewardship of the environment … Protecting clean air and clean water, conserving our natural habitats and preserving our natural reserves will remain a high priority."

[iv] Several economists emphasize that this $50 trillion figure is a gross exaggeration of economically recoverable oil and gas reserves on public lands. They note that Trump relied on a report by the Institute of Energy Research, funded by the oil and gas industry, that estimates these reserves at $31.7 trillion. That estimate assumed high oil and gas prices (i.e., $100 per barrel and $5.64 per thousand cubic feet of natural gas) thereby ignoring the price volatility that can and has resulted in far lower prices. Moreover, that report did not apply any discounting, a method typically used by economists to express future benefits in terms of present values [3].

[v] The Organization of the Petroleum Exporting Countries (OPEC) is an intergovernmental organization of 13 major oil-producing countries, including Saudi Arabia, Iran, Iraq, Kuwait, Libya, the United Arab Emirates, Venezuela and Nigeria.

The promise of jobs and prosperity for Americans served as the justification for the administration's far-reaching actions in favor of the oil and gas sector. The administration not only opened the United States to more oil and gas drilling and slashed regulations, it also facilitated a massive bailout of the oil and gas sector through Congress's March 2020 pandemic economic recovery package at a cost to taxpayers. The America First Energy Plan resonated with Trump's supporters and other Americans as well. A return to the heyday of the shale boom was attractive to many in the oil- and gas-reliant regions, who were suffering from the shale bust. A prosperous future built on the familiar path of oil and gas extraction was easier to fathom than the challenging prospects of diversifying economies from oil and gas reliance. Insulation from the upheavals in the global oil market was no less enticing with OPEC oil shocks still a scab on US vulnerability. However, the plan's promises comported with neither the past performance of the shale industry nor the operations of the global oil market.

In this chapter, I discuss why US expansion of domestic oil and gas extraction as an economic development strategy – as set out in the America First Energy Plan – is not in the best interest of the American public in the twenty-first century. Boom–bust cycles in the oil and gas sector, its poor past economic performance and the uncertainties in its future economic outlook cast major doubts on the prudence of relying on that sector to provide job security and longer-term local economic development. By the time Trump took office in 2017, numerous energy and financial analysts had widely acknowledged the weak financial foundations of the shale industry (the segment of the sector that fueled its past growth and its projected expansion) and the growing reluctance of investors to fund the shale industry. The US shale oil industry also suffered from the volatile prices in the global oil market and transferred negative economic shocks to US regions reliant on that sector.[5] The shale oil industry contracted in response to the decline in global oil prices from mid-2014 through 2016, and again during the March 2020 price war between Saudi Arabia and Russia.[6] These considerations alone, even before accounting for pollution costs from the oil and gas sector on other economic players, underscore the risks of betting US economic development on domestic oil and gas extraction.

I begin by describing how the Bush and the Obama administrations' policies on oil and gas extraction set the stage for the Trump administration (Section 2.1). I discuss how the Trump administration presented oil and gas expansion as central to US economic development to justify its actions that favored the industry (Section 2.2). Next, I explore why oil and gas extraction is a dubious economic development strategy by describing the shaky financial foundations of the shale sector (Section 2.3) and the financial bailout of the oil and gas sector in 2020 (Section 2.4). I detail the nuanced contributions of the oil and gas sector to economic development (Section 2.5) and the legal advantages that enabled the industry to shift its costs to other economic actors (Section 2.6). I conclude by noting how the industry's already substantial privileges – that is, its privatization of benefits and its socialization of costs – formed the backdrop to the Trump administration's efforts to secure even more advantages for the industry at a cost to the American people and the US economy.

2.1 US Government Prioritization of Securing Oil and Gas Supplies

The US government's efforts to secure oil supplies from domestic and foreign sources to meet Americans' energy demand have shaped its economic, environmental and foreign policies.[7,8] David Painter's *Oil and the American Century* chronicles how US domestic oil production and its subsequent exertion of control over global oil supply contributed to US economic prowess in the twentieth century.[9] He also details how US reliance on oil to fuel its economy has come at the cost of conflicts and war abroad; the derogation of domestic public health and environmental and global climate concerns; and the considerable influence of the oil and gas industry on US politics. The US energy path and its consequences have been detailed in numerous books, for instance, Meg Jacobs's *Panic at the Pump*, Steven Coll's *Private Empire* and Matthieu Auzanneau's *Oil, Power, and War*.[10–12] That path – pursued by the G. W. Bush administration (Section 2.1.1) and rebalanced to a small extent during the Obama administration (Section 2.1.2) – set the stage for the Trump administration's policies in favor of the oil and gas sector (Section 2.2).

2.1.1 The G. W. Bush Administration (January 2001– January 2009)

In 2000, oil and gas contributed the bulk, about 63 percent, of US energy consumption (Figure 2.1), with oil contributing 38 percent and gas contributing 25 percent of that consumption. The remaining sources were coal (23 percent), nuclear (8 percent) and hydropower and other renewable energy sources (6 percent).

America's rapacious oil and gas consumption, amid declining domestic production, led to rising imports of both oil and gas (Figures 2.2 and 2.3). In 2000, US imports came primarily from Canada (15 percent), Saudi Arabia (14 percent), Venezuela (14 percent), Mexico (12 percent), Nigeria (8 percent)

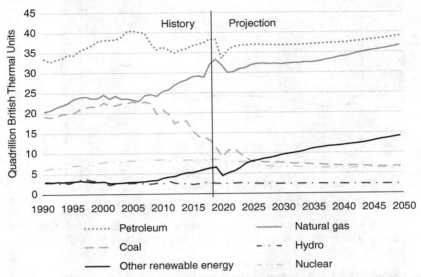

Figure 2.1 US energy consumption: History (to 2019) and projection (to 2050)
Source: Energy Information Administration, *Annual Energy Outlook*, 2020.[13]
Note: Projections were made based on policies that were in place in 2019.

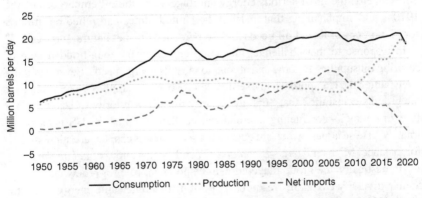

Figure 2.2 US petroleum consumption, production and net imports 1950–2020
Source: Energy Information Administration, *Petroleum & Other Liquids*, 2021.[15]

and Iraq (6 percent).[14] Rising domestic gas consumption led to rising imports (Figure 2.3), primarily from Canada.

Immediately after taking office, G. W. Bush established the National Energy Policy Development Group to advise on US energy policy. The group, which

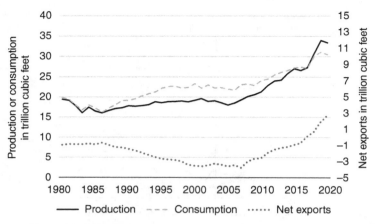

Figure 2.3 US natural gas consumption, production and net exports 1980–2020
Source: Energy Information Administration, *Natural Gas*, 2021.[16]

was headed by Vice President Dick Cheney, former CEO of oil and gas service giant Halliburton, sought out the views of many oil and gas executives while largely ignoring the views of environmental groups. The group's first and foremost concern was energy scarcity, warning that "America in the year 2001 face[d] the most serious energy shortage since the oil embargoes of the 1970s" and emphasizing that in 1999 the United States had relied on imports for 52 percent of its net oil needs and 16 percent of its net natural gas needs.[14]

In response to that challenge, the group embraced the long-trodden path of securing oil supplies at home and abroad to meet its energy needs. Several authors have faulted the US approach of exerting control over oil supplies – exemplified by the group's declaration that "Energy security must be a priority of US trade and foreign policy" – for fueling international conflicts such as the US invasion of Iraq.[17,18] These issues, which are outside the scope of this chapter, are addressed in, for instance, Michael Klare's *Blood and Oil* and Greg Muttitt's *Fuel on the Fire*.

Domestically, the group played down the potential role for renewable energy to help diversify US energy resources,[vi] positing that renewable energy was "still years away" from meeting the bulk of US energy needs and predicting that renewables would meet only 2.5 percent of US energy consumption in 2020.[14] That prediction proved pessimistic of the actual achievements in 2020.[19,vii]

[vi] Bush's lack of support for renewable energy at the federal level stood in contrast with his support as governor for wind development in Texas (Chapter 3).

[vii] According to the US Energy Information Administration, in 2020, wind provided 3 percent of US energy consumption, hydropower provided 2.6 percent and biomass provided 4.7 percent.

The group also advocated for new oil and gas drilling technologies and criticized the regulatory structure for "excessively restricting the environmentally safe production of energy."[14] At that time, companies were already experimenting with innovations in hydraulic fracturing and horizontal drilling to extract shale resources (Section 2.3). In line with the group's recommendation, the Bush administration subsequently pushed for exemptions for hydraulic fracturing from the federal environmental laws.

In 2005, Congress enacted the Energy Policy Act, which exempted hydraulic fracturing from a provision of the Safe Drinking Water Act governing underground injection of wastewater. That exemption added to an already lengthy list of exemptions from federal statutory environmental protections that Congress had granted to the oil and gas industry in its prioritization of energy needs.[20,21] The act incorporated numerous other perks for the oil and gas industry, with only a small share of expenditures going to renewable energy.

State governments, the primary regulators of oil and gas extraction, largely gave the greenlight for shale extraction to proceed. Maryland and New York, however, broke the trend by passing precautionary moratoriums on hydraulic fracturing until the public health and environmental impacts were better understood.[22,23] By the time Obama began his presidency, the industry was primed for major expansion. Having achieved success in the Barnett Shale in Texas, shale companies fanned out to exploit other shale basins throughout the United States (Map 2.1).

2.1.2 The Obama Administration (January 2009–January 2017)

When Obama took office, the shale industry was already on its way to rapid growth, driving the staggering rise in US domestic oil and gas production. Oil production from shale basins, including the Barnett, Eagle Ford, Permian and Bakken shale regions, soon exceeded that from conventional oil fields in Alaska, offshore and elsewhere (Figure 2.4). Likewise, natural gas production from shale basins, including the Marcellus Shale, outpaced natural gas production in the rest of the United States (Figure 2.5).

The Obama administration supported oil and gas expansion but attempted to balance that support with mitigating the industry's impacts. In his 2012 speech in Oklahoma, Obama boasted,

> "Under my administration, America is producing more oil today than at any time in the last eight years. Over the last three years, I've directed my administration to open up millions of acres for gas and oil exploration across 23 different states. We're opening

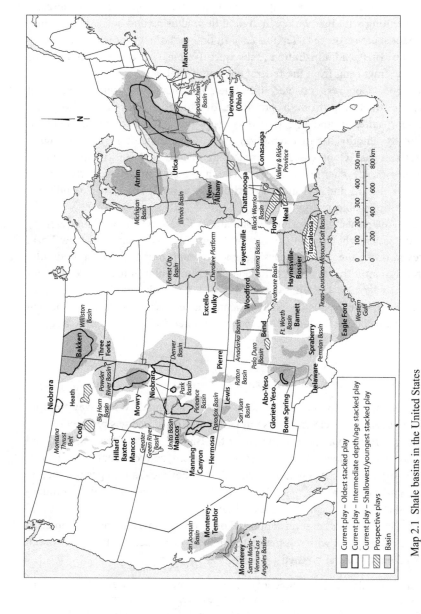

Map 2.1 Shale basins in the United States
Source: Energy Information Administration, *Shale Gas and Oil Plays, Lower 48 States*, 2016.[194]

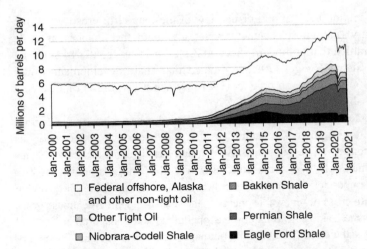

Figure 2.4 US shale oil production as a share of US oil production
Source: Energy Information Administration, *Petroleum & Other Liquids*, 2021.[15]

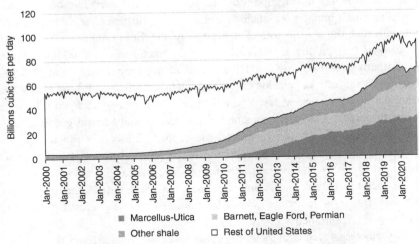

Figure 2.5 US shale gas production as a share of US natural gas production
Source: Energy Information Administration, *Natural Gas*, 2021.[16]

up more than 75 percent of our potential oil resources offshore. We've added enough new oil and gas pipeline to encircle the Earth and then some."[24],[viii]

[viii] In the same speech, Obama also noted, "But what we're also going to be doing as part of an all-of-the-above strategy is looking at how we can continually improve the utilization of renewable energy sources, new clean energy sources, and how do we become more efficient in our use of energy."

The administration held an optimistic view of the long-term prospects of the shale industry (though subsequent revelations about the finances of shale companies rattled that optimism) (Section 2.3). For instance, a 2014 study by the Congressional Budget Office estimated that shale development would increase US gross domestic product (GDP) by about two-thirds of 1 percent in 2020 and about 1 percent in 2040.[25]

Obama touted the "climate credentials" of natural gas as a bridge fuel,[26] with countries shifting from coal to natural gas first and then to renewable energy as the latter becomes more cost-competitive. Natural gas did assist the United States make its initial shift away from coal. However, this bridge argument at best glosses over important concerns and practical alternatives, including the growing cost-competitiveness of renewable energy,[27] the scale of potent greenhouse gases such as methane leaking from natural gas operations and transportation[28,29] and the technological lock-in from building out natural gas infrastructure.[30] Secretary of State Hilary Clinton spearheaded the Global Shale Gas Initiative that promoted shale and marketed US companies' shale expertise, even in countries with growing opposition in local communities to shale extraction.[31,32]

The Obama administration downplayed concerns over the health and environmental impacts of shale operations. For instance, it attempted to misrepresent the conclusion of a report by the Environmental Protection Agency (EPA) on the impacts of hydraulic fracturing on drinking water. Investigative reporters revealed that political appointees had attempted to revise the top-line conclusions that had been written by EPA scientists to state that there was no widespread evidence of negative impacts on drinking water resources.[21,33] Such a revision (abandoned when the investigative report broke) would have mischaracterized the report's findings – the failure to systematically collect publicly accessible data before and after shale operations made it difficult for researchers to determine the extent to which the lifecycle of shale operations affected drinking water resources and, indeed, the report laid out incidents in which water quality from water wells deteriorated following shale operations in their vicinity.

Nevertheless, faced with mounting evidence of the adverse impacts from the industry, the administration took several steps to mitigate those costs. The Obama administration promulgated several new regulations on the oil and gas sector, for instance, to curb the flaring of natural gas in shale oil operations on public lands and to reduce methane leaks in oil and gas operations (Chapter 4). The oil and gas industry and Republican state attorneys general filed lawsuits challenging these regulations and the courts granted a stay for a subset of these rules as cases were litigated (Chapters 4 and 7). The administration proposed to pare down tax preferences for oil companies, estimated to cost the US Treasury $4 billion per year[34] but Congress rejected those proposals.[35]

Echoing prior presidents' warnings, including those of Richard Nixon and George H. W. Bush,[36,ix] that reliance on oil alone cannot bring energy security, Obama underscored that Americans needed to "continually improve the utilization of renewable energy sources, new clean energy sources, and ... become more efficient in our use of energy."[24] The administration oversaw investments of $90 billion to promote clean energy, authorized under the 2009 American Recovery and Reinvestment Act, the economic rescue plan Congress enacted in response to the 2008 financial crisis (Chapter 11).

2.2 The Trump Administration (January 2017–January 2021)

The Trump administration drilled down the path of past administrations' support for domestic oil and gas expansion but failed to concede any downsides of such a strategy. Even the ardently pro-oil G. W. Bush had acknowledged in his 2006 State of the Union address, "Here we have a serious problem: America is addicted to oil."[37] Trump's America First Energy Plan focused unflinchingly on domestic oil and gas extraction as the cornerstone of its energy and economic strategy. It set in motion ambitious multipronged actions in support of the oil and gas sector within a system that was already bent to support that sector, including the expansion of oil and gas extraction and infrastructure in America's lands and seas (discussed in Parts II and III of this book); the undermining of the regulatory process to eliminate any restraints on the oil and gas sector (see Part IV); and an attempt to derail the private sector's actions to reduce their investments into the oil and gas sector (see Part V).

To be sure, the shale expansion brought economic benefits to the United States.[38] By January 2017, shale oil and shale gas accounted for 50 percent of total US oil production and about 63 percent of US total gas production (Figures 2.2 and 2.3). (Less than a decade before, in December 2008, shale gas and shale oil accounted for 16 percent and 12 percent, respectively, of total US gas and crude oil production.)[39] The expansion in domestic gas production, when the US gas markets were largely decoupled from world markets, reduced gas prices for US consumers, leading to savings of $200 per year for gas-consuming

---

[ix] While President George H. W. Bush supported expansion of oil extraction, he also supported energy efficiency programs. He signed into law the 1990 Clean Air Act Amendments that established the EPA's energy efficiency program known as Energy Star. He also signed into law the 1992 Energy Policy Act that implemented other energy efficiency programs.

households.[40] The expansion in gas production enabled the transformation of the United States into a net exporter of gas beginning in 2017 (Figure 2.3). The expansion in oil production also contributed to the decline in US net imports of petroleum (Figure 2.2). (High US oil consumption, which surpassed that of any other country,[x] meant that it remained a net oil importer, despite being the largest global producer of oil in 2019 and 2020.)[41–43]

The America First Energy Plan vision of oil and gas extraction as a source of jobs and prosperity for Americans hinged on a rosy assessment of the shale sector that had driven the past expansion and that was expected to drive the future expansion of the oil and gas industry. I pull back the curtain on that assessment by examining the financial foundation of the shale sector (Section 2.3), the bailout of the industry (Section 2.4) and the nuanced benefits from the industry to oil and gas workers and local communities (Section 2.5).

2.3 The Financial Foundation and Outlook of the Oil and Gas Sector

The Trump administration's optimistic predictions of expansion for the US oil and gas sector were predicated on the *continued* growth of the shale sector. I drill down to the economics of the shale sector (Section 2.3.1),[xi] its financial reckoning (Section 2.3.2) and the poor outlook for the oil and gas sector even before COVID-19 (Section 2.3.3).

2.3.1 Shale Economics: Debt Financing to Sustain the Drilling Treadmill

Shale oil and gas wells experience rapid depletion in their production. In order to maintain adequate production, companies must sustain the treadmill of drilling new wells. This rapid depletion of wells, plus the likely trajectory of shale companies drilling first in resource-rich areas before moving to less productive areas, led several energy analysts to cast doubt early in the shale era on the financial fundamentals of the shale sector.[44–46] Views that technological innovations could reduce costs sufficiently to make shale operations

---

[x] In 2020, the top global oil producers were the United States (20 percent of global production), Saudi Arabia (12 percent), Russia (11 percent), Canada (6 percent) and China (5 percent). Top consumers were the United States (20 percent of global consumption), China (14 percent), India (5 percent), Russia (4 percent) and Japan (4 percent).

[xi] Bethany McLean's *Saudi America* discusses the financial foundations of the shale sector and how the shale sector was able to attract funds from investors despite early warning signs about the uncertain financial returns and early evidence of poor returns to investors [195].

profitable eclipsed these early concerns, but, as described below, these early concerns were subsequently vindicated. Shale companies relied on a debt-financed strategy, borrowing heavily to drill new wells.[47] Companies also borrowed to buy more acreage, not only to lock-in promising reservoirs but to buoy the companies' perceived value to attract investors.[48] Debt financing to grow a company can be a legitimate financial strategy provided that the company becomes financially viable, a threshold not achieved by a number of shale companies, even by the end of 2019 (Section 2.3.3).

Independent oil and gas companies drove the early shale boom. That success led numerous other independent oil and gas companies to pursue shale drilling. From 2010 through 2011, oil giants Exxon and Chevron entered the shale sector by purchasing or pursuing joint ventures with independent oil and gas companies.[49] Other oil companies headquartered outside the United States followed suit, including Shell, Total, Statoil and BHP Billiton, as well as industrial conglomerates such as India's Reliance.[50–53]

2.3.2 Shale Bust

By the end of 2011, shale gas production had precipitated a regional gas glut and falling prices.[52] "American natural gas production growth is essentially useless at this particular point in time since you can't make any profit on North American natural gas," EOG Resources Inc. Chief Executive Mark Papa told an energy conference in late November.[52] "We are all losing our shirts today," Exxon CEO Rex Tillerson said in a talk before the Council on Foreign Relations in New York in June 2012, "We're making no money. It's all in the red." Tillerson said energy companies would not be able to continue drilling unless prices rose.[54]

The shale oil sector suffered a bust when global oil prices crashed in mid-2014 and remained low until 2016.[55,56] OPEC decided not to curb production to temper the falling oil prices but instead attempted to capitalize on the falling prices to gain market share and to push out US shale oil producers, who have higher costs.[57] Energy analysts and investors conceded that the shale sector did not generate positive cash flows even with the high oil prices at $100 per barrel in early 2014.[58]

In 2015, Steve Schlotterbeck, former CEO of shale behemoth EQT, called the shale industry an "unmitigated disaster for the buy/hold investor."[59] Similarly, in 2015, hedge-fund manager David Einhorn[xii] announced that,

---

[xii] Einhorn had made the call on the dot.com bubble [60, 61]. Frenzied investments into stocks of internet-related companies drove their skyrocketing prices. The stock market suffered a crash in 2000 when the dot.com bubble burst.

according to his analysis, 16 publicly traded shale producers from 2006 to 2014 spent $80 billion more than they received from selling oil; and companies could stay afloat only with the infusion of capital.[60,61] In 2017, Wall Street widely acknowledged poor returns to investors and called on shale companies to shift strategies to produce profits, rather than channel funds into production.[62] Between the start of 2015 and September 2018, shale companies with a combined debt of $171 billion filed for bankruptcy protection.[63]

Despite a moderate recovery, more bad news came in 2019 confirming early warnings by shale skeptics that projections of output from shale wells had been overstated.[64] In January 2019, the *Wall Street Journal* published an analysis showing consistent, sizable overestimates of output projections from shale wells.[65] In August 2019, the CEO of Pioneer Natural Resources announced that the Permian Midland Basin in Texas would likely be the only section of the basin to produce after 2025 at oil prices of $55 per barrel.[66] He conceded that wells had become less productive as operators depleted the resource-rich areas and moved to less productive areas. Well production also declined because operators drilled closely spaced wells that drew from the same reservoir.[66,67] A December 2019 report by IHS Markit confirmed first-year-production declines of 65–85 percent in wells in the Permian Basin,[68] a shale basin that had been predicted to drive growth in US oil production.[69]

By the end of 2019, many banks and traditional shale investors had cut off credit lines and shale companies had slashed production, all leading to an acceleration in the number of shale companies declaring bankruptcy.[70–72,xiii] In January 2020, North American oil and gas companies were estimated to hold $200 billion of debt that would mature by 2024, with $40 billion of that debt coming due in 2020.[63]

2.3.3 Oil and Gas Industry Outlook before COVID-19

Ample signs of uncertainties warn against the United States premising its economic development strategy on expanding domestic oil and gas extraction and on exporting oil and gas. US oil producers face projections that future global oil demand will plateau, with unabated supply from low-cost producers such as Saudi Arabia. In 2019, the International Energy Agency projected that it would plateau by 2030 due to cheaper renewable energy generation, the shift

---

[xiii] Oil and gas companies that went bankrupt numbered 21 in 2015, 58 in 2016, 15 in 2017, 21 in 2018, and 25 by August 2019 [70][72]. Under a chapter-11 bankruptcy, the company restructures its debt and finances. Under a chapter-7 bankruptcy, the company ceases operations. Both can result in losses for creditors and shareholders. For instance, in 2016, Quicksilver Resources, a Texas-based gas producer, went into chapter-11 bankruptcy protection with creditors expected to lose about $2 billion [196].

to electrified transportation, and government policies to cut greenhouse gas emissions.[73] That echoed similarly pessimistic projections from Standard & Poor Global Platts that oil consumption would peak in 2040 and from the BP Energy Outlook that oil demand already peaked in 2020.[74,75]

US natural gas producers turned to export markets to address the gas glut,[64] but US liquefied natural gas (LNG) exports face competition from cheaper exporters, such as Qatar. While the shift away from coal to natural gas sustains demand for natural gas, a number of localities in Europe are shifting directly to renewable energy, denting the demand for natural gas. For instance, the European Parliament's call to cut carbon dioxide emissions by 55 percent from 1990 levels by 2030 would mean substantial shares of the projected gas demand could be displaced by more renewables and energy efficiency.[76] The European Investment Bank announced that it would end its investment in natural gas by the end of 2021.[77]

Another strategy to absorb US natural gas production – supporting the growth of ethane cracker plants – faces overcapacity in both US and global markets. Oversupply drove down ethylene prices in the United States from $739 per metric ton in 2017 to $347 per metric ton in July 2020.[78] Oil- and gas-producing countries' build-up of production of ethane (or the substitute naphtha) compounded the global overcapacity problem.[79] The chemical divisions of oil and gas companies suffered losses, and plans for petrochemical plants in the US Gulf Coast (where the industry is well established) were shelved.[61,79–81] At the same time, global demand for plastics is expected not to grow as rapidly as a result of bans and restrictions on single-plastic use and requirements for recycled content for plastics.[82,xiv]

2.4 The 2020 Oil and Gas Bailout

The shaky financial foundations of the shale industry and its crushing debt ensnared American taxpayers through the March 2020 bailout of the oil and gas sector. That bailout was part of the economic rescue package that Congress enacted following the global pandemic (Section 2.4.1). Congress failed to legislate hard rules[xv] on the conditions for allocating bailout funds to corporations but

---

[xiv] Nevertheless, these bans on single-use plastics and requirements of recycled content are not expected to make a major dent on plastics use. Governments worldwide would need to take much more stringent actions in order to substantially curb plastics use [82].

[xv] Several members of Congress had advocated for Congress to legislate conditions for corporations to receive bailout funds including salary and employment guarantees for workers. Damon Silvers, the deputy chair of the Congressional Oversight Panel for the 2008 bailout of the US financial sector, drawing lessons from the mistakes of the past bailout, called for the establishment of a board to make decisions on the allocation of funding and of an oversight committee with subpoena powers. Regrettably, Congress did not legislate these safeguards.

instead granted broad powers to the secretary of the treasury, Steve Mnuchin, to implement the bailout.[83–85] The bailout provides a stark illustration of how a subset of wealthier Americans in the oil and gas sectors, not workers or communities, were able to capture taxpayers' funds (Section 2.4.2), thanks to the support of the administration and pro-oil members of Congress (Section 2.4.3).

2.4.1 The Coronavirus Aid, Relief, and Economic Security (CARES) Act

In March 2020, Congress enacted the Coronavirus Aid, Relief, and Economic Security Act (CARES) Act that Trump had signed into law. Even within the context of a bailout that prioritized corporations over workers and households,[xvi] the oil and gas sector's bailout was substantial.[86,87] As of December 2020, US taxpayers had purchased $475 million in corporate bonds, including those rated as junk bonds,[xvii] from the energy sector (primarily oil and gas, refinery and pipeline companies) (Table 2.1). US taxpayers also provided $828 million in loans to the oil and gas sector.[88] It received this assistance through programs funded by the CARES Act and administered by the Federal Reserve (Table 2.2). Oil and gas companies were also able to secure immediate and massive tax rebates from the CARES Act's changes in the US tax code (Table 2.2).

2.4.2 Losses to American Taxpayers and Misguided Signals in the Economy

The Federal Reserve's purchase of debt issued by the oil and gas sector has been criticized by Sarah Bloom Raskin, former Federal Reserve Board governor and former deputy secretary of the treasury; Eugene Ludwig, former comptroller of currency,[102–104] tax reform groups;[105] environmental groups;[96] and a number of congressional representatives.[106] They cite at least four adverse impacts on taxpayers and the US economy.

First, taxpayers face high risk of losses, given the poor past performance of that sector and its uncertain future prospects. In the five years prior to the

---

[xvi] The bailout cost $4 trillion. Of the $2.3 trillion that went to businesses, $651 billion went to tax breaks for companies, $454 billion went to the Federal Reserve ostensibly to stabilize markets and $670 billion went to the Paycheck Protection Program, ostensibly to help workers. Only 16 percent of the $4 trillion went to fighting the health crisis, including $253 billion to help state governments and public agencies pay for protective equipment, first responders and other pandemic impacts [86].

[xvii] For instance, Continental Resources's corporate debt was downgraded to junk by Standard & Poor's on March 27, 2020. The Federal Reserve bought $2.5 million in its bonds in June 2020 and $2 million more in July 2020 [95]. Junk bonds are debt rated by credit rating agencies as below investment grade, an indication that these bonds are highly risky.

Table 2.1 *Corporate bonds purchased from selected oil and gas companies under the Secondary Market Corporate Credit Facility under the CARES Act (as of December 2020)*

Oil and gas companies	BP Capital Market	$53 million
	Marathon Oil/Marathon Petroleum/MPLX	$33 million
	Chevron	$24 million
	Exxon Mobil	$22 million
Pipeline companies	Kinder Morgan	$33 million
	Spectra Energy Partners (subsidiary of Enbridge)	$27 million
	Western Midstream	$22 million
	Williams	$20 million
LNG terminal	Sabine Pass Liquefaction	$29 million
Oil and gas service companies	Halliburton	$10 million
	Baker Hughes	$11 million

Source: Federal Reserve Board, *Secondary Market Corporate Credit Facility Transaction Specific Disclosure Report*.[88]

bailout, the oil- and gas-drilling sector faced the largest cuts in credit ratings compared to other sectors in the US economy.[97,98] Moreover, even after the Federal Reserve's purchases of the debt of 19 oil and gas companies, credit rating agencies downgraded the short-term and long-term debt of 12 of these companies.[95] Second, the Federal Reserve's purchase provided a misleading signal to investors of confidence in the debt issued by the oil and gas sector and helped entice the flow of capital into that sector. With the Federal Reserve's purchase of their debt, several of these companies were able to borrow even more funds from the capital markets by issuing $60 million in new bonds.[95,xviii]

Third, the Federal Reserve's actions created a moral hazard in the investment market by enabling investors to reap the benefits from the initial large payout from junk bonds but to avoid losses from their risky gamble by selling off those bonds to the US Federal Reserve. Investors in junk bonds, who are typically wealthier than the average American, were able to offload the less desirable bonds from their portfolio and enjoyed a transfer of wealth from taxpayers.[103] Fourth, the Federal Reserve's holding of large amounts of oil and gas debt endangers the entire economy. The Federal Reserve itself had warned that financial institutions' holdings of substantial oil and gas assets threatens the stability of the entire US financial sector, and its expansive holdings of oil and gas debt only accentuate that instability (Chapter 11).

[xviii] Diamond Bank received a tax rebate of $9.7 million and subsequently asked a bankruptcy judge to permit that same amount as bonus payments to nine senior executives [92].

Table 2.2 *Benefits to oil and gas companies from the CARES Act*

Provisions in the CARES Act that made changes to the US tax code
Purpose: In theory, provisions providing tax relief in 2020 to corporations could have helped abate their pandemic-related financial losses and make that revenue available to keep workers paid and employed. *Beneficiaries*: These provisions permitted numerous corporations to secure immediate and significant tax rebates that came with no strings attached.[89–91] The ability to deduct losses in 2018, 2019 and 2020 from the previous five years of tax liabilities was especially useful for oil and gas companies that had made profits in earlier years but suffered sharp downturns pre-COVID-19.[92] Analysis of Securities and Exchange Commission data by the nongovernmental organization Documented revealed that a number of large oil and gas drillers, service companies and refiners took millions in tax rebates, including Marathon Petroleum Corporation ($1.1 billion), Occidental ($195 million), Valero ($117 million) and Devon ($98 million).[93] Reporting by Bloomberg[xix] reveals that several companies that had secured rebates remunerated company executives and financed stock buybacks, while shedding jobs.[92]
Secondary Market Corporate Credit Facility (SMCCF) Program
Purpose: The Fed made credit available to corporations by purchasing corporate bonds[xx] through the SMCCF.[94] The Fed rules permit the purchase of debt rated at least BBB-/Baa3 (i.e., the lowest investment-grade rating) as of March 22, 2020, but the debt rating at the time of purchase can in fact be lower. *Beneficiaries*: By September 2020, the Fed had purchased $355 million in bonds issued by oil and gas companies.[95] About 17 percent of the energy bonds purchased were highly risky junk bonds (compared to only 9 percent of such bonds purchased across sectors).[96] The Federal Reserve's purchases of debt in the energy sector were also larger relative to the energy sector's share of employment, equity market and outstanding debt compared to other sectors of the economy as represented by 1,500 companies in the Standard & Poor index.[97,98]
Main Street Lending Program
Purpose: In theory, these Federal Reserve-backed loans[xxi] were to provide bridge funds for companies that had performed well prior to COVID-19. *Beneficiaries*: Analysis by NGO BailoutWatch revealed that 46 fossil fuel companies received subsidized loans totaling $828 million or 13 percent of the total loans as of November 2020.[99]

[xix] The SMCCF makes purchases in the secondary market of corporate bonds and US listed exchange trade funds with broad exposure to the US corporate bond market. The Department of the Treasury made a $75 billion equity investment to back both the SMCCF and the Primary Market Corporate Credit Fund and the company BlackRock, which serves as the investment manager for the SMCFF.

[xx] Under the program, for most loans, the Federal Reserve purchased 95 percent of the loan from the originating bank (and thus taxpayers assumed the risks for unpaid loans) and left 5 percent of the loan held by the originating bank.

[xxi] Analysis of the loan data found that companies worth as much as $15 million received loans [93].

Table 2.2 *(cont.)*

Paycheck Protection Program (PPP)
Purpose: In theory, this program was to provide loans to "small" businesses to protect workers.[xxii]   *Beneficiaries*: About 7,100 oil, gas and petrochemical companies received funds totaling between $3 billion and $7 billion.[93] Reporting by the *Washington Post* on the PPP generally (not the oil and gas sector specifically) reveals that a number of companies that received large PPP loans did not use funds to keep workers employed[100] and that the bulk of the funds went to larger businesses.[101] About 5 percent of the program's recipients received 50 percent of the $522 billion, with 600 companies receiving $10 million each.[101]

2.4.3 Weakening Program Safeguards and Bypassing Oil and Gas Workers

Trump, Secretary of Energy Dan Brouillette[xxiii] and members of Congress channeled support from the rescue package to the oil and gas sector.[107–110] Trump tweeted "We will never let the great US Oil & Gas Industry down. I have instructed the Secretary of Energy and Secretary of the Treasury to formulate a plan which will make funds available so that these very important companies and jobs will be secured long into the future!" As revealed by media investigations, their efforts led to the opening of the Federal Reserve's programs to companies that had performed poorly prior to the pandemic. Changes to the Main Street Lending Program rules enabled more heavily indebted oil and gas companies to receive larger loans.[xxiv] The removal of the prohibition against the use of the loans to pay interest on existing loans and to refinance preexisting debt proved helpful to indebted oil and gas companies.[109–111] Contrary to its stated goal of keeping workers employed, the administration weakened the requirements on companies to retain and rehire workers (in contradiction to the precise numerical requirement set out in the CARES Act).[105]

[xxii] Those bonds amounted to 60 percent of the debt issued by the energy sector between March 2020 and September 2020.

[xxiii] Senator Ted Cruz's letter to Federal Reserve Chair Jerome Powell and Secretary of Treasury Steve Mnuchin argued for relaxation of restrictions of loans to indebted companies and the use of loans for refinancing preexisting debt. According to a Bloomberg report, Brouillette stated that he and the treasury secretary worked with the Fed "to modify its Main Street Lending Program to include more mid-size companies in order to help oil firms cope with the plunge in crude prices" [108].

[xxiv] It raised the cap for new loans from $25 million to $35 million [110] and for refinancing of existing loans from $200 million to $300 million [111].

Even as Trump's rhetoric emphasized jobs, his administration chose not to pursue economic rescue strategies that could have provided direct support to workers who lost their jobs in the oil and gas sector. Direct support of unemployed workers as an economic rescue strategy has been advocated by both conservative and progressive-leaning economists.[112] However, the administration did not fund the request from a consortium of 31 state governments to employ oil and gas sector workers to cap abandoned wells, of which there are growing numbers.[113] This type of direct job assistance can absorb unemployed workers, while ensuring the flow of income to communities highly reliant on oil and gas extraction.[114,115]

2.5 Nuanced Contribution of the Oil and Gas Sector to Local Economic Development

The Trump administration and members of Congress who championed assistance to the oil and gas sector in the rescue package emphasized its contributions to US economic development. In reality, the contribution of the oil and gas sector to jobs (Section 2.5.1) and to economic development has been more nuanced[xxv] (Sections 2.5.2–2.5.5), and these facts call for a reassessment of economic development strategies that rely on oil and gas expansion.

2.5.1 Jobs

In 2019, nearly 1.7 million Americans worked in fossil fuel industries – ranging from extraction to construction to manufacturing.[116] While this number is only about 1.1 percent of total employment, these jobs tend to be concentrated in fossil-fuel-reliant communities. In certain counties in West Texas, Oklahoma, Wyoming, North Dakota and West Virginia, for instance, about 30–50 percent of all workers are employed in fossil fuel industries.[116]

Nevertheless, three facts call for a reassessment of deepening reliance on jobs in the oil and gas sector to achieve economic prosperity for American workers. At the national level, jobs in the oil extraction sector, which are the higher-paying jobs in the sector, have declined (Figure 2.6) despite overall increase in production (Figures 2.2 and 2.3). Oil and gas companies reported that they accelerated their shift to automation as a cost-reduction strategy following the 2014–16 shale oil bust and that automation was

---

[xxv] Anthony Ladd's *Fractured Communities* documents the nuanced social impacts of shale development in local communities [197].

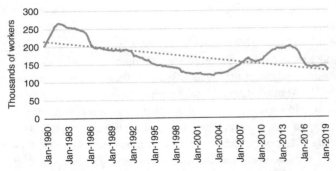

Figure 2.6 Number of jobs in oil and gas extraction (NAICS 211) 1980–2019
Source: US Bureau of Labor Statistics, "Oil and Gas Extraction," 2021.[118]

expected to reduce employment by 25 percent.[117] The new technologically oriented jobs in the oil and gas sector require different skill sets from those possessed by the earlier cohorts of oil and gas extraction workers, raising the crisis of displaced workers even within the oil and gas sector itself.[117]

Second, employment in the sector suffers the drawback of booms and busts. The boom phase of shale development, without question, created jobs. One study documented that the shale boom contributed 640,000 jobs and reduced the unemployment rate by 0.43 percentage points during the Great Recession.[119] Another study reports that the shale boom from 2005 to 2011 created 560,000 jobs in nonurban counties and average income grew by 11 percent.[120] Nevertheless, jobs at the wells peter out after the drilling phase, and job numbers rely on the treadmill of new wells (Section 2.3). The drilling phase of each well equates to roughly 13 associated jobs, whereas only 0.2–0.4 full-time-equivalent jobs per well remain in the production phase.[121,122]

Third, overall job growth in some shale regions has been disappointing despite massive production increases. For instance, a study of the Marcellus Shale from 2008 through 2019 reports little job growth in counties that were the major producers of natural gas.[123] Moreover, in areas without preexisting oil and gas operations, most of the jobs in the sector went to workers outside the counties where wells were located.[124]

2.5.2 Economic Development in Oil- and Gas-Reliant Communities

Economic development strategies should ideally aim to secure the longer-term financial health of local communities. However, several studies that span both booms and busts warn of the overall income decline and communities' overreliance on oil and gas extraction. One study of oil booms and busts in the 1970s and 1980s documents how residents in counties relying on oil and gas extraction suffered overall decline in income, with their average income decline in the busts offsetting their income increases during the booms.[125] Another study of oil booms and busts in the western United States from 1969 through 1998 finds that employment conditions and income per capita were lower than the counterfactual case had the boom not occurred.[126] Likewise, another study of the American West from 1980 through 2011 reports that specialization in oil and gas extraction led to lower per capita income and educational attainment.[127] Overdependency on oil and gas can crowd out other economic activities and lead to the loss of the systems sustaining those activities, making these hard to revive when extractive activities peter out.[122]

Oil and gas extraction provides revenue to mineral rights owners, who can earn rents from leasing out these rights to companies and earn royalties when oil and gas are produced. However, this revenue channel brings only limited benefits to local communities in numerous parts of the United States where nonresidents own the bulk of the mineral rights.[122,128] Farming households that are fortunate enough to own the underlying mineral rights have earned income that has enabled them to continue farming.[129] However, farming households have also faced negative spillovers from shale activities.[129]

Shale development has also led to inequitable distribution of benefits and costs as a result of the severance of ownership of mineral rights from that of surface rights, alongside the dominance of mineral rights over surface rights and limited legislative protections for surface owners (Section 2.6.1). Because extraction can proceed over the objection of surface owners and with minimal protections offered to them,[130,131] mineral owners reaped benefits, while surface owners bear the brunt of the costs. Even so, not all mineral owners have walked away happy from their deals; some have found that shale companies were able to impose unfavorable terms, such as deductions of significant postproduction costs, that diminished mineral owners' royalty incomes.[132]

2.5.3 Local Government Management of the Boom–Bust Cycles

A number of local governments had welcomed shale operations for the tax revenue they could bring.[133,134] However, the boom–bust of oil and gas development has posed significant challenges for local governments in providing local services to the shale industry, and in some cases the bust has left local governments holding substantial debt. The US tax system is structured such that local communities are responsible for financing the bulk of local services, such as schools, police and fire services, with only limited contributions from state governments.[133,134] Local governments, particularly in rural, less densely populated areas, face major challenges in providing local services to the sudden and large population influx.[135,136] Several local governments made the misstep of taking on projects beyond their means,[137] typically overestimating the length of the boom.[138]

The fiscal consequences of miscalculations are significant. For instance, Williston, North Dakota nearly quadrupled its debt and other liabilities from 2008 to 2014 to $158 million.[139] In response to Williston's failing tax receipts, the credit rating agency Moody downgraded Williston's general obligation bonds to junk status in 2016.[139] Sadly, Williston was repeating the mistake it had made in the last boom in the 1980s, that is, taking on debt that it struggled to but finally did pay off in the 2000s.[139]

2.5.4 State and Federal Tax Policies

State governments' tax policies – both on the collection of taxes from oil and gas companies and on the investment of those tax revenues – have fallen short of prudent management that could have secured longer-lasting benefits from oil and gas extraction.[21] State governments' short-sighted response to the boom, such as slashing tax revenue from other sources, worsened their reliance on the revenue from the oil and gas sector and led to drastic cuts in social services during the bust.

One rationale for severance taxes (i.e., a tax imposed on the removal of a natural resource) on oil and gas extraction is that it draws down a nonrenewable resource. In theory, these taxes should be accompanied by societally beneficial investments, such as in education or infrastructure, that would ensure the long-run productivity of the economy. While most states have imposed severance taxes on the extraction of oil and gas, others have failed to do so. Pennsylvania's Republican majority state legislature, for instance, has rejected the adoption of such a tax multiple times. (Pennsylvania has imposed impact fees on oil and gas operations and a substantial fraction of that revenue is directed to local governments. However, the receipt of impact fees by local governments has not always covered the damages experienced by local communities from shale activity.[140])

Only a few states, North Dakota for instance, created permanent trust funds to direct some tax revenue to longer-term investments and rainy-day savings.[141] Earnings from North Dakota's Legacy Fund helped it weather budget deficits during the shale downturn. Sadly, several states with permanent funds did not consistently direct tax revenue into those funds, for instance New Mexico. Others even raided those trust funds for current spending, for instance Alabama.[142] States such as Oklahoma and Louisiana worsened their exposure to the boom–bust cycle by reducing alternative tax revenues such as income taxes and by suspending severance-tax collection during the bust. Both states then cut spending drastically, including for schools and social services, to cope with their budget deficits,[143,144] having foregone tax collections of $1 billion in Oklahoma from 2012 through 2015[144] and $1.1 billion in Louisiana from 2010 through 2014.[145]

The federal government fell short in collecting taxes from oil and gas companies, especially compared to tax rates imposed on companies in other sectors. The oil and gas sector enjoys generous tax preferences, some that date back to the 1920s. These tax preferences cost the Department of the Treasury an estimated $41.4 billion in foregone revenue over a 10-year period.[34,146,147,xxvi] The oil and gas sector, along with the rest of the corporate sector, also benefitted from the reduction in corporate tax rates in the Tax Cuts and Jobs Act of 2017. Thanks to all these tax preferences, several major companies in oil and gas extraction and infrastructure did not pay any corporate taxes in 2018 and even received tax rebates,[148] including Chevron, Cliffs Natural Resources, Devon Energy, EOG Resources, Halliburton, Kinder Morgan, MDU Resources and Occidental Petroleum.[148]

State legislators have justified their decisions to forego or suspend severance taxes to entice oil and gas companies to stay in their states. At the federal level, Congressional representatives have defended tax preferences for the oil and gas sector to ensure domestic production. However, several studies find that oil drilling and production are fairly insensitive to state tax rates, suggesting that higher state tax rates would raise revenue without significant declines in in-state production.[149–151] Studies also report that removal of federal tax preferences would lead to only minimal reduction in US domestic oil and gas production[34,152] (Chapter 11).

2.5.5 Impairment of Quality of Life and Livelihoods

A comprehensive assessment of oil and gas extraction as an economic development strategy cannot overlook its public health and environmental impacts. These operations have competed with life-sustaining needs and other economic

---

[xxvi] In 2017, these tax preferences cost the federal government $3.2 billion from the expensing of intangible drilling costs, $1.7 billion from taking the depreciation as a percentage of revenue, and $1.1 billion from the domestic manufacturing deduction [34].

pursuits. For instance, hydraulic fracturing in the semiarid regions of the United States competes for water with agriculture and urban municipalities.[153] Oil and gas operations have also impaired economic development in a number of communities that suffered contamination and damages to their health and livelihoods.[154,155,xxvii] Several local communities across shale basins suffered contamination of their drinking water wells, traced to shale operators' spills of fracking fluids and wastewater at the surface or to the migration of methane out of poorly cased gas wells.[156–158] Large wastewater spills, which contain high concentrations of salts and in some cases heavy metals and naturally occurring radioactivity, have impaired swaths of agricultural lands.[159] Lands littered with abandoned wells have been taken out of productive use.[160] Contaminants from oil and gas extraction have been spread, in some cases *legally*, into waterways and onto land through a variety of ways, including the discharge of inadequately treated wastewater into waterways,[161,xxviii] the use of wastewater for irrigation in the American West as beneficial use,[162] and the spread of wastewater for deicing of roads.[163]

Natural gas from the shale boom, which facilitated the shift away from coal-fired power plants, did improve air quality and brought health benefits.[164,165] However, oil and gas extraction has also caused deterioration in local and regional air quality. The short setback distance separating oil and gas wells from homes and schools has exposed communities to poor air quality and to safety risks.[166] The intentional flaring of gas in operations that focus on oil extraction is commonplace, as well as methane leaks and emissions of volatile organic compounds.[167–169] A growing number of studies document adverse health impacts in areas close to shale operations in rural and suburban communities, lower-income communities, and Hispanic communities in Texas.[169–174] Scientific studies document the adverse impacts of the shale industry throughout its entire lifecycle, adding to the already substantial evidence on the adverse impacts of the conventional oil and gas industry.[175–177]

2.6 Institutions That Enabled the Externalization of Costs

The oil and gas industry has been able to externalize public health and environmental costs to the rest of Americans as a result of its privileges under US common law, state law and federal law (Sections 2.6.1–2.6.3 and detailed in my

---

xxvii State and federal regulators' failure to monitor the environment and to track contamination events means full counts of incidents are unavailable.

xxviii Pennsylvania's Department of Environment Protection initially permitted wastewater from unconventional oil and gas operations to be treated in municipal treatment plants.

edited volume *The Shale Dilemma*[21]). These privileges, secured prior to the Trump administration, meant regulatory rollbacks by the Trump administration, discussed in the next chapters, attacked a system that already offered limited protections for local communities.

2.6.1 Dominance of Mineral Rights over Surface Rights

One of the earliest instances of preferential treatment for oil and gas came through the courts as early as the 1850s. As mining proceeded, it became common for ownership of the mineral rights to be split from that of the surface rights. At the time, courts saw mining as an essential driver of economic growth, and the common law evolved to grant dominance of the mineral rights over the surface rights.[178] As discussed in Section 2.5.2, that dominance of mineral rights and the split ownership of the mineral and the surface rights in many parts of the United States[122,128] means that oil and gas extraction has bestowed benefits on mineral owners and externalized the costs to surface owners. While state legislatures have subsequently moved toward granting greater protections for surface owners, protections remain limited in most states and mineral rights still take precedence.[179] The orientation of US laws in favor of drilling has not only prioritized mineral rights owners but specifically those owners who support drilling. Forced pooling laws in some states mandate that the owners of the majority share of mineral rights located in spatial proximity can proceed with extraction over the objection of owners who hold the minority share of mineral rights.[180]

2.6.2 State Governments Preempted Local Governments

State governments have opposed efforts by some local governments to place restrictions on shale operations in their jurisdictions. State governments have generally supported extraction as they derive tax revenue and other benefits from extraction, while local communities receive some benefits but bear more of the costs.

State governments have been able to rely on laws that were enacted at a time when oil was far scarcer and one of the few sources of energy. Those laws, unsurprisingly, focused on incentivizing drilling.[181] State courts in Colorado, New Mexico and West Virginia ruled against local governments' long-term moratoria as contradicting state laws that favor oil and gas expansion.[182] States such as Texas and Oklahoma have asserted that state powers preempt local communities' powers and state lawmakers have enacted legislation to block local governments from restricting shale development.[183]

Proponents of state preemption argue that granting local powers to restrict shale development grants local governments veto power against shale development and threatens much-needed revenue streams to the rest of the state. However, the blanket preemption pursued by state governments gives little option for local communities to negotiate for more benefits, while they are forced to shoulder most of the cost. A number of local governments resorted to striking memoranda of understanding with oil and gas operations to achieve protections for local communities.[184]

Only in a few cases have local governments successfully fought battles to assert their powers. For instance, in 2013, Pennsylvania's Supreme Court, citing the Environmental Rights Amendment in Pennsylvania's constitution, ruled that the state government could not restrict the powers of local municipalities to regulate oil and gas operations within their borders.[185] While that court ruling offered protections, many local governments in Pennsylvania, which lacked the expertise to develop these zoning regulations, had not done so as late as 2017 and even 2019.[186,187] Mineral owners, who stand to lose from drilling restrictions, opposed local governments' attempts to limit drilling to the industrial zones.[187]

Communities have succeeded in making statewide changes only in a few cases. Colorado is one of the only states where the governor and legislature overhauled the state law, first, to grant powers to local governments to restrict shale operations and, second, to recast the mission of oil and gas regulators to protect public health and the environment.[182] That legislation was enacted in 2019, after a Democratic governor without strong ties to oil and gas and a Democratic majority legislature took power, and after at least two decades of legal and political battles waged against the oil and gas industry by local communities (particularly in Colorado's Front Range that hosted wells near homes and schools).

2.6.3 State Governments' Limited Regulatory Responses

Many state governments, which are the primary regulators of the oil and gas sector, did not undertake comprehensive public health and environmental assessments prior to the expansion of shale extraction. Instead, the industry was initially regulated under preexisting, inadequate conventional oil and gas regulations. The lack of publicly available baseline and monitoring data, the lack of disclosure laws, and the pervasiveness of nondisclosure agreements in settlements between companies and victims of contamination made it difficult for public health and environmental advocates to document the urgency for stricter regulatory responses.[20]

With mounting contamination incidents, state governments did step up regulations, but gaps remained. The oil and gas sector's significant political campaign contributions to politicians who hold pro-oil viewpoints[188] and the shift to many more pro-oil legislators following the shale boom[189] contributed to legislative decisions that placed greater weight on the oil and gas sector's contributions than on its cost to communities.[xxix]

The cost of remediating the abandoned wells that pockmark the US landscape demonstrates a clear case in which legislators were aware of poor past practices of the oil and gas sector.[190] Yet, legislators failed to update regulations on bonding to reflect the actual cost of remediating abandoned wells and left the legacy of this cost to communities and taxpayers. In 2016, there were 2.1–2.4 million abandoned wells that had not been capped throughout the United States.[115,191] The cost of plugging and remediating a mere 56,000 of those wells is estimated to range from $1.4 billion to $2.7 billion.[115] Abandoned wells have contaminated drinking-water sources and are likely to continue to do so.[192] Abandoned wells are also a major contributor to global greenhouse gas emissions, ranking as the tenth largest source of anthropogenic methane in the United States.[193]

2.7 Conclusion

Trump's America First Energy Plan claimed that oil and gas expansion would bring jobs to workers and prosperity to the average American and would insulate the United States from the volatility of global markets. Evidence reviewed in this chapter indicates that these claims rang hollow even in the years prior to the pandemic. The 2020 bailout of the oil and gas sector demonstrates how executives, shareholders and bondholders in the oil and gas industry benefitted handsomely, while local communities, workers and taxpayers lost dismally from the administration's financial support for the industry.

Workers and communities in the oil and gas sector built their livelihoods around oil and gas operations and their hard work energized America's progress. However, the costs they have borne from overreliance on the oil and gas

---

[xxix] It would not be surprising if patterns observed by these studies in Congress occur also in state legislatures. A study by Goldberg et al. finds that oil and gas companies reward congressional members' antienvironmental votes in one congressional session with a greater financial contribution in the next session [188]. A study by Cooper et al reveals that following the boom in the Marcellus Shale, Republicans who were more inclined to be pro-shale and less supportive of environmental protections enjoyed greater electoral success [189].

sector and its uncertain prospects call for a shift away from economic development strategies that are anchored in oil and gas extraction.

The notion that Americans have little choice but to accept a trade-off – granting financial and legal privileges for the oil and gas sector to externalize its costs to the rest of society in exchange for securing energy, jobs and economic development – is a false bargain in the twenty-first century. Many of the privileges bestowed to the industry, whether in the past under both Republican- and Democrat-led administrations and congresses, or during the more recent Trump era, are simply no longer justified. Renewable energy provides cost-competitive alternative energy resources and fossil-fuel-reliant communities have feasible, albeit challenging, pathways to diversify their economies, a topic I turn to next.

References

1. Energy Information Administration. *History of Energy Consumption in the United States, 1775–2009* (February 9, 2011). www.eia.gov/todayinenergy/detail.php?id=10.
2. Energy Information Administration. *US Energy Facts Explained: Imports and Exports* (April 27, 2020). www.eia.gov/energyexplained/us-energy-facts/imports-and-exports.php.
3. J. Proville and J. Camuzeaux. "Six Ways President Trump's Energy Plan Doesn't Add Up." *Environmental Defense Fund Blog*, 2017. http://blogs.edf.org/climate411/2017/03/23/six-ways-president-trumps-energy-plan-doesnt-add-up.
4. T. DiChristopher. "Trump's Energy Plan Overstates Benefits of More Drilling: Economists." *CNBC*, September 6, 2016. www.cnbc.com/2016/09/06/trumps-energy-plan-overstates-benefits-of-more-drilling-economists.html.
5. Columbia Center on Global Energy Policy. *Economic Volatility in Oil Producing Regions: Impacts and Federal Policy Options*. Report by D. Raimi et al. (New York: October 30, 2019). www.energypolicy.columbia.edu/research/report/economic-volatility-oil-producing-regions-impacts-and-federal-policy-options.
6. J. Yaffa. "How the Russian–Saudi Oil War Went Awry – for Putin Most of All." *New Yorker Magazine*, April 15, 2021.
7. P. Sabin. "Crisis and Continuity in US Oil Politics, 1965–1980." *Journal of American History* 99, no. 1 (June 1, 2012): 177–186. https://academic.oup.com/jah/article/99/1/177/855009.
8. D. S. Painter. "Oil and the American Century." *Journal of American History* 99, no. 1 (2012): 24–39.
9. D. Painter. *Oil and the American Century*. Baltimore, MD: Johns Hopkins University Press, 1986.
10. M. Jacobs. *Panic at the Pump: The Energy Crisis and the Transformation of American Politics in the 1970s*. New York: Hill and Wang, 2016.

11. S. Coll. *Private Empire: Exxon Mobil and American Power.* New York: Penguin Books, 2013.
12. M. Auzanneau. *Oil, Power, and War: A Dark History.* White River Junction, VT: Chelsea Green Publishing, 2018.
13. Energy Information Administration. *Annual Energy Outlook 2020* (Washington, DC: January 29, 2020). www.eia.gov/outlooks/aeo/pdf/AEO2020%20Full%20Report.pdf.
14. National Energy Policy Development Group. *National Energy Policy: Reliable, Affordable, and Environmentally Sound Energy for America's Future.* Report by D. Cheney et al. (2001). www.wtrg.com/EnergyReport/National-Energy-Policy.pdf.
15. Energy Information Administration. *Petroleum & Other Liquids* (2021). www.eia.gov/petroleum/data.php.
16. Energy Information Administration. *Natural Gas* (2021). www.eia.gov/naturalgas/data.php.
17. G. Muttitt. *Fuel on the Fire: Oil and Politics in Occupied Iraq.* New York: The New Press, 2012.
18. M. Klare. *Blood and Oil: The Dangers and Consequences of America's Growing Petroleum Dependency.* New York: Metropolitan Books, 2005.
19. Energy Information Administration. *Renewable Energy Explained* (May 20, 2021). www.eia.gov/energyexplained/renewable-sources/#:~:text=In%202020%2C%20renewable%20energy%20provided,of%20total%20U.S.%20energy%20consumption.
20. S. Gamper-Rabindran. "Information Collection, Access, and Dissemination to Support Evidence-Based Shale Gas Policies." *Energy Technology* 2, no. 12 (2014): 977–987. https://onlinelibrary.wiley.com/doi/full/10.1002/ente.201402114.
21. S. Gamper-Rabindran, ed. *The Shale Dilemma: A Global Perspective on Fracking and Shale Development.* Pittsburgh, PA: University of Pittsburgh Press, 2018.
22. M. Finkel, J. Hays and A. Law. "The Shale Gas Boom and the Need for Rational Policy." *American Journal of Public Health* 103, no. 7 (2013): 1161–1163.
23. T. Sangaramoorthy. "Maryland Is Not for Shale: Scientific and Public Anxieties of Predicting Health Impacts of Fracking." *Extractive Industries and Society* 6, no. 2 (2019): 463–470.
24. White House Office of the Press Secretary. "Remarks by the President on American-Made Energy." Obama Administration. March 22, 2012. https://obamawhitehouse.archives.gov/the-press-office/2012/03/22/remarks-president-american-made-energy.
25. Congressional Budget Office. *The Economic and Budgetary Effects of Producing Oil and Natural Gas from Shale.* Report by R. Gecan et al. (December 2014). www.cbo.gov/publication/49815.
26. White House Office of the Press Secretary. "President Barack Obama's State of the Union Address." Obama Administration. January 28, 2014. https://obamawhitehouse.archives.gov/the-press-office/2014/01/28/president-barack-obamas-state-union-address.
27. Lazard. *Levelized Cost of Energy and Levelized Cost of Storage* (October 19, 2020). www.lazard.com/perspective/levelized-cost-of-energy-and-levelized-cost-of-storage-2020.

28. A. R. Brandt et al. "Methane Leaks from North American Natural Gas Systems." *Science* 343, no. 6172 (2014): 733–735.
29. R. A. Alvarez et al. "Assessment of Methane Emissions from the US Oil and Gas Supply Chain." *Science* 361, no. 6398 (2018): 186–188. https://science.sciencemag.org/content/361/6398/186.
30. K. C. Seto et al. "Carbon Lock-In: Types, Causes, and Policy Implications." *Annual Review of Environment and Resources* 41 (2016): 425–452.
31. L. Fang and S. Horn. "Hillary Clinton's Energy Initiative Pressed Countries to Embrace Fracking, New Emails Reveal." *The Intercept*, May 23, 2016. https://theintercept.com/2016/05/23/hillary-clinton-fracking/+&cd=4&hl=en&ct=clnk&gl=ca.
32. S. Horn. "Obama Alums Are Pushing Fracked Gas Exports. That's Exactly What Trump Wants." *DeSmog Blog*, February 2, 2018. www.desmogblog.com/2018/02/02/obama-officials-trump-energy-dominance.
33. T. Scheck and S. Tong. "EPA's Late Changes to Fracking Study Downplayed Risk of Polluted Drinking Water." *American Market Place*, November 30, 2016. www.apmreports.org/story/2016/11/30/epa-changes-fracking-study.
34. G. Metcalf. "The Impact of Removing Tax Preferences for US Oil and Gas Production." *Journal of the Association of Environmental and Resource Economists* 5, no. 1 (2018): 1–37.
35. Congressional Research Service. *Oil and Natural Gas Industry Tax Issues in the FY 2014 Budget Proposal*. Report by R. Pirog. R42374 (Washington, DC: October 30, 2013). https://fas.org/sgp/crs/misc/R42374.pdf.
36. R. Nixon. "President Nixon's Nationwide Radio Address on the National Energy Situation." *New York Times*, January 20, 1974. www.nytimes.com/1974/01/20/archives/transcript-of-nixons-speech-on-energy-situation-a-call-for.html.
37. G. W. Bush. "State of the Union Address." George W. Bush White House. January 31, 2006.
38. C. F. Mason, L. A. Muehlenbachs and S. M. Olmstead. "The Economics of Shale Gas Development." *Annual Review of Resource Economics* 7, no. 1 (2015): 269–289.
39. Energy Information Administration. *Today in Energy: EIA Adds New Play Production Data to Shale Gas and Tight Oil Reports*. Principal contributors: J. Perrin and E. Geary (February 15, 2019). www.eia.gov/todayinenergy/detail.php?id=38372.
40. C. Hausman and R. Kellogg. "Welfare and Distributional Implications of Shale Gas." *Brookings Papers on Economic Activity*. Conference draft, March 2015. www.brookings.edu/wp-content/uploads/2016/07/2015a_hausman.pdf.
41. Energy Information Administration. *Frequently Asked Questions: What Countries Are the Top Producers and Consumers of Oil?* (April 1, 2021). www.eia.gov/tools/faqs/faq.php?id=709&t=6.
42. bp Energy Economics. *Energy Outlook* (2020). www.bp.com/content/dam/bp/business-sites/en/global/corporate/pdfs/energy-economics/energy-outlook/bp-energy-outlook-2020.pdf.
43. R. Rapier. "The World's Top 10 Oil Producers and Consumers." *Forbes*, June 26, 2020. www.forbes.com/sites/rrapier/2020/06/26/the-worlds-top-10-oil-producers-and-oil-consumers.
44. J. D. Hughes. "A Reality Check on the Shale Revolution." *Nature* 494 (February 20, 2013): 307–308. www.nature.com/articles/494307a.

45. Post Carbon Institute. *Drill, Baby, Drill: Can Unconventional Fuels Usher in a New Era of Energy Abundance?* Report by J. D. Hughes (Santa Rosa, California: February 19, 2013). www.postcarbon.org/publications/drill-baby-drill.
46. Energy Policy Forum. *Shale and Wall Street: Was the Decline in Natural Gas Prices Orchestrated?* Report by D. Rogers (February 2013). https://shalebubble.org/wp-content/uploads/2013/02/SWS-report-FINAL.pdf.
47. R. Elliott. "US Shale Producers Struggle on Wall Street." *Wall Street Journal*, August 4, 2019.
48. A. Loder. "Energy Giant Undone by Big Land Grab." *Washington Post*, July 8, 2012.
49. R. Gold. "Exxon Acquires Two Marcellus Shale Gas Drillers." *Wall Street Journal*, June 8, 2011. www.wsj.com/articles/SB10001424052702304392704576374103408464670.
50. Á. González and A. Flynn. "Shell Gains Boost in Shale Output with Latest Deal." *Wall Street Journal*, September 13, 2012.
51. B. Casselman. "Total Will Buy Texas Gas-Field Stake." *Wall Street Journal*, January 4, 2012.
52. D. Strumpf and R. Dezember. "Natural Gas Ends 2011 at 27-Month Low." *Wall Street Journal*, December 31, 2011.
53. G. Fouche. "Statoil Agrees $1.3 Billion US Shale Gas JV with Talisman." *Reuters*, October 10, 2010. www.reuters.com/article/us-norway-talisman-idUSTRE69913Y20101010.
54. J. A. DiColo and T. Fowler. "Exxon: 'Losing Our Shirts' on Natural Gas." *Wall Street Journal*, June 27, 2012.
55. L. Kilian. "Why Did the Price of Oil Fall after June 2014?" *Vox EU*, February 25, 2015. https://voxeu.org/article/causes-2014-oil-price-decline.
56. B. C. Prest. "Explanations for the 2014 Oil Price Decline: Supply or Demand?" *Energy Economics* 74 (2018): 63–75.
57. A. Behar and R. A. Ritz. "An Analysis of OPEC's Strategic Actions, US Shale Growth and the 2014 Oil Price Crash." International Monetary Fund working paper WP/16/131 (July 2016). www.imf.org/external/pubs/ft/wp/2016/wp16131.pdf.
58. Oxford Institute for Energy Studies. *US Shale Oil Dynamics in a Low Price Environment.* Report by T. Curtis (November 2015). www.oxfordenergy.org/wpcms/wp-content/uploads/2015/11/WPM-62.pdf.
59. H. Richards. "Is US Shale Facing an 'Unmitigated Disaster'?" *E&E News*, September 19, 2019.
60. S. Foley. "Einhorn Targets US 'Frack Addicts.'" *Financial Times*, May 4, 2015.
61. E. Crooks. "Shale Looks More Like Dotcom Boom than Lehman Debt Bubble." *Financial Times*, May 6, 2015.
62. B. Olson and L. Cook. "Wall Street Tells Frackers to Stop Tallying Barrels, Focus on Profits." *Wall Street Journal*, December 17, 2017.
63. R. Dezember. "Energy Producers Face Big Tab after Shale Bonanza." *Wall Street Journal*, January 2, 2020.
64. A. Berman. "Shale Gas Is Not a Revolution." *Forbes*, July 5, 2017. www.forbes.com/sites/arthurberman/2017/07/05/shale-gas-is-not-a-revolution.
65. B. Olson, R. Elliott and C. M. Matthews. "Fracking's Secret Projection Gap – Analysis Shows Many Wells Underperform." *Wall Street Journal*, January 3, 2019.

66. L. Hampton. "Top US Shale Producer Offers Bleak View of US Output Growth." *Reuters*, August 7, 2019. www.reuters.com/article/us-pioneer-natl-rsc-results-idUSKCN1UX1SF.
67. N. Newman. "Is the US Shale Boom Winding Down?" *Rigzone*, August 6, 2019. www.rigzone.com/news/is_the_us_shale_boom_winding_down-06-aug-2019-159498-article.
68. Staff. "Permian Basin Decline Rates Have 'Increased Dramatically' amid Ongoing Slowdown." *Journal of Petroleum Technology*, December 11, 2019. https://jpt.spe.org/permian-basin-decline-rates-have-increased-dramatically-amid-ongoing-slowdown.
69. Energy Information Administration. *Today in Energy: Permian Region Is Expected to Drive US Crude Oil Production Growth through 2019*. Report by D. Murali (August 23, 2018). www.eia.gov/todayinenergy/detail.php?id=36936.
70. R. Adams-Heard, D. Wethe and K. Gupta. "Epitome of America's Shale Gas Boom Now Warns It May Go Bust." *Bloomberg*, November 5, 2019.
71. Institute for Energy Economics and Financial Analysis. *Mounting Negative Cash Flows Highlight Struggles of Appalachian Fracked Gas Producers*. Report by K. Hipple et al. (November 2019).
72. Haynes and Boone, LLP. *Oil Patch Bankruptcy Monitor* (March 31, 2021). www.haynesboone.com/-/media/Files/Energy_Bankruptcy_Reports/Oil_Patch_Bankruptcy_Monitor.
73. International Energy Agency. *World Energy Outlook 2019* (2019). www.iea.org/reports/world-energy-outlook-2019.
74. R. Perkins. "Global Oil Demand Set to Plateau, Not Decline by 2040: IEA." *S&P Global*, October 13, 2020.
75. Engine No. 1. *Energy Transformations: Technology, Policy, Capital and the Murky Future of Oil and Gas*. Report by D. G. Victor (March 3, 2021). https://reenergizexom.com/wp-content/uploads/2021/03/Energy-Transformations-Technology-Policy-Capital-and-the-Murky-Future-of-Oil-and-Gas-March-3-2021.pdf.
76. S. Elliott. "Decarbonize or Die: Is Europe Turning Its Back on Gas?" *S&P Global Insight Blog*, February 20, 2020. www.spglobal.com/platts/en/market-insights/blogs/natural-gas/022020-decarbonize-or-die-is-europe-turning-its-back-on-gas.
77. K. Taylor. "'Gas Is Over': EU Bank Chief Signals Phaseout of Fossil Fuel Finance." *Climate Home News*, January 21, 2021. www.climatechangenews.com/2021/01/21/gas-eib-president-signals-complete-phase-unabated-fossil-fuels.
78. Deloitte Insights. *Building Resilience in Petrochemicals: Navigating Disruption and Preparing for New Opportunities*. Report by D. Dickson, D. Yankovitz and A. Hussain (2020). www2.deloitte.com/content/dam/insights/us/articles/6878_ER-I-Building-resilience-in-downstream-chemicals-article-3-Petro/DI_Building-resilience-in-petrochemicals.pdf.
79. M. Pooler. "Producing a Plastics Backlash." *Financial Times*, February 13, 2020.
80. H. Robertson. "Now for the Glut: Petrochemical Supplies Are Still Rising but Demand Growth Is Less Certain than It Was." *Petroleum Economist*, July 5, 2016. www.petroleum-economist.com/articles/midstream-downstream/refining-marketing/2016/now-for-the-glut.

81. West Virginia Center on Budget & Policy and Institute for Energy Economics and Financial Analysis. *Falling Short: Shale Development in West Virginia Fails to Deliver on Economic Promises*. Report by C. Kunkel et al. (February 2019).
82. International Energy Agency. *Chemicals*. Report by P. Levi, T. Vass, H. Mandová and A. Gouy (June 2020). www.iea.org/reports/chemicals.
83. D. Silvers. "Repeating the Mistakes of the 2008 Bailout." *American Prospect*, March 24, 2020. https://prospect.org/economy/repeating-the-mistakes-of-the-2008-bailout/.
84. E. Warren. *Concerns Regarding the Distribution of Funds to Large Corporations Under the Coronavirus Aid, Relief, and Economic Security (CARES) Act*. Submitted to S. Mnuchin, Department of the Treasury secretary, and J. Powell, Board of Governors of the Federal Reserve System. March 31, 2020.
85. J. Bivens and H. Shierholz. "Despite Some Good Provisions, the CARES Act Has Glaring Flaws and Falls Short of Fully Protecting Workers during the Coronavirus Crisis." *Working Economics Blog* (Economic Policy Institute), 2020.
86. P. Whoriskey, D. MacMillan and J. O'Connell. "'Doomed To Fail': Why a $4 Trillion Bailout Couldn't Revive the American Economy." *Washington Post*, October 5, 2020.
87. S. Gamper-Rabindran. "Fracked Communities and Taxpayers: Shale Economics in the US and Argentina." In *Oxford Handbook on Comparative Environmental Politics*, edited by J. Sowers, S. VanDeveer and E. Weinthal. Oxford University Press, 2021. Online.
88. Federal Reserve Board. *Secondary Market Corporate Credit Facility Transaction Specific Disclosure Report* (January 11, 2021). www.federalreserve.gov/monetarypolicy/smccf.htm.
89. T. Metheson. "Who Benefits from the CARES Act Tax Cuts?" *TaxVox* (Tax Policy Center at the Urban Institute & Brookings Institution), April 17, 2020. www.taxpolicycenter.org/taxvox/who-benefits-cares-act-tax-cuts.
90. S. M. Rosenthal and A. Boddupalli. "Heads I Win, Tails I Win Too: Winners from the Tax Relief for Losses in the CARES Act." *TaxVox* (Tax Policy Center at the Urban Institute & Brookings Institution), April 20, 2020. www.taxpolicycenter.org/taxvox/heads-i-win-tails-i-win-too-winners-tax-relief-losses-cares-act.
91. D. Butler, S. Mufson and D. MacMillan. "How the Cares Act Gave Millions to Energy Companies with No Strings Attached." *Washington Post*, October 6, 2020.
92. J. A. Dlouhy. "'Stealth Bailout' Shovels Millions of Dollars to Oil Companies." *Bloomberg*, May 15, 2020. www.bloomberg.com/news/articles/2020-05-15/-stealth-bailout-shovels-millions-of-dollars-to-oil-companies.
93. A. Juhasz. "Bailout: Billions of Dollars of Federal COVID-19 Relief Money Flow to the Oil Industry." *Sierra: The National Magazine of the Sierra Club*, August 26, 2020. www.sierraclub.org/sierra/bailout-billions-dollars-federal-covid-19-relief-money-flow-oil-industry.
94. S. Gilchrist et al. "The Fed Takes on Corporate Credit Risk: An Analysis of the Efficacy of the MCCF." National Bureau of Economic Research working paper 27809 (September 2020). www.nber.org/system/files/working_papers/w27809/w27809.pdf.

95. BailoutWatch, Friends of the Earth and Public Citizen. *Big Oil's $100 Billion Bender: How the US Government Provided a Safety Net for the Flagging Fossil Fuel Industry.* Report by L. Ross, A. Zibel, D. Wagner and C. Kuveke (September 2020). https://prismic-io.s3.amazonaws.com/bailout/1b1e1458-bbff-49bc-a636-f6cbd47a88af_Big+Oils+Billion+Dollar+Bender.pdf.
96. Algalita Marine Research & Education et al. *Concern over the Failure of the Board of Governors of the Federal Reserve System to Serve the Public Interest and Promote Financial Stability.* Submitted to J. Powell, chair of the Board of Governors of the Federal Reserve System. July 30, 2020.
97. InfluenceMap. *Necessary Intervention or Excessive Risk? Corporate Bond Risk Before and after COVID-19 amid the Fed's Market* (June 23, 2020). https://influencemap.org/report/Necessary-Intervention-or-Moral-Hazard-5e42adc35b315cc44a75c94af4ead29c.
98. InfluenceMap. *Update: Necessary Intervention or Excessive Risk? Corporate Bond Risk before and after COVID-19 amid the Fed's Market* (September 2020).
99. "Oil & Gas Dominates in 'Main Street' Lending Program." *BailoutWatch*, December 16, 2020. https://bailoutwatch.org/analysis/mslp-november-analysis.
100. P. Whoriskey. "PPP Was Intended to Keep Employees on the Payroll. Workers at Some Big Companies Have Yet to Be Rehired." *Washington Post*, July 27, 2020.
101. J. O'Connell et al. "More than Half of Emergency Small-Business Funds Went to Larger Businesses, New Data Shows." *Washington Post*, December 2, 2020.
102. S. B. Raskin. "Why Is the Fed Spending So Much Money on a Dying Industry?" *New York Times*, June 10, 2020.
103. G. Ludwig and S. B. Raskin. "How the Fed's Rescue Program Is Worsening Inequality." *Politico*, May 28, 2020. www.politico.com/news/agenda/2020/05/28/how-the-feds-rescue-program-is-worsening-inequality-287379.
104. R. Kuttner. "The Bailout, the Fed, and the Aftermath." *American Prospect*, April 21, 2020. https://prospect.org/economy/the-bailout-the-fed-and-the-aftermath.
105. Americans for Financial Reform Education Fund. *Comments on Primary and Secondary Market Corporate Lending Facilities, Main Street Lending Facilities, Municipal Lending Facility, and Term Asset Lending Facility.* Submitted to Staff Groups for Primary and Secondary Market Corporate Lending Facilities, Main Street Lending Facilities, Municipal Lending Facility, and Term Asset Lending Facility. April 16, 2020.
106. E. Warren. *Concerns Regarding Reports That You Are Considering Bailing Out Oil and Gas Companies Using Loans Backed by Taxpayer Funds Provided by the Coronavirus Aid, Response, and Economic Stability (CARES) Act.* Submitted to S. Mnuchin, Department of the Treasury secretary. May 6, 2020.
107. T. Cruz, US senator. *Support for the Federal Reserve and Treasury to Take Urgent Action to Ensure Access to Capital for America's Oil and Gas Industry.* Submitted to J. Powell, chair of the Board of Governors of the Federal Reserve System and S. Mnuchin, Department of the Treasury secretary. April 24, 2020.
108. S. Mohsin and A. Natter. "Energy Chief Says Fed Was Asked to Expand Lending for Oil Firms." *Bloomberg*, May 12, 2020.
109. D. Grandoni. "The Energy 202: Oil and Gas Companies Stand to Gain from Fed Loosening Coronavirus Loan Rules." *Washington Post*, May 1, 2020.

110. T. Mann and B. Mullins. "Texas Fracking Billionaires Drew Covid-19 Aid While Investing in Rivals." *Wall Street Journal*, December 27, 2020.
111. N. Timiraos. "Fed Makes Terms More Favorable for Main Street Lending Program." *Wall Street Journal*, June 8, 2020.
112. G. Hubbard et al. "Taskforce Report: Promoting Economic Recovery after COVID-19." *Aspen Economic Strategy Group*, June 16, 2020. www.economic strategygroup.org/publication/promoting-economic-recovery-after-covid-19.
113. N. Groom. "States Ask Trump Administration to Pay Laid Off Oil Workers to Plug Abandoned Wells." *Reuters*, May 6, 2020. www.reuters.com/article/us-global-oil-usa-wells-idUSKBN22I2KA.
114. J. T. L. Project. *Workers and Communities in Transition*. Report by J. M. Cha et al. (2021). www.labor4sustainability.org/files/JTLP_report2021.pdf.
115. J. Bordoff, D. Raimi and N. Nerurkar. "Green Stimulus for Oil and Gas Workers: Considering a Major Federal Effort to Plug Orphaned and Abandoned Wells." *Columbia Center on Global Energy Policy*, July 20, 2020. www.energypolicy.colum bia.edu/research/report/green-stimulus-oil-and-gas-workers-considering-major-fed eral-effort-plug-orphaned-and-abandoned.
116. A. Tomer, J. W. Kane and C. George. "How Renewable Energy Jobs Can Uplift Fossil Fuel Communities and Remake Climate Politics." *Brookings*, February 23, 2021. www.brookings.edu/research/how-renewable-energy-jobs-can-uplift-fossil -fuel-communities-and-remake-climate-politics/?preview_id=1414272.
117. C. Matthews. "Oil's New Technology Spells End of Boom for Roughnecks." *Wall Street Journal*, July 10, 2018.
118. Bureau of Labor Statistics. "Oil and Gas Extraction: NAICS 211." Updated 2021. www.bls.gov/iag/tgs/iag211.htm.
119. J. Feyrer, E. T. Mansur and B. Sacerdote. "Geographic Dispersion of Economic Shocks: Evidence from the Fracking Revolution." *American Economic Review* 107, no. 4 (April 2017): 1313–1334. https://mansur.host.dartmouth.edu/papers/ feyrer_mansur_sacerdote_frackingjobs.html.
120. P. Maniloff and R. Mastromonaco. "The Local Employment Impacts of Fracking: A National Study." *Resource and Energy Economics* 49 (2017): 62–85.
121. Marcellus Shale Education and Training Center. *Pennsylvania Statewide Marcellus Shale Workforce Needs*. Report by T. L. Brundage et al. (Williamsport, PA: 2011).
122. T. W. Kelsey, M. D. Partridge and N. E. White. "Unconventional Gas and Oil Development in the United States: Economic Experience and Policy Issues." *Applied Economic Perspectives and Policy* 38, no. 2 (2016): 191–214.
123. Ohio River Valley Institute. *Appalachia's Natural Gas Counties: Contributing More to the US Economy and Getting Less in Return*. Report by S. O'Leary (February 12, 2021). https://ohiorivervalleyinstitute.org/wp-content/uploads/ 2021/02/Frackalachia-Report-update-2_12_01.pdf.
124. R. K. Gittings and T. Roach. "Who Benefits from a Resource Boom? Evidence from the Marcellus and Utica Shale Plays." *Energy Economics* 87 (2020): 104489.
125. G. D. Jacobsen, D. N. Parker and J. B. Winikoff. "Are Resource Booms a Blessing or a Curse? Evidence from People (Not Places)." *Journal of Human Resources* (2021): 0320-10761R1.

126. G. D. Jacobsen and D. P. Parker. "The Economic Aftermath of Resource Booms: Evidence from Boomtowns in the American West." *Economic Journal* 126, no. 593 (2016): 1092–1128.
127. J. H. Haggerty et al. "Longterm Effects of Income Specialization in Oil and Gas Extraction: The US West, 1980–2011." *Energy Economics* 45 (2014): 186–195. https://doi.org/10.1016/j.eneco.2014.06.020.
128. T. Fitzgerald. "Importance of Mineral Rights and Royalty Interests for Rural Residents and Landowners." *Choices*, 2014. www.choicesmagazine.org/choices-magazine/theme-articles/is-the-natural-gas-revolution-all-its-fracked-up-to-be-for-local-economies/importance-of-mineral-rights-and-royalty-interests-for-rural-residents-and-landowners.
129. J. H. Haggerty et al. "Tradeoffs, Balancing, and Adaptation in the Agriculture-Oil and Gas Nexus: Insights from Farmers and Ranchers in the United States." *Energy Research & Social Science* 47 (2019): 84–92.
130. A. Boslett, T. Guilfoos, and C. Lang. "Valuation of the External Costs of Unconventional Oil and Gas Development: The Critical Importance of Mineral Rights Ownership." *Journal of the Association of Environmental and Resource Economists* 6, no. 3 (2019): 531–561.
131. A. R. Collins and K. Nkansah. "Divided Rights, Expanded Conflict: Split Estate Impacts on Surface Owner Perceptions of Shale Gas Drilling." *Land Economics* 91, no. 4 (November2015): 688–703.
132. C. J. Sachs, D. E. Bugden and R. C. Stedman. "Grand Theft Hydrocarbon? Post-Production Clauses and Inequity in the US Shale Gas Industry." *Extractive Industries and Society* 7, no. 4 (2020): 1443–1450.
133. R. G. Newell and D. Raimi. "The Fiscal Impacts of Increased US Oil and Gas Development on Local Governments." *Energy Policy* 117 (2018): 14–24.
134. R. G. Newell and D. Raimi. "US State and Local Oil and Gas Revenue Sources and Uses." *Energy Policy* 112 (2018): 12–18.
135. Multi-State Shale Research Collaborative. *Executive Summary: Assessing the Impacts of Shale Drilling County Case Studies* (April 10, 2014). https://docs.google.com/viewer?a=v&pid=sites&srcid=ZGVmYXVsdGRvbWFpbnxtdWx0aXN0YXRlc2hhbGV8Z3g6NGU4MjIyNWU5ZjFhZjM4Yg.
136. J. H. Haggerty et al. "Geographies of Impact and the Impacts of Geography: Unconventional Oil and Gas in the American West." *Extractive Industries and Society* 5, no. 4 (November 2018): 619–633. https://doi.org/10.1016/j.exis.2018.07.002.
137. K. K. Smith et al. "Using Shared Services to Mitigate Boomtown Impacts in the Bakken Shale Play: Resourcefulness or Over-adaptation?" *Journal of Rural and Community Development* 14, no. 2 (2019).
138. Energy & Local Economies. *The Relationship between Oil and Gas Development and Businesses in McKenzie, Richland, Sheridan, and Tioga Counties*. Report by J. Haggerty et al. (January 2018). www.montana.edu/energycommunities/documents/BusinessFinal-PDF.pdf.
139. E. Scheyder. "In North Dakota's Oil Patch, A Humbling Comedown." *Reuters Investigates*, May 18, 2016. www.reuters.com/investigates/special-report/usa-northdakota-bust.

140. A. P. Behrer and M. S. Mauter. "Allocating Damage Compensation in a Federalist System: Lessons from Spatially Resolved Air Emissions in the Marcellus." *Environmental Science & Technology* 51, no. 7 (March 3, 2017): 3600–3608. https://pubs.acs.org/doi/pdf/10.1021/acs.est.6b04886?rand=tyllxaov.
141. B. G. Rabe and R. L. Hampton. "Trusting in the Future: The Re-emergence of State Trust Funds in the Shale Era." *Energy Research & Social Science* 20 (October 2016): 117–127.
142. D. Saha and M. Muro. "Permanent Trust Funds: Funding Economic Change with Fracking Revenues." *Metropolitan Policy Program at Brookings*, April 2016. www.brookings.edu/wp-content/uploads/2016/07/Permanent-Trust-Funds-Saha-Muro-418-1.pdf.
143. Energy Information Administration. *State Severance Tax Revenues Decline As Fossil Fuel Prices Drop*. Report by R. McManmon and G. Nülle (January 12, 2016). www.eia.gov/todayinenergy/detail.php?id=24512.
144. L. Cohen and J. Schneyer. "When the Oil Boom Went Bust, Oklahoma Protected Drillers and Squeezed Schools." *Reuters*, May 17, 2016. www.reuters.com/investigates/special-report/usa-oklahoma-bust/.
145. Performance Audit Services. *Severance Tax Suspension for Horizontal Wells*. Report by D. G. Purpera et al. (August 19, 2015). https://app.lla.state.la.us/PublicReports.nsf/65C7443D8D09105F86257EA6007174D9/$FILE/00009E0B.pdf.
146. Office of Management and Budget. *FY 2013 Administration Budget*. White House. 2012.
147. Congressional Budget Office. *Options for Reducing the Deficit: 2014 to 2023*. 2013.
148. Institute on Taxation and Economic Policy. *Corporate Tax Avoidance Remains Rampant under New Tax Law*. Report by M. Gardner, S. Wamhoff, M. Martellotta and L. Roque (April 11, 2019). https://itep.org/notadime.
149. U. Chakravorty, S. Gerking and A. Leach. "State Tax Policy and Oil Production: The Role of the Severance Tax and Credits for Drilling Expenses." In *US Energy and Tax Policy*, edited by G. E. Metcalf. 305–337. Cambridge: Cambridge University Press, 2010.
150. M. Kunce et al. "State Taxation, Exploration, and Production in the US Oil Industry." *Journal of Regional Science* 43, no. 4 (2003): 749–770.
151. J. P. Brown, P. Maniloff and D. T. Manning. "Spatially Variable Taxation and Resource Extraction: The Impact of State Oil Taxes on Drilling in the US." *Journal of Environmental Economics and Management* 103 (2020): 102354.
152. Resources for the Future. *Eliminating Subsidies for Fossil Fuel Production: Implications for US Oil and Natural Gas Markets*. Report by M. Allaire and S. Brown. Issue Brief 09-10 (December 2009). https://media.rff.org/documents/RFF-IB-09-10.pdf.
153. B. R. Scanlon et al. "Will Water Issues Constrain Oil and Gas Production in the United States?" *Environmental Science & Technology* 54, no. 6 (February 16, 2020): 3510–3519. https://pubs.acs.org/doi/abs/10.1021/acs.est.9b06390.
154. R. B. Jackson et al. "The Environmental Costs and Benefits of Fracking." *Annual Review of Environment and Resources* 39 (October 2014): 327–362. https://doi.org/10.1146/annurev-environ-031113-144051.

155. J. L. Adgate, B. D. Goldstein and L. M. McKenzie. "Potential Public Health Hazards, Exposures and Health Effects from Unconventional Natural Gas Development." *Environmental Science & Technology* 48, no. 15 (February 24, 2014): 8307–8320. https://pubs.acs.org/doi/abs/10.1021/es404621d.
156. A. Vengosh et al. "A Critical Review of the Risks to Water Resources from Unconventional Shale Gas Development and Hydraulic Fracturing in the United States." *Environmental Science & Technology* 48, no. 15 (March 7, 2014): 8334–8348. https://pubs.acs.org/doi/abs/10.1021/es405118y.
157. B. E. Fontenot et al. "An Evaluation of Water Quality in Private Drinking Water Wells near Natural Gas Extraction Sites in the Barnett Shale Formation." *Environmental Science & Technology* 47, no. 17 (2013): 10032–10040.
158. R. B. Jackson et al. "Increased Stray Gas Abundance in a Subset of Drinking Water Wells near Marcellus Shale Gas Extraction." *Proceedings of the National Academy of Sciences* 110, no. 28 (July 9, 2013): 11250–11255. www.pnas.org/content/110/28/11250.short.
159. N. E. Lauer, J. S. Harkness and A. Vengosh. "Brine Spills Associated with Unconventional Oil Development in North Dakota." *Environmental Science & Technology* 50, no. 10 (April 27, 2016): 5389–5397. https://pubs.acs.org/doi/10.1021/acs.est.5b06349.
160. V. Nallur, M. R. McClung and M. D. Moran. "Potential for Reclamation of Abandoned Gas Wells to Restore Ecosystem Services in the Fayetteville Shale of Arkansas." *Environmental Management* 66 (2020): 180-190.
161. K. J. Ferrar et al. "Assessment of Effluent Contaminants from Three Facilities Discharging Marcellus Shale Wastewater to Surface Waters in Pennsylvania." *Environmental Science & Technology* 47, no. 7 (2013): 3472–3481.
162. B. McDevitt et al. "Isotopic and Element Ratios Fingerprint Salinization Impact from Beneficial Use of Oil and Gas Produced Water in the Western US." *Science of the Total Environment* 716 (May 10, 2020): 137006. https://doi.org/10.1016/j.scitotenv.2020.137006.
163. T. L. Tasker et al. "Environmental and Human Health Impacts of Spreading Oil and Gas Wastewater on Roads." *Environmental Science & Technology* 52, no. 12 (May 20, 2018): 7081–7091. https://doi.org/10.1021/acs.est.8b00716.
164. A. P. Pacsi et al. "Regional Air Quality Impacts of Increased Natural Gas Production and Use in Texas." *Environmental Science & Technology* 47, no. 7: 3521–3527.
165. R. Johnsen, J. LaRiviere and H. Wolff. "Fracking, Coal, and Air Quality." *Journal of the Association of Environmental and Resource Economists* 6, no. 5: 1001–1037.
166. M. Haley et al. "Adequacy of Current State Setbacks for Directional High-Volume Hydraulic Fracturing in the Marcellus, Barnett, and Niobrara Shale Plays." *Environmental Health Perspectives* 124, no. 9 (2016): 1323–1333.
167. C. W. Moore et al. "Air Impacts of Increased Natural Gas Acquisition, Processing, and Use: A Critical Review." *Environmental Science & Technology* 48, no. 15 (March 3, 2014): 8349–8359. https://pubs.acs.org/doi/abs/10.1021/es4053472.
168. A. Pozzer, M. G. Schultz and D. Helmig. "Impact of US Oil and Natural Gas Emission Increases on Surface Ozone Is Most Pronounced in the Central United States." *Environmental Science & Technology* 54, no. 19 (September 9, 2020): 12423–12433. https://pubs.acs.org/doi/10.1021/acs.est.9b06983.

169. G. P. Macey et al. "Air Concentrations of Volatile Compounds near Oil and Gas Production: A Community-Based Exploratory Study." *Environmental Health* 13, no. 1 (2014): 1–18.
170. Staff. "Fractured: The Body Burden of Living Near Fracking." *Environmental Health News*, 2021. www.ehn.org/fractured-series-on-fracking-pollution-2650624600/far-reaching-impacts.
171. A. C. Kroepsch et al. "Environmental Justice in Unconventional Oil and Natural Gas Drilling and Production: A Critical Review and Research Agenda." *Environmental Science & Technology* 53, no. 12 (2019): 6601–6615.
172. L. M. McKenzie et al. "Population Size, Growth, and Environmental Justice near Oil and Gas Wells in Colorado." *Environmental Science & Technology* 50, no. 21 (2016): 11471–11480.
173. J. Currie, M. Greenstone and K. Meckel. "Hydraulic Fracturing and Infant Health: New Evidence from Pennsylvania." *Science Advances* 3, no. 12 (2017). https://doi.org/10.1126/sciadv.1603021.
174. E. Hill. "Shale Gas Development and Infant Health: Evidence from Pennsylvania." *Journal of Health Economics* 61 (2018): 134–150.
175. Institute of Medicine. *Health Impact Assessment of Shale Gas Extraction: Workshop Summary*. Washington, DC: National Academies of Sciences, Engineering, and Medicine, 2014. https://doi.org/10.17226/18376.
176. D. Kaden and T. Rose. *Environmental and Health Issues in Unconventional Oil and Gas Development*. Amsterdam: Elsevier, 2016.
177. M. L. Finkel. *The Human and Environmental Impact of Fracking: How Fracturing Shale for Gas Affects Us and Our World*. Santa Barbara, CA: Praeger, 2015.
178. S. S. Ryder and P. M. Hall. "This Land Is Your Land, Maybe: A Historical Institutionalist Analysis for Contextualizing Split Estate Conflicts in US Unconventional Oil and Gas Development." *Land Use Policy* 63 (2017): 149–159. www.sciencedirect.com/science/article/abs/pii/S0264837716303301.
179. M. A. Wegener. "Balancing Rights in a New Energy Era: Will the Mineral Estate's Dominance Continue?" *Houston Law Review* 57, no. 5 (2020): 1037–1082.
180. H. G. Robertson. "Get Out from under My Land: Hydraulic Fracturing, Forced Pooling or Unitization, and the Role of the Dissenting Landowner." *Georgetown Environmental Law Review* 30, no. 4 (2018): 633–694.
181. T. K. Righetti. "The Incidental Environmental Agency." *Utah Law Review* 2020, no. 3 (2020): 685–754.
182. T. K. Righetti, H. J. Wiseman and J. W. Coleman. "The New Oil and Gas Governance." *Yale Law Journal Forum* 130 (June 29, 2020): 51–77.
183. H. J. Wiseman. "Disaggregating Preemption in Energy Law." *Harvard Environmental Law Review* 40, no. 2 (2016): 293–350.
184. A. Shaffer, S. Zilliox and J. Smith. "Memoranda of Understanding and the Social Licence to Operate in Colorado's Unconventional Energy Industry: A Study of Citizen Complaints." *Journal of Energy & Natural Resources Law* 35, no. 1 (2017): 69–85.
185. J. C. Dernbach, J. R. May and K. T. Kristl. "Robinson Township v. Commonwealth of Pennsylvania: Examination and Implications." *Rutgers University Law Review* 67 (2015): 1169–1196.

186. D. Hopey. "Many Allegheny County Cities, Municipalities Unprepared for Shale Gas Drilling Rebound." *Pittsburgh Post-Gazette*, May 30, 2017. www.post-gazette.com/local/region/2017/05/30/Marcellus-Shale-drilling-Allegheny-County-zoning-ordinance-pa/stories/201705240023.
187. S. Bojarski. "Economy Borough Residents Fret the Potential Harm from Fracking. across the Region, Municipalities Regulate Drilling Very Differently." *Public Source*, October 16, 2019. www.publicsource.org/economy-borough-residents-fret-the-potential-harm-from-fracking-across-the-region-municipalities-regulate-drilling-very-differently.
188. M. H. Goldberg et al. "Oil and Gas Companies Invest in Legislators That Vote against the Environment." *Proceedings of the National Academy of Sciences* 117, no. 10 (2020): 5111–5112.
189. J. Cooper, S. E. Kim and J. Urpelainen. "The Broad Impact of a Narrow Conflict: How Natural Resource Windfalls Shape Policy and Politics." *Journal of Politics* 80, no. 2 (2018): 630–646.
190. A. L. Mitchell and E. A. Casman. "Economic Incentives and Regulatory Framework for Shale Gas Well Site Reclamation in Pennsylvania." *Environmental Science & Technology* 45, no. 22 (2011): 9506–9514.
191. M. Kang et al. "Reducing Methane Emissions from Abandoned Oil and Gas Wells: Strategies and Costs." *Elsevier Energy Policy* 132 (2019): 594–601.
192. Ground Water Protection Council. *State Oil and Gas Agency Groundwater Investigations and Their Role in Advancing Regulatory Reforms – A Two-State Review: Ohio and Texas*. Report by S. Kell (August 2011). www.atlanticaenergy.org/pdfs/natural_gas/Environment/State%20Oil%20&%20Gas%20Agency%20Groundwater%20Investigations_US_GWProCoucil.pdf.
193. J. P. Williams, A. Regehr and M. Kang. "Methane Emissions from Abandoned Oil and Gas Wells in Canada and the United States." *Environmental Science & Technology* 55, no. 1 (2021): 563–570. https://pubs.acs.org/doi/abs/10.1021/acs.est.0c04265.
194. Energy Information Administration. *Shale Gas and Oil Plays, Lower 48 States* (June 30, 2016). www.eia.gov/maps/maps.htm.
195. B. McLean. *Saudi America: The Truth about Fracking and How It's Changing the World*. New York: Columbia Global Reports, 2018.
196. C. Helman. "The 15 Biggest Oil Bankruptcies (So Far)." *Forbes*, May 9, 2016.
197. A. Ladd (ed.). *Fractured Communities: Risk, Impacts, and Protest against Hydraulic Fracking in US Shale Regions*. New Brunswick, NJ: Rutgers University Press, 2017.

3
Renewable Energy
Setbacks, Successes and Strategies for the Energy Transition

America's reliance on oil and gas extraction to secure energy supplies and to spur economic development has exposed local and regional economies to painful cycles of economic booms and busts and has caused public health and environmental damages, as discussed in the preceding chapter. The shift to renewable energy offers an alternative, viable pathway for generating affordable energy, creating jobs and raising tax revenue for local and state governments. These benefits can help a broad spectrum of Americans when pursued with strategies that address equity concerns. Charting an economy-wide shift to renewable energy, with renewable energy powering the electricity grid and with the electrification of various sectors of the economy such as transportation, is no doubt a challenging undertaking. Progress in renewable-energy adoption in United States, while modest, reveals why this undertaking can succeed.

As of 2020, the United States derived a modest 20 percent share of its electricity from renewable energy (wind, hydropower, solar biomass and geothermal energy)[1] and 12 percent of its total energy consumption from renewable energy.[2] This progress was possible, despite the variety of political and policy advantages enjoyed by the oil and gas sector (Chapter 2) because of several factors. Wind- and solar-energy development enjoyed support across the political spectrum, gaining traction in red, purple and blue states. State governments in wind-rich oil producing states, for example Texas, Oklahoma and North Dakota, and midwestern states, for example Iowa and Kansas, have pursued wind projects to secure revenue for rural landowners and local governments. These projects also provided cheaper electricity and jobs in installation and manufacturing (with about 560 wind-related manufacturing facilities operating across the United States).[3,4] Likewise, solar projects in irradiance-rich southeastern and southwestern states offer tax revenue to local governments.[5] The cost of wind and solar energy and the cost of storage continue to decline. In response to this cost competitiveness, a number of utilities have shifted to renewable energy. Large corporations with significant energy demands are also choosing to purchase cheaper renewable energy, and cities and municipalities are setting ambitious targets for renewable energy.

The development of renewable energy has also been heavily influenced by the complex mix of federal and state policies regulating electricity markets and incentivizing renewable-energy adoption. The federal government regulates interstate electricity markets, while state governments regulate intrastate electricity markets. Since the mid-2000s, the Federal Energy Regulatory Commission (FERC), through rules promulgated to regulate interstate electricity markets, has enabled renewable energy and complementary services, such as storage and demand response,[i] to compete in electricity markets.[6] The federal government's tax credits for wind and solar development have also contributed to renewable-energy expansion.[7,8] State governments have encouraged solar and wind energy by setting mandatory targets for the shares of renewable-energy consumption in their states, known as Renewable Portfolio Standards.[9–11,ii]

Despite mostly bipartisan support for renewable-energy development and a broad set of policy tools to implement it, the energy transition has been vulnerable to political opposition. The Trump administration repealed regulations aimed at shifting the power sector toward renewable energy, proposed subsidies that would have propped up expensive coal and nuclear plants, and interfered with research from national laboratories that provide useful guidance for renewable-energy growth. While these actions were not able to reverse the growing competitiveness of renewable energy and its expansion, the United States did lose precious time in accelerating the shift to renewable energy. That acceleration is necessary for the United States to contribute to global reductions in greenhouse gas emissions to address the climate crisis (Chapter 11).

To provide context for the prospects of America's further expansion in renewable energy, I describe the factors driving its progress across US states (Section 3.1). That advancement has continued despite the Trump administration's attempts to derail the adoption of renewable energy (Section 3.2). Next, I turn to the challenges of accelerating the renewable-energy transition (Section 3.3), strategies to incorporate equity into energy transition

---

[i] Storage addresses the intermittency of renewable energy. Demand response reduces peak demand by rewarding consumers for decreasing their demand at peak times.

[ii] While these standards are less cost-effective than carbon taxes, they have received popular support across states and they have provided some certainty for investments into renewable energy [10]. On the contrary, carbon taxes pegged to the social cost of carbon have not gained traction (Chapter 11).

(Section 3.4), and the specific challenges faced by fossil-fuel-reliant communities (Section 3.5). Coalitions of beneficiaries from renewable energy have been an essential bulwark against the opposition from the incumbent energy sector (Section 3.6). I conclude by underscoring how the bipartisan support for renewable energy among voters, anchored in benefits bestowed to local communities, provides optimism for the progress of the energy transition.

3.1. Renewable-Energy Expansion in the United States: A Brief Overview

Two patterns in renewable-energy adoption in the United States provide reasons to foresee its continued progress. First, significant declines in the cost of wind and solar and the continuing decline in the cost of storage have made adoption of renewable energy cost-competitive (Section 3.1.1). Second, states across the country have made progress in renewable-energy generation, including Republican-majority states (Section 3.1.2).

3.1.1 Cost-Competitiveness of Renewable Energy

The share of renewable energy in US electricity generation has grown significantly since 2010, thanks to its declining cost. Renewable energy contributed 20 percent of US electricity generation in 2020, doubling the 10 percent share it supplied in 2010 (Figure 3.1). By 2020, its share of generation rivaled that of nuclear or coal, which contributed 20 percent and 19 percent of generation, respectively. Cheaper wind and solar, as well as cheaper natural gas, have driven the retirement of coal power plants and the announcements of planned nuclear plant retirements.[12,13,iii] Coal's share of overall generation has halved since 2010, while nuclear has remained steady at 20 percent. From 2010 through 2020, wind and solar expansion was eclipsed by that of natural gas, which almost doubled from 24 percent of generation in 2010 to 40 percent of generation in 2020. Nevertheless, as prices for wind and solar continue to fall, they have become cost-competitive with natural gas generation in some locations in the United States, as discussed in this section.

Among renewable-energy resources, wind energy has seen the largest expansion (Figure 3.2). By 2018, wind overtook hydropower to become the

---

[iii] In response to announcements of potential retirements of nuclear plants, several state governments have provided subsidies to keep those nuclear plants operating. These states include New York, Illinois, Connecticut, New Jersey and Ohio.

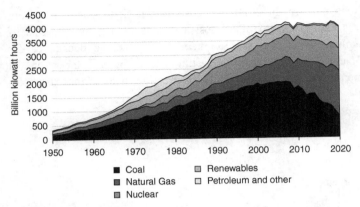

Figure 3.1 US electricity generation by major source 1950–2020
Source: Energy Information Administration, *Electricity Explained*, 2021.[1]

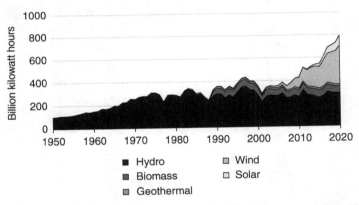

Figure 3.2 US electricity generation from renewable energy resources 1950–2020
Source: Energy Information Administration, *Electricity Explained*, 2021.[1]

largest renewable-energy contributor to electricity generation. By 2020, wind contributed 8.4 percent of electricity generation while hydropower contributed 7.3 percent. Solar contributed only 2.5 percent of generation, ahead of biomass (1.4 percent) and geothermal (0.5 percent).

Annual reports by Lazard provide the levelized cost of electricity, a metric that is widely used to make cost comparisons, albeit imperfectly, across sources of electricity.[14] The cost of wind and solar, for instance, depends on their location. Electricity is more valuable when produced during peak-demand periods. Even with these considerations, the levelized cost of wind and of solar has fallen sharply. In 2020, the cost of electricity generation from wind was less than one-third of its cost a decade earlier and the cost of solar photovoltaic generation was one-tenth of its cost a decade earlier (Figures 3.3 and 3.4). Improved forecasting, technological innovations in storage, demand response and distributed generation have permitted greater

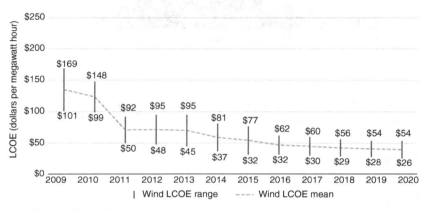

Figure 3.3 Levelized cost of electricity for wind (excluding subsidies)
Source: Lazard, "Levelized Cost of Energy and Levelized Cost of Storage," 2020.[14]

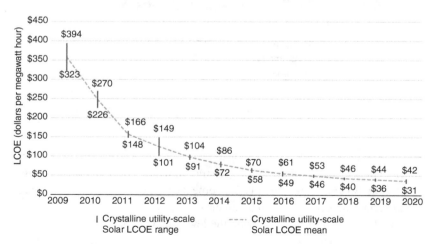

Figure 3.4 Levelized cost of electricity for solar photovoltaic (excluding subsidies)
Source: Lazard, "Levelized Cost of Energy and Levelized Cost of Storage," 2020.[14]

shares of renewable energy to be incorporated into the grid, while maintaining grid reliability.[15–18,iv]

According to Lazard's 2020 report, new construction of onshore wind and utility-scale solar photovoltaic in some locations were already cost-competitive relative to natural gas combined-cycle plants, the type of natural gas plants that produce a steady supply of electricity (Figure 3.5). Lazard's figures echo reports that as early as 2017 wind was the cheapest source of electricity generation in parts of the Midwest, while utility-scale solar was the cheapest in parts of the southwest.[19]

The decline in the price of wind and solar preceded the decline in the price of storage; the latter has taken place since around 2015. That decline in the price of energy storage, which helps address the intermittency of wind and solar, has begun to facilitate greater adoption of both. Lazard's figures indicate that new construction of wind and solar combined with storage in some locations has also become cost-competitive with natural gas peaker plants (i.e., natural gas plants that can be ramped up and down to cope with the intermittent nature of solar and wind). An analysis by Bloomberg also found that building batteries is a cheaper option for meeting peak demand for two hours or less.[20] The cost of utility-scale battery storage in the United States decreased from $2,152 per kilowatt-hour in 2015 to $625 per kilowatt-hour in 2018.[21] That decline in the cost of storage has prompted the growth of projects combining wind and solar with storage.[22]

Simultaneously, technological innovations have enabled grid operators to absorb greater shares of wind and solar into the grid, while maintaining the reliability of the electricity supply and minimizing the impact and the duration of disruptive events.[23] Lazard's figures also indicate that new construction of wind and solar is more cost-competitive than the marginal cost of generating electricity from coal and nuclear. In other words, it is cheaper to build new wind and solar facilities for electricity generation than to continue operations of existing coal and nuclear plants.

3.1.2 Renewable-Energy Adoption across US States

Renewable energy has grown across various states in the United States (Figure 3.6), building upon geographic strengths and political momentum unique to each. Benefits from renewable energy have grown across these states as well, including state and local governments' receipt of tax revenue, payments to landowners hosting wind and solar projects, and the opportunity for

---

[iv] The term grid reliability denotes that the electricity system provides users with a stable supply of electricity that meets users' demand.

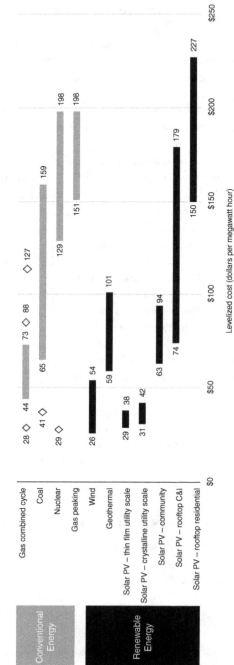

Figure 3.5 Levelized cost of electricity comparison (excluding subsidies)
Source: Lazard, "Levelized Cost of Energy and Levelized Cost of Storage," 2020.[14]

jobs in the clean-energy sector[24–26] (Table 3.1). These benefits in turn have fostered support for wind and solar in state legislatures.

Several states have also benefitted from the creation of domestic manufacturing jobs, which supply equipment to wind projects in the United States. These projects

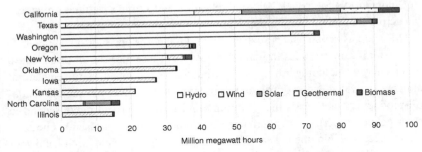

Figure 3.6 Top states in renewable electricity generation (2019)
Source: Energy Information Administration, *New York Generated the Fourth Most Electricity from Renewable Sources of Any State in 2019*, 2020.[27]

Table 3.1 *Contributions of wind, solar and storage systems in selected states in 2020*

	Power generation (GW)	Share of electricity (%)	Tax revenue (million)	Payment to farmers ranchers and landowners (million)	Clean-energy jobs
United States	171	11	$1,500	$1,100	300,000
Texas	39	19	$380	$250	36,700
California	20	6	$290	$121	58,100
Iowa	12	53	$70	$79	5,000
Oklahoma	9	35	$120	$54	3,100
Kansas	7	43	$34	$42	3,000
Illinois	7	9	$61	$46	13,400
North Carolina	5	7	$32	$27	5,600
Minnesota	5	21	$29	$35	4,500
Colorado	5	22	$16	$22	12,500
North Dakota	4	30	$13	$24	2,200

Source: American Clean Power, *Facts: Clean Power State-by-State*, 2020.[5]
Notes: Tax revenue and lease payments to landowners continue for the lifetime of the projects, unlike the revenue from oil and gas extraction that declines as wells dry up. Clean-energy jobs include jobs in the energy-efficiency sector, e.g., weatherization of homes and retrofitting of heating and cooling systems.

boast sizable domestic content, that is, materials and equipment produced in the United States, thus supporting a supply chain of American manufacturers.[3,4,28] Larger manufacturing facilities include Siemens Gamesa's facilities in Hutchinson, Kansas and Fort Madison, Iowa; General Electric's facilities in Grand Forks, North Dakota; and Vestas's facilities in Brighton and Pueblo, both in Colorado. An estimated 560 wind-related manufacturing facilities are spread out across the United States (Map 3.1).

Oil- and gas-reliant states such as Texas, Oklahoma and North Dakota have made significant progress in wind generation (Figure 3.6). In all three states, wind energy, including utility-scale wind farms, received state legislators' support as strategies to contribute to rural development. Texas, the top crude-oil-producing state, leads the nation in wind generation and ranks second in solar generation. Texas's experience also demonstrates bipartisan support for wind-energy development as a means to secure affordable electricity and spur rural development. Democrat governor Ann Richards and Republican governors George Bush and Rick Perry (alongside the state legislature during their governorships) created fertile ground for wind-energy growth. Richards initiated research on state government policies that could support wind; Bush signed into law Texas's first Renewable Portfolio Standard; and Perry signed into law even more ambitious renewable-energy targets.[30]

Map 3.1 Wind manufacturing facilities across the United States
Source: American Wind Energy Association, *Executive Summary: Wind Powers America Annual Report*, 2019.[29]

Texas offers lessons on how a government entity, by taking the lead in coordinating the build-out of necessary transmission lines paid for by ratepayers, can enable an economic environment that allows private entities to invest while promoting longer-term benefits to all ratepayers. During Perry's governorship, the state legislature enacted legislation authorizing the Public Utility Commission of Texas to coordinate with the Texas grid operator, the Electric Reliability Council of Texas (ERCOT) on the build-out of transmission lines. The commission identified Competitive Renewable Energy Zones that are high-wind-resource areas. It subsequently authorized 3,600 miles of transmission lines with 11.5 GW of capacity, which connected wind-rich West Texas and the panhandle to population centers in San Dallas–Fort Worth and San Antonio.[31]

That authorization solved the coordination issue inherent in the building of transmission lines, that is, wind developers are reluctant to invest in wind farms without the ability to transmit that electricity, while transmission-line developers cannot demonstrate the proposed line is in the public interest until wind farms contract for the use of the line. The state legislature also solved the problem of financing the $6.5 billion construction of the transmission line by spreading the cost across all ratepayers. Thanks to those investments, Texas was able to ramp up wind production, provide ratepayers with lower electricity prices,[32] bring rents to rural landowners leasing their lands, secure tax revenue for local governments and promote improved air quality.[33]

Midwest and Great Plains states, although historically less occupied with energy production as a driver in their economies, have seen significant growth in wind energy. Major producers of wind power among these states include Iowa, Illinois and Kansas. (Kansas hosts some oil and gas production.) State legislators and local communities saw wind power as a strategy to spur economic development.[34] Revenue to landowners has provided a stable income to complement farming revenue (the latter can be volatile, given the vagaries of commodity markets and droughts).[35,36] A study from Michigan found that farmers with wind turbines were more likely to invest in their farm, to buy additional farmland, to envision their farming activities continue, and to have a succession plan in place.[37] Moody's investor services reported in 2018 that wind farms' tax contributions to local governments enabled spending on infrastructure and local services and debt repayment.[38] It further described investments from companies setting up data centers that rely on wind power. Iowa alone attracted $10 billion in investments from Alphabet, Apple, Facebook and Microsoft.[38] These data centers contribute tax revenue to local governments. A study of rural counties in 12 midwestern states found that, in 2017, jobs in clean energy far outstripped those in fossil fuel power plants, extraction, refinement and transportation combined, with two exceptions: Kansas (where the numbers are comparable) and North Dakota (where fossil jobs are far more numerous).[39]

The Sun Belt states in the southeast and the southwest have expanded their solar energy, particularly North Carolina and Arizona, which rank third and fourth in solar generation in the United States after California and Texas. In North Carolina, utility-scale solar farms have contributed to tax revenue in economically challenged rural counties.[40] The state's investment-tax credit to support the growth of solar was estimated to have generated $1.93 in tax revenue for every dollar spent on the credits.[41] The state has also supported the adoption of residential solar by exempting the value of that installation from property taxes.[41] Arizona, despite generating limited shares of its electricity from solar and hydropower by 2020, shifted toward more ambitious clean-energy goals in recognition of the falling prices for solar and wind. The Republican-majority Arizona Corporation Commission, the state's public utility commission, voted to require utilities in Arizona to achieve 100 percent clean energy (including nuclear) by 2050. A study commissioned by the solar industry describes benefits from distributed solar generation to Arizona consumers, including those who receive their electricity from the utility companies' centralized generation. The utility can reduce its overall costs by reducing generation from more expensive fossil fuel power plants and save on investments in transmission lines.[42]

States such as California, New York and Washington have also pursued renewable energy, not only to secure cheaper electricity with less local pollution but also to achieve their climate goals and to cope with adverse climate impacts. As one of its strategies to cope with the more frequent and extensive wildfires induced by the climate crisis, California has been exploring the option of microgrids, with distributed renewable generation and storage, that can be isolated from the main electricity grid.[43] New York is pursuing offshore wind generation and, at the same, positioning itself as a manufacturing hub to supply offshore wind projects that have been proposed along the Atlantic coast from Massachusetts to Virginia.[44] Washington state has updated its legislation to give utilities credit, not only for the shift to renewable energy but for other investments in electric-car infrastructure.

Apart from state governments, large corporations in the United States have supported renewable energy with their direct purchase of cost-competitive wind and solar.[45] These corporations purchase renewable energy directly from wind and solar developers (retiring the renewable-energy credits) or from utilities (paying a green tariff for utilities to secure renewable-energy generation). That procurement amounted to 8.5 GW in 2018, 13.6 GW in 2019 and 11.9 GW in 2020. US corporations are expected to purchase about 44–72 GW of additional wind and solar from 2021 through 2030.[46]

3.2 The Trump Administration's Efforts to Block the Progress of Renewable Energy

The executive branch of government can encourage renewable-energy expansion, but it can just as easily prove to be an obstruction. The Trump administration's attacks on renewable-energy progress illustrate some of the strategies a pro-fossil-fuel executive branch can take to hinder the progress of the renewable-energy transition.

When Trump took office in 2017, many state governments had already embraced renewable energy to secure cheaper electricity and to promote economic development. The administration took overt steps to derail that progress, taking actions to prolong coal generation (Sections 3.2.1 and 3.2.2) to delay the progress of renewable-energy projects (Sections 3.2.3 and 3.2.4). While these efforts certainly took their toll, FERC, which regulated the interstate electricity markets, remained largely supportive of facilitating greater shares of renewable energy into the grid (Section 3.2.5).[47,v]

3.2.1 The Clean Power Plan Rule

The Obama administration promulgated the Clean Power Plan rule in 2015 under its authority to implement the Clean Air Act.[48] The rule set targets for states to cut greenhouse gas emissions and gave them the flexibility to achieve those reductions. In proposing these targets, the EPA considered the costs and technical feasibility of emission reductions from three strategies: improved efficiencies at coal plants; the shift from coal to natural gas plants; and the shift from fossil fuel generation to wind and solar. Under that rule, coal generation was projected to decline, while natural gas generation would increase before plateauing. When Trump took office in 2017, the Clean Power Plan had not taken effect, as lawsuits filed by Republican state attorneys general were underway and the Supreme Court had stayed the rule in 2016 in a 5–4 decision.[49]

---

[v] However, FERC's decisions have impeded other aspects of the energy transition. Its approval of interstate gas pipelines that fail to fully account for climate impacts contributed to the build-out of natural gas infrastructure. That infrastructure build-out can lock the United States into reliance on natural gas (Chapter 4).

Faced with the EPA's mandate to regulate greenhouse gas emissions[vi] but opposed to the regulations pursued by the prior administration, the Trump administration repealed the Clean Power Plan rule and finalized a replacement rule in 2019, that is, the Affordable Clean Energy rule.[50] The Trump administration interpreted the EPA's statutory powers under the Clean Air Act narrowly, arguing that Section 111(d) of the act permitted the EPA to regulate only at power plant level. In other words, the EPA could not take a systemwide approach to regulating the power sector nor could it consider shifting power generation toward wind and solar. The EPA's own analysis showed that the Affordable Clean Energy rule would cut carbon dioxide emissions by just 0.7–1.5 percent from 2005 levels by 2030, in contrast to the Clean Power Plan rule, which projected a 32 percent reduction from 2005 levels by 2030.[vii]

However, in January 2021, the US Court of Appeals for the DC Circuit in *American Lung Association v. EPA*[51] vacated the Affordable Clean Energy rule and remanded it to the EPA for fresh consideration. The court ruled that interpretation of the Clean Air Act by the EPA under the Trump administration – that its powers under the Clean Air Act were confined to regulating at the power-plant level – was arbitrary and capricious. While the court's decision opens the door for the EPA to take a systemwide approach to regulating emissions in the power sector, the EPA will nevertheless be constrained by the conservative majority of the Supreme Court justices. Judge Justin Walker's dissent foreshadows future legal arguments – an agency must have a clear grant of statutory authority for rules with "vast economic and political significance" – that could be used to derail EPA's climate actions (Chapter 10).

3.2.2 Proposed Subsidies for Coal and Nuclear Power

The Trump administration was willing to act against market forces in order to protect the corporate interests behind coal and nuclear generation.[52,viii] In September 2017, Secretary of Energy Rick Perry asked FERC to undertake

---

[vi] In 2007, the Supreme Court ruled 5–4 in *Massachusetts v. EPA* that the Clean Air Act mandates the EPA to regulate greenhouse gas emissions if those emissions are a threat to public health. The EPA's 2009 endangerment finding affirmed that greenhouse gas emissions were a threat to public health [194].

[vii] Despite the repeal of the Clean Power Plan rule, the power sector cut its carbon emissions to almost 31 percent below that 2005 baseline by the end of 2019, almost achieving the rule's goal of 32 percent and doing so well ahead of time [49].

[viii] Bob Murray, founder and CEO of Murray Energy, a coal-mining company that filed for bankruptcy in 2019, was reported to be a close confidant of Trump and a major donor to the Trump campaign. Murray also met with Secretary Perry and other administration officials to advocate for subsidies for coal [52].

rulemaking that would provide guaranteed revenue to coal and nuclear plants.[53] Perry claimed, disingenuously, that failures in the electricity markets were responsible for their "premature" retirement. Specifically, he claimed that markets failed to reward the important service of "resiliency" provided by the coal and nuclear plants, services that were not provided by renewable energy or natural gas.

The Department of Energy's own draft report underscored that the retirement of more expensive coal and nuclear generation, relative to renewable energy and natural gas, demonstrated that the electricity markets were indeed functioning as they should. Moreover, the claim that coal and nuclear plants added to "resilience"[ix] is factually wrong. In reality, these plants failed to provide electricity during extreme weather events, such as the 2011 and 2021 winter freezes in Texas, as well as Hurricane Sandy in the Northeast.[54,55] On the contrary, renewable energy was able to return online faster following extreme events.[56,57] The proposal would have cost ratepayers an estimated $311 million–$11.8 billion per year.[23] Bipartisan former FERC commissioners warned that the proposal, by enabling subsidized resources to drive out more cost-competitive resources, would destroy investor confidence and would risk the collapse of the market.[58] Fortunately, that proposal was so bereft of merit that even Trump-appointed FERC commissioners rejected it.

3.2.3 Obstructions to Leasing and Permits for Wind and Solar on Federal Lands and Seas

The Trump administration slow-walked the federal leasing and permitting process for wind and solar projects on federal lands and seas. According to a report by the Center for American Progress, a political appointee overrode the Bureau of Land Management Utah office's proposal for a lease sale for solar and the Bureau of Ocean Energy Management (BOEM) did not follow through with the proposed New York Bight offshore wind lease sale.[59] In August 2019, BOEM ordered a new environmental review of all proposed Atlantic offshore wind projects. That new assessment delayed the approval process for the proposed Vineyard Wind 800 MW project off the coast of Massachusetts, which is expected to be the first large-scale offshore wind project in the United States. BOEM's December 2018 draft environmental impact statement

---

[ix] Perry used the term "resilience" without defining its meaning. At that time, energy analysts focused on the reliability of the electricity supply. Subsequently, energy analysts have conducted research on the *resilience* of the electricity system, i.e., its vulnerability to threats of disruptions (e.g., from events such as hurricanes, droughts or cyberattacks) and its ability to recover from disruptions.

for Vineyard Wind had concluded that the project and seven other projects cumulatively did not pose significant risk to the environment.[60,x] The administration's decision to undertake an extensive review for offshore wind, but to abandon the review of coal leasing on public lands, raised suspicion that the action was motivated by a desire to delay the progress of offshore wind.[61]

3.2.4 Interference with Renewable-Energy Research at Federal Laboratories

Political appointees interfered with the work of career scientists at the federal national laboratories who conduct research on the shift to renewable energy. In October 2020, Representative Eddie Bernice Johnson, chairwoman of the House Committee on Science, Space, and Technology wrote to the Secretary of Energy Dan Brouillette raising her concerns about the Department of Energy's suppression, modification and delays of publications from the national laboratories pertaining to renewable energy, energy efficiency, transportation electrification and the electricity grid.[62] The department's 2018 policy requiring studies to be cleared by political appointees violated the department's own scientific integrity policy. An investigative report published in *The Atlantic* detailed how political appointees blocked dissemination and publication of the Interconnections Seam Study, undertaken by the National Renewable Energy Lab (NREL).[63] That study demonstrated how connecting the eastern and western US grids could facilitate greater reliance on renewable energy and accelerate the retirement of coal plants.[64] Political appointees perceived the study as a threat to coal and blocked the publication of the original version of the paper written by NREL experts.[63]

3.2.5 FERC Decisions That Ran Counter to Trump's Anti-renewables Agenda

FERC – the federal agency tasked with regulating wholesale electricity markets (as well as licensing hydroelectric power generation and permitting interstate natural gas pipelines) – is run by a bipartisan body of five commissioners. Although appointed by the president, no more than three commissioners can be

---

[x] Nevertheless, the project developers had to address the mitigation of impacts on commercial and recreational fishing, which were expected to be moderate to major. The project received BOEM's final approval in May 2021. The Vineyard Wind project is far larger than Rhode Island's Block Island offshore wind project (30 MW in operation since 2016) and the Coastal Virginia Offshore Wind project (12 MW for the pilot share).

of the same political party, an institutional design that can provide FERC with some insulation from political pressures.

While the Trump administration sought to undermine renewable-energy development, orders issued by the FERC commissioners during Trump's presidency by and large facilitated the integration of renewable energy into the grid.[xi] Many of these orders were directed to grid operators, known as regional transmission organizations or independent system organizations (RTOs/ISOs), which manage the electricity grid that extends across states. In February 2018, FERC issued Order 841, which ordered RTOs/ISOs to configure market rules that permit storage resources to bid into these markets to provide capacity, energy and ancillary services.[65] In September 2020, FERC approved Order 2222, which permitted distributed energy sources – such as rooftop solar, battery-storage technologies and demand response – to band together to sell some of their electricity to the grid.[66] In October 2020, FERC approved a statement that encouraged regional grid operators to consider rules to incorporate state's carbon prices, noting that carbon pricing provides a transparent market tool to reduce greenhouse gas emissions.[67] This support is significant: the New York ISO, for example, was already working on plans to include carbon pricing into its market rules to support the state of New York's plan to accelerate renewable-energy adoption.

Nevertheless, not all of FERC's decisions were in support of renewable energy. In December 2019, FERC commissioners in a 2–1 decision imposed a price floor for energy resources, including renewable-energy resources, that received state subsidies. This price floor applies when these resources bid into the capacity markets[xii] managed by PJM, the RTO that covers all or parts of 13 states and the District of Columbia.[68–70] Two FERC commissioners took the position that state programs that support renewable energy provide those resources with an unfair advantage and that price floors were necessary to ensure a fair capacity market. The majority's reasoning fails to recognize a key distortion to electricity markets – state and federal policies subsidize fossil fuels.[71,72] Commissioner Richard Glick's dissent argued against the use of FERC powers to nullify state governments' expressed policies to expand renewable-energy resources in their states.[6]

[xi] FERC's decisions on natural gas pipelines are discussed in Chapter 4.
[xii] In capacity markets, electricity generators commit to supplying specific amounts of electricity to the grid sometime in the future. Grid operators are able to better plan for grid reliability with generation resources that have committed to providing electricity when called upon to meet demand.

3.3 Accelerating the Shift to Renewable Energy

The report *Accelerating Decarbonization of the US Energy System*, from the National Academies of Sciences, Engineering, and Medicine, describes the feasibility for the United States to generate 75 percent of its electricity from renewables by 2030.[73] That decarbonization goal will help the United States meet its commitment under the Paris Climate Agreement to reduce its greenhouse gas emissions.[xiii] Decarbonization of the electric grid is among the more feasible options for more rapid cuts in greenhouse gas emissions and serves as the enabling step for decarbonizing other sectors such as transportation.[74] These two sectors – electricity generation and transportation – are the two largest sources of emissions in the United States, contributing 25 percent and 29 percent respectively of its total emissions in 2019.[75]

The report envisions rapid scaling up of wind and solar deployment, drawing down coal plants and some natural gas plants, and preserving the use of operational nuclear plants and hydroelectric facilities, as well as increasing overall electrical transmission capacity by approximately 40 percent.[73] While these goals are, no doubt, ambitious, several studies document technically and economically feasible pathways to ambitious expansion of the share of renewable energy in the US electricity grid.[76,77] Moreover, a number of studies underscore that as much as 80 percent of the decarbonization of the US grid can be achieved with existing technologies.[16,76–80,xiv]

Another study, from the University of California Berkeley and GridLab, a network of independent consultants with expertise in power systems, describes reaching 90 percent clean energy by 2035 by adding wind, solar, storage, and demand response and by relying on existing hydropower and on nuclear plants that are not planned for retirement.[77,81] The remaining 10 percent of the grid would be powered by existing natural power plants. No new gas power plants would need to be built and all coal plants would be retired. Wholesale electricity prices are projected to be 13 percent lower than that in 2019.

The plan envisions the build-out of 1,100 GW of wind and solar projects and the construction of several regional transmission lines. This target requires a sustained growth of 70 GW each year for the next 15 years, which is thrice the typical annual investment in the past.[81] Reaching that goal requires a whole

---

[xiii] In June 2017, Trump announced that the United States would withdraw from the Paris Climate Agreement and the United States formally withdrew in November 2020. However, under the Biden administration, the United States formally rejoined the Paris Agreement in February 2021.

[xiv] There is disagreement among energy analysts on how the final stages of decarbonization of the grid can be achieved [16][79, 80].

suite of policies, including but not limited to policies that facilitate the build-out of transmission lines, price the services of renewable energy and its complements in the wholesale electricity market, support the research and development of technologies that enable the integration of greater shares of wind and solar into the grid, and reform the business model of utility companies.[82]

Both studies, from the National Academies and from UC Berkeley, underscore the need for a substantial shift away from natural gas generation. Likewise, the United Nations Environment Programme's assessment of global methane emissions emphasizes how reliance on natural gas, given the reality that carbon capture and sequestration technologies are not financially feasible at a large scale, runs counter to the aspirational targets of keeping the rise in average global temperature to below 1.5°Celsius.[83]

The proposition of shifting away from natural gas, unsurprisingly, has already sparked opposition from the natural gas sector.[84] Proponents of the natural gas industry championed the energy source as the cleanest fossil fuel, capable of supporting the energy transition by facilitating the shift away from coal. Under policies in place in 2020, the US Energy Information Administration projected that natural gas would make up 40 percent of new capacity additions in the United States until 2050, and wind and solar would contribute the remaining 60 percent (Figure 3.7).[85]

However, the share of natural gas in future capacity additions may well fall short of that projection. The rapidly declining price of wind and solar, plus the

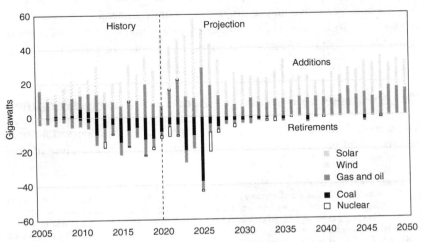

Figure 3.7 Projected capacity additions and retirements
Source: Energy Information Administration, *Annual Energy Outlook*, 2021.[85]

declining price of storage, has enabled the direct shift away from coal, bypassing natural gas. In 2021, solar and wind made up most of the new utility-scale generation capacity in the United States. The share of solar (39 percent), wind (31 percent) and battery (11 percent) in new generation capacity far exceeded that for natural gas (19 percent) and nuclear (3 percent).[86]

Reports by the Rocky Mountain Institute, a clean energy think tank, detail how the planned replacement of coal plants with 70 GW of natural gas plants would be more expensive than relying on the clean energy portfolio of wind, solar, energy efficiency and demand response.[87,88] If the planned natural gas plants were built anyway, about 90 percent of that planned natural gas generation would be more expensive to operate compared to the clean energy portfolio by 2035.[87,88] Examples of utilities shifting to clean energy portfolio solutions as opposed to building out new natural gas plants demonstrate that the former strategy can indeed prove cheaper.[19,87,88] Some regulators,[xv] for example those in Arizona and Indiana, have blocked plans for new gas plants, ruling that the utilities had not fully considered alternatives and that large new gas projects were financially risky with the rapid decline in the cost of renewable energy. However, regulators in other states have permitted the construction of large natural gas plants to proceed.[89]

The retirement of plants in the coal and nuclear sectors (Figure 3.7), which are far smaller players in the US economy than the natural gas sector, has already raised financial, socioeconomic and political hurdles in the energy transition (Section 3.5). These challenges are amplified in the case of the transition away from the natural gas sector, which provides significant revenue to state and local governments, royalty income to mineral owners and jobs (Chapter 2). To be sure, a renewable-energy transition presents challenges in equitable distribution of benefits and impacts – a topic I turn to next.

3.4 Strategies to Broaden the Benefits of Energy Transition

The economic benefits of renewable-energy projects are an important factor in securing local communities' support.[90] Simultaneously, the decline of coal

---

[xv] About half of the states have traditionally regulated electricity markets. Vertically integrated utilities own power plants and transmission and distribution lines and operate as a monopoly in a geographical location. Regulators, known as public utility commissions, control the rates the utility charges, balancing consumers' interests with utilities' interest in earning returns on capital investments and in paying for building and maintenance of transmission and distribution lines [74]. Other states have restructured markets in which companies compete in selling electricity to retail consumers and utility companies impose charges on the use of transmission and utility lines [74].

mining and coal plants has fueled understandable concern from communities that stand to lose the most economic benefits of those operations. I discuss strategies that can broaden the benefits to rural communities and low-income households (Sections 3.4.1–3.4.4).

3.4.1 Transmission Projects: Benefits to Local Communities

The build-out of transmission lines is essential to integrating greater shares of highly variable wind and solar into the electricity system.[91–93] As illustrated in Map 3.2, these lines help reduce the variability of wind and solar power supplied to the grid by linking geographically dispersed wind- and solar-generation resources and by balancing supply and demand (which vary over the day) across time zones. It also reduces the cost of electricity generation by enabling generation to be sited at the windiest and sunniest locations and by connecting to consumers in population centers. At present, the three grids in the United States – the eastern interconnection grid, the western interconnection grid and the grid in Texas – are fairly isolated systems.[91,92] Under various modeling scenarios, linking the eastern and the western interconnections would yield benefit-to-cost ratios of 1.25 or above, the typical threshold used by utility companies to justify new investment in transmission lines.[94]

While the overall benefits from these transmission lines are clear, local and state governments are often reluctant to subject their communities to transmission line development. Many of these projects, in fact, do not directly benefit the communities they transect. However, there are strategies that can compensate local communities for hosting these transmission lines and ensure their genuine participation in siting decisions. One proposed interstate transmissions line, the 800-mile Grain Belt Express line, has been opposed by rural landowners in Kansas, Missouri and Illinois at state utility commissions and in the courts.[95] Likewise, the proposed Plains and Eastern Clean Line from the Oklahoma Panhandle to Tennessee has faced court challenge from landowners in Arkansas.[95]

Community benefit agreements between the transmission line owners and local communities provide a promising strategy to extend benefits to local communities.[74,96–98] Project developers negotiate with community representatives and contract to provide benefits, such as payments to local governments to support local services and payments for conservation projects alongside transmission lines.[74,96–98] Several community benefit agreements negotiated by trusted representatives of the local community during the development of offshore wind projects offer lessons on processes that are participatory and responsive to local communities' feedback.[99,100] The Center for Rural Affairs,

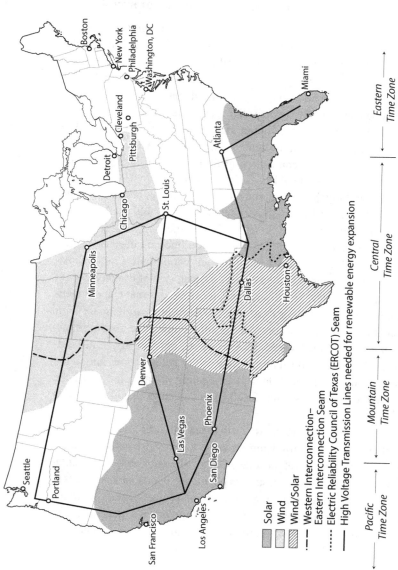

Map 3.2 Wind and solar resources and illustrative high-voltage direct current transmission lines
Source: National Renewable Energy Laboratory, *Interconnections Seam Study*, 2020.[64]
Notes: The straight lines illustrate the potential build-out of high-voltage direct current transmission lines available in a continuum across the United States and the shaded areas show the areas with greatest wind and solar potential. Technological advances have made solar and wind feasible even in the nonshaded areas. Offshore wind resources have grown in importance. The federal government has auctioned leases for offshore wind in the Atlantic Coast. In May 2021, the Biden administration opened the Pacific coast to offshore wind development.

an organization that works on renewable energy in rural America, further recommends that developers consider support for improvements in local interconnection to facilitate local small-scale wind and solar projects. Such investments can enhance local communities' recognition of local benefits that are directly connected to transmission line projects.[101]

3.4.2 Careful Contemplation of Tax Exemptions

Utility-scale wind farms and utility-scale solar farms contribute to local property taxes and have financed services such as schools and road maintenance and in some cases a reduction in property taxes for other residents.[35,102] These projects also contribute lease payments and royalties for landowners that host these projects, often complementing income for farmers and enhancing investments into farms.[103] Wind and solar projects also create jobs during both installation and operation (although to a lesser degree in the latter phase).

In light of these benefits, local and state governments have offered significant tax exemptions as enticements for developers to locate these projects within their jurisdiction. As a result, state and local governments must juggle two competing goals: offering incentives to promote projects that can contribute to the local economy and ensuring local communities do capture adequate benefits. Opponents of tax abatement programs, for instance in Texas and Kansas, have argued that school districts have foregone much-needed tax revenue as a result of these programs.[104,105]

In markets where renewable-energy generation is already cost-competitive, a careful consideration of paring down tax abatement programs is warranted to strike a balance between these two competing goals. Tax exemptions offered to both renewable-energy projects and oil- and gas-extraction projects (in states that host both types of activities) should be reconsidered in a comprehensive assessment of tax reforms. In 2015, Kansas, for instance, capped property tax abatement for new wind farms at 10 years. On the other hand, in 2019, Texas extended its tax abatement program for renewable energy for another 10 years.[xvi] Notably, neither state cut back on their subsidy programs for oil and gas.

[xvi] In June 2019, Texas extended its chapter 312 program that allows local taxation authorities, such as cities and counties, to authorize property tax abatements for manufacturing and renewable-energy projects. The corresponding program for school districts, the chapter 313 program, will be reconsidered in 2022. The value of a property granted chapter 313 tax breaks is limited to $10 million in rural areas and to between $80 million and $100 million in urban areas.

3.4.3 Encouraging Local Investors to Invest in Local Energy Generation

Many wind projects in rural regions are utility-scale wind projects owned by private, nonlocal entities.[102,106,xvii] State governments such as Minnesota have supported community-owned wind and solar projects, that is, projects owned by investors who reside in the community, to enable resident investors to gain greater shares of the benefits from these projects.[107,108] While community-owned wind farms are few relative to private-investor-owned farms, a number of studies have collated cases of successful development of community-owned farms that sell electricity to utility companies.[109,110]

Community-owned wind farms are by no means a guaranteed success. The bankruptcy, reorganization and resale of a celebrated pioneer in community wind farms in Minnesota provides a cautionary tale.[111,xviii] Smaller-scale community-owned wind farms face significant competition from utility-scale wind farms, which have more experience and resources to navigate the business and legal landscape. At the same time, community-owned wind farms have secured greater buy-in from local communities, including members of the community who are not investors in the project.[102,112,113] Mindful of these risks and the prospect of greater benefits to local communities, organizations advocating for an equitable energy transition have called for greater technical and financial assistance from federal and state governments to assist local communities in developing community-owned wind and solar farms.[114]

Minnesota offers lessons on the federal and state support needed for such projects to succeed. In 1997, Minnesota legislated an incentive of 1.5¢ per kilowatt-hour for small wind projects (2 MW or less) for the first 10 years of the project. Part of the funding for that subsidy came from Minnesota's requirement for the utility company Xcel Energy to contribute to the renewable-energy development fund that provides support for small wind projects.[107,115–117] Minnesota's laws, unlike those in several states, permit community-owned wind farms to sell into the electricity grid.

[xvii] Wind farms that are owned by groups of individuals within local communities and that sell electricity to utility companies are more widespread in Denmark and in Germany than in the United States [195].

[xviii] The 11.5 MW Minwind wind farm was owned by 360 residents in Rock County, Minnesota, who had purchased shares at $5,000 per share. The share of investment by each individual was capped at 15 percent of the project. The farm began operations in 2004, supplying electricity to the utility companies Xcel Energy and Alliant Energy under long-term contracts. However, the wind farm filed for bankruptcy in 2015 – noting it did not have the funds to repair its turbines, which had been damaged in a 2013 ice storm, and to pay a $1.9 million fine imposed by FERC for its failure to submit required filings. Another set of investor-residents from the county subsequently purchased the wind farm.

The expansion of these community-owned projects would require a number of supportive actions from federal and state government. Even the most motivated community members would benefit from technical assistance to develop such a project, to assess its feasibility and to shepherd it through the complex legal and financing landscape.[110,117] Federal support via grants as opposed to tax credits would also simplify community-owned farms' access to federal incentives. The provision of grant support in lieu of tax credits, which had been temporarily available under the 2009 American Recovery Reinvestment Act, proved to be helpful for new players to enter the wind energy sector.[109,118]

3.4.4 Facilitating Rural Electricity Cooperatives' Shift to Wind and Solar

Many rural communities receive their electricity from rural electricity cooperatives. About 800 rural electricity cooperatives (coops) serve 42 million customers in rural and suburban America.[119,120] Nonprofit distribution coops purchase that power from generation and transmission (G&T) coops and then distribute it to households. G&T coops own power-generation assets and transmission grids. These coops, which played a critical role in electrifying rural America during the New Deal in the 1930s, face hurdles to switch to wind and solar.

A subset of coops has made progress with their shift to wind and solar. For instance, in 2018, Great River Energy, a G&T coop serving 28 distribution coops in Minnesota and Wisconsin, committed to 50 percent renewable energy by 2030, noting that wind and solar were its lowest-cost option for generating power.[121] That expansion followed Great River Energy's partnership with a cooperative bank to offer leases for solar arrays to distribution coops in its network.[122] Great River Energy, however, has planned to convert one coal plant to natural gas.[123]

Nevertheless, studies show that a substantial share of distribution coops face high electricity rates[124,xix] and significant hurdles to shift to solar and wind.[119–121] Many of these coops face higher electricity rates because they are locked into long-term contracts with G&T coops that generate the bulk of their electricity from inefficient coal plants and bear debt burdens from the construction and retrofitting of those plants. Moreover, these coops' contracts

---

[xix] These high electricity costs result in poorer rural households experiencing energy poverty, with 9 percent of their budget going to electricity bills (compared to nonpoor rural households' expenditure of only 3.1 percent of their budget) [124].

with G&T coops limit their self-generation of power to a certain percent of their electricity consumption, and thus these coops would have to forego cheaper options for wind and solar. Second, G&T coops are slow to shift away from coal because of their debt due to these coal plants.

A distribution coop's individual decision to exit its contracts with a G&T coop, in order to purchase cheaper electricity and to subscribe to greater shares of renewable energy, can lead to a downward spiral for remaining coops. The G&T coop would spread the fixed cost of the system to the remaining members, leading to even higher rates for the remaining members, prompting other members to seek to exit their contracts. These effects can be mitigated though. In one case, negotiations between distribution coops and Tri-State, a G&T coop that services Colorado and New Mexico, resulted in a stabilization plan. Tri-State committed to a plan to shift some generation to renewable energy and to permit distribution coops a higher share of self-generation. That shift to renewable energy was estimated to save Tri-State $600 million by 2030.[125]

However, many G&T coops, given their exemption from many states' clean-energy mandates, may not face sufficient incentives to shift toward wind and solar. More importantly, many of them, given their debt due to their coal plants, do not have the financial capability to shift away from coal reliance. That financial challenge led to them opposing policies aimed at accelerating the energy transition.[121] RE-AMP, a network of nonprofit organizations working on energy transition in the Midwest, proposed strategies to address the financial hurdles for G&T coops to retire their coal plants.[121]

The Rural Utility Service, a unit within the US Department of Agriculture mandated to support electricity, water and broadband infrastructure in rural communities, holds debt undertaken by G&T coops to build coal plants.[xx] The Rural Utility Service could offer loan terms with lower interest payments or even forgive these loans altogether. The debt refinancing or drawdown would need to be coupled with guarantees the G&T coop would shift to wind, solar, storage or demand response.[xxi] Such proposals – given an estimated $8.5 billion outstanding loan guarantees from the Rural Utility Service to G&T coops – would no doubt require greater scrutiny, but they underscore the need for a solution to the financial hurdles that constrain rural America's transition to renewable energy.

[xx] The Center for Rural Affairs recommends that refinancing or drawdowns be limited to loans held by the Rural Utility Service and to exclude G&T coops' loans that were shifted to the private sector as a possible strategy to reduce regulatory oversight.
[xxi] Pacyniak discusses how G&T managers, whose incentives may not align with those of ratepayers, may choose not to invest in the cheapest energy option for the G&T coop [119].

3.4.5 Community-Shared Solar and Low-Income Communities

Distributed energy sources refers to renewable energy and storage projects that are smaller in scale and located closer to consumers, as contrasted with utility-scale centralized generation. These include community-shared solar programs that are owned by utilities or solar developers and offer subscriptions to purchasers, as well as community-shared solar programs that are owned by the subscribers.[126,127] Distributed generation of solar (along with storage) reduces the need for transmission build-out and for peaker plants.[128] A number of utilities have recognized that the deployment of distributed generation delivers cost savings to both utilities[xxii] and consumers by shifting reliance away from the grid during peak times and by providing power during extreme weather events such as wildfires.[129]

Community-shared solar offers economies of scale and greater accessibility to consumers than other forms of distributed generation such as rooftop solar.[130,131] These benefits have spurred a number of state governments to support the growth of community-shared solar programs.[122,132] Additionally, recognition that traditional programs have bypassed lower-income communities – alongside their potential to lead to higher-income households with rooftop solar to be cross-subsidized by lower-income households that are not able to afford such installations[131,133,134] – has prompted renewable-energy advocates to call for an expansion of programs that specifically target low- and middle-income consumers.[114,135,136]

Some of these programs require utility companies to set aside a minimum share of subscriptions for low- and middle-income households. State governments have supported these programs by providing subsidies to ensure savings to electricity bills for low-income consumers and by providing assistance for utilities to reach out to low-income communities.[132] Other programs require utility companies to locate a share of community-shared solar programs in low-income neighborhoods, thus enabling these communities to derive benefits through property tax revenues.[132] Community-shared solar and storage projects also hasten the replacement of inefficient and highly polluting peaker plants, which are disproportionately located in low-income and minority neighborhoods, bringing additional air-quality benefits to these communities.[137]

[xxii] A number of utilities opposed distributed generation, which reduces their customer base and therefore their profits. These utilities have deployed the argument that distributed generation hurts poor households, i.e., wealthier households are better able to install rooftop solar and thus shift the cost of maintaining transmission and distribution lines to nonparticipants. Baker points out that the equity concerns can be addressed with community-shared solar that genuinely serves low-income communities [135].

Organizations such as the National Association for the Advancement of Colored People, Hispanic Federation and Appalachian Voices have called for federal and state policies that enhance local community ownership of community solar programs.[114] While the community assumes more risks in these projects, it also stands to capture a greater share of the benefits.[109,114,135,138] These organizations have called for technical assistance to communities (e.g., to set up the community-ownership business model and to navigate various other legal requirements)[139] and for federal incentives tailored to community-owned projects.[109,114,135,xxiii]

These organizations have advocated for federal grants in lieu of tax incentives[xxiv] and called for the expansion of the US Department of Agriculture's Rural Energy for America Program.[114] This program, first authorized by Congress in 2002 and reauthorized in 2008, has contributed to farmers' adoption of renewable energy and of energy-efficiency projects on their farms.[39,140] The program covers up to 25 percent of the total solar or wind system cost (with a cap of $500,000 of the total cost) and provides loan guarantees for the remaining amount. The program is limited to agricultural producers that receive at least half of their gross income from agricultural operations and small businesses in eligible rural areas.

The expansion of community-owned solar (in both urban and rural areas) requires legislative changes in several states, as only 16 states have legislation that permits owner-members to take credit in their electricity bills from shared solar panels.[110] Disputes on net metering, that is, provisions for participants in community-shared solar programs to receive credits on their electricity bill for unused solar electricity that is sold to the grid, will also need to be resolved. Net metering policies need to ensure that the credits reflect the value of the electricity at the time it is sold into the grid and of the other benefits distributed generation provides to the grid. At the same time, all ratepayers will need to contribute equitably to the cost of maintaining the transmission and distribution lines. Pricing strategies such as time-of-use pricing for electricity sold by distributed generation as well as cost-sharing strategies for utilities and for

---

[xxiii] Jones and James describe how members of one community-owned solar project in Vermont were able to receive the residential investment tax credit for their investment into solar arrays located offsite [109]. These members receive credit in their residential electric bills (within the typical amounts used by a household) for the solar power generated by the offsite solar arrays.

[xxiv] Jones and James note that outright grants as incentives are more compatible with nonprofit organizations that are spearheading community-shared solar for church congregations and other civic societies [109]. On the other hand, the federal government's incentives for solar as investment tax credit favor projects by solar developers in partnership with for-profit entities with significant tax liability. For-profit entities provide the upfront capital and derive benefits from the investment tax credit.

distributed energy resources to cover the maintenance of transmission and distribution lines have been proposed as potential solutions.[141,142]

3.5 Strategies to Address Economic Disruption in Coal Communities

Achieving an equitable energy transition requires not only broadening the benefits from renewable-energy expansion but also assisting communities that are facing economic distress from the retirement of coal and nuclear plants. State and federal efforts to ease economic and social challenges in retiring coal plants (Sections 3.5.1 and 3.5.2) anticipate the challenges, discussed in Chapter 11, in the future retirement of natural gas plants, and the even broader challenges of shifting from fossil-fuel-anchored jobs to jobs compatible with a lower-carbon economy.

3.5.1 State Governments' Planning for Coal Retirements

Colorado and New Mexico provide examples of state governments setting ambitious renewable-energy goals, while taking proactive steps to address the next wave of coal plant retirements and to assist workers and communities to bridge that transition.[143,144] These programs face the challenge of securing funds to pay for these transition costs and of reaching agreements on who should bear these costs.

In 2019, the Colorado General Assembly enacted legislation pledging a "moral commitment to assist the workers and communities that have powered Colorado for generations" by supporting "a just and inclusive transition" away from coal. It created the Office of Just Transition and charged it with preparing a "final just transition plan for Colorado" by the end of 2020.[145] The legislation requires utilities that plan to retire coal plants to submit a workforce transition plan to the Office of Just Transition and the affected community at least six months prior to the planned retirement. Colorado also authorized a refinancing mechanism to assist with the retirement of coal plants.[146,xxv]

[xxv] Colorado is a regulated electricity market in which the public utility commission approves a utility's investment plans and authorizes a return to that investment borne by ratepayers [146]. The legislation permits the issuance of ratepayer-backed bonds, which ratepayers are responsible for repaying. The funds from the issuance of the bonds are used to pay the utility, which recoups some returns to its investments into the coal plant, and also to pay down its debts. The legislation also charged the Colorado public utility commission with scrutinizing proposals for refinancing and overseeing the bond issuance to protect the ratepayers and the public interest. The commission is permitted to earmark some of the funds from the bond issuance to fund assistance to workers and communities [144].

New Mexico enacted legislation with ambitious renewable-energy targets and assistance to workers and communities, but its experience demonstrates the financial hurdles in coal retirements and in assisting workers in the transition. In March 2019, New Mexico enacted the Energy Transition Act. The legislation requires investor-owned utilities and rural electric cooperatives to meet the state's goals of 50 percent renewable energy by 2030 and 80 percent by 2040, while laying strategies to finance the retirement of coal plants and to assist local communities. The legislation required some of the renewable-energy investments to be directed into the counties facing coal retirement, in order to provide electricity and to cushion the loss of tax revenue to the local governments. The legislation also authorized a refinancing strategy, whereby ratepayer-backed bonds would be issued and the owners of the San Juan coal plant would recoup part of their investments into the plant, and forego part of the anticipated returns to those investments, and the ratepayers would repay the bonds.[147]

While these programs are innovative in recognizing the need for planning for the retirement of coal plants, they also signal the challenges ahead for securing funding for worker retraining and securing bridge funds for local communities.

3.5.2 Federal Programs That Assist Coal Communities

Grassroots and labor organizations, such as Just Transition and the BlueGreen Alliance, have advocated for partnerships between federal and state governments and local communities to assist with the diversification of local economies and to provide bridge funds to finance local services.[148–150] The federal government sponsors a number of programs to assist communities reliant on coal mining and on coal and nuclear plants, but the magnitude of support remains limited relative to the retirement of many more coal and nuclear plants (Table 3.2).

In 2015, the Obama administration launched the Partnerships for Opportunity and Workforce and Economic Revitalization (POWER) program[152–154] in response to the economic distress in coal-reliant communities in Appalachia[xxvi] and the closure of coal mines and coal plants across the United States. This program includes economic assistance to coal communities under the Department of Commerce's Economic Development Administration; workforce and leadership training by the Department of Labor; support of the regional clusters to promote innovation under the Small Business Administration; and

---

[xxvi] Jobs in coal mines in the Appalachian region have been declining due to automation in mines and cheaper coal from the western United States. Mining jobs have declined further with the retirement of coal plants in response to cheaper electricity production from solar, wind and natural gas [152].

Table 3.2 *Selected federal funding for economic diversification projects in coal communities in $ millions*

	2015	2016	2017	2018	2019
Appalachian Regional Commission POWER initiative	–	50	50	50	50
Abandoned Mine Land Economic Development: Appalachia Region	–	90	105	115	115
Economic Development Administration's Assistance to Coal Communities program	10	15	30	30	30

Source: Congressional Research Service, *The POWER Initiative*, 2019.[151]

demonstration projects on workforce training and economic diversification under the Appalachian Regional Commission.[151]

Another important program is the funding for the reclamation of abandoned mine lands in order to turn these lands into productive assets. The funds are from the collection of fees from coal mines, under the Surface Mining Control and Reclamation Act of 1977. (The authority to collect these fees will expire in September 2021, unless reauthorized by Congress.)[155] Additionally, in December 2015, Congress appropriated funds for the Abandoned Mine Land Reclamation Economic Development project, which supports combining reclamation of abandoned mine lands with economic and community development in the Appalachian region.[156]

These programs, which have yielded a number of successes,[157,158] have built upon partnerships between federal and state governments and stakeholders, including grassroots organizations that have on-the-ground experience with these diversification efforts. The long road to recovery for these coal communities requires durable financing. To Congress's credit, it has continued to support these projects, despite pressure for cuts. For instance, it rejected the Trump administration's proposed cuts of $340 million, in its 2018 budget proposal, to the Appalachian Regional Commission and Economic Development Administration programs.[159]

The experiences of coal communities anticipate the challenges that communities reliant on the oil and gas sector will face as the economy shifts away from oil and gas extraction. As I discuss in Chapter 11, that transition will be far more challenging because a larger number of communities across the United States are reliant on the oil and gas sector than on the coal sector as the primary anchor in their local economies.[160] Jobs in the oil and gas sector contribute 30–50 percent of employment in some counties in West Texas, Oklahoma,

Wyoming, North Dakota and West Virginia.¹⁶⁰ Meeting the economic transition challenge will require workforce training (to build skills required in the clean-energy sector) and for local communities to pursue economic diversification strategies beyond the clean-energy sector (as jobs created and lost during the energy transition may well be in different geographical locations).

3.6 The Incumbent Energy Sector's Opposition to Renewable Energy

Despite many years of bipartisan support for renewable energy from Republican and Democratic governors and legislators and the sector's contributions to state and local economies, the sector faces political opposition. A study published in 2021 documented at least 100 ordinances adopted in 31 states blocking or restricting renewable-energy facilities in the preceding years.¹⁶¹ They include the weakening of renewable-energy standards (e.g., Ohio in 2014 and 2019), the repeal of mandatory standards (e.g., Kansas and West Virginia in 2015) and mandatory setback requirements that block the placement of wind turbines and solar panels. While local communities' perception of the lack of benefits from wind and solar underlie some of these land-use conflicts, the incumbent energy sector has played a heavy hand in strident opposition.

A dominant subset of companies in the incumbent energy sector – that is, some utility companies, the coal sector, the nuclear sector and the oil and gas sector – have attempted to block progress in renewable energy and the electrification of transportation.¹⁶²,¹⁶³,ˣˣᵛⁱⁱ The opposition from certain utility companies stems from their slow and apprehensive shift to wind and solar. The Sierra Club's 2021 study of 50 parent companiesˣˣᵛⁱⁱⁱ that own half the coal and natural gas generation capacity in the United States finds that the majority of these companies plan to build out natural gas plants, retire coal plants slowly and add only limited amounts of wind and solar by 2030.¹⁶⁴ Another study, by MSCI Inc., a consulting and financial services company,¹⁶⁵ reports that out of the 42 largest US companies that own power-generation assets, only 5 companies will achieve net-zero emissions by 2035, while 7 other companies have specified targets for achieving net-zero emissions by 2050.¹⁶⁵

ˣˣᵛⁱⁱ Companies with expertise in offshore oil drilling such as the Norwegian company Equinor and the Danish company Ørsted have built on their expertise to become leaders in offshore wind. On the contrary, US-based oil and gas companies such as Chevron and ExxonMobil have not pursued that path.

ˣˣᵛⁱⁱⁱ These include investor-owned utilities, power authorities (e.g., the Tennessee Valley Authority), G&T coops and large municipal utilities.

The incumbent energy sector, which includes Koch Industries,[xxix] a conglomerate with interests in the oil and gas sector, has funded networks of organizations that work collaboratively to oppose renewable energy.[10,166–169] The American Legislative Exchange Council, an organization of state legislators and businesses that is backed by wealthy donors, provides model bills to state legislators.[166,167] In 2012, it worked with the Heartland Institute, a libertarian climate-denial think tank, to write the model bill titled the Electricity Freedom Act.[168] That model bill, which aimed at repealing renewable-energy mandates, was shared with state legislators and subsequently championed by a subset of them.[168,170] Americans for Prosperity, an advocacy organization, lobbied in support of that model bill among state legislators and also lobbied against the reauthorization of the federal production tax credit for wind.[171] One of these organizations' early successes is the 2015 repeal of the West Virginia renewable-energy standards signed into law by Democratic Governor Earl Ray Tomblin.[172] Various astroturf organizations, pretending to be grassroots organizations that champion local community or consumer interests,[xxx] form another potent source of opposition.[173,174] Utility companies have also fought directly to oppose renewable-energy policies.[173–175] For instance, American Electric Power, FirstEnergy and other state utilities in Ohio successfully pushed to freeze Ohio's renewable portfolio standard in 2014.[175]

While Republican governors and legislators had long supported wind and solar power for their economic benefits especially to rural communities, some of their successors have taken an oppositional stance. Such a transformation can be seen, for example, in Texas, one of the pioneers and leaders in wind energy. The natural gas sector, which exercises political and economic clout in Texas, faces increasing competition from cheap wind power in Texas's electricity markets.[176] In contrast to Bush and Perry, who supported renewable-energy standards and the build-out of transmission lines that transformed Texas into a wind powerhouse, Governor Greg Abbott and a number of state legislators have opposed wind power. They blamed the reliance on wind for the Texas blackout during the February 2021 winter freeze[177,178,xxxi] and peddled

---

[xxix] Jane Mayer in *Dark Money* and Christopher Leonard in *Kochland* trace how the Koch Industries and its affiliated organizations influence public policy in the United States. Alexander Hertel-Fernandez's *State Capture* details the history and the operations of the American Legislative Exchange Council and Americans for Prosperity. The American Legislative Exchange Council, Americans for Prosperity and the Heartland Institute have received funding from Koch-affiliated organizations and oil and gas companies [166, 167, 168][196, 197].

[xxx] I discuss strategies that address local communities' and consumers' concerns in Section 3.4.

[xxxi] The blackout resulted in winners and losers. Pipeline companies, such as Kinder Morgan, reported $1 billion profit in the first quarter of 2021. It was able to sell gas at very high prices to utility companies. At the same time, municipality utilities and rural electricity cooperatives faced huge costs from the high prices of natural gas during the freeze [178].

that misinformation to advocate for legislation that would penalize wind and solar. (The Texas grid operator's assessment revealed natural gas outage was a far larger contributor to the blackout than wind and traced outages to the failure to winterize natural gas pipelines and power plants.)[179–181,xxxii]

In April 2021, the Texas Senate passed SB 1278 which requires wind and solar providers to pay other power sources to provide energy if they are not able to meet their projected supply and to pay for other load-balancing services, thus adding to their operation costs.[182] This requirement deviates from the grid operator's established practice of purchasing these ancillary services for the entire grid. Financial and technology corporations opposed the bill, which would raise electricity costs and more importantly would set a negative precedent harming the business climate in Texas. Three utility companies in Texas also opposed the bill, arguing the unjustified shifting of costs to wind and solar facilities would discourage future investments.[183] Proponents of the bill argued that imposing fees on wind and solar are necessary to counteract federal subsidies that purportedly tilted policies against natural gas.[184] In reality, 63 percent of Texas's energy subsidies go to the oil and gas sector in the form of exemptions from severance taxes.[185] That sector annually receives $2.9 billion of state subsidies[186] and $4.2 billion of federal subsidies.[71]

Beneficiaries of the renewable-energy transition have served as a bulwark against some of this opposition. Within the Texas legislature, Republicans from West Texas, whose districts host wind and solar farms and rely on their revenue,[187] have opposed anti-renewable-energy bills advocated by other Republicans. Beneficiaries of solar in North Carolina, including farmers, defeated efforts to repeal the renewable-energy standard in the Republican-controlled North Carolina legislature.[168,170,188] Likewise, beneficiaries of renewable-energy expansion, including farmers, formed alliances with the wind industry to defeat proposed repeals of the renewable-energy standard in the Republican-controlled Kansas legislature in 2013.

However, proponents of renewable energy have failed to win some of these battles. For instance, in 2015, the wind sector accepted the repeal of the renewable-energy standard in Kansas (with the wind power projected to exceed that standard regardless), in exchange for the legislature agreeing not to impose an excise tax on the sale of wind energy.[189] Nevertheless, these examples

---

[xxxii] The Texas legislature failed to mandate winterization for power plants and natural gas pipelines and to ensure those costs could be recouped. The nonprofit North American Electric Reliability Corporation had made recommendations on winterization after Texas faced a similar blackout during the 2011 freeze [181]. In addition to problems during the winter freeze, Texas also faces challenges in meeting peak demand during the summer when shortages of cooling water force thermal power plants to reduce their operations.

underscore the critical role of beneficiaries of wind and solar in sustaining the progress of the energy transition.

3.7 Conclusion

The prospects and challenges of renewable energy in America are complex, to be sure. Accelerating the energy transition will be challenging. On the bright side, a number of polls indicate bipartisan support among voters for the shift to renewable energy. For instance, a January 2021 poll reported that 66 percent of respondents said that developing sources of clean energy should be a high or very high priority for the US president and Congress.[190] That support was 58 percent among moderate Republicans and 93 percent among all Democrats. More than 80 percent of respondents not only supported incentives for consumers to adopt clean energy but also supported assistance for coal- and gas-reliant communities through job programs aimed at remediating their local environments. In a June 2020 poll among 1,000 Republicans and Republican-leaning independents, 70 percent of respondents agreed that the growth of clean energy in the United States can help it to become a leader in green economic development.[191]

Voter support for renewable energy has seen corresponding commitments from state governments to set more ambitious goals on renewables. State legislation on renewable-energy standards with specific target dates have provided investors with some certainties on the returns to their investments in wind and solar, prompting more states to ramp up their commitments and a variety of bills proposing a federal standard in the 116th Congress.[192] As of December 2020, four states set goals for 100 percent renewable energy and twelve others for 100 percent clean energy, with most states setting target dates between 2030 to 2050.[193,xxxiii] In December 2020, Congress enacted legislation that offered support for research, development and deployment for a variety of technologies, including wind, solar and storage, and that extended the federal tax credit for solar and wind projects (Chapter 11). Republican Senator Lisa Murkowski from Alaska and Democrat Senator Joe Manchin from West Virginia, two states reliant on fossil fuel extraction, had championed the bill, underscoring their support for technological innovations that bolster the energy transition. This

---

[xxxiii] Some of these states set binding goals through legislation and others set nonbinding goals through executive action. The definition of "clean energy" differs across states, with some states including fossil fuel generation equipped with carbon capture and sequestration technology.

bipartisan support (with exceptions among a subset of pro–fossil fuel politicians) is anchored in renewable energy's delivery of benefits and provides optimism on progress for the energy transition. That the United States can meet more of its energy needs from wind and solar, while providing benefits to local communities, makes feasible the challenging shift away from US overreliance on oil and gas extraction. That shift is essential to avert oil and gas extraction's adverse impacts on America's private and public lands, a topic I turn to next.

References

1. Energy Information Administration. *Electricity Explained: Electricity in the United States* (March 18, 2021). www.eia.gov/energyexplained/electricity/electricity-in-the-us.php.
2. Energy Information Administration. *Renewable Energy Explained* (May 20, 2021). www.eia.gov/energyexplained/renewable-sources.
3. Lawrence Berkeley National Lab. *Wind Technologies Market Report*. Report by M. Bolinger and R. Wiser (2020). https://emp.lbl.gov/wind-technologies-market-report.
4. Energy Information Administration. *Three Turbine Manufacturers Provide More than 75% of US Wind Capacity* (November 28, 2016). www.eia.gov/todayinenergy/detail.php?id=28912.
5. American Clean Power. *Facts: Clean Power State-By-State* (April 2020). https://cleanpower.org/facts/state-fact-sheets/.
6. R. Glick and M. Christiansen. "FERC and Climate Change." *Energy Law Journal* 40, no. 1 (2019): 1–46.
7. C. Hitaj. "Wind Power Development in the United States." *Journal of Environmental Economics and Management* 65, no. 3 (2013): 394–410.
8. National Renewable Energy Laboratory. *Impacts of Federal Tax Credit Extensions on Renewable Deployment and Power Sector Emissions*. Report by T. Mai et al. NREL/TP-6A20-65571 (Golden: February 2016). www.nrel.gov/docs/fy16osti/65571.pdf.
9. National Renewable Energy Laboratory. *Policies and Market Factors Driving Wind Power Development in the United States*. Report by L. Bird et al. NREL/TP-620–34599 (July 2003). www.nrel.gov/docs/fy03osti/34599.pdf.
10. T. P. Lyon. "Drivers and Impacts of Renewable Portfolio Standards." *Annual Review of Resource Economics* 8, no. 1 (October 2016): 141–155.
11. S. Carley et al. "Empirical Evaluation of the Stringency and Design of Renewable Portfolio Standards." *Nature Energy* 3 (July 23, 2018): 754–763. www.nature.com/articles/s41560-018-0202-4.
12. Energy Information Administration. *Nuclear and Coal Will Account for Majority of US Generating Capacity Retirements in 2021* (January 12, 2021). www.eia.gov/todayinenergy/detail.php?id=46436.

13. Energy Information Administration. *Five States Have Implemented Programs to Assist Nuclear Power Plants* (October 7, 2019). www.eia.gov/todayinenergy/detail.php?id=41534.
14. Lazard. "Levelized Cost of Energy and Levelized Cost of Storage, 2020." *Lazard Insights*, October 19, 2020. www.lazard.com/perspective/levelized-cost-of-energy-and-levelized-cost-of-storage-2020/.
15. H. Johlas, S. Witherby and J. R. Doyle. "Storage Requirements for High Grid Penetration of Wind and Solar Power for the MISO Region of North America: A Case Study." *Renewable Energy* 146 (2020): 1315–1324.
16. M. Z. Jacobson et al. "Low-Cost Solution to the Grid Reliability Problem with 100% Penetration of Intermittent Wind, Water, and Solar for All Purposes." *Proceedings of the National Academies of Sciences of the United States of America* 112, no. 49 (December 8, 2015): 15060–15065. www.pnas.org/content/112/49/15060.
17. A. A. Solomon, D. M. Kammen and D. Callaway. "Investigating the Impact of Wind–Solar Complementarities on Energy Storage Requirement and the Corresponding Supply Reliability Criteria." *Applied Energy* 168 (2016): 130–145.
18. S. C. Johnson, J. D. Rhodes and M. E. Webber. "Understanding the Impact of Non-Synchronous Wind and Solar Generation on Grid Stability and Identifying Mitigation Pathways." *Applied Energy* 262 (March 2020): 114492.
19. D. Roberts. "Clean Energy is Catching Up to Natural Gas." *Vox*, October 26, 2018. www.vox.com/energy-and-environment/2018/7/13/17551878/natural-gas-markets-renewable-energy.
20. A. Colthorpe. "BloombergNEF: 'Already Cheaper to Install New-Build Battery Storage than Peaking Plants.'" *Energy Storage News*, April 30, 2020.
21. Energy Information Administration. *Utility-Scale Battery Storage Costs Decreased Nearly 70% between 2015 and 2018* (October 23, 2020). www.eia.gov/todayinenergy/detail.php?id=45596.
22. Energy Information Administration. *Battery Storage in the United States: An Update on Market Trends*. Report by the Department of Energy (July 2020). www.eia.gov/analysis/studies/electricity/batterystorage/pdf/battery_storage.pdf.
23. S. Gamper-Rabindran. "Markets, States and the Federal Government in the Transition to Wind Energy." *Journal of Land Use and Environmental Law* 33, no. 2 (2018): 355–378. https://shanti1.weebly.com/uploads/6/2/8/6/6286936/markets_states_and_the_federal_governmen_in_the_transition_to__wind_energy.pdf.
24. D. G. De Silva, R. P. McComb and A. R. Schiller. "What Blows in with the Wind?" *Southern Economic Journal* 82, no. 3 (January 2016): 826–858.
25. B. Castleberry and J. S. Greene. "Impacts of Wind Power Development on Oklahoma's Public Schools." *Energy, Sustainability, and Society* 7, no. 34 (2017). https://link.springer.com/article/10.1186/s13705-017-0138-8.
26. E. A. H. Shqeib, E. H. Infield and H. C. Renski. "Measuring the Impacts of Wind Energy Projects on US Rural Counties' Community Services and Cost of Living." *Energy Policy* 153 (2021): 112279.
27. Energy Information Administration. *New York Generated the Fourth Most Electricity from Renewable Sources of Any State in 2019* (November 23, 2020). www.eia.gov/todayinenergy/detail.php?id=45996.

28. Natural Resource Defense Council. *American Wind Farms: Breaking Down the Benefits from Planning to Production*. Report by P. Jordan and C. Steger (September 2012). www.nrdc.org/sites/default/files/american-wind-farms-IP.pdf.
29. American Wind Energy Association. *Executive Summary: Wind Powers America Annual Report* (Washington, DC: 2019). www.powermag.com/wp-content/uploads/2020/04/awea_wpa_executivesummary2019.pdf.
30. K. Galbraith and A. Price. *The Great Texas Wind Rush: How George Bush, Ann Richards, and a Bunch of Tinkerers Helped the Oil and Gas State Win the Race to Wind Power*. Austin: University of Texas Press, 2013.
31. J. Zarnikau. "Successful Renewable Energy Development in a Competitive Electricity Market: A Texas Case Study." *Energy Policy* 39, no. 7 (2011): 3906–3913.
32. Energy Information Administration. *Fewer Wind Curtailments and Negative Power Prices Seen in Texas after Major Grid Expansion* (June 24, 2014). www.eia.gov/todayinenergy/detail.php?id=16831.
33. H. Fell, D. T. Kaffine and K. Novan. "Emissions, Transmission, and the Environmental Value of Renewable Energy." *American Economic Journal: Economic Policy* 13, no. 2 (May 2021): 241–272.
34. M. C. Slattery et al. "The Predominance of Economic Development in the Support for Large-Scale Wind Farms in the US Great Plains." *Renewable and Sustainable Energy Reviews* 16, no. 6 (2012): 3690–3701.
35. S. B. Mills. "Preserving Agriculture through Wind Energy Development: A Study of the Social, Economic, and Land Use Effects of Windfarms on Rural Landowners and Their Communities." Doctor of Philosophy dissertation, University of Michigan (2015). https://deepblue.lib.umich.edu/bitstream/handle/2027.42/111508/sbmills_1.pdf?sequence=1&isAllowed=y.
36. Polsinelli. *Annual Economic Impacts of Kansas Wind Energy, 2020 Report*. Report by A. C. Anderson et al. (Kansas City: March 22, 2021). https://sftp.polsinelli.com/publications/energy/Annual_Economic_Impacts_of_Kansas_Wind_Energy_Report.pdf.
37. S. Mills. *Farming the Wind: The Impact of Wind Energy on Farming* (2015). https://static1.squarespace.com/static/564236bce4b00b392cc6131d/t/575b315d9f7266050a4143aa/1465594217864/Sarah+Mills+Summary+Findings.pdf.
38. H. Russ. "Wind Farms Boost Tax Base for Local US Governments-Moody's." *Reuters*, May 7, 2018. www.reuters.com/article/usa-municipals-windfarms-idUSL1N1SE0WH.
39. Natural Resource Defense Council. *Clean Energy Sweeps across Rural America*. Report by A. Krishnaswami and E. Mittelman (2018). www.nrdc.org/sites/default/files/rural-clean-energy-report.pdf.
40. NC Sustainable Energy Association. *Increased North Carolina County Tax Revenue from Solar Development*. Report by C. Carson, D. Brookshire, J. Carey and D. Parker (2019). https://energync.org/wp-content/uploads/2019/07/Small_Increased-NC-County-Tax-Revenue-from-Solar-Developmentv3.pdf.
41. K. D. Walsh. "An Industry on the Precipice of Change: Maintaining Solar Energy's Competitive Advantage in North Carolina after the Expiration of Investment Tax Credits." *North Carolina Law Review* 93, no. 6 (2015): 1935.

42. Crossborder Energy and Solar Energy Industries Association. *The Benefits and Costs of Solar Distributed Generation for Arizona Public Service*. Report by R. T. Beach and P. G. McGuire (Berkeley: May 8, 2013).
43. W. Ajaz and D. Bernell. "California's Adoption of Microgrids: A Tale of Symbiotic Regimes and Energy Transitions." *Renewable and Sustainable Energy Reviews* 138 (2021): 110568.
44. J. St. John. "New York's Latest Clean Energy Push Includes 2.5GW of Offshore Wind Contracts for Equinor and BP." *Greentech Media*, January 13, 2021. www.greentechmedia.com/articles/read/new-yorks-new-green-push-includes-2.5gw-of-offshore-wind-contracts-for-equinor-and-bp.
45. Columbia Center on Global Energy Policy. *The Role of Corporate Renewable Power Purchase Agreements in Supporting US Wind and Solar Deployment*. Report by J. Kobus, A. Nasrallah and J. Guidera (March 24, 2021). www.energypolicy.columbia.edu/research/report/role-corporate-renewable-power-purchase-agreements-supporting-us-wind-and-solar-deployment.
46. K. Adler. "Corporate US Renewable Procurement Outlook: Optimism amid a Pessimistic Year." *IHS Markit*, November 22, 2020. https://ihsmarkit.com/research-analysis/corporate-us-renewable-procurement-outlook-optimism-amid.html.
47. D. Grandoni and S. Mufson. "Trump-Picked Head of Energy Panel Says He Was 'Demoted for My Independence' On Climate Change." *Washington Post*, November 6, 2020.
48. *Carbon Pollution Emission Guidelines for Existing Stationary Sources: Electric Utility Generating Units: Final Rule*. 40 Code of Federal Regulations Part 60. Environmental Protection Agency. 80 Federal Register 64661–65120 (October 23, 2015). www.govinfo.gov/content/pkg/FR-2015-10-23/pdf/2015-22842.pdf.
49. Environmental Integrity Project. *Greenhouse Gases from Power Plants 2005–2020: Rapid Decline Exceeded Goals of EPA Clean Power Plan*. Report by E. Schaeffer and T. Pelton (February 25, 2021). https://environmentalintegrity.org/wp-content/uploads/2021/02/Greenhouse-Gases-from-Power-Plants-2005-2020-report.pdf.
50. *Repeal of the Clean Power Plan; Emission Guidelines for Greenhouse Gas Emissions from Existing Electric Utility Generating Units; Revisions to Emission Guidelines Implementing Regulations: Final Rule*. 40 Code of Federal Regulations 60. Environmental Protection Agency. 84 Federal Register 32520–32584 (July 8, 2019). www.federalregister.gov/documents/2019/07/08/2019-13507/repeal-of-the-clean-power-plan-emission-guidelines-for-greenhouse-gas-emissions-from-existing.
51. *American Lung Association v. Environmental Protection Agency*, No. 19–1140 (D.C. Cir. 2021).
52. H. Northey and B. Storrow. "Bob Murray Drafted 6 Orders on Coal, Climate for Trump." *E&E News*, June 6, 2018.
53. R. Perry. *Secretary of Energy's Direction That the Federal Energy Regulatory Commission Issue Grid Resiliency Rules Pursuant to the Secretary's Authority*. Submitted to Chairman N. Chatterjee, Commissioner C. A. LaFleur and

Commissioner R. F. Powelson of the Federal Energy Regulatory Commission. September 28, 2017.

54. Federal Energy Regulatory Commission and North American Electric Reliability Corporation. *Outages and Curtailments during the Southwest Cold Weather Event of February 1–5, 2011* (August 2011). www.ferc.gov/sites/default/files/2020-04/08-16-11-report.pdf.

55. North American Electric Reliability Corporation. *Hurricane Sandy Event Analysis Report* (January 2014). www.nerc.com/pa/rrm/ea/Oct2012HurricanSandyEvntAnlyssRprtDL/Hurricane_Sandy_EAR_20140312_Final.pdf.

56. PJM Interconnection. *Analysis of Operational Events and Market Impacts During the January 2014 Cold Weather Events* (March 8, 2014). www.hydro.org/wp-content/uploads/2017/08/PJM-January-2014-report.pdf.

57. J. Moore, senior attorney and director of the Sustainable FERC Project at the Natural Resources Defense Council. *Testimony Regarding Part 2: Powering America: Defining Reliability in a Transforming Electricity Industry.* Submitted to the Hearing before House of Representatives Committee on Energy and Commerce, Subcommittee on Energy. October 3, 2017.

58. E. A. Moler et al. *Comments of the Bipartisan Former FERC Commissioners on Proposed Rule Grid Resiliency Pricing Rule.* Submitted to the Federal Energy Regulatory Commission. Docket No. RM18-1-000. October 19, 2017.

59. N. Gentile and K. Kelly. "The Trump Administration Is Stifling Renewable Energy on Public Lands and Waters." *Center for American Progress Energy and Environment*, June 25, 2020. www.americanprogress.org/issues/green/reports/2020/06/25/486852/trump-administration-stifling-renewable-energy-public-lands-waters/?_ga=2.29066236.1701253460.1620610966-706434035.1620411321.

60. Department of the Interior. *Vineyard Wind Offshore Wind Energy Project Draft Environmental Impact Statement.* Report by the Bureau of Ocean Energy Management and Office of Renewable Energy Programs (December 2018). www.boem.gov/sites/default/files/renewable-energy-program/State-Activities/MA/Vineyard-Wind/Vineyard_Wind_Draft_EIS.pdf.

61. D. Gearino and P. McKenna. "Government Delays First Big US Offshore Wind Farm. Is a Double Standard at Play?" *Inside Climate News*, August 19, 2019. https://insideclimatenews.org/news/19082019/vineyard-wind-offshore-renewable-energy-delay-boem-environmental-cumulative-review-nepa-massachusetts.

62. E. B. Johnson, chair of the Committee on Science, Space, and Technology. *Addressing the Issue of a Department of Energy Technical Analysis Being Suppressed, Modified, and Delayed by Headquarters Officials.* Submitted to D. Brouillette, secretary of Energy. October 23, 2020.

63. P. Fairley. "How a Plan to Save the Power System Disappeared." *The Atlantic*, August 20, 2020.

64. National Renewable Energy Laboratory. *Interconnections Seam Study: Overview.* Report by G. Brinkman et al. NREL/PR-6A20-78161 (Golden: October 2020). www.nrel.gov/docs/fy21osti/78161.pdf.

65. *Electric Storage Participation in Markets Operated by Regional Transmission Organizations and Independent System Operators: Final Rule.* 18 Code of Federal Regulations 35. Federal Energy Regulatory Commission, Department of Energy. 83 Federal Register 9580–9633 (March 6, 2018) www.federalregister.gov/docu

ments/2018/03/06/2018-03708/electric-storage-participation-in-markets-operated-by-regional-transmission-organizations-and.
66. Federal Energy Regulatory Commission. *FERC Order No. 2222: Fact Sheet.* September 28, 2020. www.ferc.gov/media/ferc-order-no-2222-fact-sheet.
67. Federal Energy Regulatory Commission. *FERC Proposes Policy Statement on State-Determined Carbon Pricing in Wholesale Markets.* Policy statement by Chairman N. Chatterjee. October 15, 2020. www.ferc.gov/news-events/news/ferc-proposes-policy-statement-state-determined-carbon-pricing-wholesale-markets.
68. S. Patel. "The Significance of FERC's Recent PJM MOPR Order Explained." *Power*, December 26, 2019. www.powermag.com/the-significance-of-fercs-recent-pjm-mopr-order-explained.
69. J. St. John. "FERC Denies Rehearings on PJM Capacity Orders, in a Blow to States' Renewables Plans." *Greentech Media*, April 16, 2020. www.greentechmedia.com/articles/read/ferc-denies-rehearings-on-its-pjm-capacity-rulings-opening-door-for-legal-challenges.
70. D. Roberts. "Trump's Crude Bailout of Dirty Power Plants Failed, but a Subtler Bailout Is Underway." *Vox*, March 23, 2018. www.vox.com/energy-and-environment/2018/3/23/17146028/ferc-coal-natural-gas-bailout-mopr.
71. G. Metcalf. "The Impact of Removing Tax Preferences for US Oil and Gas Production." *Journal of the Association of Environmental and Resource Economists* 5, no. 1 (2018): 1–37.
72. PennFuture. *Buried Out of Sight: Uncovering Pennsylvania's Hidden Fossil Fuel Subsidies.* Report by E. Persico, R. Altenburg and C. Simeone (February 2021). www.pennfuture.org/Files/Admin/PF_FossilFuel_Report_final_2.12.21.pdf.
73. National Academies of Sciences, Engineering, and Medicine. *Accelerating Decarbonization of the US Energy System.* Washington, DC: The National Academies Press, 2021. https://doi.org/10.17226/25932.
74. A. B. Klass. "Transmission, Distribution, and Storage: Grid Integration." In *Legal Pathways to Deep Decarbonization in the United States*, edited by M. B. Gerrard and J. C. Dernbach. 527–546. Washington, DC: Environmental Law Institute, 2019. Reprint.
75. Environmental Protection Agency. *Sources of Greenhouse Gas Emissions* (2019). www.epa.gov/ghgemissions/sources-greenhouse-gas-emissions.
76. Princeton University, Andlinger Center and High Meadows Environmental Institute. *Net-Zero America: Potential Pathways, Infrastructure, and Impacts.* Report by E. Larson et al. (December 15, 2020). https://environmenthalfcentury.princeton.edu/sites/g/files/toruqf331/files/2020-12/Princeton_NZA_Interim_Report_15_Dec_2020_FINAL.pdf.
77. Goldman School of Public Policy. *2035 The Report: Plummeting Solar, Wind, and Battery Costs Can Accelerate Our Clean Electricity Future* (June 2020). www.2035report.com/wp-content/uploads/2020/06/2035-Report.pdf?hsCtaTracking=8a85e9ea-4ed3-4ec0-b4c6-906934306ddb%7Cc68c2ac2-1db0-4d1c-82a1-65ef4daaf6c1.
78. A. E. MacDonald et al. "Future Cost-Competitive Electricity Systems and Their Impact on US CO_2 Emissions." *Nature Climate Change* 6 (January 25, 2016): 526–531.

79. M. Z. Jacobson et al. "The United States Can Keep the Grid Stable at Low Cost with 100% Clean, Renewable Energy in All Sectors Despite Inaccurate Claims." *Proceedings of the National Academies of Sciences of the United States of America* 114, no. 26 (June 17, 2017): E5021–E5023. www.pnas.org/content/114/26/E5021.
80. C. T. M. Clack et al. "Evaluation of a Proposal for Reliable Low-Cost Grid Power with 100% Wind, Water, and Solar." *Proceedings of the National Academies of Sciences of the United States of America* 114, no. 26 (June 26, 2017). www.pnas.org/content/pnas/early/2017/06/16/1610381114.full.pdf.
81. J. Spector. "90% Clean Grid by 2035 Is Not Just Feasible, but Cheaper, Study Says." *Greentech Media*, June 9, 2020. www.greentechmedia.com/articles/read/90-clean-grid-by-2035-is-not-just-feasible-but-cheaper-study-says?_ga=2.61205036.1848826001.1618838887-1786218930.1618838887.
82. Energy Innovation Policy and Technology, LLC. *Rewiring the US for Economic Recovery*. Report by S. Aggarwal and M. O'Boyle (June 2020). https://energyinnovation.org/wp-content/uploads/2020/06/90-Clean-By-2035-Policy-Memo_June-2020.pdf.
83. United Nations Environment Programme and Climate and Clean Air Coalition. *Global Methane Assessment: Benefits and Costs of Mitigating Methane Emissions* Report by A. R. Ravishankara et al. (2021). www.ccacoalition.org/en/resources/global-methane-assessment-full-report.
84. G. Meyer. "Climate Fears Stoke Natural Gas Opposition; US Utilities Protests over Increase in Emissions Prompt Groups to Issue Lower-Carbon Pledge." *Financial Times*, February 7, 2020.
85. Energy Information Administration. *Annual Energy Outlook 2021* (February 3, 2021). www.eia.gov/outlooks/aeo/electricity/sub-topic-02.php.
86. Energy Information Administration. *Renewables Account for Most New US Electricity Generating Capacity in 2021* (January 11, 2021). www.eia.gov/todayinenergy/detail.php?id=46416.
87. Rocky Mountain Institute. *The Growing Market for Clean Energy Portfolios*. Report by C. Teplin, M. Dyson, A. Engel and G. Glazer (2019). https://rmi.org/insight/clean-energy-portfolios-pipelines-and-plants.
88. Rocky Mountain Institute. *Prospects for Gas Pipelines in the Era of Clean Energy*. Report by M. Dyson, G. Glazer and C. Teplin (2019). https://rmi.org/insight/clean-energy-portfolios-pipelines-and-plants.
89. B. Plumber. "As Coal Fades in the US, Natural Gas Becomes the Climate Battleground." *New York Times*, June 26, 2019.
90. J. Fergen and J. B. Jacquet. "Beauty in Motion: Expectations, Attitudes, and Values of Wind Energy in the Rural US." *Energy Research & Social Science* 11 (2016): 133–141.
91. Energy Systems Integration Group. *Transmission Planning for 100% Clean Electricity*. Report by L. Azar et al. (February 2021). www.esig.energy/wp-content/uploads/2021/02/Transmission-Planning-White-Paper.pdf.
92. P. R. Brown and A. Botterud. "The Value of Inter-Regional Coordination and Transmission in Decarbonizing the US Electricity System." *Joule* 5, no. 1 (January 20, 2021): 115–134.

93. Americans for a Clean Energy Grid and Grid Strategies, LLC. *Transmission Projects Ready to Go: Plugging into America's Untapped Renewable Resources*. Report by M. Goggin et al. (April 2021). https://cleanenergygrid.org/wp-content/uploads/2019/04/Transmission-Projects-Ready-to-Go-Final.pdf.
94. National Renewable Energy Laboratory. *The Value of Increased HVDC Capacity between Eastern and Western US Grids: The Interconnections Seam Study*. Report by A. Bloom et al. NREL/JA-6A20-76580 (Golden: October 2020). www.nrel.gov/docs/fy21osti/76850.pdf.
95. J. Tomich. "Can Biden Transmission Order Avoid State Backlash?" *E&E News*, February 11, 2021.
96. B. Gerstle. "Giving Landowners the Power: A Democratic Approach for Assembling Transmission Corridors." *Journal of Environmental Law & Litigation* 29 (2014): 535–578.
97. R. Winn. "Landowner Compensation in Transmission Siting for Renewable Energy Facilities." *Electricity Journal* 27, no. 5 (2014): 21–30.
98. Center for Rural Affairs. *Amplifying Clean Energy with Conservation, Part Two: Leveraging Electric Transmission Lines for Stewardship*. Report by C. Smith (November 2020). www.cfra.org/sites/default/files/publications/amplifying-clean-energy-with-conservation-part-two.pdf.
99. S. C. Klain et al. "Will Communities 'Open-Up' to Offshore Wind? Lessons Learned from New England Islands in the United States." *Energy Research & Social Science* 34 (December 2017): 13–26. www.sciencedirect.com/science/article/pii/S2214629617301172.
100. J. Firestone et al. "Faring Well in Offshore Wind Power Siting? Trust, Engagement and Process Fairness in the United States." *Energy Research & Social Science* 62 (2020): 101393–101405.
101. Center for Rural Affairs. *Clean Energy Transmission Principles*. www.cfra.org/transmission.
102. J. B. Jacquet and J. Fergen. "The Vertical Patterns of Wind Energy: The Effects of Wind Farm Ownership on Rural Communities in the Prairie Pothole Region of the United States." *Journal of Rural and Community Development* 13, no. 2 (2018): 130–148.
103. S. Mills. "Wind Energy and Rural Community Sustainability." In *Handbook of Sustainability and Social Science Research*, edited by W. L. Filho, R. W. Marans and J. Callewaert. 215–225. Cham: Springer, 2018. Reprint.
104. M. McGraw and R. Hennessy. "Rush to Attract Wind Turbine Investors Leaves Kansas School Districts Shortchanged." *Investigate Midwest*, December 4, 2017. https://investigatemidwest.org/2017/12/04/rush-to-attract-wind-turbine-investors-leaves-kansas-school-districts-shortchanged/.
105. Texas Taxpayers and Research Association. *Understanding Chapter 313: School Property Tax Limitations and the Impact on State Finances* (January 11, 2017). www.ttara.org/wp-content/uploads/2018/09/UnderstandingChapter313_Final_Web_1_11_17.pdf.
106. Lawrence Berkeley National Lab. *Community Wind Power Ownership Schemes in Europe and Their Relevance to the United States*. Report by M. Bolinger. LBNL-48357; R&D Project: 57461F; TRN: US200426%%1013 (Berkeley: May 15, 2001). www.osti.gov/servlets/purl/827946-fmUZml/native.

107. M. Bolinger. "Community-Owned Wind Power Development: The Challenge of Applying the European Model in the United States, and How States Are Addressing That Challenge." Global Windpower Conference, Chicago, Lawrence Berkeley National Laboratory, March 28, 2004.
108. E. Lantz and S. Tegen. "Economic Development Impacts of Community Wind Projects: April 2009 Review and Empirical Evaluation." Windpower Conference and Exhibition, Chicago, National Renewable Energy Laboratory, April 2009.
109. K. B. Jones and M. James. "Distributed Renewables in the New Economy: Lessons from Community Solar Development in Vermont." In *Law and Policy for a New Economy: Sustainable, Just, and Democratic*, edited by M. K. Scanlan. 189–210. Northampton: Edward Elgar, 2017. Reprint.
110. Institute for Local Self-Reliance. *Report: Beyond Sharing – How Communities Can Take Ownership of Renewable Power*. Report by J. Farrell (April 26, 2016). https://ilsr.org/wp-content/uploads/2016/04/Beyond-Sharing-report-re-release-ILSR.pdf.
111. D. Shaffer. "Southwest Minnesota Wind Farm Sold Out of Bankruptcy to Bank and Local Investors." *Star Tribune*, November 25, 2015. www.startribune.com/southwest-minnesota-wind-farm-sold-out-of-bankruptcy-to-bank-and-local-investors/353931451.
112. S. Fast et al. "Lessons Learned from Ontario Wind Energy Disputes." *Nature Energy* 1, no. 15028 (January 25, 2016). www.nature.com/articles/nenergy201528.pdf.
113. C. Walker and J. Baxter. "'It's Easy to Throw Rocks at a Corporation': Wind Energy Development and Distributive Justice in Canada." *Journal of Environmental Policy & Planning* 19, no. 6 (2017): 754–768.
114. Earthjustice et al. *Building Back Better: A Roadmap to Expand Solar Access for All* (Posted at E&E News: April 4, 2021). www.eenews.net/assets/2021/04/21/document_ew_04.pdf.
115. M. A. Bolinger. "Making European-Style Community Wind Power Development Work in the US." *Renewable and Sustainable Energy Reviews* 9, no. 6 (2005): 556–575.
116. Minnesota Project. *Lessons & Concepts for Advancing Community Wind*. Report by R. Stockwell et al. (December 2009). https://d3n8a8pro7vhmx.cloudfront.net/windustry/legacy_url/1588/Advancing-Community-Wind_Dec09.pdf?1421783488.
117. Environmental Law & Policy Center. *Community Wind Financing: A Handbook by the Environmental Law & Policy Center*. Report by C. Kubert et al. (2004). www.mresearch.com/pdfs/docket4185/NG11/doc55.pdf.
118. S. Johnston. "Non-refundable Tax Credits versus Grants: The Impact of Subsidy Form on the Effectiveness of Subsidies for Renewable Energy." *Journal of the Association of Environmental and Resource Economics* 6, no. 3 (2019): 433–460.
119. G. Pacyniak. "Greening the Old New Deal: Strengthening Rural Electric Cooperative Supports and Oversight to Combat Climate Change." *Missouri Law Review* 85, no. 2 (2020): 409.
120. A. B. Klass and G. Chan. "Cooperative Clean Energy." *North Carolina Law Review* (2021).

121. Center for Rural Affairs, Clean Up the River Environment Minnesota, and We Own It. *Rural Electrification 2.0: The Transition to a Clean Energy Economy*. Report by E. Hatlestad et al. (June 2019). www.cfra.org/sites/default/files/publications/rural-electrification-2.0-the-transition-to-a-clean-energy-economy.pdf.
122. Center for Science, Technology, and Environmental Policy. *Community Shared: Solar in Minnesota Learning from the First 300 Megawatts*. Report by G. Chan et al. (March 2018). https://static1.squarespace.com/static/5b8032c35b409b4d9458387e/t/5bcd6c6ceef1a115dd9b74bd/1540189296387/Community+Shared+Solar+in+Minnesota+%28FINAL+for+web%29.pdf.
123. I. Penn. "The Next Energy Battle: Renewables vs. Natural Gas." *New York Times*, July 6, 2020.
124. American Council for an Energy Efficient Economy. *The High Cost of Energy in Rural America: Household Energy Burdens and Opportunities for Energy Efficiency*. Report by L. Ross, A. Drehobl and B. Stickles (Washington, DC: July 2018).
125. Rocky Mountain Institute. *A Low-Cost Energy Future for Western Cooperatives: Emerging Opportunities for Cooperative Electric Utilities to Pursue Clean Energy at a Cost Savings to Their Members*. Report by M. Dyson and A. Engel (August 1, 2018).
126. E. Funkhouser et al. "Business Model Innovations for Deploying Distributed Generation: The Emerging Landscape of Community Solar in the US." *Energy Research & Social Science* 10 (2015): 90–101.
127. G. Chan et al. "Design Choices and Equity Implications of Community Shared Solar." *Electricity Journal* 30, no. 9 (2017): 37–41.
128. Vibrant Clean Energy, LLC. *Executive Summary, Why Local Solar for All Costs Less: A New Roadmap for the Lowest Cost Grid*. Report by C. T. M. Clack et al. (December 1, 2020). www.vibrantcleanenergy.com/wp-content/uploads/2020/12/WhyDERs_ES_Final.pdf.
129. Utility Dive. *2020 Outlook: Utilities Will Be Pushed to Further Embrace Distributed Energy Resources*. Report by M. Bandyk (January 17, 2020). www.utilitydive.com/news/2020-outlook-utilities-will-be-pushed-to-further-embrace-distributed-energ/569613.
130. B. R. Lukanov and E. M. Krieger. "Distributed Solar and Environmental Justice: Exploring the Demographic and Socio-Economic Trends of Residential PV Adoption in California." *Energy Policy* 134 (2019): 110935.
131. D. A. Sunter, S. Castellanos and D. M. Kammen. "Disparities in Rooftop Photovoltaics Deployment in the United States by Race and Ethnicity." *Nature Sustainability* 2 (2019): 71–76.
132. National Renewable Energy Laboratory. *Design and Implementation of Community Solar Programs for Low and Moderate-Income Customers*. Report by J. Heeter et al. NREL/TP-6A20-71652 (Golden: December 2018). www.nrel.gov/docs/fy19osti/71652.pdf.
133. S. Borenstein. "Private Net Benefits of Residential Solar PV: The Role of Electricity Tariffs, Tax Incentives, and Rebates." *Journal of the Association of Environmental and Resource Economists* 4, no. S1 (2017): S85–S122.
134. E. Johnson et al. "Peak Shifting and Cross-Class Subsidization: The Impacts of Solar PV on Changes in Electricity Costs." *Energy Policy* 106 (2017): 436–444.

135. S. H. Baker. "Unlocking the Energy Commons: Expanding Community Energy Generation." In *Law and Policy for a New Economy: Sustainable, Just, and Democratic*, edited by M. K. Scanlan. 211–234. Northampton: Edward Elgar Publishing, 2017. Reprint.
136. Interstate Renewable Energy Council. *Shared Renewable Energy for Low- to Moderate-Income Consumers: Policy Guidelines and Model Provisions*. Report by E. Schroeder McConnell et al. (2016). www.irecusa.org/wp-content/uploads/2016/03/IREC-LMI-Guidelines-Model-Provisions_FINAL.pdf.
137. E. M. Krieger, J. A. Casey and S. B. C. Shonkoff. "A Framework for Siting and Dispatch of Emerging Energy Resources to Realize Environmental and Health Benefits: Case Study on Peaker Power Plant Displacement." *Energy Policy* 96 (2016): 302–313.
138. NAACP. *Module 4: Starting Community-Owned Clean Energy Projects*. (2014). www.naacp.org/wp-content/uploads/2014/03/Module-4_Starting-Community-Owned-Clean-Energy-Projects_JEP-Action-Toolkit_NAACP.pdf.
139. National Renewable Energy Laboratory. *Shared Solar: Current Landscape, Market Potential, and the Impact of Federal Securities Regulation*. Report by D. Feldman et al. NREL/TP-6A20-63892 (Golden: May 2015). www.nrel.gov/docs/fy15osti/63892.pdf.
140. Congressional Research Service. *An Overview of USDA Rural Development Programs*. Report by T. Cowan. RL31837 (Washington, DC: February 10, 2016). http://fedweb.com/wp-content/uploads/2016/12/Rural-Development-Programs.pdf.
141. Pacific Northwest National Laboratory. *Distributed Generation Valuation and Compensation*. Report by A. C. Orrell et al. White Paper PNNL-27271 (February 2018). www.districtenergy.org/HigherLogic/System/DownloadDocumentFile.ashx?DocumentFileKey=0103ebf1-2ac9-7285-b49d-e615368725b2&forceDialog=0.
142. Institute for Policy Integrity. *Managing the Future of the Electricity Grid: Distributed Generation and Net Metering*. Report by R. L. Revesz and B. Unel. Working paper 2016/1 (New York: February 2018). https://policyintegrity.org/files/publications/ManagingFutureElectricityGrid.pdf.
143. K. F. Roemer and J. H. Haggerty. "Coal Communities and the US Energy Transition: A Policy Corridors Assessment." *Energy Policy* 151 (April 2021): 112112. www.sciencedirect.com/science/article/abs/pii/S0301421520308235.
144. UNC Center for Climate, Energy, Environment and Economics. *Communities in Transition: State Responses to Energy-Sector Job Losses*. Report by E. Blumenthal (December 2019). https://law.unc.edu/wp-content/uploads/2019/12/CommunitiesinTransition2019.pdf.
145. Colorado Department of Labor and Employment. *Learn about the Legislation: House Bill 19–1314* (2020). https://cdle.colorado.gov/learn-about-the-legislation.
146. Energy Innovation Policy and Technology, LLC. *Comparing 2019 Securitization Legislation in Colorado, Montana, and New Mexico*. Report by R. Lehr and M. O'Boyle (September 2020). https://energyinnovation.org/wp-content/uploads/2020/09/Securitization-Brief_September-2020.pdf.
147. N. Long. "New Mexico Embraces Transition to 100% Clean Energy." *Natural Resource Defense Council Expert Blog*, 2019. www.nrdc.org/experts/noah-long/new-mexico-embraces-transition-100-clean-energy.

148. Just Transition Fund. *National Economic Transition: A Visionary Proposal for an Equitable Future Platform* (2020). https://nationaleconomictransition.org/platform.
149. BlueGreen Alliance. *Solidarity for Climate Action* (July 2019). www.bluegreenalliance.org/work-issue/solidarity-for-climate-action.
150. Just Transition Listening Project. *Workers and Communities in Transition: Report of the Just Transition Listening Project*. Report by J. M. Cha et al. (2021). www.labor4sustainability.org/files/JTLP_report2021.pdf.
151. Congressional Research Service. *The POWER Initiative: Energy Transition as Economic Development*. Report by M. H. Cecire. R46015 (Washington, DC: November 20, 2019). https://fas.org/sgp/crs/misc/R46015.pdf.
152. Council of State Governments. *Trends and Market Forces Shaping the Future of US Coal Industry*. Report by D. Saha (August 2017). http://knowledgecenter.csg.org/kc/system/files/CR_coal.pdf.
153. White House Office of the Press Secretary. *Fact Sheet: The Partnerships for Opportunity and Workforce and Economic Revitalization (POWER) Initiative*. Obama Administration. March 27, 2015. https://obamawhitehouse.archives.gov/the-press-office/2015/03/27/fact-sheet-partnerships-opportunity-and-workforce-and-economic-revitaliz.
154. White House Office of the Press Secretary. *Fact Sheet: Administration Announces New Economic and Workforce Development Resources for Coal Communities through POWER Initiative*. Obama Administration. August 24, 2016. https://obamawhitehouse.archives.gov/the-press-office/2016/08/24/fact-sheet-administration-announces-new-economic-and-workforce.
155. D. Brown. "Reclamation Funds Dwindle While Congress Dawdles." *E&E News*, August 19, 2019.
156. Office of Surface Mining Reclamation and Enforcement. *Report on Abandoned Mine Land Reclamation Economic Development Pilot Program (AML Pilot Program) for FY 2016–FY 2019*. OSMRE Program Support Directorate (December 18, 2020). www.osmre.gov/programs/AML/2016_2019_Annual_Report_AML_Economic_Development_Pilot_Program.pdf.
157. J. T. Sayago-Gómez et al. "Impact Evaluation of Investments in the Appalachian Region: A Reappraisal." *International Region Science Review* 41, no. 6 (2015): 601–629. https://doi.org/10.1177/0160017617713822.
158. A. Rupasingha, D. Crown and J. Pender. "Rural Business Programs and Business Performance: The Impact of the USDA's Business and Industry (B&I) Guaranteed Loan Program." *Journal of Regional Science* 59, no. 4 (September 2018): 701–722. https://onlinelibrary.wiley.com/doi/abs/10.1111/jors.12421.
159. V. Volcovici. "Trump Seeks to Ax Appalachia Economic Programs, Causing Worry in Coal Country." *Reuters*, March 16, 2017. www.reuters.com/article/us-usa-trump-budget-appalachia-idUSKBN16N2VF.
160. Brookings. *How Renewable Energy Jobs Can Uplift Fossil Fuel Communities and Remake Climate Politics*. Report by A. Tomer et al. (2021). www.brookings.edu/research/how-renewable-energy-jobs-can-uplift-fossil-fuel-communities-and-remake-climate-politics/?preview_id=1414272.
161. Columbia Sabin Center for Climate Change Law. *Opposition to Renewable Energy Facilities in the United States*. Report by K. Marsh et al. (February

2021). https://climate.law.columbia.edu/sites/default/files/content/RELDI%20report%20MBG%202.26.21%20HWA.pdf.

162. D. J. Hess. "The Politics of Niche-Regime Conflicts: Distributed Solar Energy in the United States." *Environmental Innovation and Societal Transitions* 19 (2016): 42–50.

163. L. C. Stokes and H. L. Breetz. "Politics in the US Energy Transition: Case Studies of Solar, Wind, Biofuels and Electric Vehicles Policy." *Energy Policy* 113 (2018): 76–86.

164. Sierra Club. *The Dirty Truth About Utility Climate Pledges*. Report by J. Romankiewicz et al. (January 2021). https://coal.sierraclub.org/the-problem/dirty-truth-greenwashing-utilities.

165. V. Karadzhova and U. Ashfaq, "What Biden's Climate Plan Means for US Utilities." *MSCI Blog*, 2021. www.msci.com/www/blog-posts/what-biden-s-climate-plan-means/02278416796.

166. A. Hertel-Fernandez. *State Capture: How Conservative Activists, Big Businesses, and Wealthy Donors Reshaped the American States – and the Nation*. New York: Oxford University Press, 2019.

167. M. Gallucci. "Renewable Energy Standards Target of Multi-Pronged Attack." *Inside Climate News*, March 19, 2013.

168. J. Eilperin. "Climate Skeptic Group Works to Reverse Renewable Energy Mandates." *Washington Post*, November 24, 2012.

169. E. Halper. "Koch Brothers, Big Utilities Attack Solar, Green Energy Policies." *Los Angeles Times*, April 19, 2014.

170. B. Plumer. "State Renewable Energy Laws Turn Out to Be Incredibly Hard to Repeal." *Washington Post*, August 8, 2013.

171. Americans for Prosperity. *Partner Prospectus* (Winter 2015). www.documentcloud.org/documents/2035387-merged-document-2.html.

172. Center for American Progress. *Fact Sheet: Efforts to Repeal or Weaken Renewable Energy Schedules in the States*. Report by G. Taraska and A. Cassady (March 10, 2015). www.americanprogress.org/issues/green/reports/2015/03/10/108250/fact-sheet-efforts-to-repeal-or-weaken-renewable-energy-schedules-in-the-states.

173. D. J. Hess. "Sustainability Transitions: A Political Coalition Perspective." *Research Policy* 43, no. 2 (2014): 278–283.

174. L. C. Stokes. *Short Circuiting Policy: Interest Groups and the Battle Over Clean Energy and Climate Policy in the American States*. New York: Oxford University Press, 2020.

175. Environment America Research & Policy Center and Frontier Group. *Blocking the Sun: 12 Utilities and Fossil Fuel Interests That Are Undermining American Solar Power*. Report by G. Weissman and B. Fanshaw (October 2015). https://environmentamerica.org/sites/environment/files/reports/EA_BlockingtheSun_scrn_0.pdf.

176. Institute for Energy Economics and Financial Analysis. *As Oil and Gas Wane, Texas Wind Industry Ascends*. Report by K. Cates et al. (August 2020). https://ieefa.org/wp-content/uploads/2020/08/As-Oil-and-Gas-Wane_TX-Wind-Industry-Ascends_August-2020.pdf.

177. E. Douglas. "Wind Power a Smaller Contributor to Texas Electricity Crisis than Initially Estimated, ERCOT Analysis Shows." *Texas Tribune*, April 28, 2021. www.texastribune.org/2021/04/28/texas-power-outage-wind.

178. G. Freitas Jr. and M. Chediak. "Kinder's $1 Billion Texas Crisis Gain Foreshadows More Windfalls." *Bloomberg*, April 22, 2021. www.bloomberg.com/news/articles/2021-04-22/kinder-s-1-billion-surprise-sets-stage-for-more-earnings-shocks.
179. ERCOT Public. *Update to April 6, 2021 Preliminary Report on Causes of Generator Outages and Derates During the February 2021 Extreme Cold Weather Event* (April 27, 2021). www.ercot.com/content/wcm/lists/226521/ERCOT_Winter_Storm_Generator_Outages_By_Cause_Updated_Report_4.27.21.pdf.
180. D. Cohan and K. Hayhoe. "Opinion: Texas Needed Power and Leadership. It Got Neither." *Austin American-Statesman*, February 24, 2021. www.statesman.com/story/opinion/columns/your-voice/2021/02/24/opinion-texas-needed-power-and-leadership-got-neither/4561323001.
181. Federal Energy Regulatory Commission and North American Electric Reliability Corporation. *The South Central United States Cold Weather Bulk Electric System Event of January 17, 2018* (July 2019). www.nerc.com/pa/rrm/ea/Documents/South_Central_Cold_Weather_Event_FERC-NERC-Report_20190718.pdf.
182. G. T. Galvin. "Lone Star Solar: Challenges and Opportunities in Post-blackout Texas." *National Law Review* 11, no. 132 (April 5, 2021). www.natlawreview.com/article/lone-star-solar-challenges-and-opportunities-post-blackout-texas.
183. D. Gearino. "Texas Politicians Aim to Penalize Wind and Solar in Response to Outages. Are Renewables Now Strong Enough to Defend Themselves?" *Inside Climate News*, April 17, 2021. https://webcache.googleusercontent.com/search?q=cache:HB-0BEkNYp8J:https://insideclimatenews.org/news/17042021/texas-politicians-wind-solar-natural-gas-winter-storm/+&cd=1&hl=en&ct=clnk&gl=ca.
184. J. Wallace. "Texas Senate Passes Bill Aiming to Counter Federal Subsidies for Wind and Solar Power." *Houston Chronicle*, March 29, 2021.
185. J. D. Rhodes. "Tax Credits for Wind and Solar Work. Why Are Texas Lawmakers Going After Them?" *UT News*, April 12, 2019. https://news.utexas.edu/2019/04/12/tax-credits-for-wind-and-solar-work-why-are-texas-lawmakers-going-after-them/.
186. Oil Change International. *Dirty Energy Dominance: Dependent on Denial*. Report by J. Redman et al. (October 2017). http://priceofoil.org/content/uploads/2017/10/OCI_US-Fossil-Fuel-Subs-2015-16_Final_Oct2017.pdf.
187. C. Brannstrom and M. Fry. "New Geographies of the Texas Energy Revolution." In *The Routledge Research Companion to Energy Geographies*, edited by S. Bouzarovski, M. J. Pasqualetti and V. C. Broto. 17–31. New York: Routledge, 2017. Reprint.
188. S. Mufson and T. Hamburger. "Battle Is Looming in States over Fossil Fuels." *Washington Post*, 2014.
189. D. Henry. "Kansas Set to Repeal Renewable Energy Mandate." *The Hill*, May 5, 2015. https://thehill.com/policy/energy-environment/242353-kansas-set-to-repeal-renewable-energy-mandate.
190. J. Schwartz. "Most Voters Want Bold Action on Climate, Survey Finds." *New York Times*, January 16, 2021.
191. Public Opinion Strategies and Citizens for Responsible Energy Solutions. *National Republican Voter Online Poll – GOP Candidates*. Report by N. Newhouse (June 6–16, 2020). www.citizensfor.com/wp-content/uploads/2020/06/CRES-Poll-June2020.pdf.

192. Congressional Research Service. *Clean Energy Standards: Selected Issues for the 117th Congress*. Report by A. J. Lawson. R46691 (Washington, DC: March 26, 2021). https://fas.org/sgp/crs/misc/R46691.pdf.
193. Natural Resource Defense Council. *Race to 100% Clean*. Report by S. Ptacek and A. Levin (Washington, DC: April 16, 2021). www.nrdc.org/resources/race-100-clean.
194. *Massachusetts v. EPA*, 549 US 497 (Supreme Court 2007).
195. M. Bolinger. *Community Wind Power Ownership Schemes in Europe and Their Relevance to the United States*. No. LBNL-48357. Lawrence Berkeley National Lab (Berkeley, CA, May 2001). https://emp.lbl.gov/publications/community-wind-power-ownership.
196. J. Mayer. *Dark Money: The Hidden History of the Billionaires Behind the Rise of the Radical Right*. New York: Anchor, 2017.
197. C. Leonard. *Kochland: The Secret History of Koch Industries and Corporate Power in America*. New York: Simon and Schuster, 2020.

PART II

America's Lands

4

Public and Private Lands
Extraction and Infrastructure versus Competing Economic Pursuits

The federal government plays many roles in protecting and managing land – both public and private. Its multifaceted and interconnected responsibilities in doing so translate into a variety of influences over where oil and gas drilling occurs, how environmental and public health protections are enforced, and whether infrastructure can be constructed to move extracted resources. The United States owns US public lands that span 640 million acres or 28 percent of the US land area, over which it has a range of rights and responsibilities.[1,i] This chapter's analysis of public lands focuses on the subset of public lands in the lower 48 states that are managed by the Bureau of Land Management (BLM) under the multiple-use mandate. The federal government determines how much of these lands is open to oil and gas drilling and other extractive activities and how much is reserved to provide recreation, grazing, ecosystem services and conservation. In 2017, public lands contributed 5.6 percent of US crude oil production and 9.7 percent of US natural gas production.[2] The US government's decisions also affect the use of private lands, Native American treaty and ancestral lands (Chapter 5), and public lands through its permitting decisions for fossil fuel export terminals, interstate gas pipelines and oil and gas railcars that cross these lands. These decisions on infrastructure shape the feasibility of oil and gas extraction across the US landscape.

---

[i] The complexity of the US government's ownership of the surface or/and subsurface mineral rights to these lands are discussed in Section 4.1.1. Public lands are subject to vastly different management regimes. BLM manages 244 acres of public lands. It manages a subset of these public lands under the multiple-use mandate. It also manages national monuments or wilderness areas and other lands dedicated to a particular purpose, such as conservation. The Fish and Wildlife Service manages 89 million acres as part of the National Wildlife Refuge System. The National Park Service manages 80 million acres in the National Park System. The Forest Service manages the 193 million acre National Forest System under the National Forest Management Act of 1976 [1].

The US Constitution grants Congress the authority to manage public lands via the Property Clause. In 1911, the Supreme Court in *Light v. United States*[3] interpreted Congress's responsibilities therein to include a responsibility for the US government to hold these lands "in trust for the people of the whole country."[4] Congress subsequently enacted the Federal Land Policy and Management Act of 1976 (FLPMA) instructing BLM to manage a subset of these public lands for "multiple use and sustained yield" that "will best meet the present and future needs of the American people."[4,5] Leasing and permitting decisions determine not only short-term impacts (e.g., changes in land use, local and regional air and water quality, etc.) they also shape the long-term survival of ecosystems within and beyond those lands and affect the global climate.

Congressional authority for oversight of oil and gas drilling also extends more broadly, on both public and private lands, via the Commerce Clause. The Commerce Clause grants Congress broad powers to regulate commerce between the states, which has been interpreted to include oversight of infrastructure necessary for oil and gas development. Congress enacted laws that govern federal agencies' permitting decisions on interstate gas pipelines and LNG terminals, requiring agencies to consider environmental impacts and to approve only projects that are in the public interest. The Commerce Clause has also served as the constitutional authority for nearly all bedrock environmental statutes, many of which apply to oil and gas extraction, transportation and consumption. (State governments make the determination on whether oil and gas extraction can proceed within a state and on intrastate oil and gas infrastructure. For instance, the state governments of New York and Maryland decided not to permit fracking in their states, decisions that blocked the extraction of shale resources in those states.)

The Trump administration departed from this philosophy of balancing competing land uses on public lands designated for multiple use. Instead, it focused on preferential expansion of oil and gas and other mineral extraction on public lands. Its America First Energy Plan aimed to "take advantage of the estimated $50 trillion in untapped shale, oil, and natural gas reserves, especially those on *federal lands* that the American people own" (emphasis added).[198] The administration reoriented federal agencies to prioritize the interests of the oil and gas sector by appointing leaders with ties to that sector. From 2017 through 2020, it leased 5.4 million acres to oil and gas drilling, an area the size of New Jersey.[6] Some of this acreage was comprised of lands that had already been opened to leasing but not yet leased. The rest had been closed to leasing but was opened by the Trump administration, such as lands designated as national monuments.

The administration fast-tracked permits for pipelines, including the Keystone XL and Dakota Access pipelines and the Jordan Cove LNG terminal. It also weakened the implementation of bedrock environmental laws, including the Clean Water Act (CWA) and the National Environmental Policy Act (NEPA), and promulgated regulations that force agencies to ignore scientific facts and to systematically understate the costs from oil and gas drilling and infrastructure. In its waning days in office, the administration skipped procedures in its rush to sell leases, including in the Arctic National Wildlife Refuge.[7,8] It also rushed to publish final regulations that force agencies to understate the costs from pollution.[9]

To better understand the nexus between America's lands and its energy path, I begin with a brief overview of the federal government's responsibilities over public and private land management (Section 4.1). I detail how the Trump administration reoriented agencies to accelerate oil and gas leasing on public lands, opened protected public lands to drilling and mismanaged leasing and extraction on public lands (Section 4.2). Next, I turn to how Trump's actions threatened both public and private lands through the fast-tracking of oil and gas infrastructure (Section 4.3) and the weakening of environmental regulations (Section 4.4) and how state attorney generals, tribes and environmental groups challenged these deregulatory actions. Finally, I turn to Congress's efforts – some successful and others stalled – to protect public lands and to safeguard private lands from encroachment by infrastructure projects whose public benefits are questionable at best (Section 4.5). I conclude by underscoring the urgency of congressional action to protect these lands that support lives and livelihoods at a time when clean water, healthy lands and functional ecosystems are becoming increasingly scarce.

4.1 Federal Government and Its Relationship with Public and Private Lands

To provide the political, economic and legal contexts for the Trump administration's actions, I first outline briefly the role of the federal government in managing public lands and balancing competing uses (Section 4.1.1), in permitting oil and gas infrastructure (Section 4.1.2), and in the protection of land resources in general.

4.1.1 The Federal Government's Management of Public Lands

At the founding of the United States, the federal government declared as public all lands that were not under private Euro-American ownership or ownership of the 13 states.[10,11] The US government seized lands from tribes and acquired lands from European powers that had previously belonged to tribes.[4,12,13,ii] As detailed in Chapter 5, the US government broke various treaties that had delineated reservation lands and rights-of-use on off-reservation lands, declaring some to be public lands.[iii]

Congress's early actions focused on promoting westward expansion by granting or selling off public lands to states, white homesteaders, railroad companies and land speculators.[4,iv] The federal government made expansive grants of land to western state governments upon statehood. Many of these land grants were for specific purposes such as supporting schools. Large shares of the lands within the boundaries of a number of western states are federal lands, including 63 percent in Utah, 47 percent in Wyoming, 80 percent in Nevada and 29 percent in Montana.[13] Even during the era of disposal and privatization of public lands, Congress contemporaneously chose to retain large tracts of land under federal ownership.[13]

Since the 1900s, the protection of public lands for conservation has enjoyed bipartisan support.[14–16] Republican President Theodore Roosevelt signed legislation establishing numerous national parks (but as detailed in Chapter 5, those actions have been rightly criticized for removing Native Americans from their lands). In the 1970s, Ronald Reagan reversed his initial support for the Sagebrush rebels (who opposed federal ownership of public lands) to sign legislation designating 10 million acres of public land as wilderness, through laws that won the support of a Republican-controlled Senate. In 1996, the Republican-controlled Congress passed the Omnibus Parks and Public Lands Management Act, which Clinton signed into law. That law protected national parks and rivers. The G. W. Bush administration defended Clinton's national monument designations, including the Grand Staircase Escalante, and Bush made the pioneering declaration of four major national monuments offshore in the Pacific Ocean. Protections for public lands largely continued, despite the persistent advocacy of a minority of Republicans for the transfer of public lands to state and private interests.[16]

---

ii Notably, that acquisition assumes colonial powers could legitimately sell and the US federal government could legitimately purchase lands long inhabited by sovereign Native American tribes. These include 530 million acres from France via the Louisiana Purchase, 378 million acres via a treaty with Russia for the purchase of Alaska, and hundreds of millions of acres from Great Britain, Mexico and Spain [13].

iii For instance, in 1872, Congress designated Yellowstone as the first national park, but the land was already inhabited and used by the Crows, the Blackfeet, the Bannocks and the Shoshones. In the late 1890s, Congress allotted the Uncompahgre Indians Reservation in Utah and declared lands that were not taken by homesteaders as public lands.

iv Between 1781 and 2017, Congress transferred 1.2 billion acres of public lands out of federal ownership to state governments, businesses and individuals [4]. BLM disposed of over 24 million acres of land between 1990 and 2010, an area about the size of Indiana [13].

BLM manages about 245 million acres of public land located primarily in 11 western states and Alaska. It also administers 700 million acres of subsurface mineral estate throughout the nation.[17,v] Where ownership of the surface and the subsurface mineral estates is divided, the US government's authority is vastly different depending on whether it owns the minerals (in which case it can decide whether and how to lease oil and gas) or the surface (in which case it can only impose reasonable restrictions on private owners seeking to develop their minerals). In 1976, Congress enacted the FLPMA, through which it declared that public lands would remain in federal ownership. FLPMA also imposes the responsibility of managing sustained yields of multiple uses, including energy and minerals, grazing, timber, recreation, watershed, wildlife and fish habitat, and conservation. BLM also manages national monuments, wilderness areas and other lands dedicated to a particular purpose such as conservation (sometimes overlayed with the multiple-use mandate).

As shown in Table 4.1, oil and gas leasing competes with a range of other economic activities on public lands, including recreation and ecosystem services. Oil and gas leasing generated $4.2 billion in revenue for the federal and state governments in 2019.[18] However, the value derived from oil and gas leasing has declined while that from competing economic activities and ecosystem services has increased.[19] While oil and gas leasing on public lands provides an important source of energy for the US economy, the profitability of oil and gas extraction has declined with the oversupply of oil and gas (Chapter 2) and the growing cost-effectiveness of renewable energy resources (Chapter 3). Moreover, oil and gas extraction has imposed significant costs to land values, public health, the environment and the climate. As of 2018, taxpayers were projected to face between $46 million and $333 million in estimated potential reclamation costs[vi] associated with orphaned wells and inactive wells that were at risk of becoming orphaned in that year.[20] The production and consumption of fossil fuels extracted from public lands and federal waters contribute to 23.7 percent of US carbon dioxide emissions annually and 7.3 percent of US methane annually.[21,vii]

At the same time, consumer demand has grown for conservation and recreation on public lands, including lower-impact, nonmotorized recreation

---

[v] These mineral rights include those below federal lands managed by another agency (such as the Forest Service) and those beneath private lands (e.g., the federal government often granted lands to railroads but did not convey the mineral estate).

[vi] Oil and gas operators are required to post a bond of $2,100 per well, which would be returned to them if they plugged wells that are no longer productive. These bonds do not incentivize companies to plug their wells; the bonds are dwarfed by the estimated remediation costs of wells, which range from $15,000 to $174,000, i.e., the range of estimated reclamation costs from the 5th to the 95th percentile [20].

[vii] These estimates are based on data from 2005 through 2014 [21].

Table 4.1 *Public lands: Competing economic activities*

Oil and gas extraction	
Revenue from oil and gas leasing on public lands shared by federal and state government (2019)[a]	$4.2 billion
Revenue from onshore oil and gas leasing on public lands that went to states (2019)[a]	$2.0 billion
Oil and gas production on public lands as a share of total US production (2017)[b]	5.6% of US crude oil production and 9.7% of US natural gas production[g]
Impact on the broader economy[c]	Federal onshore and offshore oil and gas production support $85.4 billion in value added, $139 billion in economic output and 607,000 jobs.
Other mineral extraction	
Coal: Impact on the economy[c]	$6.5 billion in value added, $11.5 billion in economic output and 36,000 jobs.
Nonfuel minerals: Impact on the economy[c]	$7.4 billion in value added, $12.7 billion in economic output and 45,700 jobs.
Other energy production	
Renewable energy[c]	$3.8 billion in economic output and over 15,000 jobs.
Recreation	
Revenue from visitors to national park lands[d] in the local regions	$21.0 billion was spent in local gateway regions by 328 million park visitors.
Visits to all Interior lands: Impacts to the broader economy[c]	$33.4 billion in value added, $58.1 billion in economic output and 452,000 jobs.
Water provision	
Provision of water for irrigation, municipal and industrial users[c]	$34.5 billion in value added, $63.4 billion in economic output and 466,000 jobs.
Ecosystem services[e, f]	
These include natural hazard protections (flood control and storm protections from wetlands); water protections (water purification, natural irrigation and drought mitigation);[viii] land protections (erosion control and sediment retention); pollination; pest control and carbon sink.	

[viii] Maintenance of intact ecosystems to provide ecosystem services can be cheaper than constructing treatment plants that would otherwise be needed. For example, purifying drinking water in New York City by watershed protection in upstate New York costs $1.5 billion, while

Sources: a. Congressional Research Service, *Revenues and Disbursements from Oil and Natural Gas Production on Federal Lands*, 2020;[18] b. Congressional Research Service, *US Crude Oil and Natural Gas Production in Federal and Nonfederal Areas*, 2018);[2] c. Department of the Interior, *Economic Report FY 2018*, 2019;[22] d. National Park Service and Department of the Interior, *2019 National Park Visitor Spending Effects*, 2020;[23] e. Constanza et al., *Twenty Years of Ecosystem Services*, 2017;[24] f. Environmental Protection Agency, *The Economic Benefits of Protecting Healthy Watersheds*, 2015.[25]
Note: g. The shale boom accelerated production on private lands relative to public lands as shale resources are located primarily on private lands.[26]

pursuits. The designation of national monuments on public lands generally goes hand in hand with conservation efforts and subsequent contributions to the local economy.[19] One study of 14 monument designations in the eight-state Mountain West region over a 25-year period documents an increase in the number of jobs and business establishments in areas close to those monuments.[27] A 2016 study, based on respondents' willingness to pay more taxes to protect national parks for themselves and their grandchildren, regardless of whether they visit the parks, put the value at $92 billion.[19] Ecosystem services provided by public lands, including maintaining waters and biodiversity, have become ever more important with the ecological disruptions wrought by the climate crisis.[28] Protected public lands have become a scarce asset, as the United States loses one football field of natural habitat every 30 seconds.[29,ix]

The majority of Americans support conservation on public lands. In a January 2021 survey of 3,800 registered voters across eight western states, a bipartisan research team found that a preponderance of respondents backed greater investment in conservation even in the face of budget deficits and rallied behind holding oil and gas companies responsible for the cleanup of their operations on public lands.[30] Respondents endorsed the goal of protecting 30 percent of America's public lands and seas by 2030, also known as the "30 by 30 goal."[30] Almost three-fifths of Republican respondents, four-fifths of Independent respondents and nine-tenths of Democratic respondents supported that goal. These findings comport with an earlier 2019 national survey of 1,200 respondents, sponsored by the Center for American Progress. About 86 percent of respondents backed the "30 by 30 goal," with strong support from Republicans (76 percent), Independents (88 percent), and Democrats (94 percent).[31] Voter support for public lands

constructing a water filtration plant costs about $8–$10 billion. Treatment of conventional wastewater by remediating wetlands costs $0.47 per thousand gallons of wastewater, while a wastewater treatment plant costs $3.24 per thousand gallons [157].
[ix] Between 2001 and 2017, the United States lost about 1,533,000 acres of natural areas annually [29].

is not lost on politicians. Republicans and Democrats in the House and Senate in western states campaigned on pro-public-lands platforms in the 2018 and 2020 elections,[32–34] but sadly not all followed through on these commitments (Section 4.4).

4.1.2 The Federal Government's Management of Oil and Gas Infrastructure

The United States is crisscrossed by over 1.63 million miles of fossil fuel pipelines, that is, 190,000 miles of liquid petroleum pipelines and 300,000 miles of natural gas transmission lines, with the remaining being natural gas distribution lines that bring gas from utilities to homes.[35]

The federal government approves various permits for specific oil and gas infrastructure projects (depending on the nature of each project). In doing so, federal agencies are to balance the benefits of the proposed infrastructure while considering their impact on treaty rights, public safety and the environment. The State Department approves permits for pipelines that cross international borders, such as the Keystone XL pipeline (Chapter 5). FERC makes permitting decisions on interstate gas pipelines and fossil fuel export terminals, including LNG terminals.[36] These projects can only proceed if state governments, acting under CWA authority, approve water-certification permits. The Army Corps of Engineers (the Corps) is responsible for permits for the construction of pipelines and infrastructure that cross US waters, that is, waters subject to federal jurisdiction under the CWA. In making their decisions, FERC and the Corps must also abide by other federal environmental laws, such as the NEPA and the Endangered Species Act (ESA). (State governments make permitting decisions for oil pipelines, natural gas liquids pipelines and intrastate gas pipelines.)

4.2 The Trump Administration's Actions That Adversely Affected Public Lands

The Trump administration was able to take far-reaching actions, thanks to Congress's decades of inaction and to the support from the majority in the 115th Congress (Section 4.2.1). The administration reoriented federal agencies to support accelerated oil and gas leasing on public lands (Section 4.2.2); opened public lands to extraction (Section 4.2.3); and managed leasing and extraction poorly (Section 4.2.4).

4.2.1 How Congress Set the Stage and Supported the Trump Administration

For decades, Congress largely failed to exercise its powers and duties to improve the management of public lands and to enact protective legislation for public and private lands. While this chapter focuses primarily on BLM lands managed under the multiple-use mandate, it is important to remember there are many types of public lands. John Leshy provides a useful analysis of the 435 million acres of public lands that are under some form of conservation protection. As of 2020, Congress imposed significant protections through the National Wilderness Preservation System for only 110 million acres or one-fifth of public lands, on which only Congress can authorize intensive development activities.[16] Congress offered some protections but did not completely outlaw intensive development for another 125 million acres.[16] About 200 million acres are protected by executive action. The executive branch cannot reverse some of the protections offered under executive action, such as wilderness study areas, but it can reverse other protections.[16] Trump's purported reversal of national monuments proclaimed by previous presidents, which was criticized by several legal scholars as unlawful,[37,38] faced legal challenges (Chapter 5).

Despite demands for federal agencies to fully account for the costs of the infrastructure build-out to private landowners, public health and the environment, the majority in Congress chose not to order agencies to undertake this full accounting. For instance, during the 112th Congress, the Republican-majority House voted against amendments and resolutions that called for the decision on the Keystone XL pipeline to be made only after the federal agency on pipeline safety conducted a specific review on the pipeline and only after the developer committed not to assert eminent domain powers against landowners.[39,x] On the contrary, the House voted to transfer the decision for Keystone XL from the State Department to FERC and to order FERC to approve the pipeline within 30 days.[39] (That action died without Senate support.)

When the Trump administration took office, the majority in the Republican-controlled 115th Congress actively aided and abetted the administration. Congress opened the coastal plain of the Arctic National Wildlife Refuge to oil and gas drilling through a provision in the 2017 Tax Cuts and Jobs Act and mandated at least one lease sale by 2021. That action, which

---

[x] Pipeline developers that receive a certificate for public convenience and necessity can assert eminent domain powers under federal law to take private property with compensation when landowners do not voluntarily sell easements.

followed decades of majority votes in Congress against leasing,[xi] paved the way for the Trump administration to sell leases in the refuge on January 6, 2021 (Chapter 6).[40,41]

Congress also wielded its powers under the Congressional Review Act to rescind BLM's Planning Rule 2.0, forcing BLM to return to the outdated 1983 rule.[42,xii] The Planning Rule 2.0, based on years of consultation with stakeholders, created a process for input from the general public earlier in BLM's development of land use and management plans; encouraged planning at the landscape level (in accordance with scientific principles, as opposed to more limited species- or habitat-level planning); and updated BLM's use of modern technology for the planning process, including the now standard tool of geographical information systems. Public input in the planning process is essential if BLM is to meet its statutory duties of managing public lands for the public interest. The Congressional Review Act prohibition on agencies promulgating a rule that is "substantially the same," language whose meaning has never been tested in the courts, could potentially make it difficult for BLM to modernize planning.

A number of Republicans pushed forward their anti-public-lands agenda. The Utah congressional delegation had met presidential candidate Trump to advocate for the shrinking of the Bears Ears National Monument and the Grand Staircase Escalante National Monument, a wish granted by Trump once in office[43] (Chapter 5). In January 2017, Representative Robert Bishop from Utah, the House Natural Resources Committee chair, orchestrated the passage of a budget rule in the House so that lawmakers would not need to account for the impacts on the federal budget of giving away public lands to states or private parties.[xiii] (That rule was overturned in the 116th Congress.) Representative Jason Chaffetz, also from Utah, reintroduced his bill to sell 3.3 million acres of public land. Following public outcry from conservationists and backcountry groups, Chaffetz withdrew the proposal nine days later.[44]

[xi] Congress enacted the 1980 Alaska National Interest Lands Conservation Act, which prohibited production of oil and gas in the Arctic National Wildlife Refuge, but ordered the Department of the Interior to study the impact of such production in the coastal plain. In 1987, the Department of the Interior recommended leasing of that area, despite adverse impacts on wildlife and the ecosystems from drilling. However, in 2015, the Fish and Wildlife Service recommended that major parts of the refuge be declared as wilderness.

[xii] Under the 1996 act, Congress holds extensive power to invalidate rules established by federal agencies that are finalized 60 days before the new congressional session, provided the president signs that rescission into law.

[xiii] The rule mandated that the Congressional Budget Office, when calculating the cost of disposing of federal land, cannot consider future revenues the federal government could have received from the land from energy production, recreation, grazing or other uses [44].

4.2.2 Reorienting Federal Agencies to Accelerate Oil and Gas Leasing

The Department of the Interior's stated mission is to protect America's natural resources, including public lands and seas, to undertake scientific research on the management of these resources, and to carry out the federal government's obligations toward Native Americans. The Trump administration reoriented the Interior toward prioritizing oil and gas leasing by appointing to leadership positions those with financial ties to the oil and gas sector.[45,46,xiv] The first secretary of the interior, Ryan Zinke, a former congressman from Montana, received campaign contributions from the oil and gas sector and had a record of voting against public land and environmental interests.[47,48,xv] His successor, David Bernhardt, worked as a lobbyist for oil and gas companies prior to taking office.[45,49] William Pendley, who held the position of acting director of BLM, advocated against federal ownership of public lands and made derogatory statements against Native Americans' religious practices on public lands,[xvi] prior to taking office.[50,51,xvii] Congress chose to confirm the political appointments of Zinke and Bernhardt,[xviii] even when the mere appearance of conflicts of interest can diminish Americans' trust in government serving the people. Moreover, research studies point to how the revolving door between policymakers and regulators and the regulated industry they oversee can lead to biases in favor of the regulated industries.[52–54]

The administration weakened both scientific and analytical capacity and transparency at the Interior. The administration relocated BLM headquarters

---

[xiv] Previous administrations, including the Obama administration, have also appointed leaders with ties to the regulated industry to head regulatory agencies [197].

[xv] After leaving office, Zinke went to work for oil and gas interests, defending his efforts to help companies navigate "the permitted process" as not violating the legal prohibition against top officials of a federal agency from lobbying the agency for one year. However, Virginia Canter, counsel for Citizens for Responsibility and Ethics in Washington, raised the question "if they're going out and engaging in obtaining employment from entities with interests before their former agencies, it really begs the question of whether, when they were making policy, were they doing it in the public interest or were they really doing it to enhance their post-employment opportunities?" [48].

[xvi] On September 10, 2019, Representative Debra Haaland noted in a congressional hearing on BLM, "In 2009, at a Republican breakfast forum, you were quoted mocking American Indian religious practitioners that insisted that federal lands and private property be off limits because it is holy to them – using air quotes to punctuate holy." Pendley's response was, "I was speaking as a private attorney representing private clients."

[xvii] In September 2020, the District Court for the District of Montana ruled that Pendley's appointment as acting director violated the Federal Vacancies Reform Act of 1998 and that any "function or duty" of the BLM director that was performed by Pendley has no force or effect [195].

[xviii] The Senate confirmed David Bernhardt 56 votes to 41 votes with Democrat Senators Martin Heinrich of New Mexico, Joe Manchin of West Virginia and Kyrsten Sinema of Arizona voting in support of the nomination.

to Grand Junction, Colorado, with little consultation with employees, resulting in a large fraction of employees choosing to retire or leave the agency rather than relocate.[55] Most of the BLM employees were already in field offices, and the relocation of the headquarters weakened BLM's ability to liaise with members of Congress. It also blocked scientists from undertaking research and communicating their research to the public[xix] and pressured scientists to implement approaches that systematically understate adverse environmental and climate impacts.[45,56–58,xx] Former BLM directors from the Bush and the Obama administrations criticized the administration's "stealth plan" to destroy BLM in order to "dismantle public ownership" of the land and its resources.[59,60]

In September 2018, then-Deputy Secretary Bernhardt signed Secretarial Order 3369 Promoting Open Science, which instructed scientists to consider only those studies whose data is publicly available, thereby blocking the consideration of scientific evidence from legitimate studies that rely on confidential data (e.g., epidemiological studies that rely on confidential patient data).[61] The Interior worked toward proposing a rule that would block scientists from considering legitimate scientific studies,[62,63,xxi] a move analogous to the EPA's "Science Transparency Rule" (Chapter 9).

The Interior also blocked public access to information. In December 2018, Zinke issued a secretarial memo, making the solicitor general (and by default the acting solicitor general at the time) responsible for coordinating responses to requests under the Freedom of Information Act (FOIA). In January 2019, the Interior published a proposed rule that concentrated powers with one Interior employee to make determinations on FOIA requests.[64] Public accountability groups, environmental NGOs and journalists excoriated the rule, which could enable political appointees at the Interior to control the flow of information to the public and thus prevent public oversight.[65–67] In its critique, the Center for Biological Diversity cited the 1989 Supreme Court decision in *Department of Justice v. Reporters Committee for Freedom of Press*,[68] which had underscored the importance of FOIA in giving the public the right to know "what their government is up to."

[xix] These incidents are recorded in the Silencing Science Tracker, a joint initiative of the Sabin Center for Climate Change Law and the Climate Science Legal Defense Fund.

[xx] Scientists at the US Geological Survey were instructed to shorten their analysis on climate impacts to the year 2040 instead of applying the typical analytical time frame to 2100, thereby understating the adverse impacts of climate change.

[xxi] Under FLPMA, BLM is responsible for managing public lands in a manner that protects the quality of air and the atmosphere. BLM's air-quality resource plans serve to complement regulations under the Clean Air Act to protect air quality on public lands [63]. The rule could potentially have affected BLM's consideration of epidemiology studies on air quality.

4.2.3 Opening Public Lands to Oil and Gas Leasing

Contrary to FLPMA, which requires that BLM "receive fair market value of the use of the public lands and their resources," the administration flooded the market with leases despite minimal demand. Between 2017 and 2019, the Trump administration eliminated protections for more than 13.5 million acres of public land that were once protected as national monuments or withdrawn from mineral leasing.[69] Trump's shrinking of Bears Ears National Monument and Grand Staircase Escalante National Monument are detailed in Chapter 5.

The Trump administration offered oil and gas leases for over 18.7 million acres nationwide, a far larger acreage than in previous years. It did so even when 50 percent of acreage leased by oil and gas companies was not under production, as measured in 2018 (Table 4.2). About 63–97 percent of the acreage received no bids, while a sizable amount of land was leased cheaply. Of the acreage of leases held by oil and gas companies since 1987, 30 percent were purchased at the minimum bid of $2 per acre or below the minimum bid in noncompetitive purchases.[69]

BLM, which can exercise discretion within FLPMA's multiple-use mandate, swung far in favor of oil and gas leasing. It published revised land-use plans that weakened conservation protections and protection for endangered species and the environment and opened more public lands to leasing. A significant share of leases sold were located on lands designated as high-priority zones for endangered species.[69]

4.2.4 Poor Financial Management of Oil and Gas Leasing and Extraction

Not only did the administration sell leases cheaply, it also cut royalty payments from oil and gas companies. Royalty rates for oil and gas leases on public lands at 12.5 percent were already lower than those imposed by state governments for leases on state lands, including 25 percent in Texas.[72] In spring 2020, BLM approved one-third of the 1,689 applications from oil and gas companies for the temporary reduction of royalty rates from 12.5 percent to an average of less than 1 percent. The Government Accountability Office criticized BLM for not requiring companies to show evidence that their wells would have been shut without a reduction in royalty rates, and thus potentially giving relief to companies that would have continued producing regardless.[73]

The administration rescinded regulations aimed at preventing the waste of natural gas and the loss of royalties to the Treasury and state governments. In 2018, it rescinded the 2016 Methane Rule on public lands, which had required

Table 4.2 Land leased and offered: Small shares that received bids and that undertook production

Year	2009	2010	2011	2012	2013	2014	2015	2016	2017	2018	2019
Total land leased (million acres)[a]	45	41	38	38	36	35	32	27	26	26	NA
Land leased not in production (%)[a]	72	70	68	67	65	63	60	53	50	50	NA
Acreage of leases offered for auction (million acres)[a]	2.9	3.2	4.4	6.1	5.7	4.1	2.3	1.9	11.9	5.0	7.4
Acreage not receiving any bids (%)[a]	62	76	75	76	81	82	77	53	93	70	63
Area that received bids of $2 per acre as a share of the area receiving bids (%)[b]	NA	NA	NA	NA	NA	NA	NA	NA	28	17	17

Sources: a. Bureau of Land Management, *Oil and Gas Statistics*, 2019[70]; b. Center of Western Priorities, *Dashboard: Oil and Gas Leasing*, 2021.[71]

operators to capture methane and to pay royalties on that gas. Without these rules, states, tribes and federal taxpayers lost as much as $23 million annually in royalty revenue.[74] The administration's rescission of the 2016 rule was vacated in July 2020 by the District Court for the Northern District of California for violating the Mineral Leasing Act.[75] (However, in October 2020, the District Court for the District of Wyoming vacated the 2016 Methane Rule, ruling that BLM had exceeded its authority under the Mineral Rights Act.)[76,xxii]

The administration also rescinded the 2015 BLM rule on hydraulic fracturing on public lands and tribal lands.[77] That regulation had served to provide some safeguards against contamination of land and water that would diminish the economic value of public lands. That rule, which updated the 1988 BLM regulations that had been promulgated prior to the widespread use of high-volume hydraulic fracturing, aimed to ensure that oil and gas wells are properly constructed to protect water supplies, that wastewater is properly managed, and that chemicals used in the hydraulic fracturing process are disclosed to the public.[78]

These actions of the Trump administration were in the financial interest of oil and companies, not the American people.[79] The large amount of land locked up by oil and gas companies, with minimal intention of developing these leases, can potentially hinder society from using these lands during the 10-year duration of these leases. While mineral leases do not grant the lessee exclusive rights to use the surface, the potential for extraction to commence on those lands can discourage the incorporation of these lands into conservation practices. A study by Mark DeSantis of the Center for American Progress details how companies benefit financially from acquiring cheap leases on public lands and paying very little for holding these leases for the 10-year period and never developing them.[80] Indeed, about 90 percent of leases sold at the minimum bid and 97 percent of those sold noncompetitively never enter production but are relinquished after a decade.[26,79,81,xxiii]

The purchase of leases that could be counted as "proved undeveloped" reserves and reported as assets to investors and shareholders benefits companies by raising their perceived market value.[82,xxiv] The Securities and Exchange

---

[xxii] The attorney general of California filed an appeal to the Court of Appeals for the Tenth Circuit.

[xxiii] The Government Accountability Office study found that 26 percent of competitive leases that sold with bids above $100 per acre produced oil and gas and generated royalties in the 10-year lease period. On the other hand, only 2 percent of competitive leases that sold at the minimum bid of $2 per acre and only 1 percent of noncompetitive leases did so [20].

[xxiv] DeSantis's study suggests that companies could make themselves more attractive for acquisition by purchasing leases on public lands. Following that rule change, the 68 companies in the study raised their "proved undeveloped" holdings, which in turn led to an average increase in their market valuation. Newly acquired companies typically had larger percentage of undeveloped leases [80].

Commission's[xxv] relaxation in 2008 of a rule on when reserves can be classified as "proved undeveloped" further enabled companies to count more of these undeveloped leases as assets.[83] That rule replaced the certainty test with a *reasonable* certainty test, in permitting the inclusion of those acreage for which geological data indicate a "reasonable certainty" that commercial recoverability can be met.[80]

4.3 The Trump Administration's Fast-Tracked Approval of Oil and Gas Infrastructure

Under the Trump administration, federal agencies such as FERC (Section 4.3.1) and the Corps (Section 4.3.2) fast-tracked approval of oil and gas infrastructure at the expense of landowners, the environment and human health. However, several of these decisions have come under legal challenge. Case studies of specific pipelines and LNG terminals and encroachment onto Native American lands are detailed in Chapter 5.

4.3.1 FERC's Approval of Pipelines

FERC, through its permitting of interstate gas pipelines and export terminals, plays a major role in shaping the extent to which lands are impacted by pipeline development. FERC's approval of pipelines grants developers the powers of eminent domain, as discussed in this section.

Interstate gas pipelines and export terminals may be constructed and operated only if they receive a certificate of public convenience and necessity from FERC – however, in practice, FERC has issued these certificates with lax consideration. FERC may issue a certificate if a proposed pipeline or terminal is required by "public convenience and necessity" per Natural Gas Act section 7.[84] Criticisms that FERC's decisions bend in favor of approving pipelines predate the Trump administration.[85,86] With Trump's nominations,[xxvi] Republican commissioners

---

[xxv] As recorded in the *New York Times*, Aubrey McClendon, chief executive of Chesapeake, in May 2008 underscored how the rule change would boost the market valuation of the industry, "the event in the next few years that will have the biggest impact on the valuation of the industry, especially our company, it will be the ability to recognize proven undeveloped reserves much more expansively than we can today ... obviously I think that is going to be a huge watershed event for the valuation of the industry" [82].

[xxvi] Trump appointed the following Republicans: Neil Chatterjee in 2017, Bernard McNamee in 2018 (who departed in 2020) and James Danly in 2019. Democrat Cheryl LaFleur's term ended in 2019. Democrat Allison Clements was sworn in as FERC commission in December 2020 and Republican Mark Christie in January 2021.

made up the majority of FERC commissioners from 2017 to 2020.[87,xxvii] The majority of decisions under the Republican-controlled FERC during the Trump administration systematically favored the approval of pipelines by setting a low bar to demonstrate the pipelines served the "public convenience and necessity" and by failing to incorporate climate considerations in evaluating pipeline projects.

FERC's majority decisions drew the conclusion that a pipeline is necessary as long as there is a contract between the pipeline developer and a company that contracts to transport gas via that pipeline. Dissenting commissioners and energy analysts have argued that, in reality, these contracts between affiliated companies do not provide strong evidence that the pipeline is needed.[88–92] Instead, even when demand for pipelines is weak, affiliated companies of pipeline developers and utility companies can profit from pipeline projects. The utility companies, that is, the purchasers of pipeline capacity, receive guaranteed returns (often at 14 percent) from infrastructure investment while ratepayers pay for the costs of pipeline construction.[88] Dissenting commissioners and energy analysts have argued that FERC must conduct more searching analysis to determine the true need for the pipeline, including requiring evidence of demand beyond the contracts between affiliate companies and investigating the existing and planned pipeline capacity in the region and the supply and demand for gas in the region.[88–92]

FERC's majority decisions also failed to incorporate climate considerations in their approval of pipelines and export terminals.[92] The two minority commissioners, Commissioners Cheryl LaFleur and Richard Glick argued that FERC is obliged to consider climate impacts in making its determination on whether pipelines or export terminals are in the public interest.[84,92] Glick and other legal scholars have detailed FERC's legal responsibility to consider climate impacts: first, the Natural Gas Act section 7 requires FERC to take into account environmental impacts, including climate impacts, in its assessment of whether the pipeline is in the public interest;[xxviii] second, NEPA obliges FERC to take a hard look at environmental impacts, including climate impacts; and third, FERC can, should it approve projects with climate impact, impose conditions to mitigate those impacts.[93,94] Glick underscored that FERC

[xxvii] The three Republican commissioners outvoted the two Democrat commissioners in the approval of pipelines and in rulings that hindered the growth of renewable energy. Late in his tenure, Republican Commissioner Neil Chatterjee expressed support for carbon pricing. In September 2021, Biden nominated Willie Phillips to serve as a FERC commissioner, paving the way for a 3–2 Democratic majority in FERC for the first time in Biden's administration [199].

[xxviii] Webb details the legislative history of the Natural Gas Act section 7, courts interpretation of that provision and how the commission historically considered environmental impacts but largely ignored those impacts in recent decisions [94].

has the tools to estimate climate impacts from greenhouse gas emissions and that other agencies that are not primarily regulators of greenhouse gas emissions have obligations under NEPA to consider climate impacts.[93]

FERC chose a narrow interpretation of its obligations to consider climate impacts.[94] In 2017, the Court of Appeals for the DC Circuit, in *Sierra Club v. FERC*,[95] ordered FERC, per its NEPA obligations, to estimate the Sabal Trail pipeline's impact on greenhouse gas emissions or explain more fully why it could not do so. The court ruled that FERC's decision was a "legally relevant cause" of the emissions, reasoning that FERC "could deny a ... certificate on the grounds that the pipeline would be too harmful to the environment." However, in May 2018, in its 3–2 decision along party lines, FERC issued an order that it would not consider upstream and downstream emissions from pipelines except in very narrow circumstances.[94,xxix] FERC interpreted *Sierra Club* to apply only when it knew with certainty the end users of the natural gas. FERC chose not to consider downstream emissions of greenhouse gas when the pipeline delivered gas to local distributors[94,96] and the exact consumers were not known with certainty. FERC also chose not to consider upstream emissions of greenhouse gas, unless the pipeline project transported gas from a new site where production was made possible by that project.[94]

FERC's approach has not gone uncriticized. The Court of Appeals for the DC Circuit in *Birckhead v. FERC*,[97] while dismissing the challenge against FERC on procedural grounds, noted that it was "troubled" by FERC's refusal to assess upstream and downstream emissions and that NEPA obligations "requires the Commission to at least attempt to obtain the information necessary to fulfill its statutory responsibilities."[94]

FERC's reasoning on public convenience and its treatment of landowners have also come under scrutiny by the courts.[94,98] In 2019, the Court of Appeals for the DC Circuit, in *Oberlin v. Federal Energy Regulatory Commission*,[99] remanded a pipeline certificate order to FERC, ruling that FERC did not sufficiently explain why the NEXUS pipeline, proposed in part to export natural gas to Canada, served "the public convenience and necessity" under the Natural Gas Act or was a public use under the Fifth Amendment Takings Clause. How this case unfolds is of particular interest for landowners who oppose pipeline projects (a growing number of which transport natural gas destined for export markets). Landowners have advanced the argument that these projects, which are primarily export-oriented, are not within the terms of "public convenience and necessity" under the Natural Gas Act. Past pipeline projects typically focused on delivering gas to domestic consumers.[100,101]

[xxix] In *Sierra Club*, the pipeline project delivered gas to specific electricity-generating units.

In June 2020, the Court of Appeals for the DC Circuit in *Allegheny Defense Project v. Federal Energy Regulatory Commission*[102] ruled on another criticism of FERC's treatment of landowners – its failure to respond within 30 days to landowners' requests for rehearings. Delays in FERC's decisions on rehearings, which are a necessary step prior to landowners being able to proceed to file court challenges, rob landowners of the ability to seek judicial review of FERC's decisions.[103] Pipeline developers can therefore proceed with construction and assert eminent-domain powers against landowners before the judicial review of FERC's decision. Eminent-domain powers permit oil and gas companies to file lawsuits to obtain easements when landowners do not grant them voluntarily.[104,xxx] In June 2020, FERC finalized a rule that prohibits project construction from beginning until FERC has issued pending rehearing orders, but pipeline developers can still assert eminent-domain powers against landowners during that time.[105,106]

4.3.2 Army Corps of Engineers' Approval of Infrastructure Projects

The Corps' permit authority over pipelines crossing US waterways has also become an important factor in pipeline development. Under the Trump administration, the Corps approved permits for a number of controversial pipelines, without giving adequate consideration to Native Americans' treaty rights, as detailed in Chapter 5. In addition, the Corps continued its practice of issuing general or nationwide permits for pipelines, circumventing careful environmental assessments.[107] Under section 404 of the CWA and section 10 of the Rivers and Harbors Act, the Corps can issue nationwide permits to authorize categories of activities that discharge dredged or fill material into the waters of the United States but only when those activities result in minimal individual and cumulative adverse environmental effects. Otherwise, the Corps would need to issue individual permits for each crossing following an in-depth environmental impact statement (EIS). The Corps issued the nationwide permit 12 (NWP-12) to approve the construction of oil and gas pipelines in 2012 and renewed that permit in 2016.[107,xxxi] The Corps treats each waterway crossing as a separate individual crossing and therefore systematically ignores the cumulative impact of a given pipeline on America's waterways. The NWP-12 allows projects to proceed

---

[xxx] While landowners have the legal right to contest pipeline developers' assertion of eminent-domain powers, it has been reported that poorer landowners who cannot afford legal fees are forced to accept low-ball offers from pipeline developers [104].

[xxxi] That permit is also used to govern the work of other utility companies.

without the comprehensive environmental review typically required by the CWA.

Environmental groups successfully challenged the Corps' use of that permit. In April 2020, the District Court for the District of Montana in *Northern Plains Resource Council v. US Army Corps of Engineers* suspended the NWP-12 permit.[108] The court ruled that the Corps' issuance of the permit, without consultation with Fish and Wildlife Service experts, violated the ESA. Under that act, the Corps is required to consult with service experts if there is a possibility that the NWP 12 permit and the projects it fast-tracks affects endangered and threatened species. The court ordered the Corps to review the potential impacts on endangered species. The district court also suspended the construction of all permits that rely on the NWP-12 permit while the Corps reviews that permit. However, in July 2020 the Supreme Court intervened and permitted construction to continue temporarily for pipelines other than the Keystone XL pipeline.[109]

4.4 The Trump Administration Weakened Regulations under Environmental Laws

The Trump administration promulgated regulations that weakened the implementation of bedrock environmental laws. These rollbacks align with the interests of the American Petroleum Institute, a trade association representing the oil and gas industry, which sent a wish list of deregulatory actions to the EPA's Regulatory Reform Task Force in May 2017.[110–113] Violation of environmental laws had been the basis of courts' rulings that blocked several agencies' leasing and permitting decisions (Table 4.3). Likewise, state governments had asserted their powers under the CWA to deny permission for infrastructure projects.

Below, I describe these deregulatory actions and the arguments advanced in legal challenges against these deregulatory actions mounted by several state attorneys general, Native American tribes and environmental groups. The extent to which the administration's rulemaking strategies made its deregulatory actions vulnerable to legal challenges are detailed in Part IV.

4.4.1 National Environmental Policy Act: Not Taking a Hard Look at Adverse Impacts

The 1970 NEPA aims to protect the environment by ensuring informed and transparent decision-making by federal agencies and by enabling stakeholders

Table 4.3 *Projects halted due to federal agencies violating environmental laws or state agencies denying clean water certification*

National Environmental Policy Act (NEPA)	
Pipelines	In November 2018, the District Court for the District of Montana halted the construction of the Keystone XL pipeline. It ruled that the State Department failed to complete an adequate environmental review as required under NEPA. The department did not consider cumulative greenhouse gas emissions from the pipeline and did not undertake an updated assessment of potential oil spills (Chapter 5).[114] In July 2020, the District Court for the District of Columbia ordered the Dakota Access pipeline to be shut down while the Corps revised its inadequate EIS.[115] The shutdown order was temporarily reversed by the Court of Appeals for the DC Circuit in August 2020.[116] In May 2021, the District Court for the District of Columbia ruled that the pipeline could continue to operate while the Corps revises its EIS (Chapter 5).
Leasing on public lands	In March 2019, the District Court for the District of Columbia blocked BLM from issuing drilling permits on 300,000 acres of public lands in Wyoming because BLM failed to adequately consider the impact of drilling on climate change.[117] In November 2020, the court ordered yet another redo of the assessment after finding BLM's supplementary assessment inadequate.[118] (The Wyoming leases were initially sold during the Obama administration.) In May 2020, the District Court for the District of Montana vacated 287 oil and gas leases covering 145,000 acres of land in Montana. The court criticized BLM for failing to provide analysis of the impacts of oil and gas drilling on groundwater and on climate change.[119]

Endangered Species Act (ESA)	
Pipelines	In July 2020, the Supreme Court rejected the request of the Trump administration to permit the construction of the Keystone XL pipeline that had been blocked by the District Court for the District of Montana.[120] In April 2020, that district court had suspended the construction of pipelines that relied on the NWP-12, pending the Corps' review of that permit program.[120] The district court ruled that the Corps' issuance of that nationwide permit violated the ESA. (The Supreme Court decision permitted construction to continue for other pipelines that relied on the nationwide permit, pending the Corps' review of the permit; see Chapter 5.)
Leasing on public lands	In February 2020, the District Court for the District of Montana vacated oil and gas leases in Nevada, Utah and Wyoming, covering 1 million acres, because BLM illegally shortened the period for public participation and to submit protests, contrary to the regulations governing the leasing of public lands designated as sage grouse habitat areas.[121] The ESA had created the impetus

Table 4.3 *(cont.)*

Endangered Species Act (ESA)	
	for BLM under the Obama administration to designate sage grouse habitat areas. In May 2020, the District Court for the District of Montana nullified 440 oil and gas leases, covering about 336,000 acres in Montana and Wyoming.[122] As noted above, the ESA had created the impetus for BLM during the Obama administration to propose resource-use plans that designated sage grouse protected habitats. The court ruled that BLM under the Trump administration violated FLPMA in making changes to those resource plans without abiding by the required formal amendment process.[122] That ruling makes vulnerable BLM lease sales that involved analogous modifications of resource plans.[123]
Clean Water Act (CWA): State governments denied section 401 water-quality certification for pipeline projects or export terminals	
Pipelines	In 2016, the New York Department of Environmental Conservation denied a section 401 permit for the proposed Constitution natural gas pipeline. Williams, the pipeline developer, lost its appeal in the Court of Appeals for the Second Circuit in 2017. Subsequently, in August 2019, FERC ruled that the project could proceed, deeming that the state of New York had waived its section 401 authority. In February 2020, the pipeline developer cancelled the pipeline project. In May 2019, the Oregon Department of Environmental Quality denied without prejudice a 401 permit for the proposed Jordan Cove LNG export terminal and its feeder pipeline. In March 2020, FERC conditionally approved the project, contingent on Oregon's approval of state permits. In May 2020 the developer petitioned FERC to rule that Oregon had waived its authority under section 401. Oregon protested that petition.[124] In January 2021, FERC ruled unanimously that Oregon did not waive its section 401 authority.

to hold the government accountable. It mandates federal agencies to take a hard look at the environmental impact of their actions, including their approval of projects that require federal permits. Agencies are required to undertake either an environmental assessment or a more in-depth EIS and to consider reasonable alternatives including not proceeding with the project.[125] NEPA's effectiveness depends on the extent to which agencies consider adverse environmental impact and engage with stakeholders.

In July 2020, the Council of Environmental Quality, which is in charge of overseeing federal agencies' implementation of the environmental impact assessment under NEPA, published its final rule that weakened regulations on NEPA implementation.[126] In August 2020, 23 attorneys general and environmental groups filed a lawsuit challenging the rule, laying out how the council's promulgation of the rule was beyond its statutory authority and how it undermined NEPA's goals of environmental protection, public participation and informed decision-making.[127,128]

They documented the numerous ways in which the rule contradicted the language of the statute and courts' interpretation of NEPA.[129,130] Contrary to case law – including *Sierra Club v. Babbitt*,[131] which had established that NEPA applies to actions in which federal agencies have the ability to *influence* the outcome of the project – the rule narrows NEPA's application to those actions "subject to Federal control and responsibility."[132] The rule also narrows the consideration of indirect and cumulative impacts by excluding effects that are "remote in time, geographically remote, or the product of a lengthy causal chain" and by excluding "effects that the agency has no ability to prevent due to its limited statutory authority." The rule creates barriers for public participation, for instance, by requiring public comments to meet specific requirements. It also purports to limit judicial review of agency NEPA compliance by attempting to limit when judicial review can occur, by restricting remedies for those injured by deficient NEPA review and by imposing unlawful bond requirements on parties bringing a legal challenge. The rule permits the project applicant to prepare its own EIS, even though an applicant, invested in the project going forward, does not have the incentive to take a hard look at the project's adverse impacts and to consider alternatives to not proceed with the project.

4.4.2 Endangered Species Act: Weakening Protections for Species and Habitats

The ESA[133] aims to protect endangered and threatened species,[xxxii] to enable their recovery, and to protect the ecosystems on which they depend. Protections offered to endangered and threatened species, that is, the prevention of intentional and incidental taking of those species and the destruction of their habitats, limit oil and gas operations and infrastructure on both private and public lands.[134,135]

Under the Trump administration, the Fish and Wildlife Service, within the Department of the Interior, finalized rules that made it more difficult to list

---

[xxxii] A species may be listed under the ESA as (1) "endangered," if it is at risk of extinction throughout all or a significant portion of its range; or (2) "threatened," if it is likely to become endangered throughout all or a significant portion of its range.

species as threatened or endangered and reduced protections for listed species, thus opening the door for oil and gas development and infrastructure to proceed. The Fish and Wildlife Service and the National Marine Fisheries Service, within the Department of Commerce, also weakened the ESA interagency consultation process.

In September 2019, attorneys from 17 states and environmental groups sued the administration, setting out how its revised rules contradict the ESA's goals and statutory language.[136–139] The plaintiffs also underscored that the Fish and Wildlife Service and National Marine Fisheries Service violated their obligations under NEPA for finalizing rules that have tremendous environmental impacts without undertaking an environmental review.

The first set of revisions eliminates all protections for wildlife newly designated as "threatened" under the act.[140] The revisions repeal the Blanket 4(d) Rule, which had automatically prohibited the taking of any species listed as threatened. The second set of revisions makes it more difficult to list imperiled species and to protect habitats.[141] The third set of revisions weakens the interagency consultation process, which requires all federal agencies to use their authorities to conserve endangered and threatened species and to consult with the Fish and Wildlife Service if their actions can imperil endangered or threatened species and their habitats.

4.4.3 Clean Water Act: Weakening States' Powers on Clean Water Certification

The CWA aims "to restore and maintain the chemical, physical, and biological integrity of the Nation's waters." In practice, this has translated into a comprehensive permit system overseeing many industries and land-use practices, including the oil and gas industry. The CWA, in line with principles of cooperative federalism, reserved powers for states and tribes to protect their waters. The statute specifies that states have "the primary responsibilities and rights ... to prevent, reduce, and eliminate pollution" and that states can set stricter standards than the federal baseline. Under section 401 of the CWA, applicants for federal permits for infrastructure projects, such as interstate gas pipelines and export terminals for LNG or coal, must first secure a water-certification permit from state or tribal governments.[142,143] This applies to "any discharge, per section 401(a)(1), and requires compliance with state law, per section 401(d)."

In June 2020, the EPA finalized the 401 Certification Rule,[144] which restricts when the certification is required, narrows states' powers to review projects and impose conditions on permits, and creates a procedure that raises the likelihood for states or tribal governments to inadvertently waive their section 401 authority. Contrary to cooperative federalism, the rule sets up a process in which federal agencies dictate how the states conduct their review – including specifications on timelines, procedures and information requirements from applicants. Federal agencies such as the EPA, FERC or the Corps can impose tighter review timelines of less than a year on states and the clock starts ticking even if the application does not contain all the information needed by the state to properly evaluate the application. Even if states met the timeline, federal agencies could deem that the state waived its authority under section 401 if the state does not provide to the federal agency the detailed information outlined in the final rule. (Alleging that a state inadvertently waived its authority under section 401 has been used by federal agencies and project proponents to push forward projects for which state governments denied the section 401 permit; see Table 4.3.)

A number of tribes,[xxxiii] environmental groups and a coalition of 22 attorneys general sued the EPA, arguing the rule contravenes the goals of the CWA, its statutory language and judicial interpretations.[145–148] In January 2020, Montana and Wyoming filed a complaint to the Supreme Court against Washington's assertion of its section 401 authority. In September 2017, Washington's Department of Ecology had denied with prejudice a 401 permit for the Millennium Bulk Coal Export Terminal.[149] Montana and Wyoming, whose coal would be exported through that terminal, argued that the denial violated the Commerce Clause. In contrast to Montana and Wyoming, other western states have stood in support of states' section 401 powers, which are essential for states to protect their water resources.[150] (In May 2021, the US solicitor general filed a brief stating there was no issue for the Supreme Court to resolve as the project developer had declared bankruptcy and the project was not proceeding.)

4.4.4 Clean Water Act: Narrowing the Definition of Protected Waters

The CWA specifies protections for "waters of the United States" but does not provide a definition for these waters.[151] A narrower definition of protected water would reduce the scope of oil and gas extraction and infrastructure that require permits from the Corps.[xxxiv] Particularly, whether wetlands and small

---

[xxxiii] A number of tribes have received approval from the EPA for "Treatment in the same Manner as a State for purposes of the CWA Section 401."
[xxxiv] For instance, pipeline developers need permits from the Corps under section 404 of the CWA for the discharge of dredged or fill material into waters of the United States.

tributary streams, which contribute to the integrity of traditionally protected waters, are protected under the CWA has been litigated heavily. (The CWA's protection of rivers, lakes and territorial seas, i.e., traditional protected waters, is not disputed.)

The Trump administration weakened the CWA's scope by repealing the 2015 Clean Water Rule[152] in October 2019 and promulgating the Navigable Waters Protection Rule in April 2020.[153] The Navigable Waters Protection Rule removed protections for waters that had received them under the 2015 rule, including isolated non-floodplain wetlands, floodplain wetlands that are not connected to traditionally protected waters, intermittent streams that flow only after rainfall and ephemeral streams that flow only seasonally.[154,155,xxxv]

The rule's focus on permanence and continuous surface connection to other traditionally protected waters in awarding protections to water bodies is not backed by scientific evidence. The EPA Science Advisory Board had criticized the rule (at its proposal stage), explaining that such a change "decreases protection for our Nation's waters and does not support the [CWA] objective of restoring and maintaining 'the chemical, physical and biological integrity' of these waters."[156] The board further noted how the exclusion of protections for water bodies were made "without a fully supportable scientific basis, while potentially introducing substantial new risks to human and environmental health."[156]

The rule also departed from the scientific analysis underlying the agencies' previous rule. In promulgating the 2015 Clean Water Rule,[xxxvi] the EPA completed the Connectivity Report, which examined the state of the science on the hydrogeological, biological and chemical connections between wetlands and streams to downstream waters.[157,158] Numerous experts provided their input and the Science Advisory Board peer-reviewed the report.[159,xxxvii] The EPA and the Corps failed to explain the rule's departure from its previous scientific finding, other than claiming that "science cannot dictate where to draw the line between federal and state waters" and their reliance on Justice Antonin Scalia's plurality opinion in *Rapanos v. United States*.[xxxviii]

[xxxv] Perennial streams, streams with relatively permanent surface flows and floodplains wetlands that abut traditionally protected waters remain protected.

[xxxvi] Lawsuits by 31 state attorneys general meant that the Clean Water Rule came into effect in some states but was stayed in others.

[xxxvii] The board agreed with the conclusions of the Connectivity Report that streams and flood-plains are connected physically, chemically and biologically to downstream waters. The board went further than the EPA, arguing that the science supported the conclusion that non-floodplain wetlands sustained downstream waters.

[xxxviii] As discussed in Chapter 10, the *Rapanos* 4-4-1 decision was fractured. Justice Scalia used one dictionary definition for waters to limit protections to relatively permanent surface waters and to wetlands connected to larger water bodies, while Justice Anthony Kennedy applied a more ecologically informed reasoning [196].

Notably, the EPA and the Corps fudged the economic analysis to understate the benefits of regulations (or, equivalently, the costs of weakening regulations).[155,160,xxxix] They excluded from their analysis the economic benefits from those wetlands and streams that would no longer be protected under the Navigable Waters Protection Rule.[155] Yet, economic benefits from wetlands alone was estimated by the EPA and the Corps in their 2015 analysis to amount to between $300 million and $500 million per year.[160] Other analysts estimate ecosystem services from non-floodplain wetlands to amount to $673 billion per year.[155,161]

Lawsuits challenging the rule in various district courts resulted in it being stayed in some states but in effect in others.[152] Lawsuits by environmental NGOs[162] and a coalition of 18 state attorneys general[163] have argued that the rule conflicts with the CWA goals, contravenes recommendations by the Scientific Advisory Board and departs from the agencies' previous scientific findings.[162,163] As of May 2021, the litigation challenging the Navigable Waters Protection Rule continued in the US District Court for the District of Colorado and in the US District Court for the District of South Carolina.[164]

4.4.5 Hazardous Materials Transportation Act: Permitting LNG Transport by Rail

The Hazardous Materials Transportation Act (HMTA) of 1975 aims to "protect against the risks to life, property, and the environment that are inherent in the transportation of hazardous material." Rail transportation of crude oil, a common but risky practice in the oil and gas industry, is overseen by this legislation.[165] More than 25 million Americans live within a mile of a rail line that carries crude oil. Derailments have caused explosions, oil spills and environmental contamination across the United States.[166,xl] One particularly disastrous derailment of an oil train occurred in 2013 in Lac-Mégantic, Quebec, involving crude oil transported from the Bakken Shale in North Dakota. The fire and explosion of multiple tank cars killed 47 people, destroyed 30 buildings and spilt 1.6 million gallons of oil.[167]

The Trump administration facilitated the rail transportation of crude oil and LNG across the country, alongside the construction of pipelines. Contrary to the aims of the HMTA, the administration repealed a 2015 rule that required trains

---

xxxix Boyle et al. provide a detailed comparison of the EPA and the Corps' 2015 and 2017 analyses. The 2017 analysis ignored 90 percent of the total benefits from the Clean Water Rule estimated in the earlier analysis [160].

xl Derailments have occurred, among other places, in Aliceville, Alabama; Casselton, North Dakota; Lynchburg, Virginia; and near Philadelphia in 2013; in Mount Carbon, West Virginia and Heimdal, North Dakota in 2015; and Mosier, Oregon in 2016.

carrying crude oil to use special brakes with new technology.[168] In July 2020, the administration went further by finalizing the LNG-by-rail rule that permitted for the first time in the United States regular rail transportation for LNG.[169] Previously, LNG could only be transported by truck in small quantities or by rail in smaller, specialized tanker cars with the special approval of the Federal Railroad Administration.[170,xli] In December 2019, the federal government issued a special permit authorizing LNG transport via rail traversing 255 miles, between a proposed inland LNG facility in Wyalusing, Pennsylvania and Gibbstown, New Jersey. Several 100-car trains of LNG would move daily on the route and pass through highly populated areas such as North Philadelphia.[171]

In August 2020, environmental NGOs and a coalition of 15 attorneys general filed a petition for review challenging that LNG-by-rail rule.[170,172–175] Their comments lay out how the rule contravenes the HMTA, how the Department of Transportation's Pipeline and Hazardous Materials Safety Administration (PHMSA) failed to provide scientific analysis to support its claim that LNG transportation by rail is safe, and how PHMSA violated NEPA in failing to undertake an EIS prior to issuing the rule. Notably, other federal agencies and even members of Congress have raised the alarm about the rule as well.

PHMSA's claim that the rule would not impose significant risks is undercut by its own admission that its analysis is not grounded in evidence. For instance, PHMSA states "it is difficult to estimate the failure rate of the DOT-113 tank car in derailments because railroads are not required to report incidents to PHMSA or [the Federal Railroad Administration] unless they meet a baseline threshold." The Federal Railroad Administration, the other federal agency responsible for overseeing the transportation of hazardous materials, emphasized major safety concerns from the lack of field testing of tank cars used for transporting LNG on their reliability and their crashworthiness. The National Transportation Safety Board, the federal agency that investigates transportation accidents, and the National Association of State Fire Marshals also warned of the significant risks posed by LNG rail transport to public safety.

4.5 How Can Congress Protect Public and Private Lands

Several members of Congress have pursued important legislative initiatives to protect public and private lands, including by addressing conflicts of interest in

---

[xli] The Federal Railroad Administration approved two special permits for LNG rail transportation in Alaska and Florida in 2016, even though it expressed concerns about the risks to Florida's highly populated coastal areas. It also noted how these risks were exacerbated by the congested tracks with numerous highway–rail crossings [170].

federal agencies (Section 4.5.1), improving management of leasing on public lands (Section 4.5.2) and ensuring agencies undertake more complete analyses of competing land uses (Section 4.5.3). A number of these initiatives remained stalled in the 116th Congress without the support of Senate Republicans. I detail these important unfinished tasks.

4.5.1 Ensuring Federal Appointees Serve the Public Interest

In March 2019, the House passed the For the People Act of 2019 (H.R. 1) to provide some safeguards against conflicts of interest between policymakers and regulators and those whom they regulate.[176] That bill, which was passed by 234 votes to 193 votes, was subsequently read in the Senate.[177] The bill, part of a larger package on voter access, election integrity and political spending, establishes additional conflict-of-interest and ethics provisions for employees at federal agencies and the White House. Notably, it also prohibits members of the House from serving on the board of a for-profit entity.[177] As evident from reports by the Center for Responsibility and Ethics and the Brennan Center for Justice at New York Law School, these reforms are only the tip of the iceberg of legislation needed to combat conflicts of interest in the executive and legislative branches and to improve transparency in decision-making.[176,178]

4.5.2 Improving Financial Management of Public Lands

Prudent financial management of public lands calls for a prioritization of long-term more sustainable land use. Cognizant of renewable energy production's cost-competitiveness, alongside the declining economic contribution of oil and gas, Congress has shifted to supporting the leasing of public lands for renewable energy projects. In December 2020, Congress enacted the Energy Act of 2020, which directs the secretary of agriculture and heads of federal agencies to propose national goals for renewable energy production on public lands by September 2022 and to seek to permit at least 25 GW of solar wind and geothermal energy. When undertaken with appropriate community consultations and environmental safeguards, these projects can contribute energy while maintaining the ecosystem services provided by public lands.[179]

With a significant acreage of public lands already leased for oil and gas drilling, Congress must act to improve the returns to taxpayers from these leases. Republican Senator Chuck Grassley of Iowa and Democratic Senator Tom Udall called for improved management of oil and gas leases on public lands, arguing that "oil and gas companies should pay fair market value for the public resources they extract and sell." However, their proposed bill focuses only on *new* leases. Their Senate bill S.3330, known as the Fair Returns for Public Lands Act,[180] would set a uniform federal royalty at 18.75 percent for new leases and the minimum bid at auction at $10 per acre. In September 2020, the House Natural Resources Committee passed the bill H.R. 3225, known as the Restoring Community Input and Public Protections in Oil and Gas Leasing Act, which would eliminate noncompetitive leasing, shorten lease terms to five years, raise rental rates and require minimum bids and royalty rates.[181]

The Congressional Budget Office estimated that raising the royalty rate would increase by $200 million revenue to the federal and state governments over a decade as it is phased in, while causing no or minimal decline in production.[26,182] Other studies have also pointed to reforms that could raise government revenue from leased lands and eliminate the oil and gas industry's lock-in of public lands.[182,183]

4.5.3 Improving Legislative Protections for Public Lands

Protecting public lands requires recognizing the life-and-livelihood-supporting value of public lands (which includes their provision of ecosystem services and habitat for animals and plants) and acknowledging how climate change has already worsened the deterioration of ecosystems and the extinction of species in the United States.[28] Several Congressional representatives have pushed for broad protections for public lands and seas. In 2020, only 12 percent of the land area in the United States was permanently protected, mostly in Alaska and the West, and only 26 percent of federal marine territory was permanently protected, mostly in the remote western Pacific Ocean or northwestern Hawaii. Representative Debra Haaland and Senator Tom Udall introduced the Thirty by Thirty Resolution to Save Nature in the House and in the Senate, respectively. The resolutions set a goal for the United States to protect at least 30 percent of the land and the ocean by 2030. The resolution called for partnerships with states, tribes, local communities and private landowners.[184,185] This goal, long discussed in the scientific community,[28] has also been proposed in resolutions in state legislatures, including in the Republican-majority South Carolina legislature,[186] but Republicans in Congress did not cosponsor the resolution.

The 116th Congress's enactment of two pieces of legislation that offer protections for public lands lends some optimism that congressional representatives respond to public support for conservation. Various political commentators noted how Senate majority leader Mitch McConnell supported the Great American Outdoors Act, having previously blocked numerous bills in the Senate, because Republican senators faced tight races to hold the

Senate in 2020 (including in Montana and Colorado where the majority of voters strongly support conservation).[34] In July 2020, Congress enacted the law, which permanently allocates $900 million annually to the Land and Water Conservation Fund and directs $9.5 billion to fund the National Park Service's substantial maintenance backlog. That law followed one other bright spot for public lands protection in the 116th Congress, that is, the 2019 John D. Dingell, Jr. Conservation, Management, and Recreation Act. That law permanently protects 2.4 million acres of public lands and water, designated 1.3 million acres of public lands as wilderness, expanded two national parks and withdrew 370,000 acres in Montana and Washington state from mineral development.[187]

Unfortunately, these two laws were the exception to Congress's largely bifurcated response to the administration's assault on public lands. In response to the Trump administration's rollback of environmental laws to facilitate oil and gas leasing on public lands and the infrastructure buildout, a number of Democrats sponsored bills that would reverse those rollbacks, while a number of Republicans championed bills that would cement them.[188] Democrats introduced a series of resolutions in the House and Senate calling on the Trump administration to reverse its actions that weakened the CWA, the NEPA and the ESA,[189–192] while Republicans did the reverse.

Proposed bills regarding Trump's Navigable Waters Protection Rule (Section 4.4.4) illustrate the divergence between Republicans and Democrats. For instance, in May 2020, Democrat Representative Peter DeFazio, alongside 85 other cosponsors, proposed H.R. 6745, also known as the Clean Water for All Act.[193] The bill would nullify the Navigable Waters Protection Rule. The legislation would mandate the EPA and the Corps to promulgate regulations defining those waters to include "water bodies that affect the physical, chemical, or biological integrity of waters traditionally navigable and interstate waters, based on the best available scientific evidence." It would also mandate the implementation of the CWA using that definition to prevent "the degradation of surface water quality, increased contaminant levels in drinking water sources, increased flooding-related risks to human life or property, or disproportionate adverse impacts on minority or low-income populations." On the contrary, in April 2019, Republican Representative Mac Thornberry proposed H.R. 2287, known as the Federal Regulatory Certainty for Water Act, which would make into law the provisions of the Navigable Waters Protection Rule.[194] As detailed in Chapters 8 and 9, Congress has been split in its support for legislation that would ensure federal agencies undertake a more thorough, science-based accounting of the benefits and costs from competing uses of public lands and from oil and gas infrastructure.

4.5.4 Improving Legislative Protections for Private Lands

The Trump administration's deregulatory actions put at risk not only public lands but also human lives and private property. The LNG-by-rail rule is one particularly egregious example in which the administration placed at risk the lives and property of Americans who reside close to railway tracks. In July 2020, the House passed H.R.2, known as the Moving Forward Act, which would require PHMSA and the Federal Railroad Administration to undertake a safety study, to rescind any approval for LNG transportation by rail tank car issued prior to the enactment of the act, and to prohibit any approval of such transportation prior to the completion of the study. Regrettably, the Senate failed to pass these protections.

In December 2020, the House Committee on Oversight and Reform held a hearing on pipelines and landowner rights, in which both Republican and Democrat committee members recognized the lack of procedural fairness toward landowners whose lands are in the pathway of proposed pipelines.[195] The committee acknowledged the need for FERC to impose conditions on pipeline developers to better protect landowners and communities, for instance, by requiring pipeline developers to repair harm inflicted on neighboring lands prior to beginning their pipeline operations. The committee also recognized landowners' anger against pipeline developers that cancelled their project but chose to keep the easements, rather than sell them at the same price back to landowners. Although many landowners sell easements to pipeline companies voluntarily, they do so with limited choice in the matter, given the eminent-domain powers associated with FERC-approved pipeline projects.

4.6 Conclusion

Long-term protection of public and private lands against administrations that are bent on serving oil and gas interests will require Congress's legislative action. Indeed, Congress's failure to enact protections paved the way for the Trump administration's assault on public lands. Cognizant of voters' preferences, Congress enacted two major pieces of conservation legislation in 2019 and 2020. Several members of Congress, including those who ran for office in 2020, pledged to protect public lands and to safeguard the rights, health and safety of communities.[33] Voters will need to keep vigilant watch on the actions of their representatives once in office. The Trump administration's support for drilling and infrastructure development not only intruded upon Americans' use of public and private lands but also encroached upon Native Americans' treaty lands and ancestral lands to which they hold deep religious, cultural and historical ties, a topic I turn to next.

References

1. Congressional Research Service. *The Federal Land Management Agencies.* Report by K. Hoover, coordinator, specialist in Natural Resources Policy. IF10585 (Washington, DC: February 16, 2021). https://crsreports.congress.gov/product/pdf/IF/IF10585.
2. Congressional Research Service. *US Crude Oil and Natural Gas Production in Federal and Nonfederal Areas.* Report by M. Humphries, specialist in Energy Policy. R42432 (Washington, DC: October 23, 2018). https://crsreports.congress.gov/product/pdf/R/R42432.
3. *Light v. United States*, 220 US 523, 537 (Supreme Court 1911).
4. J. R. Pidot. "Compensatory Mitigation and Public Lands." *Boston College Law Review* 61, no. 3 (March 30, 2020): 1045–1110. https://lawdigitalcommons.bc.edu/bclr/vol61/iss3/5.
5. C. F. Wilkinson. "The Public Trust Doctrine in Public Land Law." *Colorado Law Scholarly Commons* 14 (1980): 269–316. https://scholar.law.colorado.edu/cgi/viewcontent.cgi?article=2098&context=articles.
6. E. Holden, J. Tobias and A. Chang. "Revealed: The Full Extent of Trump's 'Meat Cleaver' Assault on US Wilderness." *The Guardian*, October 26, 2020. www.theguardian.com/environment/ng-interactive/2020/oct/26/revealed-trump-public-lands-oil-drilling.
7. Bureau of Land Management Alaska. "Notice of Sale to Be Issued for Coastal Plain Oil and Gas Leasing Program Dec. 7." News release, December 3, 2020. www.blm.gov/press-release/notice-sale-be-issued-coastal-plain-oil-and-gas-leasing-program-dec-7.
8. E. Lipton. "In Last Rush, Trump Grants Mining and Energy Firms Access to Public Lands." *New York Times*, December 19, 2020. www.nytimes.com/2020/12/19/us/politics/in-last-rush-trump-grants-mining-and-energy-firms-access-to-public-lands.html.
9. *Increasing Consistency and Transparency in Considering Benefits and Costs in the Clean Air Act Rulemaking Process: Final Rule.* 40 Code of Federal Regulations 83. Environmental Protection Agency. 85 Federal Register 84130–84157 (December 23, 2020). www.federalregister.gov/documents/2020/12/23/2020-27368/increasing-consistency-and-transparency-in-considering-benefits-and-costs-in-the-clean-air-act.
10. Wilderness Society and Avarna Group. "How Did Public Lands Come to Be?" In *Public Lands in the United States: Examining the Past to Build a More Equitable Future* (July 2019). www.wilderness.org/sites/default/files/media/file/Module%202%20-%20Reading.pdf.
11. Wilderness Society and Avarna Group. *Public Lands in the United States: Examining the Past to Build a More Equitable Future* (July 2019). www

.wilderness.org/sites/default/files/media/file/Public%20Lands%20in%20the%20United%20States.pdf.
12. I. Kantor. "Ethnic Cleansing and America's Creation of National Parks." *Public Land and Resources Law Review* 28 (2007): 41–64. https://scholarship.law.umt.edu/cgi/viewcontent.cgi?article=1267&context=plrlr.
13. J. C. Ruple. "The Transfer of Public Lands Movement: The Battle to Take Back Lands That Were Never Theirs." *Colorado Natural Resources Energy & Environmental Law Review* 29, no. 1 (2018): 1–78. www.colorado.edu/law/sites/default/files/attached-files/ruple_final_1-5-web_1.pdf.
14. J. L. Sax. "Why We Will Not (Should Not) Sell the Public Lands: Changing Conceptions of Private Property." *Utah Law Review*, no. 2 (1983): 313–326.
15. J. L. Sax. "The Legitimacy of Collective Values: The Case of the Public Lands." *University of Colorado Law Review* 56 (1984): 537–558.
16. J. D. Leshy. "Public Land Policy after the Trump Administration: Is This a Turning Point?" *Colorado Natural Resources, Energy & Environmental Law Review* 31, no. 3 (2020): 472–507. www.colorado.edu/law/sites/default/files/attached-files/002_leshy_final_copy.pdf.
17. Congressional Research Service. *Federal Land Ownership: Overview and Data.* Report by C. H. Vincent, specialist in Natural Resources Policy; L. A. Hanson, senior research librarian; and L. F. Bermejo, research assistant. R42346 (Washington, DC: October 23, 2020). https://fas.org/sgp/crs/misc/R42346.pdf.
18. Congressional Research Service. *Revenues and Disbursements from Oil and Natural Gas Production on Federal Lands.* Report by B. S. Tracy, analyst in Energy Policy. R46537 (Washington, DC: September 22, 2020). www.everycrsreport.com/files/2020-09-22_R46537_a7dc7a1cdb61406e0cd5344716eccaf9e960bc72.pdf.
19. L. J. Bilmes and J. B. Loomis. *Valuing US National Parks and Programs: America's Best Investment.* Abingdon, UK: Routledge, 2019.
20. US Government Accountability Office. *Oil and Gas: Bureau of Land Management Should Address Risks from Insufficient Bonds to Reclaim Wells.* Report by F. Rusco, director of Natural Resources and Environment (September 2019). www.gao.gov/assets/710/701450.pdf.
21. Department of the Interior and US Geological Survey. *Federal Lands Greenhouse Gas Emissions and Sequestration in the United States: Estimates for 2005–14.* Report by M. D. Merrill, B. M. Sleeter, P. A. Freeman, J. Liu, P. D. Warwick and B. C. Reed (Reston, VA: 2018). https://pubs.usgs.gov/sir/2018/5131/sir20185131.pdf.
22. Department of the Interior, Office of Policy Analysis. *US Department of the Interior Economic Report FY 2018* (September 30, 2019). www.doi.gov/sites/doi.gov/files/uploads/fy-2018-econ-report-final-9-30-19-v2.pdf.
23. National Park Service and Department of the Interior. *2019 National Park Visitor Spending Effects: Economic Contributions to Local Communities, States, and the Nation.* Report by C. C. Thomas and L. Koontz (Fort Collins, CO: April 2020). www.nps.gov/subjects/socialscience/vse.htm.
24. R. Costanza et al. "Twenty Years of Ecosystem Services: How Far Have We Come and How Far Do We Still Need to Go?" *Ecosystem Services* 28 (June 4, 2017): 1–16. www.robertcostanza.com/wp-content/uploads/2017/02/2017_J_Costanza-et-al.-20yrs.-EcoServices.pdf.

25. Environmental Protection Agency. *The Economic Benefits of Protecting Healthy Watersheds* (April 2012). www.epa.gov/sites/production/files/2015-10/documents/economic_benefits_factsheet3.pdf.
26. Congressional Budget Office. *Options for Increasing Federal Income from Crude Oil and Natural Gas on Federal Lands*. Report by A. Stocking and P. Beider. (Washington, DC: April 2016). www.cbo.gov/sites/default/files/114th-congress-2015-2016/reports/51421-oil_and_gas_options.pdf.
27. M. Walls, P. Lee and M. Ashenfarb. "National Monuments and Economic Growth in the American West." *Science Advances* 6, no. 12 (March 18, 2020). https://advances.sciencemag.org/content/6/12/eaay8523/tab-pdf.
28. E. Dinerstein et al. "A Global Deal for Nature: Guiding Principles, Milestones, and Targets." *Science Advances* 5, no. 4 (April 19, 2019). https://advances.sciencemag.org/content/5/4/eaaw2869.
29. Conservation Science Partners. *Executive Summary: Loss and Fragmentation of Natural Lands in the Conterminous US from 2001 to 2017*. Report by D. M. Theobald, I. Leinwand, J. J. Anderson, V. Landau and B. G. Dickson (January 18, 2019). www.csp-inc.org/public/CSP%20Disappearing%20US%20Exec%20Summary%20011819.pdf.
30. State of the Rockies Project. *Key Findings: The 2021 Survey of the Attitudes of Voters in Eight Western States*. Report by L. Weigel and D. Metz (January 2021). www.coloradocollege.edu/other/stateoftherockies/conservationinthewest/2021/2021-State-of-the-Rockies-D2a1.pdf.
31. D. Shiffman. "An Ambitious Strategy to Preserve Biodiversity." *Scientific American*, October 4, 2020. www.scientificamerican.com/article/an-ambitious-strategy-to-preserve-biodiversity.
32. Center for Western Priorities. *Winning the West: Election 2018* (January 2019). https://westernpriorities.org/wp-content/uploads/2019/01/Winning-the-West_2018.pdf.
33. Center for Western Priorities. *Winning the West: Election 2020* (December 2020). https://westernpriorities.org/wp-content/uploads/2020/12/WinningTheWest2020_CWP.pdf.
34. A. Bolton. "McConnell Gives Two Vulnerable Senators a Boost with Vote on Outdoor Recreation Bill." *The Hill*, May 21, 2020. https://thehill.com/homenews/senate/499084-mcconnell-gives-two-vulnerable-senators-a-boost-with-vote-on-recreation-bill.
35. K. Edelstein. *The Growing Web of Oil and Gas Pipelines*. FracTracker Alliance (February 28, 2019). www.fractracker.org/2019/02/the-growing-web-of-oil-and-gas-pipelines.
36. Congressional Research Service. *Pipeline Transportation of Natural Gas and Crude Oil: Federal and State Regulatory Authority*. Report by B. J. Murrill, legislative attorney. R44432 (Washington, DC: March 28, 2016). https://fas.org/sgp/crs/misc/R44432.pdf.
37. M. Squillace et al. "Presidents Lack the Authority to Abolish or Diminish National Monuments." *Virginia Law Review Online* 103 (June 9, 2017): 55–71. www.virginialawreview.org/wp-content/uploads/2020/12/Hecht%20PDF.pdf.
38. J. C. Ruple. "The Trump Administration and Lessons Not Learned from Prior National Monument Modifications." *Harvard Environmental Law Review* 43, no. 1 (2019): 1–76.

39. Congressional Research Service. *Congressional Roll Call Votes on the Keystone XL Pipeline*. Report by L. J. Cunningham, senior research librarian. R42432 (Washington, DC: June 7, 2019). https://fas.org/sgp/crs/misc/R43870.pdf.
40. *Call for Nominations and Comments for the Coastal Plain Alaska Oil and Gas Lease Sale: Notice*. 19X.LLAK930000.L13100000.EI0000.241A. Bureau of Land Management, Department of the Interior. 85 Federal Register 73292–73293 (November 17, 2020). www.federalregister.gov/documents/2020/11/17/2020-25316/call-for-nominations-and-comments-for-the-coastal-plain-alaska-oil-and-gas-lease-sale.
41. Congressional Research Service. *Arctic National Wildlife Refuge (ANWR): An Overview*. Report by L. B. Comay, M. Ratner and R. E. Crafton. RL33872 (Washington, DC: January 9, 2018). https://fas.org/sgp/crs/misc/RL33872.pdf.
42. *Resource Management Planning: Final Rule*. 43 Code of Federal Regulations Part 1600. Bureau of Land Management, Department of the Interior. 81 Federal Register 89580–89671 (December 28, 2018). www.govinfo.gov/content/pkg/FR-2016-12-12/pdf/2016-28724.pdf.
43. T. Burr. "Effort to Shrink Bears Ears National Monument Started before Donald Trump Was Elected President." *Salt Lake Tribune*, December 3, 2017. www.sltrib.com/news/politics/2017/12/03/effort-to-shrink-bears-ears-national-monument-started-before-donald-trump-was-elected-president/.
44. T. Cama. "GOP Lawmaker Withdraws Bill to Sell Federal Land." *The Hill*, February 2, 2017. https://thehill.com/policy/energy-environment/317514-gop-rep-pulls-bill-to-sell-federal-land.
45. L. Friedman and C. O'Neil. "Who Controls Trump's Environmental Policy?" *New York Times*, January 14, 2020. www.nytimes.com/interactive/2020/01/14/climate/fossil-fuel-industry-environmental-policy.html.
46. C. Davenport. "Inspector General Says Official at Interior Dept. Violated Ethics Rules." *New York Times*, December 11, 2019.
47. B. Patterson. "Zinke Donors Include Oil and Gas Firms Using Public Land." *E&E News*, January 13, 2017. www.eenews.net/stories/1060048348.
48. A. Natter and J. A. Dlouhy. "Ryan Zinke Is Now Taking Clients from Industries He Oversaw in Trump's Cabinet." *Bloomberg*, July 23, 2019. www.bloomberg.com/news/articles/2019-07-23/former-interior-chief-zinke-now-enlisting-energy-mining-clients.
49. Western Values Project. *The Curious Case of Interior Secretary David Bernhardt's Recusals* (August 2, 2019). https://westernvaluesproject.org/wp-content/uploads/2019/08/20190802-WVP-Report-Curious-Case-of-Bernhardts-Recusal.pdf.
50. W. P. Pendley. "The Federal Government Should Follow the Constitution and Sell Its Western Lands." *National Review*, January 19, 2016. www.nationalreview.com/2016/01/federal-government-should-sell-western-land-follow-constitution.
51. US House of Representatives. *Oversight Hearing on BLM Disorganization: Examining the Proposed Reorganization and Relocation of the Bureau of Land Management Headquarters to Grand Junction, Colorado*. Committee on Natural Resources. 116th Congress, 1st Sess. September 10, 2019.
52. Z. Meghani and J. Kuzma. "The 'Revolving Door' between Regulatory Agencies and Industry: A Problem That Requires Reconceptualizing Objectivity." *Journal*

of Agricultural and Environmental Ethics 24, no. 6 (September 17, 2010): 575–599.

53. J. Blanes i Vidal, M. Draca and C. Fons-Rosen. "Revolving Door Lobbyists." *American Economic Review* 102, no. 7 (December 2012): 3731–3748.
54. M. Bertrand, M. Bombardini and F. Trebbi. "Is It Whom You Know or What You Know? An Empirical Assessment of the Lobbying Process." *American Economic Review* 104, no. 12 (December 2014): 3885–3920.
55. A.-M. Fennell and F. Rusco, directors of Natural Resources and Environment. *Bureau of Land Management: Agency's Reorganization Efforts Did Not Substantially Address Key Practices for Effective Reforms*. Submitted to R. Grijalva, chairman of the Committee on Natural Resources. March 6, 2020.
56. L. Friedman. "A War against Climate Science, Waged by Washington's Rank and File." *New York Times*, July 14, 2020.
57. H. Tabuchi. "A Trump Insider Embeds Climate Denial in Scientific Research." *New York Times*, March 2, 2020.
58. G. T. Goldman et al. "Perceived Losses of Scientific Integrity under the Trump Administration: A Survey of Federal Scientists." *PLoS ONE* 15, no. 4 (April 23, 2020): e0231929. https://doi.org/https://doi.org/10.1371/journal.pone.0231929.
59. B. Abbey and J. Caswell. "The Stealth Plan to Erode Public Control of Public Lands." *Politico*, December 12, 2019. www.politico.com/news/agenda/2019/12/12/public-lands-bureau-land-management-082689;.
60. J. Freemuth and J. R. Skillen. "Moving Bureau of Land Management Headquarters to Colorado Won't Be Good for Public Lands." *The Conservation*, January 8, 2020. https://theconversation.com/moving-bureau-of-land-management-headquarters-to-colorado-wont-be-good-for-public-lands-126990.
61. D. Bernhardt. *22 Secretarial Order 3369: Promoting Open Science* (September 28, 2018). www.doi.gov/sites/doi.gov/files/elips/documents/so_3369_promoting_open_science.pdf.
62. M. Doyle. "Interior Department Moves to Impose New Rules on Use of Science in Decision-Making." *Science* (American Association for the Advancement of Science), February 27, 2020. www.sciencemag.org/news/2020/02/interior-department-moves-impose-new-rules-use-science-decision-making.
63. Bureau of Land Management. *Air Resources Management Program Strategy*. Report by K. E. Rodgers, R. A. Boyd, T. H. Alexander and L. A. Ford (February 2015). www.blm.gov/sites/blm.gov/files/AirResourceProgramStrategy.pdf.
64. *Freedom of Information Act Regulations*. 43 Code of Federal Regulations 2. Environmental Protection Agency. 83 Federal Register 67175–67180 (December 28, 2018). www.federalregister.gov/documents/2018/12/28/2018-27561/freedom-of-information-act-regulations.
65. D. McGrath. *Interior's Proposed FOIA Rule Threatens Transparency and Accountability*. American Oversight (January 29, 2019). www.americanoversight.org/interiors-proposed-foia-rule-threatens-transparency-and-accountability.
66. M. E. Townsend. open government staff attorney at the Center for Biological Diversity. *Comments on the US Department of the Interior's Freedom of Information Act Rule Revisions, Docket No. DOI-2018-0017*. Submitted to C. Cafaro, Office of Executive Secretariat and Regulatory Affairs. January 29, 2019.

67. B. D. Brown, executive director of Reporters Committee for Freedom of the Press. *Proposed Revisions to the Department of the Interior's Freedom of Information Act Regulations, RIN 1093-AA26/Docket No. DOI-2018-0017*. Submitted to Office of the Secretary, the Interior. January 28, 2019.
68. *United States Department of Justice v. Reporters Committee for Freedom of the Press*, 489 US 749 (Supreme Court 1989).
69. Wilderness Society and Center for Western Priorities. *America's Public Lands Giveaway* (April 2020). https://storymaps.arcgis.com/stories/36d517f10bb0424493e88e3d22199bb3.
70. Bureau of Land Management, Department of the Interior. *Oil and Gas Statistics* (2019). www.blm.gov/programs/energy-and-minerals/oil-and-gas/oil-and-gas-statistics.
71. Center for Western Priorities. *Dashboard: Oil and Gas Leasing* (2021). https://westernpriorities.org/dashboard-oil-gas-leasing.
72. D. R. Bucks. *A Fair Return for the American People: Increasing Oil and Gas Royalties from Federal Lands* (March 2019). https://naturalresources.house.gov/imo/media/doc/Testimony%20Attachment%20-%20Dan%20Bucks%20-%20EMR%20Leg%20Hrg%2009.24.19.pdf.
73. Government Accountability Office. *Federal Oil and Gas Revenue: Actions Needed to Improve BLM's Royalty Relief Policy*. Report by F. Rusco, director Natural Resources and Environment (October 6, 2020). www.gao.gov/assets/720/710030.pdf.
74. Department of the Interior. *Fact Sheet on Methane and Waste Prevention Rule*. (2016). www.doi.gov/sites/doi.gov/files/uploads/methane_waste_prevention_rule_factsheet.pdf.
75. *California v. Bernhardt*, 472 F. Supp. 3d 573 (N.D. Cal. 2020).
76. C. Anchondo. "'Roller Coaster': Judge Scraps Obama-Era Methane Rule." *ClimateWire*, October 9, 2020. www.eenews.net/climatewire/stories/1063715891?t=https%3A%2F%2Fwww.eenews.net%2Fstories%2F1063715891.
77. *Oil and Gas; Hydraulic Fracturing on Federal and Indian Lands; Rescission of a 2015 Rule: Final Rule*. 43 Code of Federal Regulations 3160. Bureau of Land Management, Interior. 82 Federal Register 61924–61949 (December 29, 2017) www.federalregister.gov/documents/2017/12/29/2017-28211/oil-and-gas-hydraulic-fracturing-on-federal-and-indian-lands-rescission-of-a-2015-rule.
78. *Oil and Gas; Hydraulic Fracturing on Federal and Indian Lands: Final Rule*. 43 Code of Federal Regulations 3160. Bureau of Land Management, Interior. 80 Federal Register 16127–16222 (March 26, 2015) www.federalregister.gov/documents/2015/03/26/2015-06658/oil-and-gas-hydraulic-fracturing-on-federal-and-indian-lands.
79. Taxpayers for Common Sense. *Locked Out: The Cost of Speculation on Oil and Gas Leases on Federal Land* (October 2017). www.taxpayer.net/wp-content/uploads/ported/images/downloads/LOCKED_OUT_Energy_Report.pdf.
80. Center for American Progress. *Oil and Gas Companies Gain by Stockpiling America's Federal Land*. Report by M. K. DeSantis (August 29, 2018). www.americanprogress.org/issues/green/reports/2018/08/29/455226/oil-gas-companies-gain-stockpiling-americas-federal-land.

81. Government Accountability Office. *Oil and Gas: Onshore Competitive and Noncompetitive Lease Revenues*. Report by F. Rusco, director of Natural Resources and Environment (November 19, 2020). www.gao.gov/products/GAO-21-138.
82. New York Times Staff. "Documents: Leaked Industry Emails and Reports." *New York Times*, August 2011. https://archive.nytimes.com/www.nytimes.com/interactive/us/natural-gas-drilling-down-documents-4.html.
83. *Modernization of Oil and Gas Reporting: Final Rule*. 17 Code of Federal Regulations Parts 210 and 211. Securities and Exchange Commission. 74 Federal Register 2158–2197 (January 14, 2009) www.sec.gov/rules/final/2009/33-8995fr.pdf.
84. R. Glick and M. Christiansen. "FERC and Climate Change." *Energy Law Journal* 40, no. 1 (2019). www.eenews.net/assets/2019/05/06/document_gw_02.pdf.
85. C. J. Bateman and J. T. B. Tripp. "Toward Greener FERC Regulation of the Power Industry." *Harvard Environmental Law Review* 38 (2014): 275–333.
86. A. Flyer. "FERC Compliance under NEPA: FERC's Obligation to Fully Evaluate Upstream and Downstream Environmental Impacts Associated with Siting Natural Gas Pipelines and Liquefied Natural Gas Terminals." *Georgetown International Environmental Law Review* 27 (2015): 301–319. https://gielr.files.wordpress.com/2015/04/flyer-final-pdf-27-2.pdf.
87. J. S. John. "Senate Confirms Democrat Clements, Republican Christie to FERC." *GreenTech Media*, December 1, 2020. www.greentechmedia.com/articles/read/senate-confirms-democrat-clements-republican-christie-to-ferc#.
88. Analysis Group. *FERC's Certification of New Interstate Natural Gas Facilities*. Report by S. F. Tierney (Washington, DC: November 2019). www.analysisgroup.com/globalassets/content/insights/publishing/revising_ferc_1999_pipeline_certification.pdf.
89. R. Glick, chairman of the Federal Regulatory Commission. *Dissent on PennEast Pipeline Co. LLC*. Submitted to Federal Energy Regulatory Commission. Docket No. CP15-558-001. August 10, 2018.
90. R. Glick, chairman of the Federal Regulatory Commission. *Dissent Regarding Jordan Cove Energy Project L.P. (Jordan Cove LNG Terminal)*. Submitted to Federal Energy Regulatory Commission. May 21, 2020.
91. C. A. LaFleur. *Dissent on Order Issuing Certificates and Granting Abandonment Authority (Mountain Valley Pipeline and Atlantic Coast Pipeline)*. Submitted to Federal Energy Regulatory Commission. Docket Nos. CP15-554-000 CP16-10-000. October 13, 2017.
92. C. A. LaFleur. *Testimony at the Hearing on the Natural Gas Act*. Submitted to United States House of Representatives Committee on Energy and Commerce Subcommittee on Energy. February 5, 2020.
93. B. L. McNamee, Federal Energy Regulatory Commission. *Order Issuing Certificates and Approving Abandonment*. Submitted to Federal Energy Regulatory Commission. July 17, 2020.
94. R. Webb. "Climate Change, FERC, and Natural Gas Pipelines: The Legal Basis for Considering Greenhouse Gas Emissions under Section 7 of the Natural Gas Act." *NYU Environmental Law Journal* 28, no. 2 (2019): 179–226.
95. *Sierra Club v. Federal Energy Regulatory Commission*, 867 F.3d 1357, 1374 (D.C. Cir. 2017).

96. *Final Environmental Impact Statement – Midcontinent Supply Header Interstate Pipeline Project.* Staff of the Federal Energy Regulatory Commission PF17-3, CP17-458 (2018). www.ferc.gov/final-environmental-impact-statement-midcontinent-supply-header-interstate-pipeline-project
97. *Birckhead v. Federal Energy Regulatory Commission*, 925 F.3d 510 (D.C. Cir. 2019).
98. A. B. Klass. "Eminent Domain Law As Climate Policy." *Wisconsin Law Review* 49, no. 1 (June 8, 2020): 50–83.
99. *City of Oberlin v. Federal Energy Regulatory Commission*, 937 F.3d 599, 601, 603 (D.C. Cir. 2019).
100. A. B. Klass. "The Public Use Clause in an Age of US Natural Gas Exports." *Stanford Law Review Online* 72 (April 2019): 103–111. https://review.law.stanford.edu/wp-content/uploads/sites/3/2020/03/72-Stan.-L.-Rev.-Online-Klass.pdf.
101. D. Bookbinder and M. Gibson, Niskanen Center. *Comments on the Federal Energy Regulatory Commission's Draft Environmental Impact Statement for the Jordan Cove Energy Project.* Submitted to Federal Energy Regulatory Commission. PF17-4-000. January 2020.
102. *Allegheny Defense Project v. Federal Energy Regulatory Commission*, 964 F.3d. 1 (D.C. Cir. 2020).
103. House Committee on Oversight and Reform. "Subcommittee Releases Preliminary Findings Showing FERC Pipeline Approval Process Skewed Against Landowners." News release, April 28, 2020. https://oversight.house.gov/news/press-releases/subcommittee-releases-preliminary-findings-showing-ferc-pipeline-approval.
104. R. Ewing. "Pipeline Companies Target Small Farmers and Use Eminent Domain for Private Gain." *North Carolina Central Law Review* 38, no. 2 (2016): 125–141.
105. J. Bell. "Big Changes May Be Ahead for Natural Gas Pipelines, If FERC Does Its Job." *Utility Dive*, September 16, 2020. www.utilitydive.com/news/big-changes-may-be-ahead-for-natural-gas-pipelines-if-ferc-does-its-job/585182/.
106. *Limiting Authorizations to Proceed with Construction Activities Pending Rehearing: Final Rule.* 18 Code of Federal Regulations Part 153 and 157. Federal Energy Regulatory Commission. Docket No. RM20-15-000; Order No. 871 (June 9, 2020). www.ferc.gov/sites/default/files/2020-06/RM20-15-000.pdf.
107. D. Hayes, staff attorney at Sierra Club. *Comments on the US Army Corps of Engineers' Proposal to Reissue and Modify Nationwide Permit 12, Docket No. COE-2015-0017.* Submitted to US Army Corps of Engineers (August 1, 2016).
108. *Northern Plains Resource Council v. United States Army Corps of Engineers*, No. CV-19-44-GF-BMM, F.Supp. 3d, 2020 WL 1875455 (D. Mont. April 15, 2020).
109. R. Frazin. "Overnight Energy: Supreme Court Reinstates Fast-Track Pipeline Permit Except for Keystone, Judge Declines to Reverse Dakota Access Pipeline Shutdown." *The Hill*, July 7, 2020. https://thehill.com/policy/energy-environment/506266-overnight-energy-supreme-court-reinstates-fast-track-pipeline.
110. H. J. Feldman, senior director of Regulatory and Scientific Affairs. *Re: EPA-HQ-OA-2017-0190 (82 FR 17793) Comments in Response to the EPA's Solicitation of Input from the Public to Inform Its Regulatory Reform Task Force's Evaluation of*

Existing Regulations. Submitted to S. K. Dravis, regulatory reform officer and associate administrator in the Office of Policy at the Environmental Protection Agency. May 15, 2017.
111. American Petroleum Institute. *Attachment 1: Comments on Specific Regulations* (Washington, DC: May 15, 2017). www.api.org/~/media/Files/News/Letters-Comments/2017/5-15-17-API-Comments-on-specific-regulations.pdf.
112. H. J. Feldman, senior director of Regulatory and Scientific Affairs, A. J. Black, president and CEO of Association of Oil Pipelines and D. Van Liew, vice president of IAGC. *Update to the Regulations Implementing the Procedural Provisions of the National Environmental Policy Act, 85 Fed. Reg. 1,684 (January 10, 2020)*. Submitted to E. A. Boling, Council on Environmental Quality. March 10, 2020.
113. A. J. Turner, counsel for the ESA Cross-Industry Coalition, et al. *Comments in Response to the Three Proposals from the US Fish and Wildlife Service and the National Marine Fisheries Service to Amend Their Endangered Species Act Regulations*. Submitted to US Fish and Wildlife Service and National Marine Fisheries Service. September 24, 2018.
114. *Indigenous Environmental Network v. United States Department of State*, 347 F. Supp. 3d 561 (D. Mont.) order amended and supplemented, 369 F. Supp. 3d 1045 (D. Mont. 2018), and appeal dismissed and remanded sub nom. *Indigenous Environmental Network v. United States Department of State*, No. 18-36068, 2019 WL 2542756 (9th Cir. June 6, 2019).
115. *Standing Rock Sioux Tribe. v. US Army Corps of Engineers*, 471 F. Supp. 3d 71 (D.D.C 2020), affirmed in part, revised in part sub nom. *Standing Rock Sioux Tribe v. United States Army Corps of Engineers*, 985 F.3d 1032 (D.C. Cir. 2021).
116. *Standing Rock Sioux Tribe v. US Army Corps of Engineers*, No. 20-5197, 2020 WL 4548123 (D.C. Cir. August 5, 2020).
117. *WildEarth Guardians v. Zinke*, 368 F. Supp. 3d 41 (D.D.C. 2019).
118. *WildEarth Guardians v. Bernhardt*, No. CV 16-1724 (RC), 2019 WL 3253685 (D.D.C. July 19, 2019).
119. *WildEarth Guardians v. US Bureau of Land Management*, 457 F. Supp. 3d 880 (D. Mont. 2020).
120. A. Liptak. "Supreme Court Won't Block Ruling to Halt Work on Keystone XL Pipeline." *New York Times*, July 8, 2020.
121. *Western Watersheds Project v. Zinke*, 441 F. Supp. 3d 1042 (D. Idaho 2020).
122. *Montana Wildlife Federation v. David Bernhardt*, No. CV-18-69-GF-BMM, 2020 WL 2615631 (D. Mont. May 22, 2020), appeal dismissed, No. 20-35609, 2020 WL 6194597 (9th Cir. Oct. 16, 2020).
123. M. Freeman. "Court Strikes Down Trump Administration's Sage-Grouse Directive, Canceling Hundreds of Oil and Gas Leases." News release, May 26, 2020, https://earthjustice.org/news/press/2020/court-strikes-down-trump-administrations-sage-grouse-directive-canceling-hundreds-of-oil-and-gas-leases.
124. Oregon Department of Environmental Quality. *Protest in Opposition to the Petition for Declaratory Order Filed by Jordan Cove Energy Project LP and Pacific Connector Gas Pipeline, LP*. Submitted to Federal Energy Regulatory Commission. May 2020.

125. Council on Environmental Quality. *Major Cases Interpreting the National Environmental Policy Act*. June 1997. https://ceq.doe.gov/docs/laws-regulations/Major_NEPA_Cases.pdf
126. *Update to the Regulations Implementing the Procedural Provisions of the National Environmental Policy Act: Final Rule*. 40 Code of Federal Regulations 1500–1508, 1515–1518. Council on Environmental Quality. 85 Federal Register 43304–43376 (July 16, 2020) www.federalregister.gov/documents/2020/07/16/2020-15179/update-to-the-regulations-implementing-the-procedural-provisions-of-the-national-environmental.
127. *State of California v. Council on Environmental Quality* WL 672540 (N.D. Cal. 2021).
128. *Environmental Justice Health Alliance et al. v. Council on Environmental Quality et al*, No. 1:20CV06143 (S.D.N.Y. 2020).
129. S. Buccino, senior director of Lands Division at Natural Resources Defense Council. *Comments of the Natural Resources Defense Council on CEQ's proposed Update to the Regulations Implementing the Procedural Provisions of the National Environmental Policy Act*. Submitted to M. Neumayr, chairman of Council on Environmental Quality. March 10, 2020.
130. Attorneys General of Washington, California, New York, District of Columbia, Connecticut, Delaware, Guam, Illinois, Maine, Maryland, Massachusetts, Michigan, Minnesota, Nevada, New Jersey, New Mexico, Oregon, Pennsylvania, Rhode Island, and Vermont. *Notice of Proposed Rulemaking – Update to the Regulations Implementing the Procedural Provisions of the National Environmental Policy Act, 85 Fed. Reg. 1684 (Jan. 10, 2020) Docket ID No. CEQ-2019-0003*. Submitted to E. A. Boling, associate director for the National Environmental Policy Act, and V. Z. Seale, chief of staff and general counsel for Council on Environmental Quality. March 10, 2020.
131. *Sierra Club v. Babbitt*, 65 F. 3d. 1502, 1509 (9th Cir. 1995).
132. Congressional Research Service. *The Legal Framework of the National Environmental Policy Act*. Report by N. M. Hart and L. Tsang (Washington, DC: October 22, 2020).
133. *Endangered Species Act*. 16 U.S.C. 1531–1544, 87 Stat. 884. US Fish and Wildlife Service, Department of the Interior (1973). www.fws.gov/endangered/esa-library/pdf/ESAall.pdf.
134. T. Campbell IV et al. "Protecting the Lesser Prairie Chicken under the Endangered Species Act: A Problem and an Opportunity for the Oil and Gas Industry." *Texas Environmental Law Journal* 45, no. 1 (February 2015): 31–50.
135. J. Galperin. "Trust Me, I'm a Pragmatist: A Partially Pragmatic Critique of Pragmatic Activism." *Columbia Journal of Environmental Law* 42, no. 2 (2017): 425–496.
136. *Center for Biological Diversity v. Bernhardt*, No. 19-CV-05206-JST, 2020 WL 4188091 (N.D. Cal. May 18, 2019).
137. *Complaint for Declaratory and Injunctive Relief, California v. David Bernhardt*, 472F. Supp. 3d 573 (N.D. Cal. 2020) (No. 4:18-cv-05712-YGR).
138. Attorneys General of Massachusetts, California, Maryland, New York, Oregon, Pennsylvania, Rhode Island, Vermont, Washington, and the District of Columbia. *Comments on the Fish and Wildlife Service and the National Marine Fisheries*

Service's Proposed Rules. Submitted to R. K. Zinke, secretary of the Department of the Interior and W. Ross, secretary of the Department of Commerce. September 24, 2018.
139. *Comments on Proposed Revisions to Regulations for Listing Endangered and Threatened Species and Designating Critical Habitat*. Submitted to G. Frazer, assistant director for Endangered Species Ecological Services Program and S. D. Rauch III, deputy assistant administrator for Regulatory Programs Office of Protected Resources. Docket ID: FWS–HQ–ES–2020–0047. September 3, 2020.
140. *Endangered and Threatened Wildlife and Plants; Regulations for Prohibitions to Threatened Wildlife and Plants*. 50 Code of Federal Regulations Part 17. Fish and Wildlife Service, Department of the Interior. 84 Federal Register 44753–44760 (August 27, 2019). www.regulations.gov/document?D=FWS-HQ-ES-2018-0007-69538.
141. *Endangered and Threatened Species, Listing Species and Designating Critical Habitat: Final Rule*. 50 Code of Federal Regulations Part 424. Fish and Wildlife Service, Department of the Interior. 84 Federal Register 45020–45053 (August 27, 2019). www.regulations.gov/document?D=FWS-HQ-ES-2018-0006-64025.
142. D. Duncan. "Clean Water Act Section 401: Balancing States' Rights and the Nation's Need for Energy Infrastructure." *Hastings Environmental Law Journal* 25, no. 2 (2019): 235–262.
143. Congressional Research Service. *Clean Water Act Section 401: Background and Issues*. Report by C. Copeland, specialist in Resources and Environmental Policy. 97-488 (Washington, DC: July 2, 2015). https://fas.org/sgp/crs/misc/97-488.pdf.
144. *Clean Water Act Section 401 Certification Rule: Draft of Final Rule*. 40 Code of Federal Regulations Part 121. Environmental Protection Agency. EPA-HQ-OW-2019-0405; FRL-10009-80-OW (June 2020). www.epa.gov/sites/production/files/2020-06/documents/pre-publication_version_of_the_clean_water_act_section_401_certification_rule_508.pdf.
145. *Complaint for Declaratory and Injunctive Relief, Suquamish Tribe et al. v. Andrew Wheeler and US Environmental Protection Agency* (N.D. Cal. 2019) (No. 3:20CV06137).
146. *Complaint for Declaratory and Injunctive Relief, State of California et al. v. Andrew R. Wheeler et al.*, WL 9172918 (N.D. Cal. 2020) (Nos 20-cv-04869-WHA, 20-cv-04636-WHA).
147. Attorneys General of California, Connecticut, Maryland, Maine, Massachusetts, Minnesota, New Jersey, New Mexico, New York, Oregon, Pennsylvania, Rhode Island, Vermont, Washington, and Pennsylvania Department of Environmental Protection. *Objection to the Clean Water Act Section 401 Guidance for Federal Agencies, States, and Authorized Tribes*. Submitted to A. R. Wheeler, administrator of the Environmental Protection Agency. July 25, 2019.
148. M. Nasmith, staff attorney at Earthjustice, et al. *Comments on EPA Proposed Rule Updating Regulations on Water Quality Certification*. Submitted to A. R. Wheeler, administrator of the Environmental Protection Agency. EPA–HQ–OW–2019–0405. October 21, 2019.
149. M. D. Bellon, director of Washington State Department of Ecology. *Decision on the Request for a Section 401 Water Quality Certification*. Submitted to K. Gaines, Millennium Bulk Terminals-Longview, LLC. September 26, 2017.

150. J. D. Ogsbury, executive director of Western Governors' Association, et al. *Comments in Response to the US Environmental Protection Agency's Proposed Rule, Updating Regulations on Water Quality Certification.* Submitted to A. R. Wheeler, administrator of the Environmental Protection Agency. No. EPA-HQ-OW-2019-0405. October 16, 2019.
151. Congressional Research Service Legal Sidebar. *Wading into the "Waters of the United States."* Report by S. P. Mulligan, legislative attorney. LSB10236 (Washington, DC: December 28, 2018). https://fas.org/sgp/crs/misc/LSB10236.pdf.
152. D. Y. Chung and L. P. Frost. "Dueling Navigable Waters Protection Rule Decisions Leave Uncertainty in Their Wake." *Trends* (American Bar Association), October 30, 2020. www.americanbar.org/groups/environment_energy_resources/publications/trends/2020-2021/november-december-2020/dueling-navigable-water.
153. *The Navigable Waters Protection Rule: Definition of "Waters of the United States": Final Rule.* 33 Code of Federal Regulations 328 and 40 Code of Federal Regulations 110, 112, 116, 117, 120, 122, 230, 232, 300, 302, 401. Department of the Army, Corps of Engineers, Department of Defense and Environmental Protection Agency. 85 Federal Register 22250–22342 (April 21, 2020) www.federalregister.gov/documents/2020/04/21/2020-02500/the-navigable-waters-protection-rule-definition-of-waters-of-the-united-states.
154. S. M. P. Sullivan, M. C. Rains and A. D. Rodewald. "The Proposed Change to the Definition of 'Waters of the United States' Flouts Sound Science." *Proceedings of the National Academy of Sciences of the United States of America* 116, no. 24 (June 11, 2019): 11558–11561. https://doi.org/https://doi.org/10.1073/pnas.1907489116. www.pnas.org/content/116/24/11558.
155. S. M. P. Sullivan et al. "Distorting Science, Putting Water at Risk." *Science* 369, no. 6505 (August 14, 2020): 766–768.
156. M. Honeycutt, chair of the Science Advisory Board. *Draft Commentary on the Proposed Rule Defining the Scope of Waters Federally Regulated under the Clean Water Act.* Submitted to A. R. Wheeler, administrator of the Environmental Protection Agency. EPA-SAB-20-xxx. October 16, 2019.
157. Environmental Protection Agency. *Connectivity of Streams and Wetlands to Downstream Waters: A Review and Synthesis of the Scientific Evidence* (Final Report). EPA/600/R-14/475 F (Washington, DC: 2015). https://cfpub.epa.gov/ncea/risk/recordisplay.cfm?deid=296414.
158. L. C. Alexander. "Science at the Boundaries: Scientific Support for the Clean Water Rule." *University of Chicago Press Journal* 34, no. 4 (August 7, 2015). www.journals.uchicago.edu/doi/full/10.1086/684076.
159. Science Advisory Board. *SAB Review of the Draft EPA Report Connectivity of Streams and Wetlands to Downstream Waters: A Review and Synthesis of the Scientific Evidence.* Submitted to G. McCarthy, administrator of the Environmental Protection Agency. October 17, 2014.
160. K. J. Boyle, M. J. Kotchen and V. K. Smith. "Deciphering Dueling Analyses of Clean Water Regulations." *Science* 358, no. 6359 (October 6, 2017): 49–50. https://science.sciencemag.org/content/358/6359/49.summary.
161. I. F. Creed et al. "Enhancing Protection for Vulnerable Waters." *Nature Geoscience*, no. 10 (October 2, 2017): 809–815. www.nature.com/articles/ngeo3041?WT.feed_name=subjects_law.

162. *Complaint, Conservation Law Foundation et al. v. US Environmental Agency et al.*, WL 8669769 (D. Mass. 2020) (No. 1:20-cv-10820-DPW).
163. *Complaint for Declaratory and Injunctive Relief, California. v. Wheeler*, 472 F. Supp. 3d 573 (N.D. Cal. 2020) (No 4:18-cv-05712-YGR 2020).
164. J. P. Jacobs and P. King. "Biden Races Courts for Chance to Torpedo Trump Water Rule." *E&E News*, April 28, 2021.
165. T. Lalley. "Fifteen State AGs Challenge Rule Allowing Transport of Liquified Natural Gas (LNG) by Rail." News release, NYU School of Law State Energy & Environmental Impact Center, August 18, 2020, www.law.nyu.edu/centers/state-impact/press-publications/press-releases/lng-by-rail-challenge.
166. Associated Press. "A Timeline of Recent Oil Train Crashes in the US and Canada." *Associated Press*, June 13, 2016. https://apnews.com/article/oil-spills-fires-north-dakota-accidents-canada-84b1e8273d854697b34af57bc60badc2.
167. B. Campbell. *The Lac-Mégantic Rail Disaster: Public Betrayal, Justice Denied.* Toronto, ON: Lorimer, 2018.
168. T. Cama. "Trump Officials Roll Back Obama Oil Train Safety Rule." *The Hill*, September 24, 2018. https://thehill.com/policy/energy-environment/408125-trump-admin-rolls-back-obama-oil-train-safety-rule.
169. *Hazardous Materials: Liquefied Natural Gas by Rail: Final Rule.* 49 Code of Federal Regulations 172–174, 179, 180. Pipeline and Hazardous Materials Safety Administration, Department of Transportation. 85 Federal Register 44994–45030 (July 24, 2020). www.federalregister.gov/documents/2020/07/24/2020-13604/hazardous-materials-liquefied-natural-gas-by-rail.
170. B. Marshall, staff attorney at Earthjustice. *Comments Objecting to the Proposed Rulemaking to Authorize the Transportation of Methane, Refrigerated Liquid by Rail, Docket No. PHMSA-2018-0025 (HM-264).* Submitted to D. Pearce, deputy administrator for the Pipeline and Hazardous Material Safety Administration. January 13, 2020.
171. A. Maykuth. "Plan to Send LNG Trains through Philly to S. Jersey Port Sparks Outrage from Residents, Environmentalists." *Philadelphia Inquirer*, September 20, 2020. www.inquirer.com/business/philadelphia-lng-gibbstown-new-fortress-energy-port-drbc-fracking-trains-20200920.html.
172. *Petition for Review, State of Maryland et al. v. United States Department of Transportation et al.*, D.C. No. 20-___ (Court of Appeals 2020).
173. *Petition for Review, Sierra Club et al. v. United States Department of Transportation et al.*, D.C. No. 20-___ (Court of Appeals 2020).
174. Attorneys General of Maryland, New York, California, Delaware, Illinois, Massachusetts, Michigan, Minnesota, New Jersey, North Carolina, Oregon, Pennsylvania, Rhode Island, Vermont, Washington and the District of Columbia. *Comments to Voice Their Strong Objection to the Pipeline and Hazardous Materials Safety Administration's Notice of Proposed Rulemaking.* Submitted to D. Pearce, deputy administrator for the Pipeline and Hazardous Material Safety Administration. January 13, 2020.
175. Congressional Research Service. *Rail Transportation of Liquefied Natural Gas: Safety and Regulation.* Report by P. W. Parfomak, specialist in Energy and Infrastructure Policy and J. Frittelli, specialist in Transportation Policy. R46414 (Washington, DC: July 28, 2020). https://fas.org/sgp/crs/misc/R46414.pdf.

176. R. A. Mehrbani, Spitzer fellow and senior counsel at Brennan Center for Justice. *Written Testimony at Hearing on H.R. 1: Strengthening Ethics Rules for the Executive Branch*. Submitted to US House of Representatives Committee on Oversight and Reform. February 6, 2019.
177. US Congress. House. *For the People Act of 2019*. H.R. 1, 116th Congress, 1st Sess. Introduced in House January 3, 2019.
178. Citizens for Responsibility and Ethics in Washington. *What Democracy Looks Like*. Report by J. Ahearn, C. Shaw, G. Lezra, M. Woodard and H. Hammado (Washington, DC: December 2, 2020). www.citizensforethics.org/reports-investigations/crew-reports/democracy-reform-blueprint-accountable-inclusive-ethical-government.
179. Yale Center for Business and the Environment; Wilderness Society. *Key Economic Benefits of Renewable Energy on Public Lands*. Report by N. Springer, Yale Center for Business and the Environment, and A. Daue, Wilderness Society (May 2020). www.wilderness.org/sites/default/files/media/file/CBEY_WILDERNESS_Renewable%20Energy%20Report_0.pdf.
180. S. C. Grassley and S. T. Udall. "End the Taxpayer Giveaway to Big Oil and Gas." *New York Times*, December 2, 2020. www.grassley.senate.gov/news/news-releases/grassley-udall-nyt-op-ed-end-taxpayer-giveaway-big-oil-and-gas.
181. US Congress. House. *Restoring Community Input and Public Protections in Oil and Gas Leasing Act of 2020*. H.R. 3225, 116th Congress, 2nd Sess. Introduced in House June 12, 2019.
182. Government Accountability Office. *Oil, Gas, and Coal Royalties: Raising Federal Rates Could Decrease Production on Federal Lands But Increase Federal Revenue*. Report by F. Rusco, director of Natural Resources and Environment (Washington, DC: June 2017). www.gao.gov/assets/690/685335.pdf.
183. F. Rusco, director Natural Resources and Environment. *Testimony on Federal Energy Development: Challenges to Ensuring a Fair Return for Federal Energy Resources*. Submitted to Subcommittee on Energy and Mineral Resources, House of Representatives (September 24, 2019).
184. US Congress. Senate. *Expressing the Sense of the House of Representatives that the Federal Government Should Establish a National Goal of Conserving at Least 30 Percent of the Land and Ocean of the United States by 2030*. H.R.___ 116th Congress, 2nd Sess. Introduced in Senate January 8, 2020.
185. US Congress. Senate. *A Resolution Expressing the Sense of the Senate that the Federal Government Should Establish a National Goal of Conserving at Least 30 Percent of the Land and Ocean of the United States by 2030*. S. 372 116th Congress, 1st Sess. Introduced in Senate October 22, 2019.
186. South Carolina General Assembly. State Senate. *To Establish the Goal of Protecting Thirty Percent of the State of South Carolina by Not Later than 2030*. Bill 1024, 123rd Sess. Introduced in State Senate January 21, 2020.
187. US Congress. Senate. *John D. Dingell, Jr. Conservation, Management, and Recreation Act*. S. 47, 116th Congress, 1st Sess. Introduced in Senate January 8, 2019.
188. R. Beitsch. "House GOP Seeks to Cement Trump Rollback of Bedrock Environmental Law." *The Hill*, September 22, 2020. https://thehill.com/policy/

energy-environment/517619-house-gop-seeks-to-cement-trump-rollback-of-bedrock-environmental.

189. US Congress. House. *Encouraging the Environmental Protection Agency to Maintain and Strengthen Requirements under the Clean Water Act and Reverse Ongoing Administrative Actions to Weaken This Landmark Law and Protections for United States Waters.* H.R. 797, 116th Congress, 2nd Sess. Introduced in House January 14, 2020.

190. US Congress. Senate. *A Resolution Encouraging the Administrator of the Environmental Protection Agency to Maintain and Strengthen Requirements under the Clean Water Act and Reverse Ongoing Administrative Actions to Weaken the Clean Water Act and Protections for Waters of the United States.* S. 714, 116th Congress, 2nd Sess. Introduced in Senate September 9, 2019.

191. US Congress. House. *Encouraging the Trump Administration to Maintain Protections under the National Environmental Policy Act and Reverse Ongoing Administrative Actions to Weaken this Landmark Law and Its Protections for American Communities.* H.Con.Res. 89, 116th Congress, 2nd Sess. Introduced in House February 11, 2020.

192. US Congress. Senate. *A Resolution Encouraging the Trump Administration to Maintain Protections under the National Environmental Policy Act and Reverse Ongoing Administrative Actions to Weaken This Landmark Law and Its Protections for American Communities.* S. 537, 116th Congress, 2nd Sess. Introduced in Senate March 10, 2020.

193. US Congress. House. *Clean Water for All Act.* H.R. 6745, 116th Congress, 2nd Sess. Introduced in House May 8, 2020.

194. US Congress. House. *Federal Regulatory Certainty for Water Act.* H.R. 2287, 116th Congress, 1st Sess. Introduced in House April 10, 2019.

195. US House of Representatives. *Pipelines over People: How FERC Tramples Landowner Rights in Natural Gas Projects.* Committee on Oversight and Reform. 116th Congress, 2nd sess. December 9, 2020.

195. *Bullock v. Bureau of Land Management*, Case No. 4:20-cv-00062-BMM (D. Montana September 25, 2020).

196. *Rapanos v. United States*, 547 US 715 (US Supreme Court 2006).

197. A. Cantu and R. Galbraith. *Oil and Gas Industry Dominates Federal Agency Responsible for Pipeline Approvals.* Public Accountability Initiative (2017).

198. D. J Trump. *An America First Energy Plan.* Trump White House. March 2017.

199. T. Gardner and D. Chiacu. "Biden to Nominate Democrat Willie Phillips to US Energy Regulator – White House." *Reuters*, September 9, 2021. www.reuters.com/business/energy/biden-nominate-lawyer-willie-phillips-ferc-white-house-2021-09-09.

5

Native American Lands
Respect for Tribes' Rights versus Encroachment

The Trump administration's expansion of oil and gas extraction and infrastructure marked a new wave of assault on not only public and private lands but also Indian lands.[i] In the lower 48 states,[ii] the 337 federally recognized tribes[iii] have their land base on reservations that total 56 million acres (or about 2.3 percent of the US land base).[1,2] Tribes hold hunting, fishing and gathering, and water rights in off-reservation areas across the United States. Tribes also hold lands that are no longer part of reservations.[3,iv] Additionally, tribes continue to have cultural, spiritual and historical ties to lands that they historically occupied but that are presently designated as public lands.[4] Tribes' land base is critical to their assertion of sovereignty, that is, their powers to self-govern,[5,v] to determine their own way of life and to live that life.[6] Many Indian nations' religious and cultural worldviews are tied to particular geographic sites, including their understanding of their origins and their obligations to the land and to other beings.[7–9]

---

[i] The term Indian is used in various treaties and statutes to refer to Native Americans.
[ii] Indigenous peoples in Alaska and the oil and gas industry's operations in Alaska are discussed in Chapter 6.
[iii] Federal tribal recognition acknowledges tribes' rights to certain benefits. Scholars have criticized the rationale and process used by the federal government to grant or exclude Federal tribal recognition [2].
[iv] Indian country is legally defined as all land that has been set aside by the US government for the use of Indians, including Indian reservations, dependent Indian communities and restricted private individual allotments, owned by Indians, that are no longer part of the reservations. Congress terminated a number of reservations during the Termination Era. Indian lands refer to Indian country plus lands on which Indians have rights of way or treaty-based hunting, fishing and gathering rights and public lands that are comanaged by tribes [3].
[v] Tribes exert sovereignty in their reservation and exert civil jurisdiction over natives and non-natives within reservation boundaries. Tribes' reservation lands are not subject to taxation by the federal government. Kalt and Singer traces how self-rule – e.g., making their constitution, choosing their economic pursuits, etc. – is essential for tribes' political survival and economic well-being [5].

Tribes' battles throughout history to safeguard their sovereignty and their rights to lands and resources continue to the present day.[10–14] Encroachment by oil and gas infrastructure and extraction threatens their treaty rights, their lands and resources, and their sovereignty.[15] The oil and gas sector, along with their political supporters, have sought to access oil and gas resources on Indian lands and to construct pipelines and terminals that cross Indian lands.[16] Reservation lands are estimated to include 20 percent of known natural gas and oil reserves as well as 50 percent of potential uranium reserves and 30 percent of the nation's coal reserves west of the Mississippi.[1] Numerous pipelines and LNG and fossil fuel export terminals encroach upon on tribes' reservation and off-reservation lands and ancestral lands. The federal government's historical mismanagement of extraction on reservations, including operations that began as early as the 1800s, has resulted in lost earnings,[vi] worsened tribes' public health and degraded their environment.[17–19] A few tribes have chosen to anchor their economies around oil and gas drilling (Section 5.4.1). However, views vary across and within tribes on trade-offs in pursuing oil and gas extraction.[20]

The Obama administration began a deliberative process to recognize tribes' rights and to rectify a number of past injustices. Obama established the first national monument proposed by tribes – Bears Ears National Monument – protecting a 1.4 million acre parcel of land in southeast Utah from extractive activities.[21] Following domestic and international condemnation of violence perpetuated by the North Dakota state government against water protectors at Standing Rock,[22–24] the Obama administration committed to respecting tribes' treaty rights when reviewing the easement application for the Dakota Access pipeline. The administration also refused to grant the Keystone XL permit after finding that the pipeline was not in the national interest, thus blocking the pipeline that would have crossed Indian lands. Additionally, the administration settled more than a hundred lawsuits against the federal government's misman-agement of tribes' lands, mineral rights and accounts that the government held in trust for tribes.[25,26] As part of the settlement of one of the major lawsuits, *Cobell v. Salazar*, Congress provided $2 billion for a land-buyback program to restore tribal lands.[25,27]

However, Trump and his administration reversed tribes' hard-won successes and exacerbated prior injustices, by thumbing the scales against the protection of Indian lands and in favor of oil and gas extraction and infrastructure on these lands. The Trump administration deliberately disregarded tribal governments'

[vi] The US government facilitated companies securing leases for oil, gas, coal and uranium on tribal lands on poor terms for the tribes. It also failed to collect full royalties on behalf of tribes, to manage those royalties and to police companies' theft of resources [17, 18, 19].

sovereignty and the US government's treaty obligations.[28–30,vii] In December 2017, Trump shrunk Bears Ears National Monument and Grand Staircase Escalante National Monument and opened former monument lands to oil and gas leasing.[31] His administration appointed William Perry Pendley, who had a history of advocating against Native American interests[viii] and against federal management of public lands, as acting head of BLM.[32–34] BLM subsequently approved oil and gas leasing on former monument lands. The administration approved the easement for the Dakota pipeline, permitting construction to proceed. Trump also issued a presidential permit for the Keystone XL pipeline and fast-tracked the Jordan Cove LNG terminal under his purported emergency powers during the COVID-19 pandemic.[35]

The Trump administration supported the privatization of reservations,[16] a strategy, decried by critics, that aims primarily to gain access to tribal lands and resources.[16,29,36,37] The privatization actions harken back to the US government's policies of breaking up reservations and selling tribal lands to non-natives during the Allotment and Assimilation Era from 1871 through 1934.[38,39,ix] In April 2020, the administration revoked the reservation of the Mashpee Wampanoag tribe in Massachusetts and removed these lands from federal status[x] – a move that eerily recalls the federal government's actions during the Termination Era from 1953 through 1969[40,xi] – opening the door for other revocations.[41]

[vii] The Trump administration attempted to abnegate the US government's treaty obligations to provide tribes with healthcare in exchange for their lands. It proceeded to build the US–Mexico border wall that cut through the Tohono O'odham Nation's reservation [28, 29]. Critics have denounced as racist a number of remarks by Trump that allude to Native Americans [28].

[viii] Senator Tom Udall of New Mexico, in his speech to the Senate in September 2020, excoriated Pendley's opposition against Native Americans' rights, noting, "He has disdain for Native Americans – their Tribal sovereignty and their religious practices. . . . He's fought against protecting their sacred sites on federal land. . . . He's questioned the basis of Tribal sovereignty and even Tribal recognition" [34]. Prior to taking office in July 2019, Pendley had served as the head of the nonprofit Mountain States Legal Foundation, which had challenged federal oversight of public lands [33].

[ix] The 1887 Dawes Act, which codified the conversion of Native Americans' communal land-holdings into private property allotments, sought to quash Native American sovereignty and identity (both tied to their land base), to break up communities and to legally appropriate Native Americans' land [38]. Native American lands were slashed from 138 million acres in 1887 to 48 million acres by 1934 [39]. The 1934 Indian Reorganization Act stopped the allotment process and returned some land and mineral rights to select tribes.

[x] In June 2020, the District Court for the District of Columbia ruled that the Department of the Interior's decision to revoke the reservation failed to consider extensive evidence and ordered the Interior to reconsider its decision. In May 2019, the House voted in support of a bill that would affirm the Mashpee Wampanoag tribe's reservation, but the Senate failed to move on the bill.

[xi] In the 1950s and 1960s, the federal government terminated the status of a number of tribes, sold off lands totaling 2.5 million acres, ended tribal affiliation for 12,000 Native Americans, deprived tribes of their land base, and effectively ended tribal sovereignty for terminated tribes [40].

I begin with a brief description of tribes' reservation lands, their off-reservation rights and their limited rights on public lands that are of cultural and historical significance to tribes (Section 5.1). I then discuss the unique challenges tribes face when seeking to protect their lands and resources on unceded and encumbered lands (Section 5.2) and public lands that are of cultural and historical significance (Section 5.3). Next, I return to the larger energy transition facing the United States, detailed in Part I of this book, and describe tribes' efforts in this transition (Section 5.4). I conclude by underscoring the calls by tribes for Congress to act and for the executive branch to fulfill the US federal government's obligations to protect tribes' lands and resources.

5.1 Reservations Lands, Off-Reservation Lands and Ancestral Lands

Sovereign Indian tribes once used and occupied vast swaths of land. Today, tribes' use and occupation of those lands have been restricted to a far smaller collection of reservation lands held in trust by the United States and some treaty-based off-reservation hunting, fishing and gathering rights (Section 5.1.1). Other lands historically occupied by tribes are presently designated as public lands, to which tribes' rights of use are subject to the decision of the US federal government that manages those lands (Section 5.1.2). Understanding how the United States arrived at this configuration of rights is essential to comprehending how tribes have carved out legal and political strategies to protect their lands and resources (Section 5.2).

5.1.1 On-Reservation and Off-Reservation Rights

Reservation and off-reservations rights were reserved by sovereign Indian tribes in treaties signed with the US government. After the end of the treaty-making era, these reservations were reserved by statutes, executive orders or judicial decisions.[42] (Reserved water rights are created by implication at the moment that a reservation is established by Congress, the president or the judiciary.) Although the specifics can vary across reservation lands, title to the land is generally held by the United States while the tribe or tribes hold an "occupancy title."[xii] The lands are held in trust by the federal government. In

---

[xii] The specific legal status depends on the unique treaties governing the specific reservations. For instance, the Seneca Nation holds its lands in restricted fee status. It does not regard the United States as holding the Nation's lands in trust for the tribe. The 1784 Treaty of Canandaigua reserved lands within the specified boundaries as belonging to the Seneca Nation [242].

other words, while tribes have guaranteed occupancy and use rights, the federal government's approval is needed for a number of land-use decisions. The 2012 Helping Expedite and Advance Responsible Tribal Home Ownership (HEARTH) Act provided some autonomy to tribes. Tribes whose leasing regulations have been approved by the Secretary of the Interior do not need approval for individual business and agriculture leases.[43] Nevertheless, the HEARTH Act does not cover leases for exploration, development or extraction of mineral resources.[43]

Property rights within reservations are further broken into a complex patchwork of Indian lands held under trust by the US government, fee simple Indian lands and fee simple non-Indian lands.[8,xiii] In addition to reservation lands, tribes hold treaty-based rights to hunt, fish and gather on off-reservation lands. These are property rights protected under federal law and do not require the tribe to have title to the underlying land.[44]

Article 6 of the United States Constitution declares treaties to be the supreme law of the land. In about 400 treaties signed from 1778 through 1871, tribes ceded significant lands that they had occupied. In exchange, the US government agreed to recognize the boundaries of Indian territories (as delineated in the treaties), to protect tribes' retained lands (i.e., reservations), to uphold tribes' rights to use off-reservation lands for traditional uses such as fishing, hunting and gathering, and to offer other protections.[xiv] Forty-five tribes relinquished certain aspects of their exterior sovereignty but retained their governmental powers over their territories.[45] In 1871, Congress ended treaty-making with Indians,[xv] after which Indian reservations were reserved only through executive orders, congressional acts and judicial decisions.[42,46,47,xvi]

[xiii] As Tsosie notes, the Supreme Court curtailed the ability of tribes to impose uniform regulations in their reservations by curtailing tribal jurisdiction over non-Indian property owners on the reservation [8]. The court ruled in *Brendale* that the Yakima Nation did not have jurisdiction to regulate land use on non-Indian fee land within the "open" area of the reservation and in *Montana* that the Crow Tribe did not have the jurisdiction to regulate hunting and fishing by nonmembers on fee land within the reservation [243, 244].

[xiv] In various treaties, the US government, in exchange for Indian lands, undertook obligations to provide healthcare, schooling, protection for tribes' natural resources, and other protections [37]. The fact that the United States undertook these obligations in various treaties has been underscored by the courts, including in *Worcester v. Georgia*. In 1974, the Supreme Court in *Morton v. Mancari* ruled that "the United States [has] assumed the duty of furnishing ... protection, and with it the authority to do all that was required to perform that obligation" [245].

[xv] As noted in Wilkinson and Volkman, "The legislation declared that no Indian tribe was to be acknowledged as an independent nation with whom the United States could contract by treaty" [254].

[xvi] Rey-Bear and Fletcher note that federal courts "have long held that when it comes to protecting tribal rights against non-federal interests, it makes no difference whether those rights derive from treaty, statute or executive order, unless Congress has provided otherwise" [47][246].

Approximately 9.5 million acres of reservation lands were established by treaty, and 45.5 million acres are recognized under other legal instruments.[42]

The concept that reservation and off-reservation rights are *reserved* by tribes, based on rights they possessed as original sovereigns, was affirmed by the Supreme Court in 1905 in *United States v. Winans*.[48] Tribes reserved their inherent rights as sovereigns in treaties and they retain these rights unless expressly granted or transferred away to the United States. In that case, the court ruled that a treaty reserved the tribe's "right of taking fish at all usual and accustomed places," creating a servitude on the lands relinquished to the United States that continued with the land to any future grantees. In 1908, the Supreme Court in *Winters v. United States*[49] ruled that there was an implied reservation of water rights for Gros Ventre and Assiniboine bands and other tribes in the treaty that established the Fort Belknap Reservation – thus establishing a presumption for the reservation of water rights to tribes on federally reserved lands.

A series of legal decisions transformed the rights of tribes over reservation lands into what is now known as "occupancy title," while the federal government holds the ownership title in trust for tribes. In the 1823 Supreme Court decision in *Johnson v. McIntosh*,[50] Justice Marshall declared that tribes did not hold absolute ownership title to land, but that they had only a "right of occupancy" while the US government owned the title to lands. Justice Marshall based his reasoning on the colonialist Doctrine of Discovery, that is, a European colonial power gains title to lands it "discovers."[xvii] His reasoning ignored the tribes that had long inhabited those lands – a number of which had even developed elaborate property systems.[51–53] The trust relationship between the federal government and tribal nations arose from the Supreme Court's decision in *Worcester v. Georgia*,[45] in which Justice Marshall noted that the relation between the Cherokee Nation and the United States "was that of a nation claiming and receiving the protection of one more powerful: not that of individuals abandoning their national character, and submitting as subjects to the laws of a master."[45,xviii] The decision in *Worcester v. Georgia* also clarified that that relationship between the US federal government and tribes meant that state laws did not apply in Indian country.[xix]

[xvii] The Supreme Court cited the Doctrine of Discovery as recently as 2005 in *City of Sherrill v. Oneida Indian Nation of New York*. Justice Ginsburg wrote that "Under the 'doctrine of discovery . . . ' fee title [ownership] to the lands occupied by Indians when the colonists arrived became vested in the sovereign – first the discovering European nation and later the original states and the United States" [247][255].

[xviii] Wood notes that the trust relationship has been incorrectly framed as based on a guardian–ward relationship, in reference to Justice Marshall's unfortunate use of that analogy in *Cherokee Nation v. Georgia* [45].

[xix] State laws do not apply to tribal members on a reservation unless there is an express federal statutory authorization.

Judicial decisions continued to shape Indian sovereignty and land rights with the development of the doctrine of plenary powers.[54,55,xx] That doctrine, as developed in the 1886 case *United States v. Kagama*[56] and the 1903 case *Lone Wolf v. Hitchcock*,[57] asserts that Congress has plenary powers over Indian Country and over tribal members. (Congress does not have powers over matters regarding the inherent sovereignty of the tribe such as its membership determinations.) Notably, the Supreme Court's highly deferential judicial review of Congress's legislative acts led to Congress's wide breadth of plenary powers.[54,55] In subsequent cases, the Supreme Court ruled that Congress's plenary powers are subject to the provisions of the Due Process Clause and the Just Compensation Clause.[3] It affirmed that Congress's powers over tribal affairs are subject to "pertinent constitutional restrictions" in *Delaware Tribal Business Committee v. Weeks*[58] and *United States v. the Sioux Nation*[59] in 1977 and 1980, respectively.

Even so, Congress's plenary powers have been used to justify numerous actions, such as Congress's abrogation of treaties; its violation of treaty agreements; its divestment of vast landholdings reserved for tribes; its allotment and sale of tribal lands in the Allotment Era; and its termination of federal recognition of tribes in the Termination Era.[42,54,60] In 1903, in *Lone Wolf*, the Supreme Court ruled that Congress's action of allotting and transferring land that tribes had owned as fee simple and had been guaranteed under treaty was not a constitutional taking, but an *appropriate* exercise of federal administrative power over tribal property.[60] More than a century later, in 2020, Justice Neil Gorsuch reiterated the expansiveness of Congress's plenary powers in *McGirt v. Oklahoma*,[61] noting "This Court long ago held that the Legislature wields significant constitutional authority when it comes to tribal relations, possessing even the authority to breach its own promises and treaties."[61]

How courts conceptualize the federal trust responsibility has also influenced the extent to which tribes have been able to hold the US government accountable in its management of tribes' lands and resources. Congress acknowledged, in the Indian Trust Asset Reform Act, the enforceable nature of these trust obligations, noting that

> the fiduciary responsibilities of the United States to Indians . . . are founded in part on specific commitments made through written treaties and agreements securing peace, in exchange for which Indians have surrendered claims to vast tracts of land, which

---

[xx] Newton underscores "that the Constitution does not explicitly grant the federal government general powers to regulate Indian affairs" [55]. Hoffman traces the development of this doctrine from the nineteenth century onwards [54]. She writes that "the Court finally acknowledged that congressional plenary power was not an enumerated constitutional power, making plenary power something of a congressional inherent power even though the Court generally declined to recognize that Congress possessed inherent, unenumerated powers in other contexts. According to the Court, it was simply 'valid' for Congress to legislate in this area."

provided legal consideration for permanent, ongoing performance of Federal trust duties ... [T]he foregoing historic Federal–tribal relations and understandings ... have established enduring and enforceable Federal obligations.[47]

However, courts have not always held the US government to strict standards as a trustee. For instance, in 2011, Justice Samuel Alito in *United States v. Jicarilla Apache Nation*[62] wrote that Congress can choose to shape its "trust" relationship with Indians, without taking on all the fiduciary duties associated with a private trustee in common law. Justice Sonia Sotomayor in her dissent underscored how the Alito approach could open the door for the government to manage the trust in ways incongruent and even contrary to interests of the tribe.[xxi]

Despite these decisions, tribes have achieved a number of notable successes in the courts. The assertion of treaty rights, alongside other legal arguments, has proven successful in a few cases in which tribes sought to protect lands and resources.[xxii] The 2018 Supreme Court decision in *Washington v. United States*,[63] which upheld the decision of the Court of Appeals for the Ninth Circuit, injected optimism for tribes' use of treaty rights to protect their resources and the habitat for those resources.[64,65] That decision concerned the Pacific Northwest tribes' fishing rights as reserved in the Stevens Treaties signed in 1854 and 1855. The decision defined tribes' treaty-based fishing rights to include enforceable rights to protect the fishery habitat and emphasized the state and the federal governments' duty to refrain from actions that damage the natural habitat of tribes' treaty-based natural resources.[65]

Other court decisions have underscored that the federal government's trust responsibility to tribes, alongside treaty rights, offers protections for tribes' lands and resources.[45] For instance, in *Northern Cheyenne Tribe v. Hodel*,[66] the district court rejected BLM's proposal to lease federal lands for coal development just outside the Northern Cheyenne reservation. The court ruled that "a

---

[xxi] Justice Sotomayor also noted that the courts' past decisions had applied common law principles to the federal trust relationship. The Alito majority decision would restrict the application of those principles only to cases when Congress accepted those principles.

[xxii] Asserting treaty rights is particularly difficult for tribes in two common positions. First, tribes could base their rights on one of the 150 treaties for which the Senate did not approve a resolution for ratification. Although tribes had signed these treaties with US Treaty Commissions, they are not viewed as valid under US law. Second, whether or not treaty rights were terminated during the Termination Era can be unclear for some tribes. On the one hand, the Court of Claims ruled in 1967 that the Menominee have the same hunting and fishing rights free from state regulation on their former reservation as they had before termination [248]. On the other, the Court of Appeals for the Ninth Circuit ruled in 1964 that those Klamaths maintaining their tribal membership after termination lost all special fishing and hunting privileges on the former reservation [256]. Nevertheless, that court took a different position in 1974, ruling that tribes terminated pursuant to the Klamath Termination Act retained their treaty hunting, trapping and fishing rights [257, 258].

federal agency's trust obligation to a tribe extends to actions it takes off a reservation which uniquely impact tribal members or property on a reservation." In *Northwest Sea Farms v. US Army Corps of Engineers*,[67] a district court upheld the Corps' decision that the federal trust responsibility prohibits its permitting of activities that would have more than a de minimis impact on tribes' treaty fishing rights.[45]

Tribes face at least two hurdles in asserting treaty rights to contest the actions of the federal government (or those of state governments or private parties), The first is, as discussed earlier, the judiciary's interpretation that a wide range of legislative actions by Congress (even when these adversely affect tribes) represents Congress's exercise of its plenary powers. The second hurdle is for tribes to establish that the treaty has not been abrogated. On a more encouraging note, the Supreme Court has set some requirements for concluding that Congress has abrogated treaties or extinguished treaty rights. In 1968, the Supreme Court ruled in *Menominee Tribe v. United States*[68] that express congressional authorization is required for the abrogation of Indian treaty rights. In that case, the court also underscored that tribes retain treaty-based hunting rights even when Congress terminated the tribes' status.[69] In 1999, the Supreme Court ruled in *Minnesota v. Mille Lacs Band of Chippewa Indians*[70] that the Chippewa retain hunting, fishing and gathering rights on lands, rivers and lakes that it had ceded to the federal government in the 1837 Treaty of St. Peters – despite allegations that those rights were eliminated by an executive order in 1850 and a subsequent treaty with the tribe in 1855. In 2019, the Supreme Court held in a 5–4 opinion in *Herrera v. Wyoming*[71] that a treaty reserving hunting rights on federal lands was not extinguished by Wyoming's statehood. In 2020, the Supreme Court held in another 5–4 opinion in *McGirt v. Oklahoma*[61] that Congress never dissolved its original treaty with the Muscogee (Creek) Nation, and thus the Creek Reservation located in most of eastern Oklahoma remained reservation lands.[xxiii]

5.1.2 Public Lands That Overlap with Tribes' Ancestral Lands

In addition to their reservation lands and their rights on off-reservation lands, tribes have sought to protect their ancestral lands, that is, lands with which

---

[xxiii] In an 1832 treaty between the United States and the Creek Nation, the Creek Nation ceded their lands to the east of the Mississippi River in exchange for a guarantee of the Creek country west of the Mississippi River, and an 1833 treaty delineated the boundaries of the Creek reservation in present day Oklahoma. In 1838, in the Trail of Tears, the US government then forced the members of some tribes, including the Creek, to march more than 1,000 miles to Oklahoma. Beginning in 1893, Congress allocated the reservation lands and sold lands to non-natives.

tribes continue to have cultural, religious and historical ties and that tribes had historically occupied.[9] Tribes were forcibly removed from their ancestral lands to make room for white settlers, extractive and infrastructure projects, conservation, and, ironically, the preservation of Indian historical sites.[72,73] Today, significant shares of tribes' ancestral lands are designated as public lands, owned by the United States. As discussed in Chapter 4, these lands are managed by BLM and the US Forest Service for multiple use, and by other federal agencies such as the National Park Service and the Fish and Wildlife Service.[74]

Tribes have limited property-based legal avenues to protect these lands. Tribes have rarely been able to repatriate these public lands into tribal trust lands.[4,75,xxiv] The aboriginal title doctrine,[xxv] which recognized the claims of Indian nations, as the original habitants of the land, to legal protection, has gradually narrowed to the judiciary's failure to accord protections for aboriginal title against takings by the federal government.[76,77] Early Supreme Court decisions such as the 1835 case of *Mitchel v. United States*[78] recognized that Indians' occupancy of their aboriginal lands is "as sacred as the fee simple of the Whites."[76,77,79] A number of cases affirmed that aboriginal title does not rest on treaty, state or other formal government actions, including *Cramer v. United States* in 1923 and *United States ex rel Huaipai Indians v. Santa Fe Pacific Railroad* in 1941.[80,81] As recently as 1946, Justice Vinson in *United States v. Alcea Band of Tillamooks*[82] wrote that "the Indians' right of occupancy has always been held to be sacred; something not to be taken from him except by his consent, and then upon such consideration as should be agreed upon." Vinson further underscored that aboriginal title was accorded the protection of complete ownership (against all but the sovereign). Indeed, tribes mounted successful property-based actions, such as trespass, against state governments and private parties that infringed upon aboriginal lands.[77]

However, the 1955 Supreme Court decision in *Tee-Hit-Ton Indians v. United States*[83] severely narrowed protections for aboriginal lands from takings by the federal government.[76,77] The Supreme Court ruled that aboriginal lands, unless recognized by Congress through treaty or unambiguous legislation, are not accorded protections under the Fifth Amendment Takings and Just

---

xxiv Rarely has Congress returned ancestral lands, designated as public lands, back to tribes. In 1970, the Taos Pueblo's repatriated Blue Lake, which was taken from the Pueblo in 1906 when President Theodore Roosevelt established the Taos Forest Reserve, which eventually became the Carson National Forest. President Nixon signed House Resolution 471 that restored 48,000 acres of land to the Taos Pueblo [8]. In 2019, the Western Oregon Tribal Fairness Act restored 17,000 acres of public land to the Cow Creek Band and 15,000 acres to the Confederated Tribes of Coos, Lower Umpqua and Siuslaw Indians [4].

xxv See Newton for detailed discussion of aboriginal lands [76].

Compensation Clause.[xxvi] The court reasoned that aboriginal title represents the sovereign's permission for tribes to occupy lands and entails nothing more than "mere possession not specifically recognized as ownership."[76,77] The 1988 Supreme Court decision in *Lyng v. Northwest Indian Cemetery Protective Association*[84] further exalted the US government rights to use public lands without offering protections for tribes' sacred sites. Justice Sandra Day O'Connor wrote that "Whatever rights the Indians may have to use of the area ... those rights do not divest the Government of its right to use what is, after all, *its* land."

Therefore, tribes have relied on other avenues for protecting these lands, while securing their use of these lands for religious, spiritual and cultural ceremonies. These avenues include securing presidential proclamation of national monuments and congressional designation of national parks and wilderness.[xxvii] However, vast areas of public lands of significance to tribes are designated simply as public lands where tribes are treated as one among many stakeholders.

5.1.3 Challenging Oil and Gas Infrastructure on Indian Lands

Tribes exercise the most control over proposed or existing infrastructure projects on reservation lands. Under the Indian Right-of-Way Act, the secretary of the interior's decision to grant an easement is subject to tribes' consent.[64] Several tribes have chosen not to approve easements for pipelines that have come up for renewal. For instance, in 2017, Bad River Band of the Lake Superior Tribe of Chippewa Indians in northern Wisconsin voted not to renew an easement for the Enbridge Line 5 pipeline that crosses 200 miles of its reservation.[85,86,xxviii] In 2020, the Court of Appeals for the Ninth Circuit in *Swinomish Indian Tribal Community v. BNSF Railway Company*[87] upheld a district court ruling that barred the railway company from using tracks on Swinomish Indian Tribal Community reservation land to carry crude oil from North Dakota's Bakken formation.[xxix] The company had violated the right-of-way agreement, which limited the number of railcars that could use the tracks daily.

[xxvi] In a few cases, Congress enacted legislation that permitted Alaska Native nations to sue in the federal claims courts for the taking of aboriginal property interests [76].

[xxvii] The secretary of the Interior has the authority to add some types of rivers to the National Wild and Scenic Rivers System. Both the secretary of the Interior and the secretary of Agriculture have the authority to add certain types of trails to the National Trails System [74].

[xxviii] Notably, in November 2020, the governor of Michigan, citing risks to Michigan waters from a catastrophic oil spill, revoked an easement granted by the state of Michigan for a different segment of that pipeline through the Straits of Mackinac [86].

[xxix] Railcars carrying Bakken crude oil have exploded and caused fatalities. For instance, in July 2013, a derailment and explosion in Lac-Mégantic, Quebec caused 47 fatalities, destroyed almost all downtown buildings and contaminated lands and waters [249].

Within the reservation boundaries, the checkerboard of land ownership may include fee simple lands owned by nonmembers of the tribe. Arguably, the tribal government could also block oil and gas infrastructure projects that cross fee simple land owned by nonmembers of the tribe within reservation boundaries.[64] Tribes may regulate non-Indians on non-Indian lands within the reservation if their "conduct threatens, or has some direct effect on the political integrity, the economic security, or the health and welfare of the tribe," according to the Supreme Court's 1981 decision in *Montana v. United States*.[88] Oil and gas pipelines and crude-oil-laden railcars that pass through reservations impose risks on the health and safety of tribes, thus falling under the purview of tribal oversight.[64]

However, tribes face more legal hurdles in opposing proposed oil and gas infrastructure on lands where they hold off-reservation rights. Federal agencies, per various laws governing energy, infrastructure and the environment, are responsible for permitting various aspects of these infrastructure projects. Tribes have asserted their treaty rights alongside legal challenges based on statutes that govern the approval of infrastructure projects and the Administrative Procedure Act (APA). (Section 5.2 provides a more detailed discussion of these particular types of legal challenges.) When extraction projects are proposed on public lands of cultural significance to tribes, tribes have focused their actions on statutes that govern leasing, the National Historic Preservation Act and the APA (discussed in more detail in Section 5.3). Specific lawsuits have also been brought alleging that the executive branch overstepped its power (i.e., challenges to the presidential permit granted to the Keystone XL pipeline and to the president's shrinking of Bears Ears National Monument and of Grand Staircase Escalante National Monument).

Federal agencies face additional responsibilities to consult[xxx] with tribes under statutes such as the National Environmental Policy Act and the National Historic Preservation Act. However, the Government Accountability Office has detailed inadequacies in the consultation process, including the lack of consideration of tribes' input into the final decision-making process.[89] Consultation has been further diminished by federal agencies that have extended consultation to those tribes that are recognized by states but not to those recognized by the federal

---

[xxx] The duty to consult fails short of the United Nation's General Assembly's 2007 Declaration of Rights of Indigenous Peoples, which includes the duty of states to obtain indigenous peoples' free, prior, and informed consent on projects that affect them. In 2010, Obama announced the United States' qualified support for the declaration by noting "the United States understands [the importance of a] call for a process of meaningful consultation with tribal leaders, but not necessarily the agreement of those leaders, before the actions addressed in those consultations are taken" [250].

government.[90,xxxi] For instance, the Federal Energy Regulatory Commission, which approved the Atlantic Coast pipeline in October 2017, was criticized for not undertaking in-depth consultation with the Lumbee, Coharie and Haliwa-Saponi tribes, whose contemporary and traditional areas are affected by that proposed pipeline.[90]

5.2 Legal Challenges on Unceded or Encumbered Lands

Tribes face uphill battles to challenge oil and gas infrastructure on lands where land rights are disputed. The US government's failure to honor treaties (i.e., treaties that had delineated tribes' reservation lands and off-reservation rights) and Congress's imposition of new land schemes on reservation lands and on lands where tribes hold off-reservation rights has created large swaths of land over which ownership is disputed.[xxxii] Tribes argue that many lands where treaties recognized reservation and off-reservations rights remain unceded, that is, the sovereign tribal nation possessed those lands and never ceded those lands to the United States. Tribes also argue that their unceded off-reservation rights continue to encumber lands which the US government claim to own outright and that the US government is legally bound to honor tribes' off-reservation rights. (Under US property common law, property can be owned by one party while another may still maintain an interest or right in that property – e.g., hunting, fishing and gathering rights. These lands are sometimes referred to as "encumbered.")

Tribes' arguments that reservations lands and off-reservations rights were never ceded are often based on specifications within treaties of the conditions that had to be met for the modification to the terms of the treaties, i.e., terms within the treaties to which the US government had agreed in signing those treaties. Judicial doctrine that Congress has the plenary powers to abrogate treaties jettisons tribes' legal arguments that would routinely apply under US contract law, and instead recasts the legal question to whether Congress did (or did not) abrogate the treaties and extinguish tribes' rights. Tribes' political and legal opposition to some of the most glaring actions by the Trump administration that facilitated oil and gas infrastructure on these types of lands are discussed below, including the Dakota Access pipeline (Section 5.2.1), the

---

[xxxi] For instance, the route of the Atlantic Coast Pipeline, which received approval from FERC, crossed traditional and contemporary areas of three tribes: the Lumbee, Coharie and Haliwa-Saponi [90].

[xxxii] See the discussion about *McGirt v. Oklahoma* (Section 5.1.1). Congress established the Creek reservation in eastern Oklahoma but subsequently decided to allot lands and to sell them to non-Native Americans.

Keystone XL pipeline (Section 5.2.2) and the Jordan Cove LNG terminal and related gas pipeline (Section 5.2.3).

5.2.1 Dakota Access Pipeline

The Dakota Access pipeline runs just north of the Standing Rock Sioux and Cheyenne River Sioux reservations. Construction of the pipeline began in 2015, with operations starting shortly thereafter in June 2017. It transports oil from the Bakken Shale in North Dakota through South Dakota and Iowa to an oil tank farm in Illinois. The 1,134-mile pipeline crosses the Missouri River at several points and Lake Oahe (Maps 5.1 and 5.2). Tribes on these reservations opposed the pipeline because the pipeline crosses the Missouri River, their

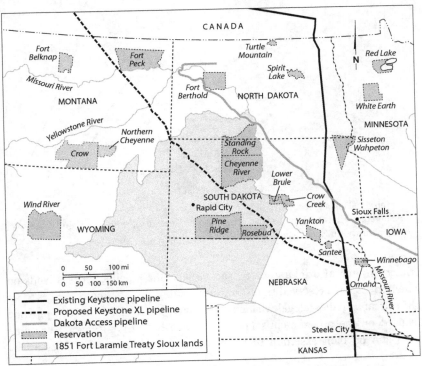

Map 5.1 Native American reservations, the Dakota Access pipeline and the proposed Keystone XL pipeline

Source: Redrawn from various maps by the Bureau of Indian Affairs,[91] US Department of State,[92] the Climate Alliance Mapping Project, Keystone Mapping Project and Indigenous Environmental Network[93] and Carl Sack.[94]

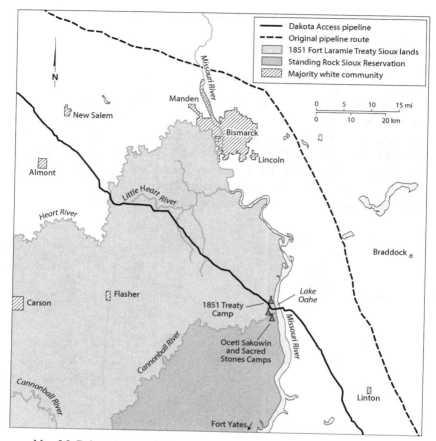

Map 5.2 Dakota Access pipeline
Source: Redrawn from various maps by the Bureau of Indian Affairs,[91] US Department of State[92] and Carl Sack.[94]

primary source of drinking water, and because the pipeline crosses lands on which tribes have reserved rights to water, fishing and hunting resources in an 1851 treaty. Tribes' legal challenges against the Dakota Access pipeline focused on the government's failure to abide by NEPA (which I examine first) and on their assertion of treaty rights (which I examine later in the section).

Even on disputed lands, tribes have some procedural rights to voice their concerns over encroaching pipelines – however, these challenges are often drawn out and there is minimal guarantee of recourse if those procedural rights are not met. The requirement to consult tribes in these instances is tied to

NEPA, which is triggered by major actions by the federal government, including the issuing of permits. In this case, pipeline construction required a 408 permit for crossing a US waterway,[xxxiii] which was reviewed by the Corps. Prior to issuing the permit in July of 2016, the Corps conducted an environmental assessment.[xxxiv] Per the environmental assessment, the pipeline would cross Lake Oahe "approximately 0.55 miles upstream from the Standing Rock Sioux Reservation and seventy miles upstream from the Cheyenne River Sioux Reservation."[95] However, following protests from tribes and criticisms from other federal agencies, the Corps announced it would undertake a more comprehensive EIS, as required under NEPA for "major projects," and consider alternative routes for the pipeline.[96] Under the Trump administration. the Corps slow-walked the environmental assessment/EIS procedure under NEPA while the pipeline developers continued construction and the pipeline eventually began operations as planned.

Tribes asserted their procedural rights to be consulted during the EA and EIS processes. They had a short-lived victory in 2017 when the District Court for the District of Columbia ordered the pipeline be shut down while the Corps revised the flawed EIS. Judge Boasberg ruled that the Corps failed to meet its obligations under NEPA when it ignored the impact of a potential oil spill on the tribes' aquatic and hunting treaty rights (augmented by ample historical examples of oil spills on similar pipelines, see Table 5.1) and when it failed to address environmental justice concerns.[15,97,98] The case was remanded back to the Corps.

However, in January 2017, the Trump administration published an executive order instructing the Corps "to review and approve in an expedited manner"[xxxv] the permit for the Dakota Access pipeline.[106] In February 2017, the Corps terminated its work on the EIS[107] and granted an easement for the pipeline under Lake Oahe.[108] Shortly afterward, the governor of South Dakota ordered the clearance of the camp in Standing Rock,[109] where water protectors had mounted teach-ins and protests.[23]

In July 2020, Boasberg again ordered that the pipeline stop operations until the Corps has completed the EIS.[110] Boasberg's decision highlighted more

---

[xxxiii] The Dakota Access pipeline did not require federal permits for most of its path where it crossed private lands.

[xxxiv] Under NEPA, federal agencies must prepare an environmental assessment of a proposal, its impacts, and alternatives. The agencies may conclude that there is a finding of no significant impact or that a more in-depth EIS is needed. The EIS provides a far more comprehensive discussion of the impacts and alternatives and a hard look at the cumulative impacts from the proposal and existing and future developments in that geographical area.

[xxxv] The executive order instructed the Corps to consider rescinding or modifying the Corps' December 2016 decision on the pipeline crossing at Lake Oahe; withdraw the notice of intent to prepare the EIS; and regard the environmental assessment issued in July 2016 as meeting the requirements of NEPA.

Table 5.1 *Selected oil spills from pipelines owned or managed by Energy Transfer Partner, TC Energy, Enbridge and their subsidiaries*

Pipelines owned and operated by Energy Transfer Partner (or its subsidiaries) Energy Transfer Partner is the developer of Dakota Access pipeline	In September 2016, the Permian Express pipeline leaked more than 33,000 gallons of oil in Texas and crews took 12 days to find the leak. Four months earlier, the PHMSA had sought a $1.3 million fine against the builder, a subsidiary of Energy Transfer Partners LP, citing subpar welding on the pipe that could lead to a leak, but company attorneys claimed that the pipe was sound.[99] In 2018, the Revolution pipeline, developed by Sunoco, a subsidiary of Energy Transfer Partner, exploded near a subdivision in Pennsylvania after a landslide. The Pennsylvania Department of Environmental Protection imposed a $31 million fine and began a federal grand jury investigation.[99] In 2018, a crude oil spill sprayed a mist of oil in an Oklahoma City suburb. State regulators issued a $1 million fine against Sunoco. Months before, a contractor had alerted the company about problems with the system to prevent corrosion in the pipe.[99] Pipelines operated by Energy Transfer Partner and its subsidiaries experienced 349 leaks, spills and other accidents from 2012 through 2016, including 19 groundwater contamination sites in Texas.[99]
The Keystone pipeline owned by TC Energy, developer of Keystone XL	In September 2017, 407,000 gallons of oil were spilled in Amherst in northeast South Dakota.[100,101] The company initially understated the spill as half of its true size. In October 2019, 383,000 gallons of oil were spilled into wetlands near the town of Edinburg in northeast North Dakota.[102]
Line 3 and Line 6B pipeline owned by Enbridge, developer of the Line 3 replacement pipeline in Minnesota that crosses the Ojibwe's treaty lands.	In March 1991, the original Line 3 pipeline spilled 1.7 million gallons in Grand Rapids, Minnesota, the largest US inland spill as of 2020.[103] In July 2010, 1 million gallons spilled from Line 6B into the Kalamazoo River in Marshall, Michigan. In 2016, the EPA fined Enbridge $61 million for that spill.[104] In 2005, Enbridge learned that the very section of the pipeline that eventually failed was cracked and corroded; they also discovered about 15,000 other defects in the 40-year-old pipeline. Enbridge decided not to excavate the pipeline to inspect it.[105]

flaws in the EIS, including "that the pipeline's leak-detection system was unlikely to work, that it was not designed to catch slow spills, that the operator's serious history of incidents had not been taken into account, and that the worst-case scenario used by the Corps [for a spill] was potentially only a fraction of what a realistic figure would be"[97] (see Table 5.1).

In August 2020 that order was reversed at least temporarily by a three-judge panel in the Court of Appeals for the District of Columbia.[111] The Corps initiated in September 2020 a more thorough, court-ordered environmental review of the pipeline that could take until March 2022. The Standing Rock Sioux tribe and environmental groups contend the pipeline is operating illegally and must be closed while the environmental review is ongoing.[112] However, in reviewing the case again in May 2021, Boasberg ruled that the pipeline could continue to operate while the Corps was undertaking the environmental review.[xxxvi] He noted that "The Court acknowledges the Tribes' plight, as well as their understandable frustration with a political process in which they all too often seem to come up just short."[113]

The Standing Rock Sioux tribe and Cheyenne River Sioux tribe argued that the pipeline violated their treaty rights as well, focusing on the land in and around Lake Oahe. In December 2016, the Department of the Interior's solicitor wrote a legal opinion largely corroborating the tribes' assertions, underscoring their treaty rights to land in and around Lake Oahe and calling for government-to-government consultations.[95] The legal opinion stated that "portions of Lake Oahe are located on both the Standing Rock and Cheyenne River Sioux Reservations, and that the Tribes retain treaty hunting, fishing, and reserved water rights in the Lake."[xxxvii]

That legal opinion also concurred with tribes' concerns that the pipeline threatened their waters. The Corps had rejected the initial pipeline route that passed 10 miles from the city of Bismarck on the grounds that the pipeline posed unacceptable risks to Bismarck's water supply[95] but concluded that the pipeline did not pose risks to the tribes' water resources. As noted in that legal opinion, "the reasons for rejecting the Bismarck route are equally (if not more) applicable to the Lake Oahe route."[95] In July 2016, the EPA had also warned that crossings of the Missouri River threatened drinking water for much of North and South Dakota and the tribal nations and that the project raised environmental justice concerns.[114] Thus, the EPA recommended the Corps consider alternative routes. As noted by Robert Bell,[115] the failure of the Corps to give ample protections to the Sioux tribe runs counter to Congress's

---

xxxvi The Court of Appeals for the DC Circuit ruled that the district court judge had applied the wrong test in his earlier decision to order the shutdown of the pipeline. The level of harm from leaving the pipeline in operation would need to rise to a significant enough level for the district court to order the shutdown of the pipeline.

xxxvii In 1944, Congress enacted the Flood Control Act and related statutes that authorized limited takings of Indian lands for specific hydroelectric and flood control dams on the Missouri River in North and South Dakota. The four statutes that applied to the Standing Rock, Cheyenne River, and Lower Brule reservations all recognized the tribes' right to "hunt and fish in and on the aforesaid shoreline and reservoir."

affirmation in the Act of 1889, section 14 that "the United States would protect the water supply of the Standing Rock Sioux, by making the Sioux a riparian proprietor to the water on, near or adjacent to the Standing Rock and the other Sioux Reservations."[xxxviii]

While the solicitor-general's legal opinion acknowledged several of the tribes' treaty rights, the Sioux tribes also contend that the pipeline, in reality, traverses a large expanse of land to which the tribes hold additional treaty rights. Specifically, the tribes assert treaty hunting rights under the 1851 and 1868 Fort Laramie Treaties on lands between the Heart River and the Cannonball River, known as Article 16 lands.[23,115–119] The 1851 Fort Laramie Treaty reserved territory for the Great Sioux Nation, with its boundary at the Heart River in North Dakota.[115,118] The subsequent 1868 Fort Laramie Treaty reserved a smaller land area for the reservation. It also reserved the Sioux's rights under the 1851 treaty, that is, the "right to hunt on any lands north of North Platte, and on the Republican Fork of the Smoky Hill river."[59] Article 16 of the 1868 treaty designated lands north of the permanent reservation as "un-ceded Indian territory."[59]

The tribes also contend that the Sioux Nation never ceded their rights to these Article 16 lands, which the Dakota Access pipeline does cross.[xxxix] The US government never secured the votes of three-fourths of the adult male tribal members that was required under the 1868 treaty for "the cession of any portion or part of the reservation." In 1877, Congress enacted a law to appropriate the Black Hills, a sacred area within the Sioux Reservation (where the US government, in defiance of the 1868 treaty, failed to remove gold miners), and to move the boundary of the Great Sioux Reservation from the Heart River to the Cannon Ball River.

Nick Estes, a professor of American Studies, and Jeffrey Ostler, a professor of History, have argued that a 1978 Indian claims court (ICC) decision and a 1980 Supreme Court decision lend support to the Sioux tribes' contention that they never ceded the Article 16 lands.[23,115,118] The 1978 ICC decision recognized that the northern boundary of the unceded Article 16 lands is the Heart River, that is, the same boundary recognized in the 1851 treaty. The 1980 Supreme Court decision, in *United States v. the Sioux Nation*,[59] upheld the ICC decision on Congress's illegal taking of the

---

[xxxviii] The act broke the 1877 Sioux Reservation into even smaller reservations.
[xxxix] In 1876, Congress enacted the sell or starve provision in its appropriations bill, which would withhold food rations unless the Indians relinquished their rights to the hunting grounds outside the reservation and ceded the Black Hills to the United States. The Manypenny Commission dispatched by Congress threatened the Sioux with starvation and with their removal from the Northern Plains to Indian territory, located in today's state of Oklahoma.

Black Hills.[xl] Estes and Ostler argue that the Supreme Court reasoning on Congress's illegal taking of the Black Hills would apply to the tribes' rights to the Article 16 lands and, in particular, to those Article 16 lands still held by the US government.[118]

5.2.2 Keystone XL Pipeline

The Keystone XL pipeline case presents another example of encroachment on disputed tribal lands. The pipeline would transport tar sands oil over 1,200 miles from Alberta, Canada (and also from the Bakken Shale in North Dakota and Montana) through Montana and South Dakota, to Steele City, Nebraska and then connect to existing pipelines that reach to the Gulf Coast. Within South Dakota, it would run southwest of the Cheyenne River Sioux Indian Reservation and north of the Pine Ridge Reservation of the Oglalla Lakota. It would also pass close to the Rosebud Indian Reservation, and, according to the tribe, it would in fact cross Rosebud's mineral estate.[120] Tribes again asserted their treaty rights to challenge the pipeline but in this case ultimately found more success from legal claims based on abuses of power by the executive branch and on violations of the APA. Pipeline construction began in April 2020 but was halted in July 2020 after the Supreme Court upheld a lower-court ruling requiring construction halt.

Tribes oppose the pipeline, underscoring how it traverses a large expanse of their treaty lands under the 1851 and 1868 Fort Laramie treaties (Map 5.3). Tribes' arguments on why their treaty claims remain valid under US law were described in Section 5.2.1. Tribes' concerns were backdropped by a larger debate as to the implications of the pipeline for climate change and international relations, pointing to the particularly polluting extraction of tar sands oil and the lack of benefit for the United States in allowing the pipeline to cross its borders. As described below, these conditions – although pivotal parts of the legal challenges to the pipeline – left the fate of their lands in the hands of Congress and the executive branch.

The Keystone pipeline, unlike the Dakota pipeline, requires a presidential permit, needed for crossing the US border (which falls under the purview of the State Department). In 2012, the State Department under the Obama administration denied the permit application by TransCanada that had been filed in 2008. Congress attempted to subvert the department's decision via the

---

[xl] Specifically, the court ruled that Congress's taking of the Black Hills in 1877 violated the Fifth Amendment of the Constitution that prohibited the taking of property without just compensation and ordered the government to pay $17.5 million for the land, plus interest. The tribes have refused money compensation and continue to insist on the return of the Black Hills [42].

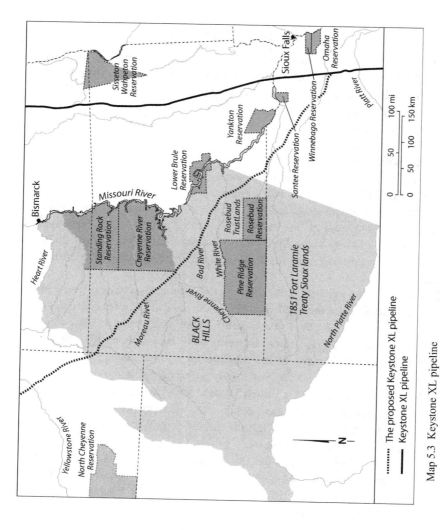

Map 5.3 Keystone XL pipeline
Source: Redrawn from various maps by the Bureau of Indian Affairs,[91] US Department of State[92] and the Climate Alliance Mapping Project, Keystone Mapping Project and Indigenous Environmental Network.[93]

Keystone Pipeline Approval Act,[121] which had been passed by the Republican-controlled House and Senate to authorize the pipeline construction. President Obama vetoed the bill in February 2015[xli] and the Senate subsequently failed to override that veto.[122] In November 2015, Obama rejected TransCanada's permit application that it had filed in 2012, citing the State Department's 2015 Record of Decision that the pipeline was not in the national interest. Specifically, the pipeline would undercut US credibility on climate diplomacy but would provide no long-term contribution to the US economy or energy security. However, this victory was short-lived.

The Trump administration disregarded the findings of the previous administration regarding both tribal and environmental concerns. In January 2017, Trump issued a memo inviting TransCanada (renamed TC Energy) to reapply for the construction permit and for the State Department to make a determination on the permit.[123] In March 2017, the State Department issued the permit with Trump's approval,[122] but a district court subsequently ruled that this permit failed to meet NEPA requirements. In March 2019, Trump issued a presidential permit authorizing the construction of the pipeline. A month later, he issued Executive Order 3867,[124] which asserted that the president has sole authority to issue, deny or amend permits for cross-border pipelines.[xlii] That same month, TC Energy began construction on the pipeline, starting in northern Montana, and projected that the pipeline would begin operations in 2023.[125,126,xliii]

In the meantime, tribes had filed several legal challenges: the first focuses on tribes' treaty rights, a second alleges that the president overstepped his power, and the third established, successfully, that the State Department violated the APA with its approval of the pipeline's construction permit. In November 2018, the District Court for the District of Montana ruled in *Indigenous Environmental Network v. Department of State*[127] that the State Department's 2017 decision, which had granted the construction permit, violated the APA. Judge Brian Morris ruled that the State Department failed to provide a reasoned explanation for its new position that the pipeline was in the national interest, contradicting its previous decision under the Obama administration. Moreover, the State

---

[xli] President Obama argued that Congress was attempting "to circumvent longstanding and proven processes for determining whether or not building and operating a cross-border pipeline serves the national interest" [251].

[xlii] That executive order rescinded previous executive orders that had set up the prior procedures for approving permits, i.e., EO 11423 issued by Lyndon Johnson in 1968 and EO 2423 issued by G. W. Bush in 2004.

[xliii] TC Energy began construction after Premier Jason Kenney of Alberta agreed to invest $1.1 billion of his province's funds to the project and to guarantee $4.2 billion of loans. Kenney acknowledged that he was aware of the political risks on that investment but argued that construction would create "facts on the ground" and make it more difficult for Biden, should he be elected, to cancel the project [126].

Department had impermissibly disregarded the 2015 decision's "prior factual findings related to climate change."[128,xliv] The ruling required the State Department to complete a supplement to the 2014 Supplement EIS that would comply with NEPA and the APA. The court noted that the State Department would need to pay greater attention to the pipeline's impacts on cultural resources, endangered species and climate change, and the impact of changed oil prices on the commercial viability of the pipeline.[129] Even though the Court of Appeals for the Ninth Circuit ruled in June 2019 that that district court's decision was moot (as Trump had revoked the State Department permit), the win remained significant in underscoring that agencies must provide a reasoned explanation for their decision and must address the factual record.

Faced with those basic requirements for agencies' decision-making, the Trump administration switched strategy and, as discussed earlier, issued the March 2019 presidential permit for the Keystone XL pipeline. In December 2019, Judge Brian Morris of the District Court for the District of Montana ruled that the tribes' challenge against the presidential permit could proceed on its merit. In the first case, *Rosebud Tribe v. Trump*,[130] Judge Morris ruled that the Rosebud Sioux Tribe and the Fort Belknap Indian Community "alleged sufficiently their treaty and statutory claims at this point in the litigation." The tribes argued that the presidential permit was unconstitutional because the United States failed to meet its fiduciary duties owed to the tribes under several treaties, including the 1851 Fort Laramie Treaty, the 1855 Lame Bull Treaty[xlv] and the 1868 Treaty of Fort Laramie, as well as duties under statutes such as NEPA.[131]

In the second case, *Indigenous Environmental Network v. Trump*,[132] Judge Morris ruled that the plaintiffs, the Indigenous Environmental Network and North Coast Rivers Alliance, pleaded plausible claims that Trump's issuance of the permit overstepped presidential powers and violated the Foreign Commerce Clause and the Property Clause.[132] The Constitution grants powers to Congress to regulate foreign commerce and to manage federal lands. Accordingly, Congress had approved a State Department review process, in

---

[xliv] The 2017 Record of Decision largely tracked the earlier 2015 Record of Decision except for the section on "Climate Change-Related Foreign Policy Considerations." The 2017 decision acknowledged that the 2015 decision concluded that the pipeline would undercut US credibility on climate diplomacy, but it stated that under the changed global context, approval of the pipeline would not undermine US foreign policy objectives. It should be noted that even the 2015 assessment has come under challenge. The 2015 assessment stated that the Keystone XL pipeline itself would not have significant impact on climate change. It reasoned that the oil from the tar sands fields would be transported through some other means and be combusted if the pipeline were not built. Auffhammer argues that without the Keystone XL pipeline and with other proposed pipelines completed, the insufficient pipeline capacity would effectively keep a substantial portion of tar sands in the ground and not combusted [252].

[xlv] The 1855 Lame Bull Treaty recognized a common area for buffalo hunting for a number of Indian nations.

which the department consults with other agencies and makes a national interest determination. Congress had demonstrated its intent to regulate cross-border pipelines under the Foreign Commerce Clause by passing legislation in 2011 that set a deadline for the State Department to make a decision on the Keystone XL pipeline permit and by passing a 2015 bill that would have approved the Keystone XL pipeline. Obama vetoed the 2015 bill.[132,xlvi]

In addition to the myriad of lawsuits described above, other legal challenges were brought based on NWP-12 and its approach of fast-track approval of pipelines without requiring pipeline developers to undertake a thorough environmental review and public consultation process.[xlvii] These challenges ultimately led to the halting of construction, at least pending a proper review of its construction permit. NEPA mandates the consideration of the cumulative impacts of a project. By taking the unlawful approach of considering each water crossing of the pipeline as a separate crossing, the Corps and pipeline developers could claim that each pipeline segment did not affect more than an acre of US waters and thus ignore the substantial cumulative impacts of the pipelines. Environmental groups challenged the Corps' issuance of the NWP-12 permit for pipelines generally and for the Keystone XL pipeline permit specifically.[133] In April 2020, the District Court for Montana ruled in *Northern Plains Resource Council v. US Army Corps of Engineers*,[134] that the Corps' NWP-12 permit process violated the Endangered Species Act. The court vacated the permit and prohibited the Corps from using this fast-tracked approval process for Keystone XL and other pipeline projects. In July 2020, the Supreme Court denied a request from the Corps and TC Energy that would allow for the Keystone XL pipeline to continue construction, pending the review of its NWP-12 permit.[135] (The Supreme Court did, however, permit construction to continue for other new oil and gas pipelines that relied on the NWP-12 permit program, pending the review of that program.)

5.2.3 Jordan Cove LNG Project

The proposed Jordan Cove LNG Terminal and the 230-mile Connector pipeline in Oregon demonstrates another encroachment on tribes' treaty rights, that is,

---

[xlvi] Congress in a provision under the Temporary Payroll Tax Cut Continuation Act of 2011 set a 60-day deadline for the State Department to make a decision on the Keystone XL pipeline permit.

[xlvii] Under Section 404 of the CWA, the Corps review permits pipelines that cross US waters. The Corps authorized the NWP-12 program in 2012 and reauthorized the program in 2017. Environmental groups objected to the program, which enabled the Corps and pipeline developers to avoid thorough environmental review and public input processes for proposed pipelines.

fishing rights. This LNG terminal project may well be a harbinger of other fossil fuel terminal projects as fossil fuel producers in the United States and Canada look to export their products to Asia.[136] The project, developed by the Canadian firm Pembina, aimed to export LNG to Asia using gas sourced from either Canada or the western United States. Tribes in Oregon and California have opposed the project, citing the adverse impacts on their traditional fishing areas.[137]

In March 2020, FERC made a 2–1 decision approving the Jordan Cove project, prompting the Confederated Tribes of Coos, Lower Umpqua and Siuslaw Indians and Cow Creek Band of Umpqua Tribe of Indians to file a lawsuit against FERC's decision. (FERC, which had two new commissioners, appointed by Trump, reversed its earlier 2016 decision to deny approval to the project.) However, in January 2021, FERC made a decision that may prove fatal for the pipeline project. FERC unanimously upheld Oregon's decision under the CWA not to approve water permits for the project.[138]

While the Jordan Cove developers announced the pause of that project in April 2021, tribes in the Pacific Northwest face the possibility of other proposed infrastructure projects in future. Legal scholars have discussed how tribes in the Pacific Northwest can potentially succeed in opposing oil and gas infrastructure projects by asserting their treaty fishing rights.[15,44,139] A number of cases illustrate the past success of this strategy. In the past, treaty fishing rights have compelled the Corps to deny permits for export terminals. In 2016, the Corps denied permits under the CWA and the Rivers and Harbors Act for the proposed Gateway Pacific Coal Terminal Washington State. The Corps explained that the proposed project would have more than a de minimis impact on the Lummi Nation's treaty fishing rights.[15,139] The Corps noted that the consideration of treaty rights is independent of the NEPA process.[15] A district court in an earlier case, *Northwest Sea Farm v. US Corps of Engineers*,[67] upheld the Corps' denial of a permit to protect treaty fishing rights.[15] The court underscored that the federal government and subsequently the Corps have a fiduciary duty "to ensure that Indian treaty rights are given full effect" and "treaty rights are not abrogated or impinged upon absent an Act of Congress."

Treaty fishing rights have also been cited by the states of Washington and Oregon as reasons for denying state permits for proposed export terminals.[15,139,xlviii] In 2017, the state of Washington denied a section

---

[xlviii] Despite these positive developments among state governments' decisions, Blumm and Litwak advise tribes to remain vigilant about backsliding in state governments' interpretation of treaties [139]. The state of Washington (and the state of Oregon) challenged tribes' treaty fishing rights in *Washington v. United States*, while simultaneously citing these rights when denying approvals for proposed export terminal projects [63].

401 permit for the proposed Millennium Bulk Terminal project, which would export coal from the Powder River Basin of Wyoming and Montana to Asia. Washington state cited the terminal's interference with the treaty fishing rights of the Confederated Tribes of the Umatilla Indian Reservation.[15] That year, Washington state also denied approval for the proposed Tesoro Savage Oil terminal, which would be one of the largest oil-export terminals in the Northwest, citing treaty fishing rights of the Confederated Tribes of the Umatilla Indian Reservation and the Confederated Tribes and Bands of the Yakama Nation.[15] Three years earlier, in 2014, Oregon denied a state permit for a proposed coal storage and barge loading project that would interfere with tribes' treaty fishing rights.[15]

5.3 Legal Challenges on Ancestral Lands Designated As Public Lands

Tribes have strived to protect against extractive activities lands to which they have cultural and historical ties, and they face specific challenges when those lands are designated as public land.[xlix] As discussed in Section 5.1.2, a large share of public lands subject to the authority and oversight of BLM and US Forest Service are managed under the multiple use mandate for the benefit of Americans present and future – leaving tribal interests to compete with a variety of other stakeholder interests. Property-based legal avenues for recourse for tribes are sparse as tribes, under US law, do not have occupancy and user rights on these lands. Rather, tribes must rely in part on presidential intervention and congressional action.

Bears Ears National Monument reveals how tribes have used the presidential declaration of National Monuments to protect these lands, but without congressional declaration of national parks or wilderness, these sites remain vulnerable to oil and gas extraction (Section 5.3.1). BLM's leasing of oil and gas on the Chaco Canyon landscape further illustrates how, without congressional protections, public lands that host tribes' sacred cultural sites are vulnerable to extractive activities (Section 5.3.2).

5.3.1 Bears Ears National Monument

Bears Ears is the first national monument proposed by tribes (in this case, the Intertribal Commission representing the Navajo, the Ute Mountain Ute, the Ute

---

[xlix] Tribes have strived to protect lands to which they have cultural and historical ties. Tribes face specific challenges when those lands are designated as public lands and private parties seek to pursue extractive operations on those lands [72].

Tribe of Uintah and the Ouray, Hopi and Zuni tribes). The Intertribal Commission reframed the use of the Antiquities Act to protect lands where tribes, applying their traditional knowledge, would undertake collaborative management of the monument along with BLM and be allowed to continue their cultural practices.[72,140] This approach reframes the past use of the national monument designations by presidents and the national park designations by Congress. Traditionally, these tools were used to preserve Indian lands and culture as historical relics or to preserve wilderness as uninhabited lands. The US government removed Native Americans from their treaty and ancestral lands and restricted their hunting and gathering rights in the creation of national parks.[72,141–143,l]

The monument declaration defeated the attempt by a subset of Utah politicians to transfer Indian lands and public lands to the state of Utah. In July 2016, Congressmen Robert Bishop and Jason Chaffetz introduced their Utah Public Land Initiative bill,[144] which would transfer to the state of Utah 100,000 acres of public land that fall within the tribe's historical Uncompahgre Reservation (subject to ongoing legal dispute) and would make management changes to another 200,000 acres of reservation lands.[145–147,li] The Intertribal Coalition, the Department of the Interior under the Obama administration and the minority members on the Committee on Natural Resources criticized the proposed transfer of Ute lands and federal lands to the state of Utah and the prioritization of extractive activities on public lands designated for multiple use.[147,lii]

The Intertribal Coalition submitted the Bears Ears National Monument proposal, covering 1.9 million acres, to Obama in October 2015, based on extensive cultural and archaeological work (Map 5.4).[148,149] The Obama administration engaged with state officials, congressional delegations and the Utah governor on both the Public Lands Initiative and on the Intertribal Coalition's proposal.[150] The Public Lands Initiative bill failed to win support in the House. In December 2016, Obama proclaimed the Bears Ears National Monument on 1.35 million acres, excluding some areas with uranium mining and resources.[21]

With Trump taking over the presidency, Congressmen Bishop and Chaffetz and Utah Senator Orrin Hatch found support.[154,155] In April 2017, Trump

---

[l] For instance, the creation of the Mesa National Park in 1906, which focused on preserving antiquities, removed lands from the Ute [141]. The creation of the Grand Canyon National Park removed lands from the Havasupai, Hualapai, Paiute, Dine, Zuni and Hopi [143].

[li] Tribes learned about this transfer only after the Public Lands Initiative proposal was published [145, 146].

[lii] During the hearing, Neil Kornze from BLM noted that "tribes deserve and must have a meaningful seat at the table in managing the Bears Ears Area." The bill defines management of conservation areas in ways that undermined protection of those areas. It would transfer management of oil and gas activities on federal lands to the state of Utah and could require the disposal of public lands.

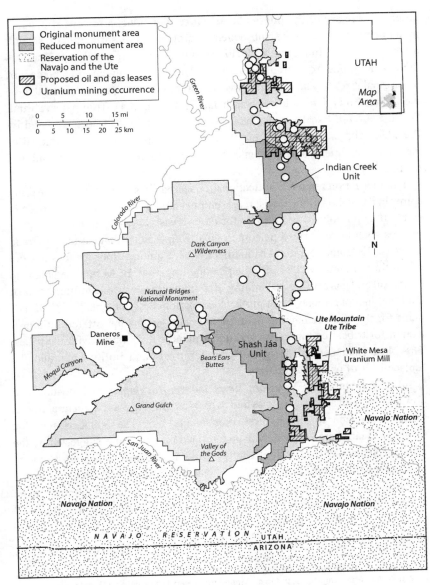

Map 5.4 Bears Ears National Monument
Source: Redrawn from various maps by the Department of the Interior,[151] the Center for Biological Diversity[152] and the Grand Canyon National Trust.[153]

ordered Secretary of the Interior Ryan Zinke to conduct a review of national monuments,[156] and Zinke subsequently proposed the shrinking of several monuments.[157,liii] In December 2017, Trump, acting purportedly upon powers in the Antiquities Act, shrunk Bears Ears National Monument to 15 percent of its original size.[158] The monument was split into two separate sections, that is, the Indian Creek Unit and the Shash Jáa Unit (Map 5.4). Trump also shrunk Grand Staircase Escalante National Monument, which had been declared by President Clinton in 1999,[159] to 50 percent its original size. The removal of national monument status opens these lands to mining and oil and gas drilling.[160]

Trump's action removed protections for sacred sites, including the towering spires in the Valley of the Gods, representing ancient Navajo warriors frozen in stone. It also removed protections for other areas valued by Indians and non-Indians such as the prized paleontological Kaiparowits Plateau in the Grand Staircase Escalante National Monument, where unique dinosaur fossils were found.[161] In February 2020, despite pending litigation, BLM finalized plans to permit oil and gas drilling and mining of uranium on nearly one million acres of land that had previously been protected as national monuments.[162,163] BLM's plans for the areas remaining under monument designation offered little protection for tribes' cultural sites and undermined collaboration with the tribes. The plans also permitted right-of-way for roads and utility lines and the replacement of native vegetation.[164,165]

Investigations by the *New York Times* and the *Washington Post* documented how access to oil and gas reserves and to uranium mining played a central role in Trump's decision to shrink these monuments.[31,166] The boundaries of the shrunken Bears Ears National Monument lined up with those suggested by the office of Senator Orrin Hatch of Utah.[31] Those boundaries excluded areas with high potential for uranium mining.[167,liv] Notably, Energy Fuels Resources Inc., which operates the White Mesa uranium mill and the Daneros uranium mine, lobbied the Trump administration[lv] to shrink the national monument.[166,168]

[liii] These include Bears Ears National Monument, Grand Staircase Escalante National Monument, Cascade Siskiyou in Oregon and California and Gold Butte in Nevada. Zinke also recommended permitting more grazing, ranching, fishing and hunting, and other activities in Katahdin Woods and Waters in Maine, in Organ Mountains-Desert Peaks and Rio Grande, in Northeast Canyons and Seamounts Marine National Monument, in Pacific Remote Island Marine National Monument and in Rose Atoll Marine National Monument.

[liv] The proposed expansion of uranium mining triggered tremendous opposition from tribes, given the continued devastation to public health and the environmental damage from uranium mining, including in the Navajo Nation, from the 1940s through to 1970 [19].

[lv] Andrew Wheeler, who subsequently served as EPA deputy administrator and administrator, listed Energy Fuels Resources Inc. as his client in the first two quarters of 2017 and met with Secretary of the Interior Zinke, who was in charge of reviewing the monuments [166][168].

Access to coal reserves motivated the decision to shrink Grand Staircase Escalante National Monument.[31]

Native American tribes, as well as conservation and environmental groups and the outdoor retailer Patagonia, filed lawsuits challenging Trump's shrinkage of the monuments.[169,170] In September 2019, Judge Tanya Chutkan of the District Court for the District of Columbia ruled that these cases[lvi] could proceed on their merits.[171] Tribes and other plaintiffs have argued that Trump overstepped his executive powers under the Antiquities Act in shrinking the two monuments. They contend that a president cannot revoke or diminish the national monuments proclaimed by past presidents and only Congress can revoke or diminish monuments.[169] Under the Constitution's property clause, Congress holds the powers to manage federal lands and the president, on the matter of public lands, can only exercise powers that Congress has delegated to the president.

A number of legal scholars argue that the text of the Antiquities Act's makes it clear that Congress gave presidents a "one way" power to protect these archaeological and historical sites.[172,173] The act specifies presidents' powers to proclaim a monument, but it is silent on presidents' powers to revoke or diminish monuments. Furthermore, these scholars argue that their analysis of legislative history and reports on the Federal Land Policy and Management Act of 1976 demonstrate that Congress reserved its powers to remove protections.[lvii] Courts have upheld presidential use of the Antiquities Act to protect large landscapes as "objects of historic or scientific interest." For instance, President Roosevelt's proclamation of Grand Canyon National Monument and President H. W. Bush's proclamation of Giant Sequoia National Monument were upheld.[172,173] Courts have largely deferred to presidents' decisions on the appropriate size of the land to be protected as a national monument.[lviii] Lawsuits against President Clinton's proclamation of Grand Escalante Staircase National Monument and President G. W. Bush's proclamation of Papahānaumokuākea National Monument, both of which attempted to reduce the size of monuments, were not successful.[174]

[lvi] The three cases against the Bears Ears National Monument shrinkage were consolidated, i.e., the first filed by the five tribes that proposed the monument; the second, by Utah Diné Bikéyah; and the third by the National Resources Defense Council and other environmental and conservation groups. The two cases against the Grand Staircase Escalante National Monument shrinkage were consolidated as well, i.e., the first filed by the Wilderness Society and other environmental and conservation groups, and the second filed by the Grand Staircase Escalante Partners, the Society of Vertebrate Paleontology and the Conservation Lands Foundation.

[lvii] The report on the House bill states that "[The bill] would also specifically reserve to the Congress the authority to modify and revoke withdrawals for national monuments created under the Antiquities Act" [172, 173].

[lviii] The few cases in which presidents reduced the size of monuments primarily occurred when resources were needed during the war years [173].

Tribes, environmental groups and conservation groups have challenged the validity of BLM's Resource Management Plans (RMPs) for the diminished Bears Ears and Grand Staircase Escalante National Monuments, both of which were drawn up while William Perry Pendley served as the unlawful director of BLM.[175] David Bernhardt appointed Pendley as acting director of BLM and Pendley then extended his own tenure.[176] (Prior to taking office in July 2019, Pendley served as the head of the Mountain States Legal Foundation, which challenged federal oversight of public lands.)[33] In September 2020, Judge Brian Morris in the District Court for Montana ruled that Pendley's appointment as acting head to a position that required Senate confirmation was unlawful and that any "function or duty" of the BLM director that Pendley performed has no force and must be set aside.[176] Challenges on the validity of BLM plans, based on Pendley's unlawful tenure at BLM, have also been filed by other state governments (Chapter 4).

5.3.2 Chaco Canyon National Historic Park

The Chaco Canyon National Historical Park and the greater Chaco region in northwest New Mexico is the ancestral home of many Southwestern tribes, including the Puebloans and the Navajos. Chaco Canyon, the hub of the Puebloan civilization between the ninth and the twelfth centuries and a relic of Puebloan architecture, was named a United Nations World Heritage Site in 1987. Tribes continue to conduct cultural, spiritual and religious ceremonies in the Chaco region. The region also faces encroachment from oil and gas development.[177,178] Already 91 percent of public lands surrounding Chaco Canyon National Historical Park has been leased by BLM to the oil and gas industry.[54]

The Trump administration repeatedly tried to sell oil and gas leases in the areas surrounding the national historical park. On May 29, 2019, Secretary of the Interior David Bernhardt agreed to issue a one-year moratorium on oil and gas leasing near Chaco Canyon to allow BLM to draft a new RMP.[179] However, BLM failed to undertake genuine consultation with tribes. In February 2020, BLM released a draft amendment to the RMP and the EIS for the Chaco region.[180] That plan could add more than 3,000 new oil and gas wells to the area. Oil and gas operations in the region have already caused tribes and other local communities to suffer from severe air pollution and caused damage to cultural sites.[181] The RMP was released for public comment at a time when the Navajo Nation, All Pueblo Council of Governors, members of the tribes and other stakeholders were addressing the COVID-19 public health emergency.[182] In May 2020, BLM and the Bureau of Indian Affairs announced a virtual consultation with tribes,[183] despite the ongoing public health emergency and the lack of broadband access for many tribal members.

The Chaco case also illustrates the legal hurdles in using the National Historic Preservation Act of 1966 (NHPA) to protect tribes' historical and cultural landscape.[9,181] In May 2019, the Court of Appeals for the Tenth Circuit ruled in *Diné Citizens v. Bernhardt* that BLM did not violate NHPA in approving approximately 350 application permits for drilling in the Chaco region.[259] The court ruled that federal agencies have the authority under NHPA to consider more expansive definitions of the area of potential effect, but they are not required to do so. Therefore, BLM could ignore the large-scale impacts of oil and gas extraction in the Chaco region, including the air pollution and noise from the industrial activities that diminished the historical landscape. The court reiterated that NHPA section 106 "is a procedural statute requiring government agencies to stop, look, and listen before proceeding when their action will affect national historical assets." BLM took a narrow approach in defining the area of potential effect as narrow buffers around archaeological sites; and with that approach, BLM concluded that archaeological sites and artifacts could "yield important historic information regardless of whether [they are] in a pristine location or surrounded by development ... So long as the site itself remains undisturbed; setting is not an important aspect of its integrity."[181]

The Trump administration's actions to open public lands that encompass tribes' cultural and historical landscapes to oil and gas drilling and uranium mining is not limited to the Chaco region. Instead these actions extended to, among others, the Ruby Mountains in Nevada, sacred to the Te-Moak Western Shoshone tribes;[184] lands around the Grand Canyon, Arizona, sacred to the Hualapai, Hopi and Navajo tribes;[185] and lands between Canyonlands National Park and Hovenweep National Monument in Utah and Colorado, sacred to Pueblo, Zuni, Hopi, Navajo and Ute tribes.[186]

5.4 Energy Transition on Reservations

Not only did the Trump administration undermine tribes' ability to challenge unwanted oil and gas infrastructure and extraction on Indian lands, it also weakened federal agencies' support for renewable energy on tribal lands.[187,lix] State governments' divergence in their energy pathways, discussed in Part I of

---

[lix] Among the federal agency divisions that work with tribes to promote renewable energy projects are the Department of Energy's Tribal Energy Program and the Department of the Interior's Indian Energy and Economic Development unit within the Bureau of Indian Affairs.

this book, is echoed in tribes' divergence in their energy pathway. The economies of a few tribes are anchored in oil and gas extraction (Section 5.4.1), while others are shifting toward renewable energy (Section 5.4.2). Tribes' diverse energy strategies have led to their contrasting stances on federal policy debates, such as the permitting of pipelines and export terminals and the regulations on oil and gas extraction.[188,lx]

5.4.1 Oil and Gas Extraction on Reservations

The Department of the Interior, which exercises oversight over Indian lands, has historically supported development of oil, gas, coal and uranium.[189] Historically, the Bureau of Indian Affairs' mismanagement and poor oversight of private companies has caused tribes to suffer underpayment of royalties and harm to their health and environment.[lxi] The 1982 Indian Mineral Development Act enabled tribes to enter into energy extraction agreements, authorizing tribes to negotiate leases and other agreements. Nevertheless, these agreements are subject to oversight by the Bureau of Indian Affairs and require approval by the secretary of the Interior.[190,lxii] Armed with greater autonomy in charting their energy pathways, some tribes have chosen to support oil and gas extraction. Oil and gas extraction provides significant revenue to a dozen or more tribes (Figure 5.1),[191,lxiii] but it has imposed adverse impacts on tribes' health, air, water and lands (Table 5.2).

The Southern Ute tribe stands out as one of the few tribes that has successfully leveraged extractive activities to build a diversified economic base. The

---

[lx] For instance, while 57 tribes in the Pacific Northwest opposed all proposals for fossil fuel transportation and export, the Ute Tribal Business Committee supported the Jordan Cove Energy proposal. Diné Citizens against Ruining Our Environment and the Fort Berthold Protectors of Water and Earth Rights joined the lawsuit to oppose the Trump administration's repeal of the hydraulic fracturing regulation on public and tribal lands. On the contrary, the Ute Indian Tribe of the Uintah and Ouray Reservation joined a lawsuit opposing the Obama administration's promulgation of the hydraulic fracturing regulations on public and tribal lands.

[lxi] For instance, the Senate launched an investigation into Koch Oil, a subsidiary of Koch Industries, one of the largest purchasers of oil from Indian reservations. The investigations revealed that the company had understated the amount of oil taken and thus underpaid royalties to the tribes. In 1999, in a civil suit against Koch Oil, the jury returned the verdict that Koch Oil was guilty of underpaying royalties on oil from Indian reservations and public lands. In 2001, Koch Industries agreed to a settlement [253].

[lxii] The 2005 Indian Tribal Energy Development and Self Determination Act created a mechanism that could give tribes more control over energy development. Tribes that enter into Tribal Energy Resource Agreements (TERAs) with the secretary of the Interior would no longer need approval for energy agreements. Kronk Warner writes that tribes have chosen not to enter into TERAs because they impose requirements on tribes while waiving federal liability associated with the federal trust responsibility [190].

[lxiii] According to the Bureau of Indian Affairs' 2013 report, reservations with existing or emerging shale plays include Blackfeet and Fort Peck in Montana, Wind River in Wyoming, Northern Ute in Utah, Ute Mountain Ute in Colorado, Navajo in Arizona, Jicarilla in New Mexico, Osage in Oklahoma, Tunica Biloxi in Louisiana and Fort Berthold in North Dakota.

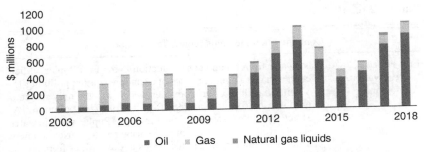

Figure 5.1 Revenue from oil, gas and natural gas liquids extraction on Native American lands
Source: Department of the Interior, Natural Resources Revenue Data, https://revenuedata.doi.gov/query-data.
Notes: According to the Natural Resources Revenue Data website, the figures above represent "The amount of money collected by the federal government from energy and mineral extraction on Native American lands." Revenue includes royalties, rents and other revenue.

Table 5.2 *Selected pollution incidents from oil and gas extraction on Indian lands*

	Contamination incidents reported post-2006
All tribal lands	Air pollution: Annual methane emissions from fugitive, vented and flared emissions are estimated at 15.6 billion cubic feet, based on estimates in 2013. This translates to $9 million in foregone royalties per year, assuming a gas price of $4 per thousand cubic feet and a 12.5 percent royalty rate.[206]
Fort Berthold Reservation, North Dakota	In 2014, 1 million gallons of wastewater produced by unconventional oil drilling spilled into a ravine filled with natural springs and then seeped into Bears Den Bay about a quarter mile upstream from a drinking water intake on Lake Sakakawea. The intake location on the Missouri River serves as the source of drinking water for the MHA Nation.[207] A 2016 study found high levels of salts, lead, other toxic contaminants and radium in sites downstream from the spills.[208]
Wind River Reservation, Wyoming	Wastewater from oil extraction was being released onto tribal lands from 2005.[209] That wastewater contained hydrochloric acid and naphthalene and carcinogens such as benzene and ethyl benzene. Under the CWA, wastewater can be released west of the 98th meridian for "beneficial use" provided it meets specific requirements, conditions not met in this case.[209]

Table 5.2 *(cont.)*

	Contamination incidents reported post-2006
Fort Peck Reservation, Montana	Wastewater from oil and gas production in the East Poplar oil field contaminated 18 square miles of a shallow aquifer, polluting 15–37 billion gallons of groundwater, which was the only source of groundwater for the Fort Peck Reservation. That pollution affected private water wells and the city of Poplar's public water-supply wells.[210] Consequently, the tribe had to construct a new drinking water pipeline in the 2000s to draw water from the Missouri River.[211] The contamination resulted from oil companies' disposal of wastewater into unlined pits in the 1950s and 1960s and from using poorly constructed injection wells in the 1970s.[212]
Navajo Nation communities of Aneth and Montezuma Creek, Utah	Mobil Exploration and Producing US, Inc. caused 83 identified oil spills along the banks of the San Juan River and its tributaries and released oil and wastewater without a permit into the San Juan River. In 2004, Mobil reached a $5.5 million settlement with the EPA to undertake cleanup operations. In 2005, Mobil reached a settlement with the EPA and the Department of Justice for $1 million for Clean Air Act violations at Mobil's oil production facility near Aneth. In 2002, Texaco Exploration and Production, Inc. reached a settlement with the EPA and the Department of Justice for $0.85 million on charges over oil spills.[213]
Sac and Fox lands, Oklahoma	Oil and gas operations, including the disposal of oil and gas wastewater and the injection of that wastewater for oil and gas recovery, caused extensive damage beyond remediation to the Vamoosa-Ada aquifer, the only source of drinking water for the Sac and Fox lands. The company Tenneco's settlement in 1997 took place long after the tribes first began complaining about the poor drinking water in the 1950s and studies in the 1960s that linked oil and gas operations to the contamination of the aquifer.[214]

tribe began to conduct its own royalty accounting and auditing to overcome historical mismanagement by the Bureau of Indian Affairs;[192,193,lxiv] it renegotiated leases with oil and gas companies; and it imposed severance tax on extraction.[192] The Southern Ute Nation Tribal Council formed its own corporations and expanded operations from extraction to distribution of oil and gas to consumers on and off the reservation.[194] By 2000, the Tribal Council had

---

[lxiv] Based on information from Southern Ute auditors, in 2010, the Department of the Interior fined BP America $5.2 million for underreporting the amount of gas it had been producing on Southern Ute lands.

created a growth fund that invested in a variety of sectors in Colorado and other states.[194] Oil and gas revenue has enabled the tribe to provide each tribal member with health insurance and a guarantee for college education.[193] Poverty rates of 18.2 percent among tribal members on the Southern Ute reservation are lower than the average poverty rates of 26.8 percent among Native Americans in the United States.[195,lxv] Despite these benefits, extraction has come at the cost of air pollution,[196] abandoned wells on their reservation,[197] and disagreement among the tribal members on their energy pathway.[197]

Unlike the Southern Ute, other tribes have suffered from the boom–bust economy associated with oil and gas extraction. For instance, the Mandan, Hidatsa and Arikara Nation, also known as the MHA Nation, of the Fort Berthold Indian Reservation,[lxvi] allowed extraction on their lands during the shale boom. By 2019, production in the reservation amounted to 370,000 barrels per day or one-quarter of North Dakota's production.[198] The MHA government received production and extraction tax and royalties ($450 million in 2013).[lxvii] That production also contributed to the decline in poverty rates and unemployment and to an increase in mean household income.[200] However, the MHA nation's reliance on oil extraction for 90 percent of its revenue resulted in severe economic shocks from the shale bust in 2015 and the oil price crash in 2020.[199] Oil and gas operations have also resulted in social problems, including the dramatic rates of sexual assault, rape and domestic violence by male oil-field workers against female tribal members.[201,202] Tribal members criticized the poor management of oil and gas extraction[lxviii] and voted in new leaders who promised greater transparency on the management of oil and gas leases and decision-making by the tribe's business council, as well as to enforce stricter environmental oversight.[194,203–205]

5.4.2 Renewable Energy on Reservations

Recognition of the adverse economic, health and environmental impacts of oil and gas extraction have led some tribes to turn to renewable energy. Smaller-scale

---

[lxv] These poverty rates are still higher than the 8 percent poverty rate for non-Native Americans on the Southern Ute reservation and the 14.6 percent poverty rate in the United States.

[lxvi] The US government, breaking numerous treaties and taking ever more lands, moved these three distinct tribes onto the reservation. In 1944, Congress approved the building of the Garrison Dam, which flooded one-fourth of the reservation including prime agricultural land [194].

[lxvii] In 2013, oil production on the reservation amounted to 170,000 barrels per day. The tribal government received $249 million in oil tax revenue. Oil firms paid $99 million in royalties to the tribal government and companies in the drilling and trucking services paid additional licensing fees for their operations. Individual owners of mineral rights on the reservation received $350 million [200].

[lxviii] In 2013, members of the MHA Nation sued some tribal leaders for leasing drilling rights at low-ball prices [203].

Table 5.3 *Selected utility-scale renewable energy projects on reservations*

The Jicarilla Apache Nation in north-central New Mexico developed a 50 MW utility-scale solar farm and a companion 20 MW/80 MWh battery-storage unit, which began construction in October 2020. The power is sold to Public Service Company of New Mexico, which supplies the city government of Albuquerque. The project benefitted from the retirement of the San Juan coal power plant. Renewable energy projects help diversify the tribe's economy, with 90 percent of the tribe's government operations in 2012 reliant on funds from oil and gas development.[216]

The Southern Paiute hosts the 250 MW Moapa Southern Paiute Solar Park in Clark County, Nevada. The project, online since 2017, supplies the Los Angeles Department of Water and Power. Two other projects were announced in 2019, the Southern Bighorn Solar & Storage with 300 MW of solar and 135 MW of battery storage on the Moapa River Indian Reservation and another project with 200 MW of solar and 75 MW of battery storage on the Moapa Band of Paiutes Indian Reservation.

The Navajo Tribal Utility Authority, which is owned by the Navajo Nation, developed the 55 MW Kayenta Solar Facility in northeastern Arizona in 2008. The project, developed in partnership with Salt River Project, an Arizona state utility company, supplies power to Phoenix. In 2021, the Navajo Tribal Utility Authority began construction of a solar project in San Juan County, Utah that will supply electricity to the Utah Associated Municipal Power System.

The Campo Kumeyaay Nation in eastern San Diego County has hosted a 50 MW, 25-turbine wind project since 2005 that serves the city of San Diego. However, the tribe opposed a larger, 250 MW project in 2014. In April 2020, the 240 MW project with sixty 4.2 MW turbines, which was approved by the Department of the Interior, faced opposition from some tribal members.

Sources: Institute for Energy Economics and Financial Analysis, 2019;[227] Maruca, 2019;[187] and Reuters, 2019.[228]

renewable energy projects have been an effective strategy for tribes to address energy poverty on reservations.[37,215] About 14.2 percent of Native American households have no access to electricity, that is, 10 times the national average.[189] Tribes have faced challenges in developing utility-scale renewable energy projects, with only a few tribes achieving success as of 2020 (Table 5.3). Tribes whose reservations are close enough to population centers can respond to the increase in demand for renewable energy, resulting from state government mandates for electricity sourced from renewable resources and the retirement of coal-fired power plants.[189,lxix]

[lxix] In 2018, California tightened its renewable portfolio standard to 60 percent by 2030 and 100 percent by 2045. In 2019, New Mexico tightened its standards to 50 percent by 2030, 80 percent by 2040 and 100 percent by 2045. That same year, Nevada tightened its standards to 50 percent renewables by 2030 and 100 percent by 2050.

Tribes face a number of hurdles to renewable energy development, including laws governing leasing and taxation on reservations. The Bureau of Indian Affairs' delayed decision on the leasing of these lands held under trust for tribes by the US government had threatened the viability of many renewable energy projects.[216] The 2012 HEARTH Act was an important step in empowering tribal governments to make autonomous decisions on leasing.[187] Under the act, tribal nations can approve leases for business and other purposes, after their regulations governing leasing are approved by the secretary of the Interior. As of December 2020, about 60 tribes have received approval for their regulations and can thus make independent leasing decisions.

Congress has yet to address at least two other hurdles. First, it has yet to clarify via legislation that non-Indian entities developing renewable energy on reservation land are not subject to taxation by state or local governments that could disincentivize such projects.[lxx] That clarification would prevent projects from being subject to dual taxation by the tribe and by the state government, as well prevent pressure on tribes to relinquish potential tax revenue to ensure the project remains financially feasible. Congress would need to enact incentive programs, including direct loans or grants, to support renewable energy projects that are fully owned by tribes. Congressional incentives for wind and solar, which are based on federal tax credits, do not provide direct benefits to tribes that are exempt from federal taxes.[187,215,217,218] Congress's Production Tax Credit to support wind and the Investment Tax Credit to support solar have forced tribes, as well as other nonprofits that develop wind or solar, to partner with taxable entities in order to take advantage of these credits. This approach reduces the potential benefits that tribes could secure from solar and wind projects. Congress had recognized that procurement by federal agencies, which are large consumers of energy and which host sites close to reservations, can provide the demand to support tribal renewable energy projects.[219] However, federal agencies, such as the Department of Energy, the Department of Defense and the General Service Administration, have been slow to implement preferential procurement of renewable energy from entities that are tribally owned, despite their powers under the 2005 Energy Policy Act.[219]

The experience of the Navajo Nation exemplifies the benefits, but also the challenges, faced by tribes in switching to renewable energy from their past reliance on fossil fuels. In 2018, the Navajo Nation elected green platform candidates for its president and vice president. In April 2019, President

---

[lxx] In 1989, the Supreme Court in *Cotton Petroleum Corp. v. New Mexico* permitted the state government to impose a severance tax on a non-Indian extraction company. The decision permits state governments to impose taxes on non-Indian businesses operating on lands leased from the tribe [241].

Jonathan Nez announced the Navajo Nation's goal to transition to renewable energy, which includes providing off-grid solar to the 15,000 Navajo households still without electricity and developing additional utility-scale solar energy projects to take advantage of New Mexico's and California's legislation to source all its electricity from renewable energy.[220] That announcement revitalized efforts toward a green energy pathway, first conceptualized in the 2009 Navajo Green Jobs legislation.[221] The Navajo Tribal Utility Authority developed the 55 MW Kayenta Solar project that now powers 36,000 homes, which came online in two phases in 2017 and 2019. In 2020, Navajo Power, a Navajo entrepreneur's solar startup, signed a co-development deal with sPower to generate 200 MW of solar energy on Navajo land by the end of 2023.[222] The switch to solar aided in ending reliance on the Navajo Generating Station, a coal-fired power plant that had contributed to air pollution and poor health on the reservation before shutting down in November 2019.

Despite the promises of renewable energy for a more sustainable energy pathway, Len Necefer, Navajo member, energy researcher and professor of Native American Studies, underscores the importance of tailoring the scale and location of the projects to gain support of tribal members and of genuine consultation with tribal members.[223] Development of renewable energy projects, whether on- or off-reservation, necessitates the construction of transmission lines that rely on right-of-way through the reservations. The placement of transmission lines would need to be compatible with tribe's land use and the right-of-way through the reservation would need to generate fair and appropriate payments for the tribe.[224]

The experience of the Navajo Nation also illustrates the continued pull for scarce tribal resources from fossil-fuel-oriented enterprises. The Nez leadership rejected the Navajo Nation Oil and Gas Company's request for a $40 million bailout from the Navajo tribe[225] and also opposed investments into coal megaprojects. Despite that rejection, in October 2019, the Navajo Transitional Energy Company, which is owned by the Navajo Nation but semiautonomous in its decision-making, acquired three large mines in Wyoming and Montana from bankrupt Cloud Peak Energy for about $100 million and assumed all cleanup costs.[226] That decision replicated another ill-fated decision in 2013 by the Navajo Nation to purchase a 60-year-old coalmine at $85 million and to extend a 50-year lease with a major coal-fired power plant; both projects eventually shutdown.[221] Nevertheless, as other state governments commit to energy transition, as discussed in Chapter 3, the opportunities for affordable electricity, jobs and revenues, alongside the decline of fossil fuels, provides continued impetus for the energy transition.

5.5 How Congress and the Executive Branch Can Support Tribes' Protection of Their Lands

Reversing centuries of injustices against tribes will require a deep reorientation of the US government's relationship to tribes. Tribes and the US Commission on Civil Rights have called on the US government to honor tribes' treaty rights, to honor obligations to tribes whose lands were taken without treaties,[lxxi] and to meet trust responsibilities to tribes.[37,229–231] Tribes have also called for the reassessment of legal doctrines that have sanctioned Congress's abrogation of treaties and that transformed tribes' land ownership to holding occupancy titles.[232]

While such a reorientation may take time to permeate the US government, immediate steps can be taken by Congress to meet its obligations to protect tribes' lands and resources. Congress, using its powers under the Property Clause, can enact laws to grant more permanent protections from extractive activities for national monuments and other public lands of cultural significance to tribes. A number of legislative initiatives in the 116th Congress spearheaded by Representative Debra Haaland would offer protections to tribal lands. In October 2019, the House successfully passed HR 2181, the Chaco Cultural Heritage Area Protection Act, by 245–174. The bill would set up a 10-mile buffer zone around Chaco Canyon National Historical Park. About 300,000 acres of public lands in that buffer would be withdrawn from oil and gas leasing, while private land and mineral rights would remain unaffected. Senator Tom Udall and Senator Martin Heinrich, Democrats of New Mexico, sponsored the companion Senate bill 1079.[233] The Navajo Nation, the All Pueblo Council of Governors, New Mexico Governor Michelle Lujan Grisham and the entire New Mexico congressional delegation supported that legislation. In October 2019, the House also passed the Grand Canyon Centennial Protection Act by 236–185 votes, which would withdraw 1 million acres of federal lands in Arizona from mineral leasing.

Another proposed piece of legislation would offer permanent protections for several national monuments. In January 2019, Representative Haaland and Representative Ruben Gallego, Democrats from Arizona, reintroduced the Bears Ears Expansion and Respect for Sovereignty Act.[234] The bill, which received the support of 132 cosponsors, would restore the protection for Bears Ears National Monument and require engagement with tribes during the management of the monument's lands. The bill would also expand the monument to the full 1.9 million acres as initially proposed by the Intertribal Coalition. In

---

[lxxi] That some tribes do not have a ratified treaty with the US government is a vagary of the colonial conquest [231].

February 2019, Senator Udall sponsored the related Senate bill S.367, America's Natural Treasures of Immeasurable Quality Unite, Inspire, and Together Improve the Economies of States (ANTIQUITIES) Act of 2019.[235] The bill, with 25 cosponsors, would expand protection for Bears Ears National Monument to 1.9 million acres. That bill would also declare Congress's support for 52 national monuments declared by previous Republican and Democratic presidents and would reinforce the intentions of the Antiquities Act of 1906 that only Congress can diminish national monuments.

Native American political representation has given greater salience to protection of tribes' interests not only at the federal level but also at the local level. For instance, with tribes winning two of the three seats as San Juan commissioners, the commission switched from opposing the Intertribal Commission proposal of Bears Ears National Monument to strident support for it.[236] Nevertheless, Republicans did not support these legislative initiatives at the federal level. With Native American votes becoming pivotal in a number of elections, including the tribes' votes that contributed to Biden's win in Arizona and in Wisconsin in 2020, politicians from both sides of the aisle may well become more responsive to tribes' concerns.[237]

The Trump administration's reversal of the milestones achieved by the Obama administration underscores the importance of Congress's more permanent legislative protections. At the same time, the executive branch plays a critical role in meeting the US government's obligations to tribes. President Biden has promised a reset in the US executive branch's relationship with tribes. The Biden–Harris Plan for Tribal Nations has committed "to upholding the US's trust responsibility to tribal nations, strengthening the Nation-to-Nation relationship between the United States and Indian tribes, and working to empower tribal nations to govern their own communities and make their own decision."[238] The plan acknowledged that "tribal homelands are at the heart of tribal sovereignty and self-governance" and the Biden administration would "restore lands and protect the natural and cultural resources within them, while honoring the role of tribal governments in protecting those resources."

The appointment of Representative Haaland as the first Native American secretary of the Interior, as noted by Raúl Grijalva, chairman of the House committee that oversees the Interior, "brings history full circle at the Department of Interior."[239] Secretary Haaland's appointment instilled hope that leaders who had strong track records of protecting tribes' lands and resources can help reset the US government's relationship with tribes.[lxxii]

[lxxii] The Senate voted 51–40 to confirm Haaland, with only four Republican Senators breaking away from the party line, i.e., senators Lisa Murkowski and Dan Sullivan of Alaska, Susan Collins of Maine, and Lindsey Graham of South Carolina.

Biden's executive order on January 20, 2021 revoked the presidential permit for the Keystone XL pipeline.[240] (In June 2021, the pipeline developer announced the cancellation of the Keystone XL pipeline project.) He also ordered a pause to new oil and gas leasing on public lands, pending an in-depth review by the Department of the Interior of that leasing program. Many tribes and environmental groups have called on Biden to restore the original boundaries of Bears Ears and Grand Staircase Escalante national monuments, as promised in the Biden–Harris Plan for Tribal Nations.[lxxiii]

The broader commitment of federal agencies to undertake genuine consultation with tribes and to conduct more thorough science- and evidence-based analysis would avoid the biased assessments of the past that understated the adverse impacts from oil and gas infrastructure and extractive projects. As discussed in Chapter 4, these include climate-related assessments (e.g., BLM's and FERC's consideration of the projects' contributions to greenhouse gas emissions); other environmental assessment (e.g., the Corps' consideration of cumulative impact of the entire pipeline on water bodies); and financial assessment (e.g., FERC's consideration of whether projects are necessary to meet consumer demand and the potential for these projects to become stranded assets). The Biden administration's promise to support tribes' clean energy projects[238] would provide the impetus for reducing the barriers to clean energy development within reservations and for securing greater consultation with tribes for on-reservation or off-reservation renewable energy projects.

The furious pace at which the Trump administration sought to facilitate oil and gas extraction and the staggering extent to which it ran roughshod over Americans opposed to drilling have ravaged not only America's lands but also its seas, to which I turn next.

References

1. Revenue Watch Institute. *United States Native American Lands and Natural Resource Development*. Report by M. Grogan, R. Morse and A. Youpee-Roll (2011). https://resourcegovernance.org/sites/default/files/documents/rwi_native_american_lands_2011.pdf.
2. A. Koenig and J. Stein. "Federalism and the State Recognition of Native American Tribes: A Survey of State-Recognized Tribes and State Recognition Processes Across the United States." *Santa Clara Law Review* 48, no. 79 (2008): 82–83.

[lxxiii] The Biden proclamation would be consistent with the reading that the Antiquities Act empowers presidents to provide protections but not to revoke protections.

3. S. Day. "Implications of Tribal Sovereignty, Federal Trust Responsibility, and Congressional Plenary Authority for Native American Lands." In *The Environmental Politics & Policy of Western Public Lands*, edited by E. A. Wolters and B. S. Steel. Corvallis: Oregon State University Press, 2020.
4. A. V. Smith. "Federal Lands Are Becoming Tribal Lands Again." *Mother Jones*, August 17, 2019. www.motherjones.com/politics/2019/08/federal-lands-are-becoming-tribal-lands-again.
5. J. P. Kalt and J. W. Singer. "Myths and Realities of Tribal Sovereignty: The Law and Economics of Indian Self-Rule." *Native Issues Research Symposium*, no. RWP04-016 (March 2004).
6. A. J. Cobb. "Understanding Tribal Sovereignty: Definitions, Conceptualizations, and Interpretations." *American Studies* 46, no. 3/4 (2005): 115–132.
7. R. Tsosie. "Tribal Environmental Policy in an Era of Self-Determination: The Role of Ethics, Economics, and Traditional Ecological Knowledge." *Vermont Law Review* 21 (1996): 225–234.
8. R. Tsosie. "Land, Culture, and Community: Reflections on Native Sovereignty and Property in America." *Indiana Law Review* 34, no. 4 (2001): 1291–1312.
9. H. M. Hoffman and M. Mills. *A Third Way: Decolonizing the Laws of Indigenous Cultural Protection*. Cambridge, UK: Cambridge University Press, 2020.
10. C. F. Wilkinson. *Blood Struggle: The Rise of Modern Indian Nations*. New York: W.W. Norton & Company, 2006.
11. D. E. Wilkins and K. T. Lomawaima. *Uneven Ground: American Indian Sovereignty and Federal Law*. Norman: University of Oklahoma Press, 2001.
12. C.E. Goldberg et al. *American Indian Law: Native Nations and the Federal System*. LexisNexis, 2015.
13. R. Dunbar-Ortiz. *An Indigenous Peoples' History of the United States*. Boston, MA: Beacon Press, 2014.
14. N. Blackhawk. *Violence over the Land: Indians and Empires in the Early American West*. Cambridge, MA: Harvard University Press, 2006.
15. E. A. Kronk Warner. "Tribal Treaty Rights: A Powerful Tool in Challenges to Energy Infrastructure." *Connecticut Law Review* 51 (2019): 843–888.
16. V. Volcovici. "Trump Advisors Aim to Privatize Oil-Rich Indian Reservations." *Reuters*, December 5, 2016. www.reuters.com/article/us-usa-trump-tribes-insight-idUSKBN13U1B1.
17. M. Ambler. *Breaking the Iron Bonds: Indian Control of Energy Development*. Lawrence: University Press of Kansas, 1990.
18. J. R. Allison III. *Sovereignty for Survival: American Energy Development and Indian Self-Determination*. New Haven, CT: Yale University Press, 2015.
19. T. B. Voyles. *Wastelanding: Legacies of Uranium Mining in Navajo Country*. Minneapolis: University of Minnesota Press, 2015.
20. R. Tsosie. "Climate Change, Sustainability, and Globalization: Charting the Future of Indigenous Environmental Self-Determination." *Environment & Energy Law & Policy* 4, no. 2 (2009): 188–255.
21. B. Obama. *Presidential Proclamation: Establishment of the Bears Ears National Monument*. White House Office of the Press Secretary. 2016. https://obamawhitehouse.archives.gov/the-press-office/2016/12/28/proclamation-establishment-bears-ears-national-monument.

22. K. P. Whyte. "The Dakota Access Pipeline, Environmental Injustice, and US Settler Colonialism." In *The Nature of Hope: Grassroots Organizing, Environmental Justice, and Political Change*, edited by C. Miller and J. Crane. 320–337. Louisville: University of Colorado Press, 2019.
23. N. Estes. *Our History Is the Future: Standing Rock versus the Dakota Access Pipeline, and the Long Tradition of Indigenous Resistance*. London: Verso Books, 2019.
24. K. P. Loor. "Tear Gas + Water Hoses + Dispersal Orders: The Fourth Amendment Endorses Brutality in Protest Policing." *Boston University Law Review* 100 (May 2020): 817–848.
25. L. Tuell. "The Obama Administration and Indian Law: A Pledge to Build a True Nation-to-Nation Relationship." *Federal Lawyer*, April 2016, 44–48. www.fedbar.org/wp-content/uploads/2016/04/Obama-pdf-1.pdf.
26. Executive Office of the President. *A Renewed Era of Federal-Tribal Relations*. Obama Administration. White House Tribal Nations Conference (January 2017). https://obamawhitehouse.archives.gov/sites/default/files/docs/whncaa_report.pdf.
27. Congressional Research Service. *The Indian Trust Fund Litigation: An Overview of Cobell v. Salazar*. Report by T. Garvey. RL34628 (July 13, 2010). www.everycrsreport.com/files/20100713_RL34628_d9b8f6b0182b2147928f22d7a214a825294e9e1f.pdf.
28. Cultural Survival Staff. "Presidents Day 2020: 11 Ways Trump Dishonors Native Americans & How Natives Fight Back." *Cultural Survival*, February 17, 2020. www.culturalsurvival.org/news/presidents-day-2020-11-ways-trump-dishonors-native-americans-how-natives-fight-back.
29. T. Perez. "Op-Ed: Trump Is Breaking the Federal Government's Promises to Native Americans." *Los Angeles Times*, August 7, 2017. www.latimes.com/opinion/op-ed/la-oe-perez-native-american-indians-trump-20170807-story.html.
30. T. Udall. *Memo: Fact Checking the Trump Administration's Attempt to Re-write Its Native American Record, False Promises to Tribes*. US Senate Committee on Indian Affairs. October 23, 2020.
31. E. Lipton and L. Friedman. "Oil Was Central in Decision to Shrink Bears Ears Monument." *New York Times*, March 2, 2018. www.nytimes.com/2018/03/02/climate/bears-ears-national-monument.html.
32. US House of Representatives. *Oversight Hearing on BLM Disorganization: Examining the Proposed Reorganization and Relocation of the Bureau of Land Management Headquarters to Grand Junction, Colorado*. Committee on Natural Resources. 116th Congress, 1st sess. September 10, 2019.
33. B. Maffly. "Feds' Top Land Manager Remains the Attorney for Two Utah Counties in a Grand-Staircase Monument Lawsuit." *Salt Lake Tribune*, August 31, 2019. www.sltrib.com/news/environment/2019/08/31/feds-top-land-manager.
34. US Senate. "Proceedings and Debates of the 116th Congress, Second Session." *Congressional Record* 166, no. 159 (September 15, 2020). www.govinfo.gov/content/pkg/CREC-2020-09-15/pdf/CREC-2020-09-15-senate.pdf.
35. M. Brown. "Drilling, Mines, Other Projects Hastened by Trump Order." *AP News*, September 2, 2020. https://apnews.com/article/cb443d7b291f62ae649ba2acd80fe9b7.

36. K. A. Carpenter and A. R. Riley. "Privatizing the Reservation?" *Stanford Law Review* 71 (2019): 791–878. https://scholar.law.colorado.edu/cgi/viewcontent.cgi?article=2345&context=articles.
37. US Commission on Civil Rights. *Broken Promises: Continuing Federal Funding Shortfall for Native Americans* (Washington, DC: December 2018). www.usccr.gov/pubs/2018/12-20-Broken-Promises.pdf.
38. V. Deloria, Jr. and C. M. Lytle. *American Indians, American Justice*. Austin: University of Texas Press, 1983.
39. N. J. Newton, ed. *Cohen's Handbook of Federal Indian Law*. New York: LexisNexis, 2012.
40. C. F. Wilkinson and E. R. Briggs. "The Evolution of the Termination Policy." *American Indian Law Review* 5 (1977): 139–184. https://scholar.law.colorado.edu/cgi/viewcontent.cgi?article=2122&context=articles.
41. R. Taylor. "Trump Administration Revokes Reservation Status for Mashpee Wampanoag Tribe Amid Coronavirus Crisis." *Vox*, April 2, 2020. www.vox.com/identities/2020/4/2/21204113/mashpee-wampanoag-tribe-trump-reservation-native-land.
42. A. R. Riley. "The History of Native American Lands and the Supreme Court." *Journal of the Supreme Court* 38, no. 3 (2014): 369–385. https://doi.org/https://doi.org/10.1111/j.1540-5818.2013.12024.x.
43. Government Accountability Office. *Indian Programs: Interior Should Address Factors Hindering Tribal Administration of Federal Programs*. Report by F. Rusco, director of Natural Resources and Environment (January 2019). www.gao.gov/assets/700/696330.pdf.
44. J. Wolfley. "Embracing Engagement: The Challenges and Opportunities for the Energy Industry and Tribal Nations on Projects Affecting Tribal Rights and Off Reservation Lands." *Vermont Journal of Environmental Law* 19, no. 2 (2018): 115–163.
45. M. C. Wood. "Indian Trust Responsibility: Protecting Tribal Lands and Resources Through Claims of Injunctive Relief against Federal Agencies." *Tulsa Law Review* 39, no. 2 (Winter 2003): 355–368.
46. K. M. Riley. "Congress, Tribal Recognition, and Legislative-Administrative Multiplicity." *Indiana Law Review* 91, no. 3 (2016): 955–1021.
47. D. Rey-Bear and M. L. M. Fletcher. "We Need Protection from Our Protectors: The Nature, Issues, and Future of the Federal Trust Responsibility." *Michigan Journal of Environmental & Administrative Law* 6, no. 2 (2017): 397–462.
48. *United States v. Winans*, 198 US 371 (Supreme Court 1905).
49. *Winters v. United States*, 207 US 564 (Supreme Court 1908).
50. *Johnson v. McIntosh*, 21 US 543 (Supreme Court 1823).
51. R. A. Williams. "Columbus's Legacy: Law As an Instrument of Racial Discrimination against Indigenous Peoples' Rights of Self-Determination." *Arizona Journal of International and Comparative Law* 8, no. 2 (1991): 51–76.
52. E. Kades. "The Dark Side of Efficiency: Johnson v. M'Intosh and the Expropriation of American Indian Lands." *University of Pennsylvania Law Review* 148, no. 4 (2000): 1065–1190.
53. J. L. Seifert. "The Myth of Johnson v. M'Intosh." *UCLA Law Review* 52 (2004): 289–332.

54. H. M. Hoffman. "Congressional Plenary Power and Indigenous Environmental Stewardship: The Limits of Environmental Federalism." *Oregon Law Review* 97, no. 353 (2019): 354–396. https://scholarsbank.uoregon.edu/xmlui/bitstream/handle/1794/24696/Hoffman_OLR97%282%29.pdf?sequence=1&isAllowed=y.
55. N. J. Newton. "Federal Power Over Indians: Its Sources, Scope, and Limitations." *University of Pennsylvania Law Review* 132, no. 2 (January 1984): 195–288.
56. *United States v. Kagama*, 118 US 375 (Supreme Court 1886).
57. *Lone Wolf v. Hitchcock*, 187 US 553 (Supreme Court 1903).
58. *Tribal Business Committee v. Weeks*, 448 US 371 (Supreme Court 1977).
59. *United States v. Sioux Nation of Indians*, 448 US 371 (Supreme Court 1980).
60. S. L. Leeds. "By Eminent Domain or Some Other Name: A Tribal Perspective on Taking Land." *Tulsa Law Review* 41, no. 1 (Fall 2005): 51–78.
61. *McGirt v. Oklahoma*, 587 US ___ (Supreme Court 2020).
62. *United States v. Jicarilla Apache Nation*, 564 US 162 (Supreme Court 2011).
63. *Washington v. United States*, 584 US ___ (Supreme Court 2018).
64. M. C. Wood. "Tribal Tools & Legal Levers for Halting Fossil Fuel Transport & Exports Through the Pacific Northwest." *American Indian Law Journal* 7, no. 1 (2018): 249–357. https://digitalcommons.law.seattleu.edu/ailj/vol7/iss1/5.
65. R. Du Bey, A. S. Fuller and E. Miner. "Tribal Treaty Rights and Natural Resource Protection: The Next Chapter United States v. Washington – The Culverts Case." *American Indian Law Journal* 7, no. 2 (2019): 54–72.
66. *North Cheyenne Tribe v. Hodel*, 12 Indian L. Rptr. 3065, 3066 Indian L. Rptr 3065, 3066 (D. Mont. 1985).
67. *Northwest Sea Farm v. US Army Corps of Engineers*, 931 F. Supp. 1515 (W.D. Wash. 1996).
68. *Menominee Tribe of Indians v. United States*, 391 US 404 (Supreme Court 1968).
69. M. C. Walch. "Terminating the Indian Termination Policy." *Stanford Law Review* 35, no. 6 (1999): 1181–1215.
70. *Minnesota v. Mille Lacs Band of Chippewa Indians*, 526 US 172 (Supreme Court 1999).
71. *Herrera v. Wyoming*, 587 US ___ (Supreme Court 2019).
72. S. Krakoff. "Public Lands, Conservation, and the Possibility of Justice." *Harvard Civil Rights – Civil Liberties Law Review* (2018): 213–258.
73. S. Krakoff. "Not Yet America's Best Idea: Law, Inequality, and Grand Canyon National Park." *University of Colorado Law Review* 91 (2020): 559–648.
74. Congressional Research Service. *Federal Land Designations: A Brief Guide*. Report by L. B. Comay, R. E. Crafton, C. H. Vincent and K. Hoover. R45340 (Washington, DC: October 11, 2018). https://fas.org/sgp/crs/misc/R45340.pdf.
75. R. Tsosie. "The Conflict between the 'Public Trust' and the 'Indian Trust' Doctrines: Federal Public Land Policy and Native Nations." *Tulsa Law Review* 39, no. 271 (2003): 309–310.
76. N. J. Newton. "At the Whim of the Sovereign: Aboriginal Title Reconsidered." *Hastings Law Journal* 31 (1980): 1215–1285. https://scholarship.law.nd.edu/law_faculty_scholarship/1199.
77. C. G. Berkey. *Legal Challenges Regarding Native Land Ownership*. Continuing Legal Education discussion paper. July 8, 2013. www.berkeywilliams.com/wp/wp-content/uploads/2013/08/Native-Land-Ownership-CLE-Paper-7.8.2013.pdf.

78. *Mitchel v. United States*, 34 US (9 Pet.) 711, 746 (Supreme Court 1835).
79. M. C. Blumm. "Why Aboriginal Title Is a Fee Simple Absolute." *Lewis & Clark Review* 15, no. 4 (2012): 975–993. https://law.lclark.edu/live/files/10655-lcb154art4blummpdf.
80. *Cramer v. United States*, 261 US 219 (Supreme Court 1923).
81. *United States ex rel Huaipai Indians v. Santa Fe Pacific Railroad*, 314 US 339 (Supreme Court 1941).
82. *United States v. Alcea Band of Tillamooks*, 329 US 40, 46 (Supreme Court 1946).
83. *Tee-Hit-Ton Indians v. United States*, 348 US 272 (Supreme Court 1955).
84. *Lyng v. Northwest Indian Cemetery Protective Association*, 485 US 439 (Supreme Court 1988).
85. P. McKenna. "Wisconsin Tribe Votes to Evict Oil Pipeline from Its Reservation." *Inside Climate News*, January 16, 2017. https://insideclimatenews.org/news/16012017/dakota-access-pipeline-standing-rock-enbridge-line-5-native-american-protest.
86. Associated Press. "Michigan Governor Seeks Shutdown of Enbridge Pipeline in Great Lakes." *CBC News*, November 13, 2020. www.cbc.ca/news/canada/windsor/michigan-gov-shutdown-great-lake-pipe-line-1.5801015.
87. *Swinomish Indian Tribal Community v. BNSF Railway Company*, 951 F.3d 1142 (9th Cir. 2020).
88. *Montana v. United States*, 450 US 544, 565 (Supreme Court 1981).
89. Government Accountability Office. *Tribal Consultation: Additional Federal Actions Needed for Infrastructure Projects*. Report by A.-M. Fennell (March 2019). www.gao.gov/assets/700/697694.pdf.
90. R. E. Emanuel and D. E. Wilkins. "Breaching Barriers: The Fight for Indigenous Participation in Water Governance." *Water* 12, no. 8 (July 25, 2020): 2113–2150.
91. Bureau of Indian Affairs. *Indian Lands of Federally Recognized Tribes of the United States*. Department of the Interior (2016). www.bia.gov/sites/bia.gov/files/assets/public/webteam/pdf/idc1-028635.pdf
92. Department of State. *Factsheet: The Keystone XL Pipeline*. Bureau of Public Affairs (2011).
93. Climate Alliance Mapping Project. *Keystone XL Pipeline Map*. Collaboration between Indigenous Environmental Network and Keystone Mapping Project and Climate Alliance Mapping Project. https://climatealliancemap.org/kxl.
94. C. Sack. "Map: The Black Snake in Sioux County Showing the Dakota Access Pipeline Reroute through Former Sioux Lands and Its Consequences." *HuffPost*, February 11, 2016. www.huffpost.com/entry/a-nodapl-map_b_581a0623e4b014443087af35.
95. H. C. Tompkins. *Tribal Treaty and Environmental Statutory Implications of the Dakota Access Pipeline*. Submitted to secretary of the Interior. No. 1:16-cv-1534-JEB. December 4, 2016.
96. *Notice of Intent to Prepare an Environmental Impact Statement in Connection with Dakota Access, LLC's Request for an Easement to Cross Lake Oahe, North Dakota*. Department of the Army, Department of Defense. 82 Federal Register 5543–5544 (January 18, 2017). www.federalregister.gov/documents/2017/01/18/2017-00937/notice-of-intent-to-prepare-an-environmental-impact-statement-in-connection-with-dakota-access-llcs.

97. Memorandum Opinion. *Standing Rock Sioux Tribe, et al. v. US Army Corps of Engineers, et al.* District Court for the District of Columbia Case No. 1:16-cv-01534-JEB: Doc. 496 (D.D.C. Mar. 25, 2020). https://earthjustice.org/sites/default/files/files/standing-rock-sj.pdf
98. M. Faitsch. "'Highest Responsibility and Trust': The National Environmental Policy Act & the Dakota Access Pipeline." *Connecticut Law Review* 51, no. 4 (2019): 1043–1072.
99. M. Soraghan. "Pipelines: Trail of Spills Haunts Dakota Access Developer." *E&E News*, May 26, 2020.
100. Associated Press Staff. "Keystone Pipeline Spill in South Dakota Twice As Big As First Thought." *Associated Press*, April 7, 2018. www.argusleader.com/story/news/crime/2018/04/07/keystone-pipeline-spill-south-dakota-twice-big-first-thought/496613002.
101. S. Mufson and C. Mooney. "Keystone Pipeline Spills about 210,000 Gallons of Oil in South Dakota." *Washington Post*, November 17, 2017.
102. E. S. Rueb and N. Chokshi. "Keystone Pipeline Leaks 383,000 Gallons of Oil in North Dakota." *New York Times*, October 31, 2019.
103. Greenpeace. *Dangerous Pipelines: Enbridge's History of Spills Threatens Minnesota Waters*. Report by T. Donaghy (November 2018).
104. E. Lipton. "A Lobbyist, A Condo Deal, A Green Light." *New York Times*, April 3, 2018.
105. L. Song and E. McGowan. "Federal Agency Blames 'Complete Breakdown of Safety at Enbridge' for 2010 Oil Spill." *Inside Climate News*, July 10, 2012. https://insideclimatenews.org/news/10072012/national-transportation-safety-board-ntsb-kalamazoo-enbridge-6b-pipeline-marshall-michigan.
106. D. J. Trump. *Memorandum for the Secretary of the Army on the Construction of the Dakota Access Pipeline*. Office of the Press Secretary, 2017. https://assets.documentcloud.org/documents/3410448/Construction-of-the-Dakota-Access-Pipeline.pdf
107. *Notice of Termination of the Intent to Prepare an Environmental Impact Statement in Connection with Dakota Access, LLC's Request for an Easement to Cross Lake Oahe, North Dakota*. Department of the Army, Department of Defense. 82 Federal Register 11021 (February 17, 2017). www.federalregister.gov/documents/2017/02/17/2017-03204/notice-of-termination-of-the-intent-to-prepare-an-environmental-impact-statement-in-connection-with.
108. Congressional Research Service. *Army Corps Easement Process and Dakota Access Pipeline Easement Status*. Report by N. T. Carter (February 14, 2017). https://fas.org/sgp/crs/misc/IN10644.pdf.
109. M. Berman. "Dakota Pipeline Protest Camp: Ten Arrested, Dozens More Believed to Remain after Evacuation Deadline." *Washington Post*, February 23, 2017.
110. J. Fortin and L. Friedman. "Dakota Pipeline Is Ordered to Shut Down During Environmental Review." *New York Times*, July 7, 2020.
111. R. Frazin. "Court Cancels Shutdown of Dakota Access Pipeline." *The Hill*, August 5, 2020.
112. J. Blum. "Federal Appeals Court Sets Nov. 4 Court Date for Dakota Access Pipeline Fight." *S&P Global*, September 18, 2020. www.spglobal.com/platts/en/

market-insights/latest-news/natural-gas/091820-federal-appeals-court-sets-nov-4-court-date-for-dakota-access-pipeline-fight.

113. A. Sick. "Judge Allows Dakota Access Pipeline to Keep Operating." *Bismarck Tribune*, May 22, 2021.
114. P. McKenna. "Dakota Pipeline Was Approved by Army Corps Over Objections of Three Federal Agencies." *Inside Climate News*, August 30, 2016. https://insideclimatenews.org/news/30082016/dakota-access-pipeline-standing-rock-sioux-army-corps-engineers-approval-environment.
115. R. A. Bell. "The Fort Laramie Treaty of 1868 and the Sioux: Is the United States Honoring the Agreements It Made?" *Indigenous Policy Journal* 28, no. 3 (2018): 1–13.
116. J. P. LaVelle. "Rescuing Paha Sapa: Achieving Environmental Justice by Restoring the Great Grasslands and Returning the Sacred Black Hills to the Great Sioux Nation." *Great Plains Natural Resources Journal* 4 (2001): 40–101.
117. A. Rome. "Black Snake on the Periphery: The Dakota Access Pipeline and Tribal Jurisdictional Sovereignty." *North Dakota Law Review* 93, no. 1 (2018): 57–86. https://law.und.edu/_files/docs/ndlr/pdf/issues/93/1/93ndlr57.pdf.
118. Public Seminar. *The Supreme Law of the Land: Standing Rock and the Dakota Access Pipeline*. Report by J. Ostler and N. Estes (February 3, 2017). https://publicseminar.org/2017/02/the-supreme-law-of-the-land/.
119. *Standing Rock Sioux Tribe and Cheyenne River Sioux Tribe v. United States Army Corps of Engineers and Dakota Access LLC*, WL 7189653 (D.D.C. 2016).
120. *Rosebud Sioux Tribe v. Donald J. Trump*, WL 1456413 (D. Mont. 2020).
121. US Congress. Senate. *Keystone XL Pipeline Approval Act*. 114th Cong. 1st sess. Introduced in Senate February 11, 2015.
122. H. M. Babcock. "Issuance of the Keystone XL Permit: Presidential Prerogative or Presidential 'Chutzpah.'" *Montana Law Review* 81, no. 1 (2020): 5–57. https://scholarship.law.georgetown.edu/cgi/viewcontent.cgi?article=3290&context=facpub.
123. D. J. Trump. *Presidential Memorandum Regarding Construction of the Keystone XL Pipeline*. 2017. www.whitehouse.gov/presidential-actions/presidential-memorandum-regarding-construction-keystone-xl-pipeline.
124. *Issuance of Permits with Respect to Facilities and Land Transportation Crossings at the International Boundaries of the United States*. E.O. 13867. Executive Office of the President. 84 Federal Register 15491–15493 (April 10, 2019) www.federalregister.gov/documents/2019/04/15/2019-07645/issuance-of-permits-with-respect-to-facilities-and-land-transportation-crossings-at-the.
125. Canadian Press. "TC Energy to Start Building Keystone XL Pipeline after Alberta Government Invests $1.1B US." *CBC News*, March 31, 2020. www.cbc.ca/news/canada/calgary/tc-energy-keystone-xl-pipeline-1.5515850.
126. R. Tuttle and M. Bellusci. "Covid-19 May Finish Keystone XL for Good, with a Little Help from Biden." *World Oil*, October 27, 2020. www.worldoil.com/news/2020/10/27/covid-19-may-finish-keystone-xl-for-good-with-a-little-help-from-biden.
127. *Indigenous Environmental Network v. Department of State*, 347 F. Supp. 3d 561 (D. Mont. 2018).
128. "Indigenous Environmental Network v. Department of State." *Harvard Law Review* 132, no. 8 (2019): 2368. https://harvardlawreview.org/2019/06/indigenous-environmental-network-v-department-of-state.

129. A. Chiu, J. Eilperon and B. Dennis. "Sending Rebuke, Judge Halts Keystone XL Pipeline." *Washington Post*, November 10, 2018.
130. *Rosebud Sioux Tribe v. Trump*, 428 F. Supp. 3d. 282 (D. Mont. 2019).
131. *Rosebud Sioux Tribe and Fort Belknap Indian Community v. Donald J. Trump, Michael R. Pompeo, David Hale, US Department of Interior, David L. Bernhardt, Transcanada Corporation, and Transcanada Keystone Pipeline, LP.*, WL 2373054 (D. Mont. 2019).
132. *Indigenous Environmental Network v. Trump*, 428 F. Supp. 3d 296 (D. Mont. 2019).
133. D. Hayes et al. *Comments on the US Army Corps of Engineers' Proposal to Reissue and Modify Nationwide Permit 12, Docket No. COE-2015–0017*. Submitted to US Army Corps of Engineers. August 1, 2016.
134. *Northern Plains Resource Council v. US Army Corps of Engineers*, 460 F. Supp. 3d. 1030 (D. Mont. 2020).
135. R. Frazin. "Supreme Court Reinstates Fast-Track Pipeline Permitting Except for Keystone XL." *The Hill*, July 6, 2020. https://thehill.com/policy/energy-environment/506112-supreme-court-reinstates-fast-track-pipeline-permitting-except-for.
136. N. H. Farah and C. Anchondo. "If Lawsuits Don't Kill Ore. LNG Terminal, Pandemic Might." *E&E News*, September 9, 2020.
137. G. Dembicki. "They're Trying to Take My House." *The Tyee*, September 8, 2020. https://thetyee.ca/News/2020/09/08/Theyre-Trying-To-Take-My-House.
138. Reuters Staff. "US FERC Delivers Blow to Oregon LNG Terminal, Upholds State's Permit Denial." *Reuters*, January 19, 2021.
139. M. C. Blumm and J. B. Litwak. "Democratizing Treaty Fishing Rights: Denying Fossil-Fuel Exports in the Pacific Northwest." *Colorado Natural Resources, Energy, & Environmental Law Review* 30, no. 1 (Winter 2019): 1–34.
140. *Edge of Morning: Native Voices Speak for the Bears Ears*. Salt Lake City, UT: Torrey House, 2017.
141. R. H. Keller and M. F. Turek. *American Indians and National Parks*. Tucson: University of Arizona Press, 1999.
142. M. D. Spence. *Dispossessing the Wilderness: Indian Removal and the Making of the National Parks*. New York: Oxford University Press, 1999.
143. S. Hirst. *I Am the Grand Canyon: The Story of the Havasupai People*. Grand Canyon: Grand Canyon Association, 2006.
144. US Congress. House. *Utah Public Lands Initiative Act*. H.R. 5780, 114th Congress, 2nd Sess. Introduced in House July 14, 2016.
145. S. Chapoose et al. "Op-Ed: PLI Would Be the First Indian Land Grab in 100 Years." *Salt Lake Tribune*, September 17, 2016. https://archive.sltrib.com/article.php?id=4358069&itype=CMSID.
146. B. Maffly. "Ute Tribe Rejects State's Plan for Reservation Lands." *Salt Lake Tribune*, September 23, 2016. https://archive.sltrib.com/article.php?id=4387082&itype=CMSID.
147. US House of Representatives. *Hearing on H.R. 5780, to Provide Greater Conservation, Recreation, Economic Development and Local Management of Federal Lands in Utah, and for Other Purposes, "Utah Public Lands Initiative Act."* Subcommittee on Federal Lands of the Committee on Natural Resources. 114th Cong. 2nd sess. September 14, 2016.

148. Bears Ears Inter-Tribal Coalition. *Proposal to President Barack Obama for the Creation of Bears Ears National Monument*. Bears Ears Inter-Tribal Coalition: A Partnership of the Hopi, Navajo, Uintah and Ouray Ute, Ute Mountain Ute, and Zuni Government. October 15, 2015. https://utahdinebikeyah.org/full-proposal.
149. Archaeology Southwest et al. *Request That the Department of Interior Continue to Support the Designation of the Bears Ears National Monument*. Submitted to R. K. Zinke, secretary of the Interior. March 3, 2017.
150. E. E. Cummings, ranking member of the House Committee on Oversight and Government Reform. "Documents Obtained by Oversight Committee Refute Republican Claims That Obama Administration Did Not Consult on Bears Ears Monument Designation." News release, April 13, 2017, https://democrats-oversight.house.gov/news/press-releases/documents-obtained-by-oversight-committee-refute-republican-claims-that-obama.
151. Department of the Interior. *Bears Ears National Monument Boundary Modification*. 2017. www.doi.gov/sites/doi.gov/files/uploads/benm_12012017.pdf.
152. Center for Biological Diversity. *Oil and Gas Expressions of Interest in Bears Ears*. Report by K. Clauser (2017). http://insideenergy.org/2017/09/01/oil-gas-eyes-bears-ears-fringes.
153. Grand Canyon National Trust. *Active Mining Claims around Bears Ears National Monument*. Report by S. Smith (2018). www.grandcanyontrust.org/map-active-mining-claims-around-bears-ears-national-monument-february-2018.
154. T. Burr. "Effort to Shrink Bears Ears National Monument Started before Donald Trump Was Elected President." *Salt Lake Tribune*, December 3, 2017. www.sltrib.com/news/politics/2017/12/03/effort-to-shrink-bears-ears-national-monument-started-before-donald-trump-was-elected-president/+&cd=19&hl=en&ct=clnk&gl=ca.
155. K. Siegler. "Utah Representative Wants Bears Ears Gone and He Wants Trump to Do It." *NPR*, February 5, 2017. www.npr.org/2017/02/05/513492389/utah-representative-wants-bears-ears-gone-and-he-wants-trump-to-do-it.
156. D. J. Trump. *Presidential Executive Order on the Review of Designations Under the Antiquities Act*. 2017. www.whitehouse.gov/presidential-actions/presidential-executive-order-review-designations-antiquities-act.
157. R. K. Zinke. *Final Report Summarizing Findings of the Review of Designations under the Antiquities Act*. Submitted to D. J. Trump. December 2017. www.doi.gov/sites/doi.gov/files/uploads/revised_final_report.pdf.
158. D. J. Trump. *Presidential Proclamation Modifying the Bears Ears National Monument*. 2017. www.whitehouse.gov/presidential-actions/presidential-proclamation-modifying-bears-ears-national-monument.
159. B. Clinton. *Proclamation 6920 – Establishment of the Grand Staircase-Escalante National Monument*. 1996. www.govinfo.gov/content/pkg/WCPD-1996-09-23/pdf/WCPD-1996-09-23-Pg1788.pdf.
160. J. C. Ruple, M. Henderson and C. Caitlin. "Up for Grabs – The State of Fossils Protection in (Recently) Unprotected National Monuments." *Georgetown Law Review Online*, October 5, 2018. www.law.georgetown.edu/environmental-law-review/blog/up-for-grabs-the-state-of-fossils-protection-in-recently-unprotected-national-monuments.

161. *The Wilderness Society, et al. v. Donald J. Trump, et al.; Grand Staircase Escalante Partners, et al. v. Donald J. Trump, et al.*, WL 1904809 (D.D.C. 2020).
162. Bureau of Land Management. "BLM Restores Access with a Blueprint for Managing National Monuments and Public Lands in Utah." News release, February 6, 2020. www.blm.gov/press-release/blm-restores-access-blueprint-managing-national-monuments-and-public-lands-utah.
163. C. Davenport. "Trump Opens National Monument Land to Energy Exploration." *New York Times*, February 6, 2020. www.nytimes.com/2020/02/06/climate/trump-grand-staircase-monument.html.
164. Bureau of Land Management, Department of the Interior. *Bears Ears National Monument: Proposed Monument Management Plans and Final Environmental Impact Statement*. Report by S. Jáa and Indian Creek Unites (July 2019). https://eplanning.blm.gov/public_projects/lup/94460/20000105/250000108/Volume1_Chapters_1-4_Bears_Ears_Proposed_MMPs-Final_EIS.pdf.
165. *Records of Decision and Approved Monument Management Plans for the Bears Ears National Monument Indian Creek and Shash Jáa Units, Utah: Notice of Availability*. Docket No. 20X 1109AF LLUT930000 L16100000.DR0000. LXSSJ0650000. Bureau of Land Management, Interior and Forest Service, USDA. 85 Federal Register 9800–9801 (February 20, 2020). www.federalregister.gov/documents/2020/02/20/2020-03375/notice-of-availability-of-the-records-of-decision-and-approved-monument-management-plans-for-the.
166. J. Eilperin. "Uranium Firm Sought Bears Ears Cut." *Washington Post*, December 10, 2017.
167. Natural Resource Defense Council. *Uranium Development Potential in Bears Ears National Monument*. 2017.
168. A. Weiss. *Documents: When a Uranium Mining Company Lobbied for Bears Ears National Monument Action, the Trump Administration Acted*. Center for Western Priorities. 2017.
169. *Hopi Tribe, et al. and Utah Dine Bikeyah, et al. v. Donald J. Trump; Natural Resources Defense Council, Inc., et al. v. Donald J. Trump*, WL 7943150 (D.D.C. 2019).
170. *The Wilderness Society, et al. v. Donald J. Trump, et al.; Grand Staircase Escalante Partners, et al. v. Donald J. Trump, et al.*, WL 7902967 (D.D.C. 2019).
171. *Hopi Tribe v. Trump*, No. 17-CV-2590 (TSC), 2019 WL 2494161 (D.D.C. Mar. 20, 2019).
172. M. Squillace et al. "Presidents Lack the Authority to Abolish or Diminish National Monuments." *Virginia Law Review Online* 103 (2017): 55–71. www.virginialawreview.org/sites/virginialawreview.org/files/Hecht%20PDF.pdf.
173. J. C. Ruple. "The Trump Administration and Lessons Not Learned from Prior National Monument Modifications." *Harvard Environmental Law Review* 43 (2019): 1–76.
174. Congressional Research Service. *National Monuments and the Antiquities Act*. Report by C. H. Vincent, specialist in Natural Resources Policy (November 30, 2018). https://fas.org/sgp/crs/misc/R41330.pdf.
175. N. Groom. "Trump Public Lands Agenda Threatened by New Court Ruling." *Reuters*, September 29, 2020. www.reuters.com/article/us-usa-drilling-lawsuit-idUSKBN26K3NK.

176. R. Frazin. "Court Removes Pendley from Role As Public Lands Chief." *The Hill*, September 25, 2020. https://thehill.com/policy/energy-environment/518376-court-removes-pendley-from-role-as-public-lands-chief.
177. US House of Representatives. Report 116–224: Chaco Cultural Heritage Area Protection Act of 2019. Report by R. M. Grijalva. 116th Congress 1st Sess. (Washington, DC: 2019). www.govinfo.gov/content/pkg/CRPT-116hrpt224/html/CRPT-116hrpt224.htm.
178. National Parks Conservation Association. *Spoiled Parks: The 12 National Parks Most Threatened by Oil and Gas Development*. www.npca.org/reports/oil-and-gas-report.
179. S. Streater. "Bernhardt Commits to Leasing Moratorium Near Chaco Canyon." *E&E News*, May 29, 2019.
180. Bureau of Land Management, Department of the Interior. *Farmington Mancos-Gallup Draft Resource Management Plan Amendment and Environmental Impact Statement*. Report by T. Spisak and B. W. Stevens (February 28, 2020). https://eplanning.blm.gov/public_projects/lup/68107/20013477/250018467/FMG_DraftRMPA-EIS_Vol-1_508.pdf.
181. S. Grant. "Aggregate Airs: Atmospheres of Oil and Gas in the Greater Chaco." *Engaging Science, Technology, and Society* 6 (2020): 534–554.
182. M. Kakol, D. Upson and A. Sood. "Susceptibility of Southwestern American Indian Tribes to Coronavirus Disease 2019 (COVID-19)." *Journal of Rural Health* 37, no. 1 (January 2021): 197–199.
183. Bureau of Land Management, Department of the Interior. "BLM and BIA to Host Five Virtual Public Meetings for the Farmington Mancos-Gallup Resource Management Plan Amendment and Environmental Impact Statement." News release, April 29, 2020, www.blm.gov/press-release/blm-and-bia-host-five-virtual-public-meetings-farmington-mancos-gallup-resource.
184. T. Higgins. "With Push to Drill in Nevada, a Failure to Consult Native Peoples." *Earth Island Journal*, April 3, 2019. www.earthisland.org/journal/index.php/articles/entry/drill-nevada-failure-to-consult-native-peoples.
185. US House of Representatives. *National Parks, Forests, and Public Lands Subcommittee Legislative Hearing on H.R. 1373 and H.R. 2181*. Subcommittee on National Parks, Forests, and Public Lands. 116th Congress, 1st sess. June 5, 2019.
186. J. Osborne. "Trump Turns Oil Firms Loose on the American West." *Houston Chronicle*, April 17, 2018. www.houstonchronicle.com/business/article/Is-Trump-selling-America-s-wilderness-to-energy-12840533.php.
187. M. Maruca. "From Exploitation to Equity: Building Native-Owned Renewable Energy Generation in Indian Country." *William & Mary Environmental Law and Policy Review* 43, no. 2 (2019): 391–500. https://scholarship.law.wm.edu/wmelpr/vol43/iss2/3.
188. B. Patterson. "Fossil Fuels: Tribes Divided over Unlocking Energy Wealth." *E&E News*, November 16, 2016.
189. Sandia National Laboratories. *Identifying Barriers and Pathways for Success for Renewable Energy Development on American Indian Lands*. Report by T. E. Jones and L. E. Necefer (November 2016). www.energy.gov/sites/prod/files/2017/05/f34/Sandia_Report_2016-311J.pdf.

190. E. A. Kronk Warner. "Tribal Energy Resource Agreements: The Unintended 'Great Mischief for Indian Energy Development' and the Resulting Need for Reform." *Pace Environmental Law Review* 29, no. 3 (October 11, 2012): 811–859.
191. Bureau of Indian Affairs. *Oil and Gas Outlook in Indian Country* (2013).
192. J. Thompson. "The Ute Paradox." *High Country News*, July 12, 2010. www.hcn.org/issues/42.12/the-ute-paradox.
193. James M. Olguin. *The GAO Report on Indian Energy Development: Poor Management by BIA Has Hindered Development on Indian Lands*. Submitted to US Senate Committee on Indian Affairs. October 21, 2015.
194. G. E. B. Thompson. "The Double-Edged Sword of Sovereignty by the Barrel: How Native Nations Can Wield Environmental Justice in the Fight against the Harms of Fracking." *UCLA Law Review* 63, no. 6 (August 2016): 1818–1860.
195. Center for Indian Country Development. "Southern Ute Reservation." Federal Reserve Bank of Minneapolis, 2017. www.minneapolisfed.org/indiancountry/resources/reservation-profiles/southern-ute-reservation.
196. Southern Ute Indian Tribe. *Final Report for 2017 Southern Ute Indian Tribe Comprehensive Emissions Inventory for Criteria Pollutants, Hazardous Air Pollutants, and Greenhouse Gases* (2019).
197. A. Crouse. "Idle Oil, Gas Wells Threaten Indian Tribes while Energy Companies, Regulators Do Little." *Investigate West*, September 5, 2018.
198. Energy Information Administration. "North Dakota: State Profile and Energy Estimates." April 16, 2020. www.eia.gov/state/analysis.php?sid=ND#113.
199. Native Business Staff. "Chairman Fox: Oil Crisis Threatens to 'Knockout' MHA Nation's Economy." *Native Business*, April 30, 2020. www.nativebusinessmag.com/chairman-fox-oil-crisis-threatens-to-knockout-mha-nations-economy/+&cd=5&hl=en&ct=clnk&gl=ca.
200. P. Davies. "Homeland of Opportunity: The Bakken Oil Boom Has Brought Unprecedented Prosperity – and Daunting Challenges – to the Fort Berthold Indian Reservation." *Fed Gazette*, October 2014. www.minneapolisfed.org/~/media/files/pubs/fedgaz/14-10/fedgazette_oct2014_homeland_of_opportunity.pdf.
201. K. Finn et al. "Responsible Resource Development and Prevention of Sex Trafficking: Safeguarding Native Women and Children on the Fort Berthold Reservation." *Harvard Journal of Law & Gender* 40, no. 1 (2017): 1–51.
202. L. Grisafi. "Living in the Blast Zone: Sexual Violence Piped onto Native Land by Extractive Industries." *Columbia Journal of Law and Social Problems* 53, no. 4 (2020): 509–539.
203. A. Lustgarten. "Land Grab Cheats North Dakota Tribes out of $1 Billion, Suits Allege." *ProPublica*, February 23, 2013. www.propublica.org/article/land-grab-cheats-north-dakota-tribes-out-of-1-billion-suits-allege.
204. E. Scheyder. "Can American Indian Reformers Slow an Oil Boom?" *Reuters*, November 3, 2014. www.reuters.com/article/idUKKBN0IN0AR20141103.
205. J. Wold Tice. "Under the Earthlodge: Extraction of the MHA Nation." *Lehigh University Theses and Dissertations* (December 8, 2016). https://core.ac.uk/reader/228656910.
206. ICF International. *Onshore Petroleum and Natural Gas Operations on Federal and Tribal Lands in the United States: Analysis of Emissions and Abatement*

Opportunities (September 16, 2015). www.edf.org/sites/default/files/content/federal_and_tribal_land_analysis_presentation_091615.pdf.

207. L. Konkel. "Salting the Earth: The Environmental Impact of Oil and Gas Wastewater Spills." *Environmental Health Perspectives* 124, no. 12 (December 1, 2016). https://ehp.niehs.nih.gov/doi/full/10.1289/ehp.124-A230.

208. N. E. Lauer, J.S. Harkness and A. Vengosh. "Brine Spills Associated with Unconventional Oil Development in North Dakota." *Environmental Science & Technology* 50, no. 10 (May 17, 2016): 5389–5397. https://doi.org/10.1021/acs.est.5b06349.

209. E. Shogren. "Loophole Lets Toxic Oil Water Flow Over Indian Land." *NPR*, November 15, 2012. www.npr.org/2012/11/15/164688735/loophole-lets-toxic-oil-water-flow-over-indian-land.

210. Department of the Interior and US Geological Survey. *Delineation of Brine Contamination in and near the East Poplar Oil Field, Fort Peck Indian Reservation, Northeastern Montana, 2004–09.* Report by J. N. Thamke and B. D. Smith (April 2, 2014). https://pubs.usgs.gov/sir/2014/5024.

211. D. K. Zoanni. "Traditional Knowledge Systems and Tribal Water Governance on Fort Peck Indian Reservation, MT." *Montana State University Theses* (November 2017). https://scholarworks.montana.edu/xmlui/bitstream/handle/1/14058/ZoanniD1217.pdf?sequence=4&isAllowed=y.

212. K. Conde. "The Damage Done: A Montana Community Braces for the Next Oil Boom while Still Dealing with the Devastating Effects of the Last One." *Missoula Independent*, February 20, 2014.

213. E. L. Quintana. "Identification of Man-Made Hazards in Aneth Chapter, Navajo Nation, Utah." *University of New Mexico Thesis* (2012). www.mobt3ath.com/uplode/book/book-73532.pdf.

214. J. V. Royster. "Oil and Water in the Indian Country." *Natural Resources Journal* 37, no. 2 (Spring 1997): 457–490. https://digitalrepository.unm.edu/cgi/viewcontent.cgi?article=1694&context=nrj.

215. P. M. Thomas. *Testimony Addressing the Urgent Needs of Our Tribal Communities.* Submitted to US House of Representatives Committee on Energy and Commerce. July 8, 2020.

216. Government Accountability Office. *Indian Energy Development: Poor Management by BIA Has Hindered Energy Development on Indian Lands.* Report by F. Rusco, director of Natural Resources and Environment (June 8, 2015). www.gao.gov/products/GAO-15-502.

217. J. Barrasso, senator from Wyoming. *Testimony on the GAO Report on Indian Energy Development: Poor Management by BIA Has Hindered Development on Indian Lands.* Submitted to US Senate Committee on Indian Affairs. October 21, 2015.

218. L. Jack, chairman of the board of directors at Oceti Sakowin Power Authority. *Testimony about the Oceti Sakowin Power Authority.* Submitted to US House Subcommittee on Energy and Mineral Resources. April 2019.

219. Government Accountability Office. *Tribal Energy: Opportunities Exist to Increase Federal Agencies' Use of the Tribal Preference Authority.* Report by F. Rusco, director of Natural Resources and Environment (April 2019). www.gao.gov/assets/700/698642.pdf.

220. L. Morales. "Navajo Nation President Proclaims Renewable Energy Top Priority." *Arizona Public Media*, April 3, 2019. www.azpm.org/p/home-articles-news/2019/4/3/148995-navajo-nation-president-proclaims-renewable-energy-top-priority.
221. A. Curley. "A Failed Green Future: Navajo Green Jobs and Energy 'Transition' in the Navajo Nation." *Geoforum* 88 (January 2018): 57–65. www.sciencedirect.com/science/article/pii/S001671851730324X.
222. J. Spector. "Developer sPower Teams Up with Navajo Power to Replace Coal Plant with Solar." *Greentech Media News*, May 6, 2020. www.greentechmedia.com/articles/read/navajo-power-spower-funds-raised-solar.
223. L. Necefer, G. Wong-Parodi and M. J. Small. "Governing Energy in Conflicted Resource Contexts: Culture, Cost, and Carbon in the Decision-Making Criteria of the Navajo Nation." *Energy Research & Social Science* 70 (2020). www.sciencedirect.com/science/article/pii/S2214629620302899#b0105.
224. Department of Energy and Department of the Interior. *Report to Congress: Energy Policy Act of 2005, Section 1813 Indian Land Rights-of-Way Study* (May 2007). www.energy.gov/sites/prod/files/oeprod/DocumentsandMedia/EPAct_1813_Final.pdf.
225. B. Donovan. "Oil & Gas Seeks $40 Million Bailout but Prez Is Barrier." *Navajo Times*, March 29, 2018. https://navajotimes.com/reznews/oil-gas-seeks-40-million-bailout-but-prez-is-barrier.
226. C. Helman. "The Navajo Wanted to Go Green and Then This Surprise Deal Made Them America's Third Largest Miner." *Forbes*, June 10, 2020. www.forbes.com/sites/christopherhelman/2020/06/10/the-navajo-wanted-to-go-green-then-this-surprise-deal-made-them-americas-3rd-largest-coal-miner/?sh=64e8d9ff2eee.
227. Institute for Energy Economics and Financial Analysis. *Tribal Utility-Scale Solar Initiatives Advance Across Southwest US*. Report by K. Cates, IEEFA research editor, and D. Wamsted, IEEFA editor/analyst (October 2019). http://ieefa.org/wp-content/uploads/2019/10/Tribal-Utility-Scale-Solar-Initiatives-Advance-Across-SW-US_October-2019.pdf.
228. Reuters Staff. "Nevada Utility Announces Three Major Solar Projects with Battery Storage." *Reuters*, June 25, 2019. www.reuters.com/article/us-usa-nevada-solar/nevada-utility-announces-three-major-solar-projects-with-battery-storage-idUSKCN1TQ2H5.
229. V. Deloria, Jr. *Behind the Trail of Broken Treaties: An Indian Declaration of Independence*. Austin: University of Texas Press, 1985.
230. M. L. M. Fletcher. "The Iron Cold of the Marshall Trilogy." *North Dakota Law Review* 82 (2006): 628–696.
231. S. Davis. "American Colonialism and Constitutional Redemption." *California Law Review* 105, no. 6 (December 2017): 1751–1806.
232. W. Eco-Hawk. *In the Courts of the Conqueror: The 10 Worst Indian Law Cases Ever Decided*. Golden: Fulcrum Publishing, 2012.
233. US Congress. Senate. *Chaco Cultural Heritage Area Protection Act of 2019*. S. 1079, 116th Congress, 1st Sess. Introduced in Senate April 9, 2019.
234. US Congress. House. *Bears Ears Expansion and Respect for Sovereignty Act*. H.R. 871, 116th Congress, 1st Sess. Introduced in House January 30, 2019.

235. US Congress. Senate. *America's Natural Treasures of Immeasurable Quality Unite, Inspire, and Together Improve the Economies of States Act.* S. 367, 116th Congress, 1st Sess. Introduced in Senate February 7, 2019.
236. H. Grabar. "The Battle for San Juan County, Utah." *Slate*, August 25, 2020.
237. K. Phillips. "It's Time to Recognize the Forgotten Americans Who Helped Elect Joe Biden." *Washington Post*, November 9, 2020.
238. J. Biden and K. Harris. *Biden–Harris Plan for Tribal Nations* (2020). https://joebiden.com/tribalnations.
239. R. Beitsch. "House Democrats Push Biden to pick Haaland As Next Interior Secretary." *The Hill*, November 20, 2020. https://thehill.com/policy/energy-environment/526853-house-democrats-push-biden-to-pick-haaland-as-next-interior.
240. J. Biden. *E.O. 13990 Protecting Public Health and the Environment and Restoring Science to Tackle the Climate Crisis.* Executive Office of the President. January 20, 2021. www.federalregister.gov/documents/2021/01/25/2021-01765/protecting-public-health-and-the-environment-and-restoring-science-to-tackle-the-climate-crisis.
241. *Cotton Petroleum Corp. v. New Mexico*, 490 US 163 (Supreme Court 1989).
242. T. Gates, president of the Seneca Nation. "Letter to the Bureau of Indian Affairs." August 30, 2017. www.BureauofIndianAffairs.gov/sites/BureauofIndianAffairs.gov/files/assets/as-ia/raca/pdf/13%20-%20Seneca%20Nation.pdf.
243. *Brendale v. Confederated Tribes and Bands of Yakima*, 492 US 408 (Supreme Court 1989).
244. *Montana v. United States*, 450 US 544 (Supreme Court 1981).
245. *Morton v. Mancari*, 417 US 535 (Supreme Court 1974).
246. *Parravano v. Babbitt*, 70 F.3d 539, 545 (9th Cir. 1995).
247. R. J. Miller. "The Doctrine of Discovery in American Indian Law." *Idaho Law Review* 42 (2005): 1–122.
248. *Menominee Tribe of Indians v. United States*, 388 F.2d 998, 1005–06 (Ct. Cl. 1967).
249. B. Campbell. *The Lac-Mégantic Rail Disaster: Public Betrayal, Justice Denied.* Toronto, ON: Lorimer, 2018.
250. B. Obama. "Remarks by the President at the White House Tribal Nations Conference." December 16, 2010. https://obamawhitehouse.archives.gov/the-press-office/2010/12/16/remarks-president-white-house-tribal-nations-conference.
251. B. Obama. *Veto Message to the Senate: S. 1, Keystone XL Pipeline Approval Act.* 2015 WL 758544. February 24, 2015.
252. M. Auffhammer. "Why I Think Not Building Keystone XL Will Likely Leave a Billion Barrels Worth of Bitumen in the Ground." *Hass Energy Institute Blog*, March 24, 2014. https://energyathaas.wordpress.com/2014/03/24/it-just-doesnt-add-up-why-i-think-not-building-keystone-xl-will-likely-leave-a-billion-barrels-worth-of-bitumen-in-the-ground.
253. C. Leonard. *The Secret History of Koch Industries and Corporate Power in America.* New York: Simon & Schuster, 2019.
254. C. F. Wilkinson and J. M. Volkman. "Judicial Review of Indian Treaty Abrogation: As Long As Water Flows, or Grass Grows upon the Earth – How Long a Time Is That." *California Law Review* 63 (1975): 601–661.

255. *City of Sherrill v. Oneida Indian Nation of New York*, 544 US 197 (Supreme Court 2005).
256. *Klamath and Modoc Tribes v. Maison*, 338 F.2d. 620 (9th Cir. 1964).
257. *Kimball v. Callahan*, 493 F.2d 564 (9th Cir. 1974).
258. M. Pearson. "Hunting Rights: Retention of Treaty Rights after Termination: Kimball v. Callahan." *American Indian Law Review* 4, no. 1 (1976): 121–133.
259. *Diné Citizens v. Bernhardt*, No. 18-2089 (10th Cir. 2019).

PART III

America's Seas

6

Oceans

Drilling versus Competing Use of Coasts and Seas

Although most US oil and gas extraction occurs on land, offshore extraction is substantial. Almost a quarter of US domestic oil comes from beneath the seabed; the proportion of US gas that comes from the ocean is smaller, at one-twentieth, but still far from negligible. The oil and gas industry is not the only economic sector, however, that depends on the sea. Tourism, recreation and fishing sectors thrive on healthy coasts and oceans. For them, the oil and gas industry is not only a competitor for access but also an existential danger because an oil spill such as the BP spill that occurred in the Gulf of Mexico in 2010 can kill the local fishing industry and drive away tourists.

The task of balancing the interests of the oil and gas industry with those of other economic sectors rests on the federal government. While some consultation with state governments is required, the final decisions shaping this balance are made in Washington – decisions such as which parts of US waters are open to drilling, what precautions should be taken to reduce the risk of pollution, and how to allocate the benefits of offshore extraction. America's need for oil has weighed heavily on past presidents, their administrations and the members of Congress tasked with balancing these interests. Some have tried hard to protect the environment and other economic sectors from the adverse effects of oil and gas extraction, whereas others have unapologetically sought to tip the legislative and regulatory balance heavily in favor of oil and gas. However, no prior administration shifted that balance in favor of oil and gas as far the Trump administration did.

In light of what preceding chapters have described about the administration's approach to onshore oil and gas production, it may come as no surprise to learn that its approach to offshore natural resource management is heavily weighted in favor of the oil and gas industry. Even so, the extent of the administration's favoritism is striking, especially when compared not only with the preceding Obama administration but also with earlier Republican administrations.

Trump and his appointees proposed the expansion of drilling into areas of the Pacific, Atlantic and Arctic Oceans that have seen little or no drilling. Trump's allies in Congress approved drilling in the Arctic National Wildlife Refuge – an expansion that comes at a time when companies already lease enormous stretches of America's seas but have not pursued production on 80 percent of that leased acreage.[1] The administration sold leases at flash-sale prices, slashed royalty rates and weakened safeguards against companies abandoning nonproductive oil rigs and other infrastructure. The Trump administration sought to bring windfalls, not for American taxpayers but for the American oil and gas industry.

This one-sided approach has found favor with some state governments but has been strongly opposed by others. As this chapter explains, offshore oil and gas exploitation is a mixed blessing – hence the need for a balanced national policy. While drilling has brought significant economic benefits to some states, it has subjected local economies to boom–bust cycles and inflicted environmental degradation that has imperiled many people's livelihoods. In other parts of the country, state governments view the prospect of offshore drilling in a largely negative light, seeing it as not only environmentally but also economically dangerous, and have staunchly opposed drilling in federal waters off their coasts. Rather than address their concerns, the Trump administration and some members of Congress sought to blunt state-level opposition, proposing to curb states' input into the management of federal waters and to penalize states that oppose offshore drilling.

This chapter delves into the Trump administration's financial management of offshore drilling; Chapter 7 turns to its management of safety in offshore drilling. I begin by describing the contributions of offshore drilling to the US economy (Section 6.1). Next, I delve into executive and congressional actions under the Trump administration to expand drilling (Section 6.2); explain why these actions were not economically prudent (Section 6.3); and describe how state governments responded to these proposed expansions (Section 6.4). Section 6.5 recounts how some members of Congress, state governments and NGOs have challenged drilling expansion and pushed for improved financial management of ocean resources.

6.1 The Contribution and Management of the Offshore Oil and Gas Sector

The offshore oil and gas sector is a major contributor to both the US energy supply and the US economy. In 2016, offshore oil and gas operations generated $30 billion for the US economy and an additional $2.7 billion for the US Treasury.[2]

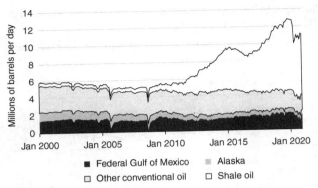

Figure 6.1 Crude oil production in the United States (2000–20)
Sources: Energy Information Administration, *Short-Term Energy Outlook Data Browser*, 2020;[7] Energy Information Administration, *Drilling Productivity Report*, 2020.[8]
Note: The Permian and Eagle Ford, which are primarily in Texas, and the Bakken, which is primarily in North Dakota, are the top shale-producing basins in the United States.

Crude oil production in the Gulf of Mexico contributes a substantial share of overall US production (Figure 6.1). Even with the expansion of shale oil production in the lower 48 states, production in the federally owned portion of the Gulf of Mexico contributed 18.2 percent of US crude oil production in 2016, second only to the contribution from the state of Texas.[3] The Gulf of Mexico also accounts for 5 percent of total US dry gas production.[4]

The state of Alaska accounted for 5.5 percent of US crude oil production in 2016.[3] This production takes place both onshore and offshore. In contrast to the overall growth in production in the Gulf of Mexico, production has declined in Alaska. By 2016, production was about a quarter of its peak production in 1988.[5] Alaska ranks third in the nation in natural gas extraction, but most of this gas does not reach markets because gas pipelines are not financially viable.[6] Instead, large volumes of gas produced during oil extraction are reinjected into oil fields to help maintain the pressure of the underground oil reservoirs and thus keep up crude oil production rates.[6]

The federal waters available for drilling are managed by the Department of the Interior under the authority of the 1953 Outer Continental Shelf Lands Act (OCSLA). Federal waters extend from 3 nautical miles off states' coastlines to 200 nautical miles seaward. State waters extend 3 nautical miles from the coast, with the exception of Texas and Florida, whose waters extend 9 nautical miles into the Gulf of Mexico.

Table 6.1 *Contribution of tourism and fisheries and the oil and gas industry to the ocean economy in 2015*

	Tourism, fishing, aquaculture, seafood processing		Oil and gas	
	Employment (% of the ocean economy)	GDP ($ billion)	Employment (% of the ocean economy)	GDP ($ billion)
Gulf of Mexico region	57	12	21	73
Texas	25	3	52	91
Louisiana	48	18	20	49
Alabama	64	<1	1	36
Mississippi	52	<1	<1	35
Florida	84	63	none	none
Mid-Atlantic region	73	61	none	none
North Carolina	92	66	none	none
South Carolina	91	84	none	none
Georgia	68	52	none	none
Northeast region	77	66	none	none
West Coast region	76	48	none	none

Source: National Ocean and Atmospheric Administration, *NOAA Report on the US Ocean and Great Lakes Economy*, 2017.[9]
Note: The source does not tabulate smaller figures.

OCSLA recognizes the need to manage oceans for the many different sectors operating in the "ocean economy," a term that denotes ocean-based economic and conservation activities that rely on the goods, services and ecological sustenance provided by marine ecosystems. As seen in Table 6.1, the oil and gas industry competes with tourism and fisheries in the Gulf of Mexico and in Alaska. The federal government sets out which areas will be available for potential oil and gas production in a five-year leasing plan known as the National Outer Continental Shelf Plan.

6.2 The Drive by Trump and Congress to Open US Seas to Drilling

The Trump administration inherited an imperfect but functioning statutory framework of regulatory oversight over offshore drilling. OCSLA requires that management of federal waters be "conducted in a manner which considers economic, social, and environmental values of the renewable and nonrenewable resources contained in the outer continental shelf, and the potential impact of oil and gas exploration on other resource values of the outer continental shelf and the marine, coastal, and human environments." The 1978 OCSLA Amendments set out an environmental review process for oil leasing, exploration and development, together with a process for the federal government to consult with state governments. Balancing competing interests and values was recognized as an essential component of an economically efficient and inclusive regulatory framework.

Together, the Trump administration and the 115th Congress whittled away at this balanced framework in various ways. In April 2017, Trump announced the America First Offshore Energy Strategy.[10] Nine months later, in January 2018, the Department of the Interior published its draft plan on the leasing of the outer continental shelf (OCS) for the years 2019–24.[11] This plan – the first of three stages before the secretary of the Interior publishes the final plan for implementation[i] – marked a major shift from the existing 2017–22 National OCS Plan (as summarized in Table 6.2).[12] The proposed draft plan, as seen in Map 6.1, would open leasing in federally owned areas of the Atlantic, the Pacific and most of the Arctic. It would also open leasing in those parts of the central and eastern Gulf of Mexico when the congressional moratorium on drilling in those areas expires in June 2022.[13]

The proposed expansion would encroach into areas that have hosted little or no drilling. Very little production currently takes place in the Pacific Ocean. Production in the federal waters in the Arctic Ocean is limited to two facilities: the Northstar facility in the Beaufort Sea, which is located in both state and federal waters,[5,14] and the Eni facility on the state-owned Spy Island, which uses extended drilling techniques to reach federal submerged waters in the Beaufort Sea.[15] Drilling in Alaska's state waters in the Arctic is limited to artificial islands in the Beaufort Sea and in the Cook Inlet. The federal waters in the Atlantic Ocean and the eastern Gulf of Mexico have no oil and gas production at all.

Trump's executive order also directed the Department of Commerce to review all designated national marine sanctuaries and marine national monuments, as well as any planned expansions of them. Following a review,

---

[i] In developing the National OCS Plan, the Department of the Interior proceeds in three stages: first, it publishes a proposed draft plan; second, it publishes a proposed final draft plan; and third, it publishes the final draft plan. The Department of the Interior is obliged to receive and to respond to public comments on each of the three drafts.

Table 6.2 *Proposed opening of seas to leasing under the 2019–24 National OCS Plan*

2017–22 National OCS Plan	Proposed draft 2019–24 National OCS Plan
10 lease sales. None in the parts of the central and eastern Gulf of Mexico where Congress enacted a moratorium until June 2022.	12 lease sales (10 lease sales in the western Gulf and in parts of the central and eastern Gulf that are not subject to the 2022 moratorium; 2 sales after 2022 in the areas subject to the moratorium).
1 lease sale in the Cook Inlet.	9 lease sales (3 in the Chukchi Sea; 3 in the Beaufort Sea; 2 in the Cook Inlet; 1 sale in other program areas off the Alaskan coast). No sales in the North Aleutian Basin Planning Area.
No lease sales since 1984.	7 lease sales (2 each for Northern California, Central California and Southern California; 1 for Washington/Oregon).
No lease sales since 1983 and no existing leases.	6 lease sales (3 each for the Mid- and South Atlantic; 2 for the North Atlantic; 1 for the Straits of Florida).

Source: Bureau of Ocean Management, *2019–2024 National Outer Continental Shelf Oil and Gas Leasing Draft Proposed Program*, 2018;[11] Bureau of Ocean Management, *2017–2022 Outer Continental Shelf Oil and Gas Leasing Proposed Final Program*, 2016.[12]

Commerce Secretary Wilbur Ross recommended reversing protection of coral reefs and atolls in the Pacific, seamounts and canyons in the Atlantic, and feeding grounds for whales and sharks off California.[16]

The Trump administration's draft leasing plan serves as the opening gambit in a multistage process before oil companies can commence drilling. The four stages of this process – development of the five-year plan, planning for specific lease sales, approval of exploration plans and approval of production plans – all require the Department of the Interior to undertake environmental assessments and invite public comment. Required levels of specificity increase with each stage.[12]

The ambition to open more of the sea to drilling did not originate with the Trump administration. The Obama administration, for instance, planned to open up drilling in the southeast Atlantic in early 2010 but shelved its plans when faced with the catastrophic BP oil spill.[17] It revisited the idea in 2015 but backed down when faced with opposition from the Pentagon,[18,19] economic sectors such as fishing and tourism, and many local communities.[20] What distinguished the Trump administration's drive to open up more of the oceans

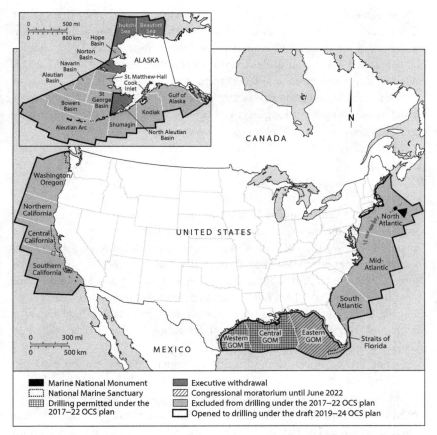

Map 6.1 Proposed expansion of offshore drilling under the proposed draft 2019–2024 National OCS Plan
Sources: Bureau of Ocean Energy Management, *2019–2024 National Outer Continental Shelf Oil and Gas Leasing Draft Proposed Program*, 2018;[11] Bureau of Ocean Energy Management, *2017–2022 Outer Continental Shelf Oil and Gas Leasing Proposed Final Program*, 2016;[12] Congressional Research Service, *Five-Year Offshore Oil and Gas Leasing Program for 2019–2024*, 2019.[13]
Note: GOM denotes the Gulf of Mexico.

to drilling was the extent of the oceans it wanted to open and its determination to ignore opposition. Republicans in Congress from coastal states such as North Carolina, South Carolina, Georgia and Florida, whose voters were opposed to offshore drilling, were largely unwilling to cooperate with Democrats to undertake bipartisan efforts in Congress to block the Trump administration's actions.

In December 2017, in step with the Trump administration's expansion of offshore drilling, Congress lifted its 40-year prohibition on drilling in the coastal plain of the Arctic National Wildlife Refuge (ANWR). Congress mandated the Department of the Interior to hold two lease sales before 2027, each covering a minimum of 400,000 acres. The department moved quickly to attempt to lock in one lease sale before the 2020 elections,[21] skirting environmental oversight by publishing an environmental *assessment* rather than a more comprehensive environmental impact *statement*.[22]

Alaska's senators and congressional representatives, such as Alaska's state government, supported drilling in ANWR, largely because the royalty revenue would be split equally between the state of Alaska and the federal government.[23] Supporters of drilling largely dismissed legitimate concerns about adverse environmental and economic impacts.[24] Senator Lisa Murkowski (R-Alaska) used the budget reconciliation process to push the Senate into passing the 2017 Tax Cut and Jobs Act with the ANWR provision. Even some Republican legislators requested that the ANWR provision be dropped from the tax bill. Representative David Reichert and 11 other House Republicans noted that "[ANWR] stands as a symbol of our nation's strong and enduring natural legacy ... Any development footprint in the refuge stands to disrupt this fragile, critically important landscape."[25] Such pleas failed to persuade a majority on Capitol Hill.

In Congress, a number of legislators who support offshore drilling sought to enact legislation that would penalize states that oppose offshore drilling. Several Republicans on the House Natural Resources Committee, including Representative Robert Bishop from Utah and Representative Paul Gosar from Arizona, supported drilling expansion. (Their states, Utah and Arizona, do not face direct costs from offshore drilling.) They proposed a draft discussion bill, the State Management of Federal Lands and Waters Act,[26,27] that would impose financial penalties on states that oppose drilling off their coasts, while rewarding states that favor it with revenue sharing. Bishop and Gosar justified these penalties as reimbursement to taxpayers, arguing that foregoing drilling would "have the effect of curtailing benefits from these taxpayer owned and high value assets."[26] Proponents also framed these penalties as states compensating the federal government for not being able to exercise its property right to develop offshore leases.

In reality, enacting these penalties into law would effectively institute drilling as the priority use of federal seas. It would be a major departure from the OSCLA approach of considering multiple uses and their impacts, including the impact of offshore drilling on other resources and on the environment. It would also disregard the long-established practice of states and the federal government cooperating to strike a balance among competing uses.

The Trump administration did not finalize its OCS draft plan prior to the end of the Trump presidency, thus leaving the 2017–22 OCS plan to govern leasing until June 2022.[ii] However, that draft plan serves a warning on how far a pro-oil administration is willing and able to go in expanding offshore drilling if Congress does not take steps to legislate protections for these oceans.

6.3 The Economics of Offshore Drilling Expansion

Management of federal ocean resources (i.e., national assets held by the federal government in trust for Americans today and for future generations) requires the executive branch and Congress to pay attention to at least two concerns. One is striking a balance between permitting offshore drilling and protecting the oceans and coasts. This entails not only managing economic activities such as tourism, fishing and recreation but also safeguarding the marine ecosystem for its intrinsic value. The other is protecting the financial interests of taxpayers by securing returns from leases awarded to the oil and gas companies and by guarding against liabilities from abandoned oil and gas infrastructure. Unfortunately, the Trump administration and the 115th Congress failed on both counts.

Until the advent of the Trump presidency, most administrations had sought to balance the risks and costs of drilling with competing economic activities. In 1991, for instance, President George H. W. Bush made parts of the Pacific and the Atlantic coasts off-limits to offshore drilling for a decade, including the North Atlantic, the Mid-Atlantic, southwest Florida, California, Washington, Oregon and southern Alaska. President Clinton extended that withdrawal for another decade.

However, not all administrations have exhibited such concern for the health of the oceans. President George W. Bush, for instance, in 2008 lifted the ban his father had imposed in 1991. The same year, Congress let expire its 27-year-old moratorium on drilling in the Atlantic and Pacific coasts as well as a small portion of the eastern Gulf of Mexico, having enacted it in 1981 and renewed it annually up to that point.

George W. Bush's successor in the White House tried harder to reconcile competing interests. The Obama administration's 2017–22 OCS plan permitted drilling in areas with large oil reserves but safeguarded other parts of the oceans and coasts for tourism, fishing and ecosystem protection. It offered lease sales in the western Gulf of Mexico, where there are substantial reserves (Table 6.3), but it did not offer new leases in the Atlantic, the

---

[ii] New lease sales after June 2022 can only be undertaken after a new five-year National OCS Plan is finalized.

Table 6.3 *Oil and gas reserves classified as undiscovered economically recoverable reserves in the OCS*

	Estimated reserves under US EIA price projections		Estimated years of consumption under US EIA price projections		Estimated reserves under lower assumed prices		Estimated reserves under higher assumed prices	
Commodity	Oil	Gas	Oil	Gas	Oil	Gas	Oil	Gas
Price	$100	$5.3	$100	$5.3	$60	$3.2	$160	$8.5
Gulf of Mexico OCS	42.9	92.0	6.0	3.3	39.6	74.7	44.8	103.5
Central Gulf	29.6	57.8	4.1	2.1	27.3	46.7	30.8	65.2
Western Gulf	10.2	27.2	1.4	1.0	9.4	21.8	10.7	30.5
Eastern Gulf	3.1	7.0	0.4	0.3	2.9	6.1	3.2	7.7
Straits of Florida	0.01	0.001	0.0004	0.01	0.01	0.01	0.01	0.01
Arctic OCS	17.3	33.6	2.4	1.2	8.4	9.4	22.0	60.4
Chukchi	9.3	22.6	1.3	0.8	2.9	4.3	12.6	40.6
Beaufort	6.1	8.1	0.8	0.3	4.0	4.2	7.1	12.6
Pacific OCS	7.3	9.4	1.0	0.3	6.5	8.3	7.9	10.4
Atlantic OCS	4.0	13.0	0.6	0.5	3.8	8.4	4.2	17.2

Source: Bureau of Ocean Management, *Assessment of Undiscovered Oil and Gas Resources of the Nation's Outer Continental Shelf*, 2017.[29]
Notes: The US Geological Service provides estimates of undiscovered, technically recoverable resources that have yet to be found (drilled) but, if found, could be produced using currently available technology and industry practices. Oil reserves are in billions of barrels. Gas reserves are in trillion cubic feet. Price of oil is per barrel. Price of gas is per million cubic feet. The Energy Information Administration projects oil prices at $93 per barrel in 2030, $105 per barrel in 2040 and $108 per barrel in 2050.[28]

Pacific, the eastern Gulf of Mexico or the Straits of Florida, where oil and gas reserves are limited and where coastal regions rely heavily on tourism and fishing. More than three-quarters of the employment and more than half of the gross domestic product of the ocean economy along the Atlantic and Pacific coasts come from tourism and fishing (as indicated in Table 6.1). As noted above, the Obama administration twice considered but withdrew draft proposals to open drilling in the southeast Atlantic coast, first when faced with the catastrophic BP oil spill and second when confronted by opposition from various quarters.

Drilling in the Arctic Sea has historically been held back by a mixture of financial, economic and ecological risks. Shell abandoned its Arctic project in the Chukchi Sea in 2015, citing poor results from its Burger J well.[30] Other companies, including BP, Chevron and Exxon, had abandoned their projects earlier.[30] The environmental impact assessment prepared for the 2017–22 OCS plan predicted that drilling operations in the Arctic would likely result in one large spill (~3,300 barrels) from a platform in each of the Chukchi and Beaufort Seas and up to four spills (~3,800 barrels) from pipelines in each sea.[31] Ultimately, the plan restricted expansion to only one sale in the Cook Inlet.

In contrast to the 2017–22 OCS plan, the Trump administration proposed to open up drilling even in those areas with minimal reserves that are highly reliant on tourism and fishing (Tables 6.1 and 6.3). The Trump administration sought to lock in lease sales and low royalty rates in contracts that would bind the US government for decades. The administration offered leases throughout the Gulf, rather than by sections of the Gulf, thus flooding the market with leases and pushing prices downwards.[32] Only a tiny fraction – 0.67 percent – of the acreage offered by the Trump administration in the Gulf of Mexico was sold (Table 6.4). And these sales fetched low prices, ranging from $153 to $238 per acre. The low demand and low prices echo the declining prices for leases even during the Obama administration (as shown in Table 6.4). Joe Balash, the assistant secretary for land and minerals management at the Department of the Interior, offered some insights into the administration's motives for holding these lease sales that brought scant benefit to the US Treasury and taxpayers. As reported by the media, Balash told oil industry representatives that the administration made a point of including contractual terms that favored oil drilling when leasing new coastal waters. He noted those terms will be difficult if not impossible for future administrations to rewrite or revoke given the sanctity of contract law in the United States.[33]

Further, the Trump administration cut royalty rates (which the George W. Bush administration had raised to boost federal revenue),[34] arguing that lower rates were necessary to incentivize companies to buy leases. In July 2017, the Trump administration reduced royalty rates in leases offered for sale in shallow water from 18.75 percent to 12.5 percent.[35] Later, in February 2018, its advisors recommended reducing royalty rates for deepwaters, also from 18.75 percent to 12.5 percent, without publishing a detailed report justifying this reduction (although, as noted below, this recommendation was withdrawn later in 2018).[36,37]

These cuts undervalue the option value of these resources to society.[38] Option value denotes the value of preserving the ability to make a choice on

Table 6.4 Gulf of Mexico offshore oil and gas leases, 2012–20

Date	Planning area	Acreage offered (millions of acres)	Percentage of offered acres sold	Average bid per acre sold	Oil price at time of lease (per barrel)
2020: March	Gulfwide	78	0.5	$234	$27
2019: August	Gulfwide	78	1.1	$190	$59
2019: March	Gulfwide	79	1.6	$194	$59
2018: August	Gulfwide	78	1.0	$222	$68
2018: March	Gulfwide	77	1.1	$153	$65
2017: August	Gulfwide	76	0.7	$238	$47
2017: March	Central Gulf	49	1.9	$301	$47
2016: August	Western Gulf	24	0.6	$131	$46
2016: March	Central Gulf	44	1.6	$225	$38
2016: March	Eastern Gulf	0.6	zero	NA	$38
2015: August	Western Gulf	22	0.9	$119	$41
2015: March	Central Gulf	41	2.2	$583	$45
2014: August	Western	22	2.0	$253	$96
2014: March	Central	40	4.3	$502	$101
2014: March	Eastern	0.5	zero	NA	$101
2013: August	Western	21	1.5	$340	$110
2013: March	Central	39	4.5	$705	$93
2011: November	Western	21	3.1	$205	$86

Sources: Cooney and Kustin, *Trump's Pick to Run Interior Looms Large behind Ocean Sell-Off*, 2019;[32] Bureau of Ocean Energy Management, *Outer Continental Shelf Lease Sale Statistics*, 2019;[41] Department of the Interior, "Gulf of Mexico Lease Sale Yields More Than $159 Million in High Bids," 2019;[42] Groom, "Could Have been Substantially Worse," 2020.[43]

Note: The average bid was high in March 2020, despite low oil prices, because of an exceptionally high bid by BHP Billiton for a tract in the Green Canyon area.[43]

how to use a resource in the future. The sale of a lease grants the oil company the option of whether to pursue drilling within 10 years or to let the lease lapse. Postponing the sale of these leases, rather than selling them now at bargain rates, would preserve society's ability to decide if and when to sell in the future, when the demand for oil and gas may be higher and prices could thus also be raised. Alternatively, society may in the future place greater value on conservation as conservation areas grow scarcer and technological innovations lead to a shift away from oil and gas. The Court of Appeals for the DC Circuit in *Center for Sustainable Economy v. Jewell*[39] recognized the importance of the Department of the Interior incorporating this option value in formulating the 2017–22 National OCS Plan. "There is," opined the court, "a tangible present economic benefit to delaying the decision to drill for fossil fuels to preserve the opportunity to see what new technologies develop and what new information comes to light."

Likewise, the administration's actions to reduce royalty rates undercuts Americans' ability to seize the benefits from offshore drilling. The US Gulf of Mexico, according to the Government Accountability Office,[40] is among the more attractive investment options for oil and gas companies. The United States is far more politically stable and more protective of foreign investors' property rights than are the OPEC countries.[40] The Gulf of Mexico possesses sizable oil and gas reservoirs and, of course, is close to US oil and gas markets.[40] Yet, prior to the G. W. Bush administration's royalty hike, the US government took only a modest share of the value of oil and gas production from publicly owned oil and gas resources.[40] Indeed, the US government's "take" was among the lowest government takes in the world.

The 115th Congress fared no better than the Trump administration when it came to recognizing the long-term economic value of the ANWR coastal plain. BLM published the final environmental impact assessment on drilling in ANWR in September 2019,[44] paving the way for the administration to sell leases for drilling in the refuge. In January 2021, the Trump administration sold oil and gas leases for 575,000 acres in ANWR.[27,iii] Similar to the challenges raised earlier with respect to contractual rights in offshore leasing, the sale of drilling rights in ANWR is dangerously difficult to undo. Property rights and contract law severely limit the ability of future congressional or presidential attempts to stop development in the refuge.

Drilling in ANWR would irreversibly destroy this unique, pristine ecosystem while producing, at most, the equivalent of just one year's total US annual

---

[iii] Biden's executive order on January 20, 2021 placed a moratorium on oil- and gas-related activities in ANWR.

consumption of oil.[45,46] ANWR is one of the few intact ecosystems left in the world. It is home to diverse species and contains pristine remote landscapes.[47] The coastal plain, known as the "American Serengeti," serves as calving grounds for caribou herds, dens for polar bears and habitats for migratory birds.[47] In 2015, the US Fish and Wildlife Service, based on the findings of its completed report, recommended that Congress designate 12.3 million acres of the refuge, including the coastal plain, as wilderness.[47] That designation would prohibit resource exploitation, including oil and gas extraction, in the refuge.

The ANWR coastal plain, prior to the lifting of the moratorium, had been the only section of the coastal plain ecosystem that was not open to drilling (Map 6.2). The rest of the coastal plain, from the west section, which contains the National Petroleum Reserve-Alaska, to the central section, which includes the Prudhoe Bay Oil Field, is open to drilling with minimal exceptions.[48–50] In keeping with its disregard for the environment and local communities, the Trump administration proposed to open to drilling the section of the National Petroleum Reserve-Alaska that had been withdrawn in 2013 to protect caribou, migratory birds and other resources important to Native Alaskans.[51,52]

Proponents justify opening drilling in the ANWR coastal plain by claiming that the resulting revenue will reduce Alaska's and the federal government's budget deficits. However, given the limited reserves in the ANWR coastal plain and the time needed to develop new oil fields, revenue from drilling would not make a sizable dent in those deficits. The estimated $35 million in state revenue for the entire oil production in ANWR would be a drop in the bucket (a 1 percent drop) of Alaska's budget deficit, which stood at $3.5 billion in 2017.[134] The federal revenue of $124 billion would likewise do little to diminish the estimated $1.7 trillion in the federal deficit by 2027.[53,134]

6.4 States' Responses to Drilling

The divergent responses among states to offshore drilling reflect the uneven benefits and costs from drilling. The Atlantic and Pacific coastal states have opposed the proposed expansion (Section 6.4.1) while the US Gulf Coast states (Section 6.4.2) and Alaska (Section 6.4.3) have supported it.

6.4.1 Atlantic and Pacific States

States that are highly dependent on tourism and fishing – Florida, Georgia, South Carolina and North Carolina, in addition to states on the West coast and

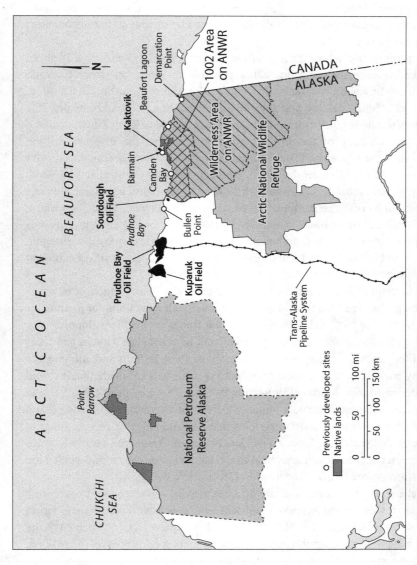

Map 6.2 The Alaska National Wildlife Refuge and the National Petroleum Reserve-Alaska
Source: Alaska Department of Natural Resources, Division of Oil and Gas (n.d.) and US Geological Survey, 2003.[136]

coastal states in the Northeast – have staunchly opposed drilling. Offshore drilling imposes significant economic risks. The costs of oil spills rest on the shoulders of coastal communities, while royalties from federal offshore drilling accrue mainly to the federal government. Not surprisingly, these states have made various attempts to resist the threat posed by drilling to their economies and their people.

Opposition to offshore drilling cuts across party lines. Governors of 10 states on the Atlantic seaboard, including Republican governors from Georgia, Florida, Maryland, Massachusetts, New Hampshire and South Carolina, opposed offshore drilling and seismic airgun blasting.[54–58] Additionally, 221 bipartisan state legislators from 17 coastal states wrote to Ryan Zinke, who was then the secretary of the Interior, protesting against the offshore drilling expansion.[59] Legislators from nine states from both parties announced joint efforts to block offshore drilling.[60] Georgia, Florida, California, Maryland, Delaware and New Jersey banned drilling in state waters to emphasize their opposition to offshore drilling.[60,61] New Jersey, Delaware, New Hampshire, California and other states passed laws prohibiting new oil and gas pipelines and other infrastructure from being built on states' seas, thereby discouraging drilling in both state and federal seas by making the transport of oil extracted offshore prohibitively expensive.[62]

In December 2018, nine state attorneys general and a coalition of environmental groups sued the federal government to block its granting of permits for seismic testing.[63–65] Seismic testing is the first step in oil exploration and drilling. South Carolina's attorney general and 16 of the state's towns and cities also filed lawsuits.[66] They argued that granting these permits and allowing the testing would result in harassment of marine mammals so severe that it would violate the Marine Mammal Protection Act, the ESA and the NEPA. Seismic testing, which uses extremely loud and persistent airgun blasts to map oil and gas resources on the seabed, disorients marine mammals that rely on sonar to navigate and search for food.[65] The Department of Commerce's refusal to turn over documents on how it arrived at its decision to issue the permits stalled the trial from proceeding as scheduled in December 2019.[135]

States also challenged the administration's thinly veiled attempt to curb their rights to review oil and gas development in federal waters off their coasts – rights that are enshrined in the 1972 Coastal Zone Management Act. In March 2019, the Trump administration published an advance notice of proposed rulemaking, seeking comments on how the federal regulatory process can be made "more efficient across all stages of OCS oil and gas projects from leasing to development" and on how "to streamline" the process "to provide industry with greater predictability when making large [offshore] investments."[67] The notice scarcely conceals the administration's goal to block states' longstanding powers under the act to raise objections to offshore energy plans (even if outside state waters) that are inconsistent with the state's federally approved coastal management plan.[68] At present, even though a state cannot veto a federal agency's decision to offer oil and gas lease sales, it may seek mediation with the secretary of Commerce or challenge a federal agency's decision in federal court.

A coalition of nine state attorneys general and environmental groups submitted comments emphasizing that the protection of states' rights to review is consistent with the act's framework of cooperative federalism.[69–71] This doctrine mandates that the federal and state governments share authority and responsibility for adopting, implementing and enforcing environmental protection standards. Federal standards serve as the minimum for the stringency of regulations and states can set more stringent standards.[69] However, as detailed in Chapter 10, the administration destroyed the approach of cooperative federalism by proposing a plethora of rules that curb states' rights to review oil and gas projects, both onshore and offshore, and by taking the unprecedented step of suing the state of California for setting more stringent environmental standards.

Nevertheless, the strident opposition to drilling prompted Secretary of the Interior David Bernhardt to announce in April 2019 that the Trump administration was suspending expansion "indefinitely."[35] Media reports suggested this response to opposition to offshore drilling was a ploy to make voters assume the administration's plans had been discarded when in reality expansion had only been postponed until after the 2020 elections.[35] This suspicion was reinforced by the Department of the Interior continuing other actions to facilitate drilling, such as granting permits for seismic testing.[72]

6.4.2 Gulf Coast States: The Case of Louisiana

In contrast to Atlantic and Pacific states, four of the five US Gulf Coast states – Louisiana, Texas, Alabama and Mississippi – support oil and gas expansion, because they need the jobs and tax revenue it promises.[73] (Florida, which has a significant Gulf coastline, does not support expansion because it depends heavily on tourism and fishing.) In doing so, however, these states risk experiencing the drawbacks as well as the advantages of drilling. To illustrate this double-edged sword, this section delves into the case of Louisiana.

Louisiana has been the site of oil and gas extraction for many decades. Louisiana hosted oil and gas extraction in its coastal wetlands and inshore bays starting in the 1930s and peaking in the 1970s. Subsequently, oil and gas extraction moved to shallow and deeper waters off Louisiana. The state benefitted economically in several ways. It received royalties from leases in state waters, from leases in federal waters within 3–6 miles from the state's coastline and from specific federal leases under the 2006 Gulf of Mexico Energy Security Act. (Louisiana also enjoyed a boom in onshore shale gas extraction, beginning in 2008, from the Haynesville Shale in the northern part of the state.) In 2015, the oil and gas industry accounted directly for 4.5 percent of state employment and more than 9 percent of state payroll, while the industry's tax and royalties payments amounted to 10 percent of all tax revenue in Louisiana.[74]

While these benefits are sizable, Louisiana has not managed its finances prudently. It did not set aside permanent trust funds to invest its oil and gas revenue for future generations.[75] Instead, it worsened its state budget balance by cutting income taxes in 2002 and 2008 and left future generations to deal with the long-term costs. The state government, despite mounting evidence of damage to the fishing industry, failed to take action to protect oyster grounds, nearshore reefs, nurseries for fisheries and other productive areas.[76] Failure to mitigate the environmental impact of oil and gas operations also contributed to Louisiana's extensive wetland losses in the Mississippi Delta and along the Gulf coastline.[76] Loss of wetlands – and Louisiana has been losing land the size of a football field every one hundred minutes[77] – worsens Louisiana's vulnerability to coastal flooding and storm surges, a vulnerability that will become more acute with a rise in sea level and more extreme weather.[78]

Scientists place the primary blame for accelerating wetland loss in interior south-central Louisiana between 1970 and 1990 on the rapid extraction of large volumes of oil and gas.[79],[iv] Researchers, including even those sponsored by oil and gas companies,[80] have revealed how high rates of extraction of oil and gas lowered the pressure below ground and hastened land subsidence.[81,82] The dredging of canals, primarily for pipelines and for access to drilling sites, caused significant saltwater intrusion and the death of freshwater wetland plants. Additionally, wastewater from oil and gas production and oil spills damaged wetland plants.

Louisiana faces a staggering price tag of $50 billion to restore its wetlands and coasts, as set out in its Louisiana 2007 Comprehensive Master Plan for a Sustainable Coast. Louisiana has cobbled together only a tiny fraction of these funds from the $5 billion payout for natural resource damages from the BP oil spill.[83] The costs of inaction are equally staggering. Over the next 25 years, if – as is projected by scientists – the sea level rises 0.14 meters within that time, Louisiana's loss of land is estimated to cost $2 billion in forgone

---

[iv] Other factors that contributed to these losses are the construction of levees that blocked the Mississippi River from replenishing the delta with sediment and the drainage from agriculture that damaged the delta ecosystem [131, 132].

output and another $2 billion to replace physical capital. Storm damage would cost an additional $2–$3 billion in forgone output and $8–$16 billion to replace physical capital.[84]

In 2019, 12 parishes in Louisiana and the city of New Orleans filed separate lawsuits against oil and gas companies, seeking to hold them accountable for damages caused by their operations and to obtain funding for restoration. These lawsuits, construed narrowly to focus on violations of state laws, contend that oil and gas companies violated provisions of the state's Coastal Zone Management Act of 1978 by failing to restore damage caused by canals and spoil banks, failing to clean up hazardous waste during drilling operations, and failing to seek permits prior to working in wetlands. One oil company, Freeport-McMoRan and its affiliates, agreed to settle and to pay the state of Louisiana $100 million in cash and environmental credits over the next several years.[80] The company, which denied any liability, nevertheless stated its decision to settle was driven by the urgency for coastal restoration. In March 2021, Louisiana's Attorney General Jeff Landry approved that settlement. However, several state legislators criticized the settlement as a "shakedown" of oil companies that threatened jobs and declared that they would not support the legislation needed for the settlement to take effect.[27]

Louisiana's reliance on the oil and gas industry continues to complicate its struggle to finance its wetland and coast restoration. Taxpayers have borne the brunt of the cost of restoration projects, even including repairs to property owned by oil and gas companies.[77] In a 2018 poll by the *Times-Picayune*, 72 percent of respondents said the oil and gas industry should help pay for coastal restoration along with the government, while 18 percent said the industry should bear the cost alone.[77] Only half said they were willing to pay higher taxes to restore the coast. Governor John Bel Edwards walked a fine line, touting the industry's key role in the Bayou state's economy but also endorsing parishes' lawsuits and the Freeport-McMoRan settlement that generated funds for restoration.[85,86] Edwards justified his position, noting, "Oil and gas is a huge component of our economy, and we want them to be successful. But, we also want them to act responsibly when they explore for and produce hydrocarbons."[77] Edwards earned the wrath of the oil industry and their political supporters but narrowly won his second gubernatorial term in 2019 against the Republican candidate endorsed and funded by, among others, the oil and gas industry.[87]

6.4.3 Alaska

Alaska, like Louisiana, relies heavily on the oil industry. It also faces significant financial and environmental costs from drilling. The uneven distribution of benefits and costs underlies the divergence of opinions among Alaskans on expanding drilling in the Arctic Ocean (and ANWR).

A staggering nine-tenths of Alaska's operating budget is financed by royalties from oil production on state and federal lands and waters, property taxes from oil production structures and pipelines, and corporate taxes.[88] In addition, the Alaska Permanent Fund, funded by royalties from oil extraction, pays annual dividends to each Alaskan. This dividend, about $1,000 per person in 2017, provides a lifeline to many Alaskans, including the 20 percent of Alaska households whose incomes are $35,000 a year or less.[89] That dividend reduced the average poverty rate from 11.4 percent in 2011 to 9.3 percent in 2015.[90]

However, falling oil production and low oil prices since 2015 have slashed Alaska's revenue.[91] Declining oil production further threatens Alaska's other key revenue source, property tax from the Trans-Alaska Pipeline System.[92] The pipeline risks being decommissioned if the oil flow drops below 350,000 barrels per day and total oil production revenue from Alaska's North Slope drops below $5 billion per year.[92] Alaska worsened its reliance on oil royalties by eliminating its income and sales taxes. As a result of those fiscal actions and declining oil revenue, Alaska faces severe budget deficits. The deficit ballooned to $3.5 billion, or two-thirds of the state's budget, in 2016.[88]

Proponents see drilling in the Arctic and in ANWR as a way to save the Trans-Alaska Pipeline System and to generate revenue for Alaska. However, as described earlier, drilling in the Arctic and in ANWR, even in the best-case scenario, would yield benefits only in a decade, while creating tremendous risks for Alaska's fishing, tourism and subsistence sectors and for its ecosystems. A 2003 study by the National Academy of Sciences underscores the likely costs to livelihoods and the ecology. By then, the livelihoods of subsistence hunters had already been adversely impacted.[93] That study's projections – most infrastructure from oil would be left abandoned and most habitats in the region would be left unrestored – have proven to be prescient.[93]

At the same time, alternative strategies for solving the budget gap – such as reinstituting the income and sales taxes and strengthening the economy through economic diversification – present their own set of challenges,[91] a fact that underlines the dilemma facing Alaska.

The split in Alaska on drilling in ANWR, both within communities and across communities,[23] reflects the uneven benefits and costs from drilling and divergent views on economic development pathways. The Arctic Slope Regional Corporation (ASRC), a for-profit Alaska Native regional corporation whose members are primarily the Iñupiat people of the North Slope,[94] supports drilling. The ASRC owns the subsurface rights to more than 92,000 acres of land within the coastal plain of the Arctic Refuge. The ASRC joined forces with the Kaktovik Iñupiat Corporation, which owns the surface rights to those lands,[95] to apply for a permit for seismic testing in the entire coastal plain. Testing may begin in winter 2020.[96] Nevertheless, Iñupiats diverge in their views on drilling. Support is stronger among residents in the larger settlements away from the coastal plain, such as the North Slope's largest city, Utqiagvik, where the ASRC is headquartered.[97] Support for drilling is weaker among villagers in the coastal plain, many of whom rely on subsistence living and on tourism.[97]

Staunchly opposed to drilling are the Gwich'in, who live and hunt in the southern part of ANWR. They argue that drilling threatens the Porcupine caribou herd that is central to the Gwich'in diet and culture.[95] Canadian tribal and provincial governments have also voiced their opposition to drilling in ANWR because the caribou herd, which calves in the ANWR coastal plain, migrates between northeastern Alaska and northwestern Canada. The Canadian First Nations, the Yukon provincial government and the Northwest Territories provincial government stress the reliance of numerous tribes on the herd and point to the 1987 US–Canadian treaty in which parties pledged to conserve the Porcupine caribou herd and its habitat[98] and to undertake consultations prior to pursuing activities that could threaten the herd and its habitat.

In Alaska's southwestern coast, united community opposition against drilling[99] compelled Governor Bill Walker and Alaska's congressional delegation – who were otherwise supportive of drilling – to request that the Trump administration abandon its proposal to lease areas stretching from the Bering Strait to the Gulf of Alaska.[100,101] As noted by the Bering Sea Elders Group, an association of elders appointed by 39 tribes in the Yukon-Kuskokwim and Bering Strait regions, these waters support fisheries and subsistence hunting.[99] Governor Walker also requested that the Kaktovik Whaling Area, Barrow Whaling Area and the 25-mile coastal buffer in the Chukchi Seas be excluded from federal drilling plans, as was the case under the previous administration, to protect communities reliant on subsistence fishing and hunting.[100]

Despite the deep division among Alaskans on expanding drilling, even the most ardent supporters of drilling, such as Senator Murkowski, acknowledge Alaska's dilemma of relying heavily on the oil industry while facing costs from drilling and from the climate crisis.[102] In a 2019 op-ed in the *Washington Post*, Senator Murkowski and Senator Joe Manchin, a Democrat who represents West Virginia, urged Congress to address the climate crisis: "Rising temperatures and diminishing sea ice on Alaska's shores are affecting our fisheries and forcing some remote communities to seek partial or total relocation. ... We must do more to pursue low- and zero-carbon technologies that will continue to lower emissions."[103]

6.5 Who Can Improve the Management of America's Oceans?

Unless and until the executive changes its policies and attitude, the wise stewardship of America's marine resources/oceans would have to be undertaken by a combination of Congress, state governments and civil society. The balance struck in the past by Congress – permitting offshore drilling in the Gulf and enacting a moratorium in the Pacific and Atlantic OCS – has become ever more important to Americans' economic and natural wealth as coastal states' economic dependence on healthy oceans deepens and as the public increasingly appreciates the value of healthy coasts and oceans. The split in states' views on expanding offshore drilling – with Atlantic and Pacific states against and the US Gulf coast states, Alaska and a number of inland states in favor – has proved to be a hurdle in the legislative efforts to restore this balance. Nevertheless, growing opposition among voters to offshore drilling may lead eventually to a shift in opinion on Capitol Hill.

In January 2019, Francis Rooney, a Republican representative from Florida, proposed H.R. 205, Protecting and Securing Florida's Coastline Act, warning that "drilling posed an existential threat to Florida."[104] The proposed bill, which was cosponsored by all but one of the other Floridian House representatives, would permanently extend the moratorium in the eastern Gulf of Mexico beyond 2022, which is when it is set to expire under the 2006 Gulf of Mexico Energy Security Act. The House passed Rooney's bill by 248 to 180 votes, including 22 votes from Republicans. Republican senators Marco Rubio and Rick Scott from Florida sponsored a less ambitious amendment to the Senate energy bill that would have extended the moratorium until 2027 only.[100] Regrettably, even this narrow proposal did not win the support of the majority of Republicans. Two other bills that passed the House, albeit with the support of only 12 and 4 Republicans, respectively, did not win support in the Senate.[v] These bills would have prohibited the Department of the Interior from offering leases in the Pacific and Atlantic OCS and the Bureau of Land Management from offering leases in ANWR.

While these bills did not win Senate support in the 116th Congress, growing voter opposition to offshore drilling may well shift the position of congressional representatives from coastal states. A May 2018 survey conducted by the University of Maryland revealed that 60 percent of respondents opposed the expansion in the Atlantic, Pacific and Arctic Seas. Among respondents from

---

[v] These bills are H.R. 1941, the Coastal and Marine Economies Protection Act, which passed by 226 to 183 votes with 12 Republicans voting for it and H.R. 1146, the Arctic Cultural and Coastal Plain Protection Act, which passed by 221 to 187 votes with 4 Republicans voting for it.

the 15 coastal states that requested the federal government not to drill in federal seas that abut their state waters, 88 percent of Democrats supported their state's request, as did 50 percent of Republicans.[105] A survey conducted at the proposal stage of the Trump administration's 2019–25 OCS plan demonstrated how executive action contradicted the views of the public. The October 2019 survey by the *Washington Post* and the Kaiser Family Foundation reported that 53 percent of respondents from the United States favored a rollback of offshore drilling, while 32 percent favored drilling remain as is.[106] Republican respondents largely shared this view, with 45 percent of them favoring a rollback and 26 percent saying it should remain as is.

The 2018 midterm elections witnessed the success of candidates running on anti-drilling platforms. For instance, Representative Joe Cunningham, who opposed drilling, flipped a long-held Republican seat in North Carolina. Additionally, candidates who had supported drilling in the past, such as Rick Scott, former Republican governor of Florida, switched to touting his anti-drilling stance to snatch a narrow win in the race for one of Florida's senate seats over his Democratic rival Bill Nelson, who had consistently opposed drilling. Floridians also showed support in state elections for a ban on drilling in state waters. Other politicians similarly seemed to respond to a growing tide of public opposition to drilling. Senator Tom Tillis of North Carolina, for instance, softened his support for drilling, while Senator Lindsey Graham of South Carolina advocated for the right of states to request federal government not to drill in federal seas that abut their seas.

Oceana, a nonprofit ocean conservation group, documented voters' opposition to drilling expansion. It compiled letters from over 370 municipalities and over 2,200 elected local, state and federal officials sent to the federal government opposing offshore drilling and seismic airgun blasting. It also documented opposition from local chambers of commerce, tourism and restaurant associations, and an alliance representing over 46,000 businesses and 500,000 fishing families from Florida to Maine. Likewise, the Southern Environmental Law Center documented widespread opposition to drilling among coastal communities in states with large numbers of Republican voters, including Virginia, North Carolina, South Carolina and Georgia.

While legislative efforts stalled, the 116th Congress, unlike its predecessor, deserves credit for executing its oversight duties of the executive branch's actions. Representative Raúl Grijalva and Senator Maria Cantwell, the top Democrats on the committees overseeing the Department of the Interior, pushed back against Secretary of the Interior Zinke's proposal to reduce royalty rates in deepwater drilling from 18.75 to 12.75 percent.[107] Zinke abandoned that proposal in October 2018.[108]

Representatives Grijalva and Alan Lowenthal, chairs of the Committee of Natural Resources and the Subcommittee on Energy and Mineral Resources, respectively, initiated investigations into the Trump administration's decision to abandon reforms aimed at protecting taxpayers from bearing the costs of decommissioning unproductive oil and gas infrastructure.[109,110] In April 2019, the administration halted plans to impose stricter criteria on companies' financial strength to qualify for waivers and to limit the amounts that companies can self-insure.[110] According to available data, as of May 2017, the federal government holds financial assurance for only 7.6 percent of the $38.2 billion in expected decommissioning costs, despite requirements under OCSLA for companies to cover the true costs of the decommissioning and for the federal government to collect bonds accordingly.[111]

Congressional oversight of the Department of the Interior's actions proved to be critical during the COVID-19 crisis. Companies sought royalty relief, citing financial woes from the crisis, and won the sympathies of 14 Republican representatives and 12 Republican senators, who pleaded their case to Secretary of the Interior Bernhardt.[112–114] Grijalva and 15 other Democrats warned Bernhardt against granting relief in these cases, arguing that such an action would violate Department of the Interior regulations.[115] They cautioned that "a national emergency is not an opportunity to do favors for the oil and gas industry, especially favors that have no legal basis, policy rationale, or societal benefit."[115]

Atlantic and Pacific state governments worked across states and across the partisan divide to dissuade the administration from expanding drilling. As described above, governors and state legislators directed their protests against the administration while state legislators enacted state laws banning oil and gas infrastructure in state waters. These states could broaden their coalition by highlighting how the administration's playbook of curbing state's rights to enable drilling expansion has the potential to adversely affect all states. For instance, governors and legislatures from both coastal and Western states united in their opposition to the administration's proposed rule to curb states' powers enshrined in section 401 of the CWA, which empowers states to review infrastructure projects that are subject to federal permits.[116] As detailed in Chapter 10, states have used those powers in the past to deny state water permits to projects such as interstate gas pipelines that threaten states' water supply, thereby making it harder for such projects to proceed in their states. A number of governors and state legislators from Western states with Republican majorities recognized how that proposed rule, while it would help extractive states to send their gas to markets, would also curb their staunchly guarded powers to protect their water resources.

Environmental groups such as Earthjustice have filed lawsuits as one strategy to protect the seas. These lawsuits, which most often point to the failure to follow administrative procedures, seek to hold the administration accountable to the rule of law. They also buy time for a reversal of policies, by Congress or a future administration, by delaying the selling of new leases.

Bernhardt acknowledged that lawsuits scuttled the administration's rush to drilling. In March 2019, the US District Court in Alaska ruled that Trump's reversal of protections on the ecologically rich Atlantic canyon and parts of the Arctic Ocean were illegal. (The court ruled that OSCLA section 12(a) granted powers to the president to withdraw parts of the OCS from leasing, but it did not grant powers to the president to revoke prior withdrawals. Only Congress, said the court, can revoke previous withdrawals.) Bernhardt admitted that the decision and the long appeals process "may be discombobulating" to the administration's drilling plan and announced that his department would halt the push forward with its proposed 2019–25 OCS plan,[117] although it continued to issues permits for seismic testing. Bernhardt also admitted that concerns over lawsuits forced the administration to proceed more carefully and thus more slowly with planned lease sales in ANWR.[118] According to media reporting, oil executives and Republicans in Congress had wanted the lease sales to take place before the 2020 elections in case "a Democrat [were to] regain the White House and mothball the plan indefinitely."[118]

Environmental groups – including Healthy Gulf, the Sierra Club and the Center for Biological Diversity – filed lawsuits against the administration challenging Gulf-wide sales in 2018 and 2019.[119,120] Past challenges under previous administrations delayed exploratory drilling, raising the costs to companies considering questionable projects. For instance, environmental groups won court rulings in 2010 and 2014 by demonstrating the flaws in the executive branch's conduct of environmental impact assessments prior to the sale of leases. The court ordered the federal government to redo its environmental impact assessment of the leases in the Chukchi Sea sold in 2008,[121,122] delaying exploratory drilling until 2015.

In the case of the Trump administration's sale of leases in the Gulf, plaintiffs laid out several errors in the environmental impact analyses. Specifically, the plaintiffs noted that the federal government failed to account for revisions to the two rules enacted after the BP oil spill. These rules were introduced with the intention of minimizing the risks of well blowouts and oil spills, but the Trump administration weakened those regulations in 2019. The Bureau of Ocean and Energy Management conceded that its analyses failed to consider the 2019 revisions to the regulations but claimed that those revisions did not compromise safety. (The next chapter details how the revisions rolled back regulations that were tailored to address the safety gaps that led to the BP oil spill.) Environmental groups, along with Native Alaskans, also filed a lawsuit against the Department of the Interior for failing to respond to freedom of information requests about how it drew up plans for oil leasing in ANWR, poised for sales by the end of 2020.[123,124]

6.6 Conclusion

As in the case of deregulation of the onshore oil and gas industry, the Trump administration's proposed expansion of drilling and its rollback of safety in the offshore oil and gas industry went so far that a subset of the industry worried about potential blowback from the public. Despite strong support for the administration from the National Ocean Industries Association, the trade group for the offshore oil and gas industry, media reports show some members of the industry were circumspect. They pointed not only to the financial hurdles of building the infrastructure from the ground up to drill in the Arctic, Atlantic and Pacific OCS but also to the uncertainties and delays that litigation launched by state governments and environmental groups would present.[125]

They worry that offshore expansion alongside the rollback of safety regulations would result in a massive disaster, akin to or worse than the BP oil spill.[126,127] Such a disaster would damage the industry's social license to operate, which is already under strain, as seen in the growth of litigation against oil and gas companies from state governments and local governments.[128,129] Concerns of blowback from the public have led several banks to announce independently that they would not finance projects in the Arctic, including ANWR.[vi] These banks include JP Morgan, Wells Fargo, UK-based Barclays Bank and Lloyds Bank, and Swiss-based UBS.[130]

Social pressure, however, seldom determines companies' plans for offshore drilling expansion; at most, social pressure may be a contributory factor in industry decision-making in those high-profile projects that capture the public's attention. Similarly, state governments have limited influence and can at most cajole a pro-drilling executive branch into exempting individual states' coastlines from offshore drilling in federal seas. Coalitions of state governments,

---

[vi] Pro-drilling members of Congress hit back at financial institutions that pledged not to invest in fossil fuel projects. In their letter to President Trump, they argued that "Wall Street's big banks, for example, should not be able to reap the benefits of participating in federally guaranteed loan programs laid out in the CARES Act, such as the Paycheck Protection Program or the trillion dollar Federal Reserve facility lending programs" [133].

local governments and environmental groups can at most delay the commencement of offshore drilling.

Ultimately, the power to safeguard the oceans and coasts rests with Congress. It would thus be wise for congressional representatives, particularly in coastal states, to recognize that the tide is turning against the expansion of offshore drilling. Voters, as well as governors, state attorneys general and state legislators, are voicing increasing opposition to such an expansion. Candidates running against pro-drilling Republicans who support curbs on states' rights may well find such support is a chink in their opponents' political armor. The economies of all coastal states, no matter whether most of their voters are Republican or Democrat, are at risk from reckless expansion of drilling, and only Congress can provide the protection they need. The urgency of congressional action extends not only to ending the threat of drilling expansion but also to preventing the dismantling of safety regulations in offshore drilling, a topic addressed in the next chapter.

References

1. Congressional Research Service. *US Crude Oil and Natural Gas Production in Federal and Nonfederal Areas*. Report by M. Humphries. R42432 (Washington, DC: 2016). https://fas.org/sgp/crs/misc/R42432.pdf.
2. Bureau of Ocean Energy Management. *Offshore Oil and Gas Economic Contributions* (Washington, DC: 2016). www.boem.gov/NP-Economic-Benefits.
3. Energy Information Administration. *Short-Term Energy Outlook (STEO)* (Washington, DC: 2017). www.eia.gov/outlooks/steo/archives/jul17.pdf.
4. Energy Information Administration, *Gulf of Mexico Fact Sheet* (Washington, DC).
5. Marine Mammal Commission. *Energy Development in the Arctic* (Washington, DC). www.mmc.gov/priority-topics/arctic/arctic-oil-and-gas-development-and-marine-mammals.
6. Energy Information Administration. *Profile Analysis: State of Alaska* (Washington, DC: October 19, 2017). www.eia.gov/state/analysis.php?sid=AK.
7. Energy Information Administration. *Short-Term Energy Outlook Data Browser* (Washington, DC: May 12, 2020). www.eia.gov/outlooks/steo/data/browser.
8. Energy Information Administration. *Drilling Productivity Report* (Washington, DC: April 2020). www.eia.gov/petroleum/drilling/pdf/dpr-full.pdf.
9. National Oceanic and Atmospheric Administration, Office for Coastal Management. *NOAA Report on the US Ocean and Great Lakes Economy* (Charleston, SC: 2017). https://coast.noaa.gov/data/digitalcoast/pdf/econ-report.pdf.
10. D. J. Trump. "Remarks by President Trump at Signing of Executive Order on an America First Offshore Energy Strategy." White House. 2017. www.whitehouse

.gov/briefings-statements/remarks-president-trump-signing-executive-order-america-first-offshore-energy-strategy.
11. Bureau of Ocean Energy Management. *2019–2024 National Outer Continental Shelf Oil and Gas Leasing Draft Proposed Program* (Washington, DC: January 2018). www.boem.gov/sites/default/files/oil-and-gas-energy-program/Leasing/Five-Year-Program/2019-2024/DPP/NP-Draft-Proposed-Program-2019-2024.pdf.
12. Bureau of Ocean Energy Management. *2017–2022 Outer Continental Shelf Oil and Gas Leasing Proposed Final Program* (Washington, DC: November 2016). www.boem.gov/sites/default/files/oil-and-gas-energy-program/Leasing/Five-Year-Program/2017-2022/2017-2022-OCS-Oil-and-Gas-Leasing-PFP.pdf.
13. Congressional Research Service. *Five-Year Offshore Oil and Gas Leasing Program for 2019–2024: Status and Issues in Brief.* Report by L. B. Comay. R44692 (August 6, 2019). https://fas.org/sgp/crs/misc/R44692.pdf.
14. E. Allison and B. Mandler. *Oil and Gas in the US Arctic.* American Geosciences Institute (Alexandria, VA: 2018). www.americangeosciences.org/geoscience-currents/oil-and-gas-us-arctic.
15. D. Joling. "Eni Receives Federal Remit for US Arctic Offshore Drilling." *Associated Press*, November 28, 2017. https://apnews.com/7331852e5c0642debf6b47ac552e0375/Eni-receives-federal-permit-for-US-Arctic-offshore-drilling.
16. B. Babbitt. "Trump Is Vandalizing Our Wild Heritage." *New York Times*, December 1, 2017.
17. K. Hendricks and R. H. Porter. "Auctioning Resource Rights." *Annual Review of Resource Economics* 6 (2014): 175–190. https://ssc.wisc.edu/~hendrick/publications/AuctioningResourceRights.pdf.
18. US Department of the Interior. "Interior Department Announces Draft Strategy for Offshore Oil and Gas Leasing." News release, January 27, 2015, www.doi.gov/news/pressreleases/interior-department-announces-draft-strategy-for-offshore-oil-and-gas-leasing.
19. D. Fears. "Pentagon Objections Prompt Changes to Oil-Drilling Plan for Atlantic Coast." *Washington Post*, March 15, 2016.
20. Department of the Interior. "Interior Department Announces Next Step in Offshore Oil and Gas Leasing Planning Process for 2017-2022." News release, March 15, 2016. www.doi.gov/pressreleases/interior-department-announces-next-step-offshore-oil-and-gas-leasing-planning-process.
21. H. Fountain and S. Eder. "In the Blink of an Eye, a Hunt for Oil Threatens Pristine Alaska." *New York Times*, December 3, 2018.
22. Bureau of Land Management. *Coastal Plain Oil and Gas Leasing Program Environmental Impact Statement.* Department of the Interior, DOI-BLM-AK-0000-2018-0002-EIS (September 2019). https://eplanning.blm.gov/epl-front-office/projects/nepa/102555/20003762/250004418/Volume_1_ExecSummary_Ch1-3_References_Glossary.pdf.
23. Congressional Research Service. *Arctic National Wildlife Refuge (ANWR): An Overview.* Report by L. B. Comay and M. Ratner. RL33872 (Washington, DC: 2018). https://fas.org/sgp/crs/misc/RL33872.pdf.
24. C. M. Albano and B. G. Dickson. *A Landscape-Level Analysis of Ecological Values of the Arctic National Wildlife Refuge.* Conservation Science Partners. Report submitted to the Center for American Progress (Truckee, CA: 2017).

25. L. McPherson. "12 House Republicans Sign Letter Opposing Arctic Drilling." *Roll Call*, November 30, 2017. www.rollcall.com/2017/11/30/12-house-republicans-sign-letter-opposing-arctic-drilling.
26. D. Grandoni. "House GOP Targets States That Resist Offshore Drilling." *Washington Post*, June 15, 2018.
27. US Congress. House. *To Amend the Mineral Leasing Act and the Outer Continental Shelf Lands Act to Enhance State Management of Federal Lands and Waters, and for Other Purposes*. 115th Cong., 2nd Sess. Introduced June 14, 2018.
28. Energy Information Administration. *Annual Energy Outlook 2019*. Department of Energy (Washington, DC: 2019). www.eia.gov/outlooks/aeo/pdf/aeo2019.pdf.
29. Bureau of Ocean Energy Management. *Assessment of Undiscovered Oil and Gas Resources of the Nation's Outer Continental Shelf* (2017). www.boem.gov/sites/default/files/oil-and-gas-energy-program/Resource-Evaluation/Resource-Assessment/BOEM-National-Assessment-Fact-Sheet.pdf.
30. W. Koch. "3 Reasons Why Shell Halted Drilling in the Arctic." *National Geographic*, September 28, 2015. www.nationalgeographic.com/news/energy/2015/09/150928-3-reasons-shell-halted-drilling-in-the-arctic.
31. L. Murkowski, senator. "Alaska Delegation Reacts to Stunning Arctic Withdrawal." News release, December 20, 2019. www.murkowski.senate.gov/press/release/alaska-delegation-reacts-to-stunning-arctic-withdrawal.
32. M. Cooney and M. E. Kustin. *Trump's Pick to Run Interior Looms Large behind Ocean Sell-Off*. Center for American Progress (2019). www.americanprogress.org/issues/green/news/2019/03/21/467545/trumps-pick-run-interior-looms-large-behind-ocean-sell-off.
33. J. A. Dlouhy. "Trump Has a Plan to Preserve Benefits for Oil Drillers in Future." *Bloomberg*, March 21, 2019.
34. Government Accountability Office. *Oil and Gas Resources: Actions Needed for Interior to Better Ensure a Fair Return*. Report by F. Rusco. GAO-14-250 (Washington, DC: 2013). www.gao.gov/assets/660/659515.pdf.
35. J. A. Dlouhy. "Trump to Push Offshore Oil Expansion to after 2020 Election, Sources Say." *Bloomberg*, April 25, 2019. www.bloomberg.com/news/articles/2019-04-25/trump-said-to-push-offshore-oil-expansion-to-after-2020-election+&cd=3&hl=en&ct=clnk&gl=us.
36. T. Cama. "Panel Recommends Zinke Cut Offshore Drilling Royalty Rates by a Third." *The Hill*, February 28, 2018. https://thehill.com/policy/energy-environment/376156-panel-votes-to-recommend-one-third-cut-to-offshore-drilling-royalty.
37. Department of the Interior. "Planning, Analysis, & Competitiveness Subcommittee Meeting Summary." News release, February 2, 2018, www.doi.gov/sites/doi.gov/files/uploads/pac_meeting_summary_2.2.18_final.pdf.
38. M. A. Livermore. "Patience Is an Economic Virtue: Real Options, Natural Resources, and Offshore Oil." *University of Colorado Law Review* 84, no. 3 (2012): 581–650.
39. *Center for Sustainable Economy v. Jewell*, 779 F.3d 588 (D.C. Cir. 2015).
40. Government Accountability Office. *Oil and Gas Royalties: A Comparison for the Share of Revenue Received from Oil and Gas Production by the Federal Government and Other Resource Owners*. Report by M. E. Gaffigan (Washington, DC: 2007). www.gao.gov/new.items/d07676r.pdf.

41. Bureau of Ocean Energy Management. *Outer Continental Shelf Lease Sale Statistics*. Report by P. R. Bryars. Office of Leasing and Plans, Gulf of Mexico OCS Region (New Orleans, LA: 2019). www.boem.gov/sites/default/files/documents/about-boem/Swilertablecompletereport19DEC19_1.pdf.
42. Department of the Interior. "Gulf of Mexico Lease Sale Yields More Than $159 Million in High Bids, Continues Upward Trend under Trump Administration." News release, August 21, 2019. www.doi.gov/pressreleases/gulf-mexico-lease-sale-yields-more-159-million-high-bids-continues-upward-trend-under.
43. N. Groom. "'Could Have Been Substantially Worse': US Offshore Oil Lease Sale Weakest since 2016." *Reuters*, March 18, 2020.
44. H. Fountain. "EIA Completed: Interior Dept. Takes Next Step toward Sale of Drilling Leases in Arctic Refuge." *New York Times*, September 13, 2019.
45. US Geological Survey. *Arctic National Wildlife Refuge, 1002 Area, Petroleum Assessment, 1998, Including Economic Analysis*. Report 0028–01 (Reston, VA: 1998). https://pubs.usgs.gov/fs/fs-0028-01/fs-0028-01.htm.
46. Energy Information Administration. *Oil Crude and Petroleum Products Explained: Use of Oil* (Washington, DC: September 19, 2017). www.eia.gov/energyexplained/index.cfm?page=oil_use.
47. US Fish and Wildlife Service. *Arctic National Wildlife Refuge Revised Comprehensive Conservation Plan: Final Environmental Impact Statement*. Report by Arctic Refuge and the Alaska Region of the US Fish and Wildlife Service in cooperation with the National Aeronautics and Space Administration (Fairbanks, AK: January 2015). www.fws.gov/home/arctic-ccp/pdfs/Executive_Summary_Jan2015.pdf.
48. Staff. "ConocoPhillips Faces Lawsuit over Alaska Exploratory Drilling." *Kallanish Energy News*, March 7, 2019. www.kallanishenergy.com/2019/03/07/conocophillips-faces-lawsuit-over-alaska-exploratory-drilling.
49. Energy Information Administration. *Oil Exploration in the US Arctic Continues Despite Current Price Environment*. Report by T. Yen and L. Singer (Washington, DC: 2015). www.eia.gov/todayinenergy/detail.php?id=21632.
50. J. Pelley. "Will Drilling for Oil Disrupt the Arctic National Wildlife Refuge?" *Environmental Science & Technology* 35, no. 11 (2001): 240–247. https://pubs.acs.org/doi/pdfplus/10.1021/es0123756.
51. Department of the Interior. "Secretary Salazar Finalizes Plan for Additional Development, Wildlife Protections in 23 Million Acre National Petroleum Reserve-Alaska." News release, February 21, 2013. www.doi.gov/news/pressreleases/secretary-salazar-finalizes-plan-for-additional-development-wildlife-protections-in-23-million-acre-national-petroleum-reserve-alaska.
52. Bureau of Land Management. *NPR-A Leases* (2019). www.blm.gov/sites/blm.gov/files/uploads/OilandGas_Alaska_NPR-A_LeaseReport_March2019.pdf.
53. Congressional Budget Office. "HR1 Deficits and Debt." Letter by K. Hall (Washington, DC: November 8, 2017). www.cbo.gov/system/files/115th-congress-2017-2018/costestimate/hr1deficitsanddebt.pdf.
54. R. Cooper et al. *Ten Atlantic Governors Oppose Seismic Airgun Surveys and Oil and Gas Drilling off our Coasts*. Submitted to W. L. Ross, secretary of Commerce and R. Zinke, secretary of the Interior. December 2018.

55. L. Hogan et al. *Joint Opposition to the Leasing, Exploration, Development and Production of Oil and Gas in the Atlantic Ocean.* Submitted to R. Zinke, secretary of the Interior. January 17, 2018.
56. A. Ropeik. "Gov. Sununu Signals Opposition to Offshore Oil Drilling in North Atlantic." *New Hampshire Public Radio*, 2018. www.nhpr.org/post/gov-sununu-signals-opposition-offshore-oil-drilling-north-atlantic#stream/0.
57. B. Petersen. "South Carolina Gov. Henry McMaster Wants Offshore 'No Drill' Oil Exemption Just Like Florida." *Post and Courier*, January 10, 2018. www.postandcourier.com/news/south-carolina-gov-henry-mcmaster-wants-offshore-no-drill-oil/article_b4162f12-f614-11e7-b5bc-1b205b230114.html.
58. J. Deaton. "Op-ed: Coastal Republicans Align with Democrats on Offshore Drilling." *PBS*, May 24, 2019. www.pbs.org/wnet/peril-and-promise/2019/05/coastal-republicans-align-with-democrats.
59. K. Ranker et al. *Opposition against the Proposed National Outer Continental Shelf Oil and Gas Leasing Program for 2019–2024.* Submitted to R. Zinke, secretary of the Interior. March 2018.
60. National Caucus of Environmental Legislators. "Legislators from Nine States Announce Coordinated Effort to Block Offshore Drilling." News release, 2019. www.ncel.net/2019/01/08/legislators-from-eight-states-announce-coordinated-effort-to-block-offshore-drilling.
61. M. Landers. "Georgia Passes Anti-drilling Resolution." *Savannah Now*, April 2, 2019. www.savannahnow.com/news/20190402/georgia-passes-anti-drilling-resolution.
62. A. M. Phillips and R. Xia. "Trump Might Limit States' Say in Offshore Drilling Plan. Here's How." *Los Angeles Times*, March 21, 2019. www.latimes.com/politics/la-na-pol-trump-offshore-drilling-states-coastal-act-20190321-story.html.
63. *South Carolina Coastal Conservation League v. Ross*, No. 2:18-cv-03326 (District Court of South Carolina 2019).
64. T. Gardner. "Nine US States Seek to Stop Trump Administration's Atlantic Oil Testing." *Reuters*, December 20, 2018. www.reuters.com/article/us-usa-drilling-lawsuit/nine-u-s-states-seek-to-stop-trump-administrations-atlantic-oil-testing-idUSKCN1OJ2MV.
65. H. McLeod. "Conservation Groups Sue Trump Administration over Atlantic Oil Testing." *Reuters* (London, UK), December 11, 2018. www.reuters.com/article/us-usa-drilling-lawsuit/conservation-groups-sue-trump-administration-over-atlantic-oil-testing-idUSKBN1OA2F2.
66. M. Kinnard. "S. Carolina Joins Offshore Drilling Lawsuit." *Associated Press*, January 7, 2019. https://apnews.com/2180d3710cfb4edea851fd99e8810b8d.
67. *Procedural Changes to the Coastal Zone Management Act Federal Consistency Process: Proposed Rule.* 15 Code of Federal Regulations 930. Office for Coastal Management, National Ocean Service, National Oceanic Atmospheric Administration, Department of Commerce. 84 Federal Register 8628–8633 (March 11, 2019). www.federalregister.gov/documents/2019/03/11/2019-04199/procedural-changes-to-the-coastal-zone-management-act-federal-consistency-process.
68. L. Krop. "Defending State's Rights under the Coastal Zone Management Act – State of California v. Norton." *Sustainable Development Law & Policy* 8, no. 1 (Fall 2007): 54–58, 86–87. https://digitalcommons.wcl.american.edu

/cgi/viewcontent.cgi?referer=https://www.google.com/&httpsredir=1&article=1160&context=sdlp.
69. R. L. Glicksman. "The Firm Constitutional Foundation and Shaky Political Future of Environmental Cooperative Federalism." In *Controversies in American Federalism and Public Policy*, edited by C. P. Banks. 132–150. London: Routledge, 2018.
70. L. James, attorney general for the State of New York, et al. *Comments on Procedural Changes to the Coastal Zone Management Act Federal Consistency Process ANPR*. Submitted to K. Kehoe, Federal Consistency Specialist, Office for Coastal Management (April 25, 2019).
71. Natural Resources Defense Council. *Comments on NOAA-NOS-2018-2017, Advanced Notice of Proposed Rulemaking Regarding Procedural Changes to the Coastal Zone Management Act Consistency Process*. April 25, 2019.
72. V. Volcovici and N. Groom. "US Still Processing Atlantic Seismic Permits Despite Drilling Plan Delay." *Reuters*, April 29, 2019. www.reuters.com/article/us-usa-oil-offshore/u-s-still-processing-atlantic-seismic-permits-despite-drilling-plan-delay-idUSKCN1S526M.
73. P. LePage, governor of Maine et al. *Request for Comments on the 2019–2024 Draft Proposed Outer Continental Shelf Oil & Gas Leasing Program*. Submitted to R. Zinke, secretary of the Interior. March 9, 2018.
74. G. B. Upton, Jr. *Oil Prices and the Louisiana Budget Crisis: Culprit or Scapegoat?* Louisiana State University (October 24, 2016). www.lsu.edu/ces/publications/2016/Upton_10-2016_Oil_and_Gas_and_the_Louisiana_Economy_FINAL.pdf.
75. B. G. Rabe and R. L. Hampton. "Trusting in the Future: The Re-emergence of State Trust Funds in the Shale Era." *Energy Research & Social Science* 20 (2016): 117–127.
76. O. A. Houck. "The Reckoning: Oil and Gas Development in the Louisiana Coastal Zone." *Tulane Environmental Law Journal* 28, no. 2 (2015): 185–296.
77. K. Sack. and J. Schwartz. "Left to the Tides and Fighting for Time." *New York Times*, February 25, 2018.
78. US Global Change Research Program. *Fourth National Climate Assessment: The Climate Science Special* Report, vol. 1 (Washington, DC: 2018). www.globalchange.gov/nca4.
79. R. A. Morton, N. A. Buster and M. D. Krohn. "Subsurface Controls on Historical Subsidence Rates and Associated Wetland Loss in Southcentral Louisiana." *Gulf Coast Assessment of Geological Societies Transactions* 52 (2002): 767–778.
80. N. Bogel-Burroughs. "Louisiana's Coast Is Vanishing. Can a Mining Company's $100 Million Offer Help Save It?" *New York Times*, September 26, 2019.
81. R. A. Morton, J. C. Bernier and J. A. Barras. "Evidence of Regional Subsidence and Associated Interior Wetland Loss Induced by Hydrocarbon Production, Gulf Coast Region, USA." *Environmental Geology* 50, no. 261 (March 23, 2006): 261–274. https://doi.org/10.1007/s00254-006-0207-3.
82. A. S. Kolker, M. A. Allison and S. Hameed. "An Evaluation of Subsidence Rates and Sea-level Variability in the Northern Gulf of Mexico." *Geophysical Research Letters* 38, no. 21 (November 11, 2011): 1–6. https://agupubs.onlinelibrary.wiley.com/doi/10.1029/2011GL049458.

83. Ocean Conservancy. *Ocean Conservancy's Analysis of the Settlement Agreement with BP* (New Orleans, LA: October 2015). https://oceanconservancy.org/wp-content/uploads/2015/07/bp-settlement-fact-sheet-1.pdf.
84. S. R. Barnes et al. "Economic Evaluation of Coastal Land Loss in Louisiana." *Journal of Ocean and Coastal Economics* 4, no. 1 (2017): 1–40. https://cbe.miis.edu/cgi/viewcontent.cgi?referer=https://scholar.google.com/&httpsredir=1&article=1062&context=joce.
85. S. Karlin. "Louisiana Governor Candidates Promise to Cut Taxes, End Lawsuits at 'Oil and Natural Gas Industry Day.'" *The Advocate*, May 1, 2019. www.theadvocate.com/baton_rouge/news/politics/elections/article_bf4fd2dc-6c12-11e9-b9c2-6b630c21a2df.html.
86. H. Richards. "La. Oil Lawsuits: A Climate Reckoning or Red Herring?" *E&E News: Energywire*, October 15, 2019. www.eenews.net/stories/1061280705.
87. C. Burkes. "Industry Invested in Louisiana Governor's Race." *Business Report Greater Baton Rouge*, October 17, 2019. www.businessreport.com/newsletters/industry-invested-in-louisiana-governors-race.
88. R. Waldholz. "Alaska Faces Budget Deficit As Crude Oil Prices Slide." *National Public Radio*, January 19, 2016. www.npr.org/2016/01/19/463551045/alaska-faces-budget-deficit-as-crude-oil-prices-slide.
89. M. Guettabi. *What Do We Know about the Effects of the Alaska Permanent Fund Dividend?* University of Alaska Anchorage Institute of Social and Economic Research (Anchorage, AK: May 20, 2019).
90. M. Berman. "Resource Rents, Universal Basic Income, and Poverty among Alaska's Indigenous Peoples." *World Development* 106 (2018): 161–172.
91. G. Knapp, M. Berman and M. Guettabi. *Short-Run Economic Impacts of Alaska Fiscal Options*. University of Alaska Anchorage Institute of Social and Economic Research (Anchorage, AK: March 30, 2016). www.iser.uaa.alaska.edu/Publications/2016_03_30-ShortrunEconomicImpactsOfAlaskaFiscalOptions.pdf.
92. Energy Information Administration. *Projected Alaska North Slope Oil Production at Risk beyond 2025 If Oil Prices Drop Sharply* (Washington, DC: September 14, 2012). www.eia.gov/todayinenergy/detail.php?id=7970.
93. National Research Council. *Cumulative Environmental Effects of Oil and Gas Activities on Alaska's North Slope*. Washington, DC: National Academy Press, 2003. https://doi.org/10.17226/10639.
94. S. Hardin and J. Rowland-Shea. *The Most Powerful Arctic Oil Lobby Group You've Never Heard Of*. Center for American Progress (Washington, DC: 2018). www.americanprogress.org/issues/green/reports/2018/08/09/454309/powerful-arctic-oil-lobby-group-youve-never-heard.
95. M. K. Hobson. "Alaska Native Communities Clash over ANWR bill." *E&E News*, December 4, 2017.
96. A. DeMarban. "No Seismic Exploration for Oil in ANWR This Winter, Company Says." *Anchorage Daily News*, February 23, 2019. www.adn.com/business-economy/energy/2019/02/23/no-seismic-exploration-for-oil-in-anwr-this-winter-company-says.
97. G. Scruggs. "Polar Opposites: The Remote Alaskan Village Divided over Oil Drilling." *Reuters*, April 24, 2019. www.reuters.com/article/us-usa-environment-

oil-feature/polar-opposites-the-remote-alaskan-village-divided-over-oil-drilling-idUSKCN1S01B5.
98. P. Morin. "Caribou and Drilling in ANWR." *Radio Canada International Eye on the Arctic*, February 25, 2019. www.rcinet.ca/eye-on-the-arctic/2019/02/25/caribou-anwr-conservation-report-porcupine-hunting-indigenous-yukon-nwt-alaska.
99. H. Lincoln, chair, Bering Sea Elder Group et al. *Opposition to the Trump Administration's Inclusion of the Northern Bering Sea in Its Five-Year Outer Continental Shelf Offshore Leasing Program*. January 5, 2018.
100. S. Contorno. "Rubio, Scott Call for 10-Year Ban on Florida Offshore Drilling." *Tampa Bay Times*, March 4, 2020.
101. L. Murkowski, D. Sullivan and D. Young. *Comments on the Draft Proposed Program for OCS Leasing for 2019-2025*. Submitted to R. Zinke, secretary of the Interior. January 26, 2018.
102. Energy Policy Institute. "A Conversation with Senator Lisa Murkowski." University of Chicago, 2019. https://epic.uchicago.edu/events/event/a-conversation-with-senator-lisa-murkowski/.
103. L. Murkowski and J. Manchin. "Tackling Climate Change from Both Sides of the Aisle." *Washington Post*, March 10, 2019.
104. US Congress. House. *Protecting and Securing Florida's Coastline Act*. H.R. 205 116th Cong., 2nd session. Introduced in House January 3, 2019.
105. M. Green. "Majority of Voters Oppose Trump Offshore Drilling Plan: Poll." *The Hill* May 8, 2018. https://thehill.com/policy/energy-environment/386695-60-percent-of-voters-oppose-the-trump-administrations-offshore.
106. D. Fears and S. Clement. "Most Americans Support Reducing or Maintaining Oil and Gas Drilling." *Washington Post*, October 26, 2019.
107. M. Cantwell, ranking member of the Senate Committee on Energy and Natural Resources, and R. Grijalva, ranking member of the House Committee on Natural Resources. *The Royalty Rate of Oil and Gas Leasing Proposed by Royalty Policy Commission*. Submitted to R. Zinke, secretary of the Interior. February 27, 2018.
108. C. Davis. "Interior to Maintain Deepwater Oil, Gas Royalties Rate at 18.75%." *Natural Gas Intelligence*, April 18, 2018. www.naturalgasintel.com/articles/114074-interior-to-maintain-deepwater-oil-gas-royalties-rate-at-1875.
109. *Notice of Availability of Notice to Lessees and Operators of Federal Oil and Gas, and Sulfur Leases, and Holders of Pipeline Right-of-Way and Right-of-Use and Easement Grants in the Outer Continental Shelf – Requiring Additional Security*. 30 Code of Federal Regulations Parts 550 and 556. Bureau of Ocean Energy Management, Department of the Interior. 81 Federal Register 137 (July 18, 2016).
110. R. M. Grijalva, chair of the Committee on Natural Resources and A. S. Lowenthal, chair of the Subcommittee on Energy and Mineral Resources. *BOEM Decision to Halt Reform Requiring Greater Financial Assurances from Companies to Cover Decommissioning Costs*. Submitted to D. Bernhardt, secretary of the Interior. July 30, 2019.
111. Government Accountability Office. *Offshore Oil and Gas Resources: Actions Needed to Better Protect against Billions of Dollars in Federal Exposure to Decommissioning Liabilities*. Report by F. Rusco. GAO-16-40 (Washington, DC: 2015). www.gao.gov/assets/680/674353.pdf.

112. *Offshore Royalty Relief: Status During the COVID-19 Pandemic.* Government Accountability Office N11380 (2019). https://crsreports.congress.gov/product/pdf/IN/IN11380.
113. D. Crenshaw et al. *Temporary Reduction in Royalties for the Oil and Gas Industry* Submitted to D. Bernhardt, secretary of the Interior. March 20, 2020.
114. J. Barrasso et al. *Reduction or Suspension in Royalty Payment for the Oil and Gas Industry.* Submitted to D. Bernhardt, secretary of the Interior. March 30, 2020.
115. R. M. Grijalva, chair of the Committee on Natural Resources, et al. *Opposition against the Granting of Royalty Relief for Oil Extraction on Federal Lands and Waters.* Submitted to D. Bernhardt, secretary of the Interior. April 6, 2020.
116. J. Ogsbury, executive director of Western Governors' Association et al. *Updating Regulations on Water Quality Certification.* Docket ID: EPA-HQ-OW-2019-0405. Submitted to A. Wheeler, EPA administrator. 2019.
117. T. Puko. "Trump's Offshore Oil-Drilling Plan Sidelined Indefinitely: Interior Secretary David Bernhardt Cites Recent Court Decision Blocking Arctic Drilling." *Wall Street Journal*, April 25, 2019. www.wsj.com/articles/trumps-offshore-oil-drilling-plan-sidelined-indefinitely-11556208950.
118. D. Grandoni. "Oil Leasing in Arctic Refuge Delayed to Bolster Legal Case." *Washington Post*, January 16, 2020.
119. *Healthy Gulf v. Bernhardt*, No. 1:19-cv-00707 (D.C. Cir. 2019).
120. *Gulf Restoration Network v. Ryan Zinke*, No. 1:18-cv-01674 (D.C. Cir. 2020).
121. Y. Rosen. "Court Order Halts Chukchi Oil and Gas Activity." *Reuters*, July 22, 2010. https://in.reuters.com/article/oil-drilling-court-idINN2117895220100722.
122. *Native Village of Point Hope v. Jewell*, No. 12-35287 (9th Cir. 2014).
123. *Gwich'in Steering Committee et al. v. US Dept of the Interior et al.*, No. 19-208 (District Court of Alaska 2019).
124. Y. Rosen. "Alaska Native Lawsuit Accuses Feds of Hiding Arctic Refuge Oil Impact Information." *Reuters*, July 31, 2019. www.reuters.com/article/us-alaska-lawsuit-oil/alaska-natives-accuse-white-house-of-hiding-arctic-oil-impact-information-idUSKCN1UQ2NG.
125. B. Patterson and Z. Colman. "Trump Opens Vast Waters to Offshore Drilling: But Will They Come?" *ClimateWire*, January 5, 2018. www.scientificamerican.com/article/trump-opens-vast-waters-to-offshore-drilling/.
126. A. Harder. "Big Energy's Surprise Warning: Trump, Slow Down Deregulation." *Axios*, May 15, 2017. www.axios.com/big-energys-surprise-warning-trump-slow-down-deregulation-1513302239-894581f4-3d4d-49ee-be4d-133954b2d024.html.
127. B. Lefebvre. "Oil and Gas Allies to Trump: Slow Down." *Politico*, August 21, 2017. www.politico.com/story/2017/08/25/oil-and-gas-allies-want-trump-to-slow-down-242008.
128. A. C. Lin and M. Burger. "State Public Nuisance Claims and Climate Change Adaptation." *Pace Environmental Law Review* 36 (2018).
129. R. Frank. "Charting the Progress of the Latest Chapter in American Climate Change Litigation." *Legal Planet*, December 21, 2019. https://legal-planet.org/2019/12/21/charting-the-progress-of-the-latest-chapter-in-american-climate-change-litigation/.

130. N. Herz. "Alaska Feels the Brunt As Investors Promise Retreat on Fossil Fuels." *Alaska Public Media*, February 26, 2020. www.npr.org/2020/02/26/809210657/alaska-feels-the-brunt-as-investors-promise-retreat-on-fossil-fuels.
131. J.-Y. Ko, J. A. Barras, J. W. Day and G. P. Kemp. "Impacts of Oil and Gas Activities on Coastal Wetland Loss in the Mississippi Delta." *Ocean and Coastal Management* 47 (2004): 597–623.
132. N. N. Rabalais, R. E. Turner and W. J. Wiseman, Jr. "Gulf of Mexico Hypoxia, aka 'The Dead Zone.'" *Annual Review of Ecology and Systematics* 33, no. 1 (2002): 235–263.
133. D. Sullivan et al. *Major American Financial Institutions Continue Unfairly to Pick Energy Winners and Losers.* Submitted to President D. Trump. May 7, 2020.
134. M. J. Kotchen and N. E. Burger. "Should We Drill in the Arctic National Wildlife Refuge? An Economic Perspective." *Energy Policy* 35, no. 9 (2007): 4720–4729.
135. B. Peterson. "SC Trial over Seismic Blast Test for Offshore Oil Stalled by US Commerce Delay." *Post and Courier*, November 23, 2019. www.postandcourier.com/news/sc-trial-over-seismic-blast-test-for-offshore-oil-stalled-by-us-commerce-delay/article_153fe78c-0ae1-11ea-bb30-7b10495733cd.html.
136. US Geological Survey. *Economics of Undiscovered Oil in the Federal Lands of the National Petroleum Reserve, Alaska.* Report by E. Attanasi. Report 03-044. (2003). https://pubs.usgs.gov/of/2003/of03-044/fig1.htm

7

Backtracking on Safety
Risking Another BP Oil Spill

The previous chapter delved into the threats to America's oceans and coasts from the poor management of offshore leasing and the push for massive expansion of offshore drilling. To safeguard the oceans and coasts, it is equally important to consider how many risks society is willing to accept from offshore drilling and how much precaution society expects the industry to take to mitigate these risks. The potential for well blowouts and major spills from offshore drilling has multiplied as the industry has extended its geographical and technological frontiers.[1] Drilling operations in the Gulf of Mexico have expanded to ultra-deepwaters of 5,000 feet or more. Exploratory drilling operations have stretched to remote waters in the Arctic seas off northern Alaska, where harsh weather and sea conditions make operations even more hazardous and oil spill cleanup impossible. Chances of poor overall risk management in complex offshore drilling operations are heightened as companies that own offshore oil and gas wells reduce their in-house expertise and outsource critical activities to a myriad of contractors.[1]

The 2010 BP well blowout in the Gulf of Mexico, the largest accidental oil spill in the world (as of May 2021), epitomizes the gravity and scale of risks associated with offshore drilling. Not only did the well blowout cause deaths and injuries, communities along 1,100 miles of contaminated coastline continue to struggle with the economic and environmental devastation even a decade after the spill.[2] Sadly, Congress has failed to guard against the myriad of risks from offshore drilling. Despite their condemnation of BP's poor risk management and regulators' weak oversight, members of Congress failed to enact offshore safety legislation recommended by safety experts. This inaction paved the way for the administration to weaken safety regulations promulgated by the Obama administration (just as congressional failure to renew the moratoria on offshore drilling in the Atlantic and Pacific seas opened the door for the Trump administration to pursue drilling expansion).

Given Congress's failure to codify safety measures into meaningful constraints on the executive branch's management of offshore resources, the Trump administration was free to toss precaution aside. President Trump's appointees at the Department of the Interior and at the Bureau of Safety and Environmental Enforcement, notably Secretary David Bernhardt and Director Scott Angelle, had a long history of challenging offshore safety regulations prior to taking office.[3,4] With these appointees at the helm, the administration weakened companies' responsibilities for risk management, slashed safety regulations and scuttled regulatory oversight of offshore drilling. It stripped away the few protections that the Obama administration had promulgated in response to the BP well blowout and to Shell's mishaps during its exploratory drilling in the Arctic seas.[1,5]

The extensive regulatory rollbacks alarmed even some in the oil and gas industry, who shared their views with journalists, "It's not helpful if regulations are streamlined so as to allow something to happen – say, a methane explosion or a spill – and we'd be painted with it as an entire industry."[6,7] Governors and legislators from states with Pacific and Atlantic coastlines, fearing adverse economic impacts to their states from spills, vehemently opposed the expansion of offshore drilling and the rollback of regulations.[8–10]

I begin with an overview of the growing risks in offshore drilling (Section 7.1) and the lessons learned from past disasters and chronic spills (Section 7.2). I delve into how the Obama administration responded to these lessons by initiating reforms (Section 7.3) and how the Trump administration's reversal of these reforms exacerbates the risks (Section 7.4). I conclude by discussing how bipartisan congressional members proposed legislation to improve safety in offshore drilling. Unfortunately, despite the shifting tides against offshore drilling in coastal states, the majority of Republicans and those Democrats from states reliant on oil extraction failed to support these initiatives, putting at risk Americans' lives, livelihoods, coasts and oceans.

7.1 Growing Safety Concerns in Offshore Drilling

The American public pays attention to the once-in-a-decade well blowouts or spills. However, accidents and near misses with devastating consequences occur much more frequently but fail to grab headlines. Auditors in the oil and gas industry underscore that these events and the failure of blowout preventers (a device often treated as the last line of defense against a well blowout) *are not rare events.*[11,12] Since 1969, US waters have suffered 44 major spills that released more than 10,000 barrels per incidence.[13,14] The number of near

Table 7.1 *Accidents and near misses in the offshore drilling industry*

Year	Spills	Fires & explosions	Gas releases	Loss of well control	Injuries	Fatalities	Evacuations & musters	Lifting accidents
2018	19	80	19	1	171	1	82	111
2017	10	73	16	0	150	0	53	126
2016	16	86	17	2	150	2	50	155
2015	24	105	21	3	206	1	70	163
2014	21	135	21	5	285	2	52	210
2013	24	116	21	8	276	4	68	197
2012	30	132	27	3	280	1	48	167
2011	4	113	17	5	221	3	36	110
2010	9	134	20	4	253	12	31	118
2009	7	148	33	7	260	4	55	243
2008	33	141	22	7	263	12	43	185
2007	7	145	14	6	322	5	33	180

Source: Bureau of Safety and Environmental Enforcement, *Offshore Incident Statistics*, 2018.[15]
Note: The 2018 data corresponds to the calendar year, while data for other years correspond to fiscal years.

misses that could lead to major catastrophes, such as fires and explosions, losses of control at the well and gas releases, are summarized in Table 7.1.

These accidents are likely to occur with greater frequency in the Gulf of Mexico for at least four reasons. First, the shift to production of oil and gas in deeper waters raises safety risks. According to a study on drilling in the Gulf of Mexico, an increase of drilling depth of 100 feet raises the chance for an adverse event (e.g., well blowouts, injuries and oil spills) by 8.5 percent.[16] As seen in Figure 7.1 and Figure 7.2, production in the Gulf of Mexico's deepwaters (more than 200 meters) has surpassed that in shallow water. The deepest well drilled (measuring the depth of both sediment and water column) increased from just over 3,000 meters in 1975 to 10,600 meters by 2017.[17]

Second, safety experts warn that well operators have weakened their capabilities to ensure overall safety of their operations. To cut costs, they reduced their in-house expertise in various aspects of drilling operations and deepened their reliance on contractors to carry out critical tasks in a compartmentalized manner.[1,19] With reduced in-house expertise, they have less ability to evaluate the expertise of the contractors, delineate clear lines of responsibility and communicate or assess the cumulative systems-level risks from interdependent decisions and tasks.[1,19] These shortcomings are evident in BP's and Shell's mismanagement of offshore drilling discussed in Section 7.2.

Third, the aging wells, drilling platforms and pipelines are prone to malfunction. As of 2017, the Department of the Interior listed 240 platforms in the

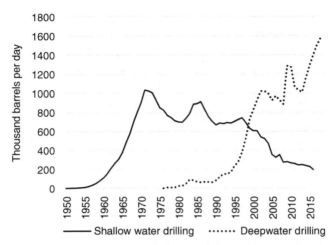

Figure 7.1 Oil production in deepwaters exceeded that in shallow waters
Source: Bureau of Safety and Environmental Enforcement, *Data Center: Production Data Online Query*, 2018.[18]
Note: Deepwaters refer to depths at or greater than 200 meters.

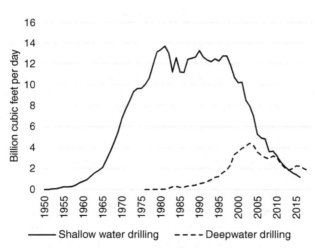

Figure 7.2 Gas production in deepwaters exceeded that in shallow waters
Source: Bureau of Safety and Environmental Enforcement, *Data Center: Production Data Online Query*, 2018.[18]
Note: Deepwaters refer to depths at or greater than 200 meters.

shallow waters of the Gulf (which serve over 2,000 oil and gas wells) as "idle iron." This designation means that this damaged infrastructure poses significant environmental and safety hazards.[3] Independent companies with less experience in drilling and remediation own the majority of these wells.[3] In 2017, about 17 percent of the 36 platforms and well operations that were inspected had oil and gas leaks that can lead to fire and many had no leak-detection systems or only malfunctioning ones.[3] Because companies may become bankrupt or even defunct prior to paying restitution for accidents, they face little incentive to address safety issues. In one egregious case, Black Elk Energy was convicted of eight felony violations related to an explosion on its platform that killed three workers and injured several others,[3] but the penalty levied after it went defunct may not be fully paid.[20]

Fourth, more intense hurricanes, as a result of climate change, damage platforms that are in production and those that are idle but have not been removed.[21,22] As seen in Figure 7.3, the majority of spills from 2001 through 2015 were caused by hurricanes. In 2006, Hurricane Katrina contributed to the spill of over seven million gallons of oil from production and storage facilities into Gulf Coast waterways.[24] In 2004, Hurricane Ivan caused an underground mudslide that damaged oil wells. The wells' owner, Taylor Energy, failed to take effective remediation action and, as a result, oil leaked into the Gulf for over 14 years – totaling to an estimated 1.5–3.5 million barrels.[25]

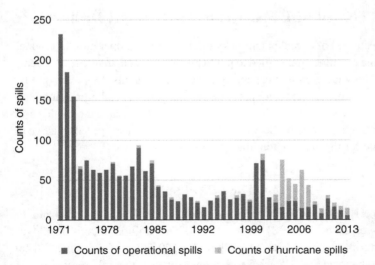

Figure 7.3 The frequency of oil spills due to hurricanes and operational problems
Source: ABS Consulting Inc., *2016 Update of Occurrence Rates for Offshore Oil Spills*, 2016.[23]

Accidents are also likely to grow in the US Arctic seas with the expansion of offshore drilling into more remote waters, that is, into the Chukchi and the Beaufort seas. The Arctic conditions heighten the risk of spills from shipping and drilling operations and limit oil-spill response. The Arctic zone experiences rough seas with ice floes, stormy and rapidly changing weather, poor visibility with summer fog, and minimal daylight in the winter.[5] The Bureau of Ocean Energy Management's 2014 environmental impact assessment for the Chukchi Sea leases projected that at least 800 small spills would occur throughout its operations.[26] It also estimated a 75 percent chance of a large oil spill (more than 1,000 barrels) would occur in the Arctic over a 77-year time horizon.[26]

A 2014 National Research Council report warned that the US Coast Guard and oil companies are not equipped to respond to an oil spill in the Arctic.[5] The lack of infrastructure, difficult weather and sea conditions would hamper efforts to contain a spill.[5] In 2017, Admiral Paul Zukunft, the then-commandant of the Coast Guard, who had served as the federal on-scene coordinator for the Deepwater Horizon oil spill, reiterated the near-impossibility of responding effectively and safely to an oil spill in the Arctic.[27,28] Rear Admiral Jonathan White, former chief Navy oceanographer, also underscored that there are still no proven methods of cleaning up an oil spill in ice.[28]

7.2 Flawed Risk Management in Offshore Drilling

Within the context of complex and risky offshore drilling operations, companies have time and time again failed to put in place prudent management systems even when they have limited capability to contain oil spills. Lax oversight by regulators weakens the external pressure on companies to take precautionary measures. The lessons from the BP spill (Section 7.2.1) and the mishaps in Shell's drilling in the Arctic (Section 7.2.2) are worth revisiting as they serve as the rationale for the Obama administration's reforms.

7.2.1 The BP Oil Spill

In 2010, the BP well blowout and oil rig explosion caused 11 deaths and 17 injuries. Over the course of 87 days, a total of 210 million gallons of oil was released into the Gulf of Mexico, polluting 1,100 miles of coastline.[29] That catastrophe happened as BP tried to temporarily seal an exploratory well 41 miles off the Louisiana coast. The improperly plugged well allowed hydrocarbons to shoot upward and the crew subsequently lost control of the well. The blowout preventer, the last line of defense, failed to seal the underwater well, leading to the well blowout and explosions on the rig.

Accident investigators excoriated BP's top management for failing to prioritize a safety culture but instead focusing on cost-cutting measures.[1,12] BP's approach resulted in other high-profile BP accidents including the 2005 Texas oil refinery fire and the 2006 Alaskan pipeline spill.[30,31] BP's top management did not clearly define the roles and responsibilities of the various players: BP, the well owner responsible for operations overall; Transocean, the rig owner and drilling contractor; and Halliburton, which supplied the cement used at the well.[1,12] Nor did it set up effective communication channels within BP and between BP and its contractors.[1,12]

BP's management did not put in a formal decision-making framework to assess risks or create an environment for workers to prioritize safety. Without that framework, the engineers and the crews, pressured by the cost of delays, opted for less meticulous options without fully appreciating the magnitude of additional risks they were taking on. Workers hesitated to raise safety concerns out of fear of retaliation.[30] Serious warning signs did not spur a redesign in operations. For example, several weeks prior to the blowout, workers observed dangerous "kicks," that is, unplanned cases of gas under high pressure entering the well.[19,32] BP's management also failed to implement inspection strategies to detect that the blowout preventer failed to function.[12]

BP's poor risk management was compounded by the regulatory agency's poor oversight. The then-regulator, the Mineral Management Service (MMS), was tasked with the dual function of generating revenue from offshore drilling and of regulating the industry. These conflicting roles led the MMS to lean far in the direction of promoting leasing over proper oversight of the industry.[1] For instance, the MMS did not fully implement the requirements under the National Environmental Policy Act (NEPA) when approving BP's drilling and production plans in the Gulf of Mexico.[11,33] NEPA requires federal agencies to prepare environmental impact statements that force the agencies to take a hard look at the environmental impacts of proposed projects and at alternatives.

Instead, the MMS granted BP a *categorical exemption* from NEPA review for all plans on drilling and production in the western Gulf of Mexico. It approved BP's production and drilling plans without probing BP's assertion that no adverse impacts were expected.[33] The occurrence of significant offshore well blowouts at an average of once every 48 months in the Gulf of Mexico almost certainly called for MMS's greater scrutiny of BP's claim.[11] MMS limited its wholly inadequate review to the environmental impact statements at the earlier stages of planning for the 2007–10 national OCS leasing plan and the western Gulf lease sales, despite the lack of information from these assessments to evaluate well-specific risks.[11,33]

Yet another shocking lesson from the spill is companies' limited capability to regain control of well gushers and contain spills. BP's multiple attempts, using strategies oil companies claimed would work in the plans they submitted prior to drilling, all failed.[34] Rex Tillerson, the then-CEO of ExxonMobil, admitted in a House Congressional Hearing on the BP oil spill "The point is we have to take every step to prevent these things [oil spills] from happening, because when they happen, we are not well equipped to prevent any and all damage. There will be damage. There is no response capability that will ensure that you won't have an impact."[35] Stopping the oil gusher took 87 days, with a team of academic and government scientists, led by Steven Chu, a Nobel Prize-winning physicist and the then-secretary of Energy, working with BP personnel.[34]

7.2.2 Shell's Arctic Mishaps

Shell's numerous mishaps in its Arctic exploration in 2012 raise red flags on the ability of oil companies to conduct safe drilling and to contain spills in the Arctic. In December of 2012, Shell's drilling rig broke loose and ran aground near Kodiak, Alaska during a risky 1,780 mile tow from Unalaska to Seattle.[36] That near miss did not turn into a catastrophe because the Coast Guard was able to mount a heroic 700-person search-and-rescue operation.

Just like BP, Shell failed to manage overall risks in its complex operations by involving a web of contractors and having limited capabilities to respond to oil spills. Investigators excoriated Shell for its series of failures, including its prioritization of cost-cutting over safety.[36-38] For instance, Shell decided to tow its drilling rig late in the season, despite rough sea conditions and against expert advice. Investigators noted that the move out of Alaskan waters, while dangerous, could reduce Shell's tax liabilities to the state of Alaska.[36] Not only did Shell fail to carefully assess the difficult Arctic conditions, it also failed to use appropriate ships and to put in place contingency plans. It even employed contractors who were inexperienced in operating in winter Arctic waters.[36]

Shell also failed to manage its contractors, from those that developed the oil spill equipment to those that undertook the drilling operations.[37] In 2012, its $400 million containment dome sunk and became "crushed like a beer can," while its undersea robots that were to assist in spill cleanup became entangled in the underwater cables.[39] In 2015, its icebreaker that was to carry the capping stack, a key safety device in the event of a well gusher, suffered a puncture and the drilling schedule had to be delayed.[40]

7.3 Reforms under the Obama Administration in Response to BP's Oil Spill and Shell's Mishaps

Following the BP oil spill and Shell's Arctic mishaps, accident investigators, safety experts and government agencies spent a great deal of effort to understand exactly what went wrong and how to avoid such disasters in the future. Investigative teams and bipartisan panels gave clear recommendations to Congress on the law to amend or to enact to address regulatory gaps. The House acted on several recommendations, passing a key bill that would strengthen safety requirements in offshore drilling, ensure compensation of victims of oil spills and remove the $75 million liability cap set in the 1990 Oil Pollution Act.[41] That cap set a ceiling on companies' liability for oil spill damages that counterproductively shifted the burden of the costs of oil spills, beyond that meager cap, to American taxpayers. The House also passed a bill that would provide protections to whistleblowers in the offshore industry who report safety lapses.[42] Sadly, the Senate failed to support these bills. The oil lobby's stranglehold on Congress, through its campaign contributions,[43] proved too strong to break, even as the historic damage from the largest oil spill in American waters to date wreaked havoc across the Gulf.[44,45] Not all was lost as the Obama administration took heed of these recommendations to reorganize the overall regulatory approach (Section 7.3.1) and to strengthen regulations (Sections 7.3.2 and 7.3.3).

7.3.1 Reforming the Regulatory Approach

The Department of the Interior reorganized the MMS into two distinct agencies. The safety tasks were entrusted to the Bureau of Safety and Environmental Enforcement (BSEE), while the revenue generation tasks were assigned to the Bureau of Ocean Energy Management.

The Department of the Interior also made an important shift to a goal-oriented safety and environmental management systems approach (SEMS). The SEMS approach places the burden on companies to demonstrate that they have taken steps to minimize risk in their operations.[46] This approach had been adopted in the United Kingdom and Norway, both major oil and gas producers in the North Sea, in response to the 1988 Piper Alpha disaster.[46] Key aspects of this approach are enhancing risk management across multiple players involved in complex operations and vigilance against the gradual, unnoticed decline in performance standards that can lead to catastrophic failures.[47–49]

Under this approach, the well operator is mandated "to assess the risks associated with a specific operation, develop a coordinated plan to manage those risks, integrate *all involved contractors* in a safety management system, and take responsibility for developing and managing the risk management process" and to do so continually.[1] Regulations serve as the backstop, specifying the minimum performance and technical standards rather than driving companies' actions.[50] Companies are required to hire *independent third-party auditors* to evaluate their risk management systems.[50] These auditors could potentially uncover key deficiencies, missed by operators, contractors and regulators, and prompt their correction – such as the deficiencies in the blowout preventer that contributed to the BP oil spill.[51]

Even so, BSEE and the Coast Guard underscored that the shift to SEMS in the United States does not eliminate the need for regulatory oversight. SEMS worked well in the United Kingdom and in Norway, which have either nationally owned oil companies held accountable to the general public or larger experienced companies whose reputation in the eyes of the general public matters for their continued ability to develop their large acreage.[52] On the contrary, the US offshore drilling industry includes less experienced smaller companies for which active oversight by regulators on their operations remains critical.[52]

7.3.2 Well Control and Blowout Preventer Rule

In addition to reforming the overall regulatory approach, the BSEE promulgated two major rules. The first is the Well Control Rule, consisting of the Well Control and Blowout Preventer Rule and the companion Production Safety Systems Rule. The Well Control and Blowout Preventer Rule finalized in 2016 sets out a number of requirements on the real-time monitoring of well operations, on the technical requirements for safety equipment, and on third-party verification on those technical requirements.[53] Table 7.2 provides examples of how provisions in the rule respond to errors that contributed to the BP well blowout and to other findings of the investigations.

Vice Admiral Brian Salerno, the director of BSEE at the time, justified the Well Control Rule by pointing to how companies failed to reduce the frequency of loss of well control even several years after the BP catastrophe. In his 2015 congressional testimony, he noted,

> The Well Control Rule synthesizes and incorporates ... these recommendations [from the investigations] in an effort to reduce risks across all phases of drilling operations. ... The need for the Well Control Rule is demonstrated by the fact that loss of well control incidents are happening at the same rate five years after the Macondo blowout as they were before. In 2013 and 2014, there were eight and seven loss of well control incidents per year, respectively, a rate on par with

Table 7.2 *The Well Control and Blowout Preventer Rule provisions and rationale*

Well Control and Blowout Preventer Rule provisions[53]	Errors that led to the BP well blowout[1,12]
Real-time monitoring of well operations is required. This ensures that additional observers, including those onshore, can monitor the well during critical operations, detect abnormalities and provide technical assistance.	The crew explained away the abnormalities in the cement tests and wrongly concluded the well was properly sealed. They did not notice warning signs that oil and gas had entered the well through the compromised cement barrier. During the critical phase when the crew struggled to regain control of the well, they had trouble reaching experts onshore for technical assistance.
Blowout preventer systems must include technologies that centers the well pipe during shearing operations and must have dual shear rams that enhances the possibility that the well pipe is successfully cut.	The blowout preventer failed because the shear ram was not able to cut the well pipe that had become off-center within the blowout preventer. BOPs are likely to fail unless they address the buckling of drill pipes that is likely to occur during a blowout.
Well Control and Blowout Preventer Rule provisions[53]	Other facts uncovered by the investigations[12]
A third party, certified by BSEE, is required to complete an annual assessment on the mechanical integrity of the blowout preventer. Physical inspection of the blowout preventer is required every five years.	Transocean and BP did not put in place testing practices that would have identified and corrected deficiencies in the blowout preventer. Accident investigators found that the blowout preventer components were not correctly wired and one of batteries to power the shear rams was drained.
Third parties are required to certify that the shearing capability of these systems meet specific technical criteria.	The shearing rams in the BP blowout preventer could not shear drill pipes of the diameter and strength as specified by its manufacturer.

pre-Macondo losses of well control. Some of these [loss of well control] incidents have resulted in blowouts.[i] ... It is abundantly clear that despite post-Macondo improvements in safety and technological advancements, there are still issues that must be addressed in order to see an appreciable decrease in dangerous loss of well control incidents.[54]

---

[i] According to Salerno, in the 2013 Walter Oil and Gas incident, the loss of well control resulted in an explosion and in a fire that burned for 72 hours and jeopardized the lives of 44 workers who were, fortunately, safely evacuated.

7.3.3 The Arctic Rule

Another regulation promulgated by the Department of the Interior, also in 2016, is the Arctic Rule.[55] The rule, recognizing the heightened challenges of operating in the harsh and remote Arctic environment, sets out a number of requirements on operators' capabilities to work under these conditions and to respond to the loss of well control and to spills in a timely way. Table 7.3 provides examples of how provisions of the Arctic Rule correspond to the lessons learned from Shell's Arctic mishaps.

7.4 The Trump Administration's Reversal of Safety Reforms in Offshore Drilling

The Trump administration laid the groundwork for its reversal of safety reforms by appointing administrators with résumés of accomplished advocacy for the oil and gas industry.[56] Trump appointed David Bernhardt as head of the transition team at the Department of the Interior, then as deputy secretary, as acting secretary and finally as secretary of the Interior. Prior to taking office, Bernhardt worked as a lobbyist and lawyer for oil and gas companies and their trade groups, including for the National Ocean Industries Association (NOIA) and for the Independent Petroleum Association of America. In those positions, Bernhardt supported expansive drilling, successfully representing NOIA interests in a lawsuit in which environmental groups sought to block the Obama administration's issuance of new offshore leases.[57] He also supported narrow interpretations of agencies' regulatory powers, arguing against the Interior's power to hold contractors liable under OCSLA. Once in office, Bernhardt spearheaded the Interior's work to expand drilling and to weaken safety regulations.[58,59] The Project on Government Oversight, a nonpartisan independent watchdog established in 1981 to expose corruption, alleged that Bernhardt oversaw oil and gas drilling issues that are of interest to his former clients. This conflict of interest was allowed under the Trump administration's weak recusal policy[60] and the weak ethics oversight in the Department of the Interior.[61] Concerns that Bernhardt may have participated in decision-making that favored his former client while he was acting secretary of the Interior even prompted senators Elizabeth Warren and Richard Blumenthal to ask for an ethics investigation into his actions.[62]

President Trump also ended the norm of appointing marine safety experts to lead BSEE, an action the National Commission had warned against. The two BSEE directors immediately before the Trump administration took office,

Table 7.3 *The Arctic Rule rationale and provisions*

Arctic Rule provisions	Rationale
Companies must capture all oil-based drilling mud and cuttings. The regional supervisor has discretionary authority to require that operators capture all water-based mud and associated cuttings.	Discharge of these materials can affect water quality and organisms in the ocean and ocean floor and thus adversely affect Alaskan native communities that rely on subsistence hunting and fishing.
Companies must submit an Integrated Operating Plan on various aspects of their operations that demonstrate they have taken into account the risks of Arctic operations.	Shell's operations failed to take into account specific Arctic conditions and underestimated the time needed for various operations.
Companies must explain how they would exercise oversight over their contractors.	Shell failed to ensure that its contractors provided ships and equipment that met safety standards and that they conducted operations safely.
Companies must demonstrate their capabilities to predict, detect and respond to changing ice, sea and weather conditions.	The Arctic weather and sea conditions can quickly change. Shell's drill ship had to disengage after commencing drilling when an ice floe was detected as due to arrive in its well site within three days.
Companies must demonstrate that they have the ability to regain control of a well and to contain a well gusher within a short period. They must have in the vicinity of their drilling operations equipment such as a capping stack, a cap and flow system and a containment dome.	The BP disaster demonstrated the catastrophic impacts of a well blowout and a well gusher. Key safety devices are not readily available in the Arctic region nor can they be easily and quickly transported to the Arctic region.
Companies must demonstrate that they have access to a relief rig in the vicinity so that they can drill a relief well and cap the original and the relief well before the seasonal ice returns to the drill site and no longer than 45 days after the loss of well control.	Access to a relief rig is an important precaution because the BP well gusher could only be stopped by drilling a relief well. The relief operations would need to be completed before the return of the winter sea ice that would hamper relief operations. Leaving a well gusher or oil spill unaddressed until the following summer would have catastrophic consequences on the Arctic ecosystem.

Source: Bureau of Safety and Environmental Enforcement, *Oil and Gas and Sulfur Operations on the Outer Continental Shelf-Requirements for Exploratory Drilling on the Arctic Outer Continental Shelf: Final Rule*, 2016.[55]

Coast Guard Rear Admiral James Watson and Coast Guard Vice Admiral Brian Salerno, built strong collaborations between BSEE and the Coast Guard, which play key roles in safeguarding the coasts and seas. The Trump-appointee Director Scott Angelle, who trained as a petroleum land manager and had served as the director of Louisiana's Department of Natural Resources, took the approach of prioritizing leasing over safety – the very approach censured by the commission. He mimicked the ill-fated path of the MMS and opened the door for the agency to be captured by the regulated industry. In 2017, Angelle announced, "efforts underway at [the] Bureau ... to create a positive investment culture: regulatory reform ... and reviewing permit processes to increase efficiency."[63] Angelle is reported to have said, "To the degree this industry wants to be part of the discussion, tell me where you want me to be and we will be there."[3]

With anti-regulatory appointees at the helm of the Interior and BSEE, the administration slashed away at past reforms that the oil industry and offshore industry trade groups had opposed.[64,65,66] It rendered the SEMS approach for risk management in the offshore drilling industry ineffective, removing its two pillars of independent third-party audits and mandatory reporting. In September 2018, it weakened the Production Safety Systems Rule by eliminating the requirement that an independent third party certify that the safety equipment would function as designed.[67] Third parties serving as extra eyes may have detected the failures in the safety systems that escaped the attention of BP and Transocean prior to the BP well blowout.[32,51]

The revised rule also eliminated the mandatory requirement that well operators submit to BSEE reports about failures in the safety and pollution prevention equipment and in the blowout preventer. Mandatory reporting of equipment failures (even those that do not result in spills) helps improve safety in the industry. For instance, in 2016, on learning about failures with the massive bolts in undersea safety equipment, including in blowout preventers, BSEE notified the entire industry.[68] The industry recognized the systemic nature of the problem and began to work on corrective action. The problem was no doubt significant, as noted by a BSEE official to a reporter: "'If your smallest component fails, you can't expect a sophisticated many-million-dollar piece of equipment' to hold fast and prevent a leak."[68] The mandatory reporting requirement had enabled BSEE to compile the data and make it public for a broad range of stakeholders, such as NGOs and shareholders of companies, that used the information to exert pressure on companies to strive for safer performance.[12]

In May 2019, the administration severely weakened the Well Control and Blowout Preventer Rule, removing the very provisions recommended by the

National Commission.[69] Removal of the requirement for the drilling crew to send real-time data from the well to observers onshore eliminates the additional set of eyes that could detect warning signs, alert the crew and assist with responses. That precaution of direct contact with additional personnel onshore could have prevented the BP crew from explaining away the abnormalities in their tests and instead helped the crew realize that the well was not properly plugged. Likewise, severely weakening the testing requirements for the blowout preventer cripples the very last line of defense in a well blowout. BP and Transocean had failed to put in place testing practices that could have detected problems in the blowout preventer, including its dead battery and mis-wired components.[70,71] Testing requirements mandated under the Well Control Rule had been effective in detecting problems. Testing procedures detected 60 percent of the equipment failures experienced by 18 companies that operate 90 percent of the new wells in the Gulf of Mexico.[70,71]

Similarly, the administration's proposed revisions to the Arctic Rule ignored the importance of ending drilling operations early enough in the season that any response to spills can be taken within the same summer season.[37] Instead, the proposed revisions would permit operators to drill into the hydrocarbon layers later in the Arctic drilling season and even when there would no longer be sufficient time to drill a relief well within the same season. (The drilling of a relief well was the only strategy that worked to stop the BP oil gusher.) These revisions dangerously risked leaving spills unaddressed during the long winter season and to be resumed only the following year. Other proposed revisions would permit delays in the response efforts by no longer requiring containment devices to arrive within seven days to the site of the spill. That seven-day provision had been promulgated out of recognition that containment devices are not stationed close to the remote Arctic. (The Trump administration did not finalize the Arctic Rule and the Biden administration withdrew that rule in May 2021.)

In justifying these rollbacks, the administration claimed that regulations impeded the growth of the offshore industry and that the rollbacks saved compliance costs without raising safety risk. Neither claims stand up to scrutiny.

As emphasized by numerous safety experts, the rollbacks of the very provisions designed to address safety gaps that had led to two catastrophes (summarized in Tables 7.2 and 7.3) would almost certainly exacerbate the risks in offshore drilling. Those who served on the bipartisan National Commission held the unanimous view that the administration's regulatory rollback compromised safety.[72] Ten state attorneys general and environmental groups submitted comments to BSEE underscoring these points.[73–75]

The administration's claims that safety would not be compromised contradict the views of BSEE engineers. Investigations by the *Wall Street Journal* revealed that BSEE administrators overruled the engineers' opposition to changes in the rule.[76] Based on interviews of BSEE staff and reviews of memos and email, the *Journal* further alleged that Angelle tried to cover up the opposition by directing a BSEE engineer to delete text from the original memos that had documented how the agency's decision to make changes to the rule contradicted guidance from the agency's own engineers. According to the *Journal*'s review of the original memos, "BSEE engineers wanted 'no change to the testing frequency' of critical safety equipment and ... the staff 'does not agree with industry' that an industry-crafted protocol for managing well pressure was sufficient in all situations." Both items were deleted from the final memos. The review also revealed that a BSEE engineer wrote in emails that "Mr. Angelle 'verbally instructed' the team to endorse an option to relax the drilling-margin [pressure] minimum, as oil-industry trade groups were urging" and "to move ahead with language adopting the 21-day testing requirement."

The administration's other justifications for regulatory rollbacks are similarly disingenuous. Its claim that regulations stifled production runs counter to the reality of record-breaking oil production in the Gulf of Mexico in 2017.[77] While the administration is correct in noting regulations' imposition of compliance costs on industry, from society's perspective, these regulations make economic sense. The savings on compliance costs to industry, shown in Table 7.4, are miniscule in comparison to the costs of cleanup, not to mention the loss of lives, the disruption to the lives of coastal communities and the damage to the ecosystems. These savings in compliance costs are also dwarfed by the value Americans place in averting another disaster akin to the BP oil spill or the Exxon Valdez spill in Alaska.

Not only did the administration roll back regulations, it also reduced inspections and weakened enforcement. According to an analysis by the Center for American Progress of BSEE data, the number of inspections fell by 30 percent between the last three years of the Obama administration (2014–16) and the first three years of the Trump administration (2017–19).[4] The 38 percent reduction in enforcement actions in that time period, the study noted, indicated BSEE's soft-pedaling of enforcement rather than greater voluntary compliance.[84] The administration also pulled the plug on an ongoing study aimed at improving the effectiveness of BSEE's oil and gas inspection program.[85] That study, by the National Academies of Sciences, Engineering and Medicine, examined the extent to which independent third parties and remote real-time monitoring could enhance safety inspections.[86]

Table 7.4 *Comparison of the large benefits from averting disasters to the more limited costs of compliance*

Costs of compliance	Benefits from averting disasters
Well Control and Blowout Preventer Rule[78]	
Compliance costs over 10 years are estimated at $790 million discounted at 3 percent or $686 million discounted at 7 percent.[78]	BP's costs for the Gulf oil spill stood at $61.6 billion, including payouts to victims, cleanup costs and legal costs.[79] Researchers find that American households are willing to pay a total $17.2 billion to avoid a repeat of the BP disaster.[80]
BSEE's revisions to the Well Control and Blowout Preventer Rule	
Savings from revisions to the rule over 10 years are estimated at 81 million discounted at 3 percent or 28 million discounted at 7 percent.[81]	See the BP case above.
Arctic Rule[55]	
Compliance costs over 10 years are estimated at $2.05 billion discounted at 3 percent or $1.74 billion discounted at 7 percent.[55]	Exxon's costs for the Exxon Valdez spill stood at $3.8 billion. Exxon spent $2.1 billion in cleanup efforts, settled a civil action by the United States and Alaska for $900 million, and paid $500 million in punitive damages and $303 million in compensatory damages to private parties.[82] Researchers estimate the public's willingness to pay to avoid another Exxon Valdez Spill between $4.9 billion to $7.2 billion dollars.[83]

7.5 How to Make Safety a Priority in Offshore Drilling

Faced with an executive branch that slashed regulations aimed to promote safety amid new and growing risks in the industry, Congress, whose constitutional duty is to manage and protect federal lands and waters, faced a threefold challenge: to enact long overdue safety laws, to take preemptive actions against new threats and to tighten its oversight of the executive branch. Sadly, congressional failure to block the rollback of offshore safety regulations and to enact updated safety laws very much mirrors its failure to curb the administration's expansion of drilling. The bipartisan push for stricter laws by representatives of Pacific and Atlantic states, including Florida, failed to secure sufficient votes in the House and the Senate. Pro-drilling members of Congress, including Democrats from oil-dependent states, while continuing to give lip service to the need to undertake offshore extraction safely, did not support putting in place laws to achieve those goals. However, members of Congress may well shift their views, with voters' opposition to drilling flipping races in favor of anti-drilling candidates, as discussed in Chapter 6.

Several members of Congress made a number of valiant efforts on the legislative front that are worth discussing. In 2019, representatives Vern Buchanan and Francis Rooney, Republicans from Florida, together with Democratic colleagues Charlie Crist and Debbie Wasserman Schultz from Florida, David Price from North Carolina and Nanette Diaz Barragán and Jared Huffman from California cosponsored the Safe Coasts, Oceans, and Seaside Towns Act.[87] The proposed legislation would combat the administration's regulatory rollback by codifying into law the original Well Control and Blowout Preventer Rule and the Production Safety Systems Rule, as promulgated by the Obama administration. A similar piece of legislation in the previous congress won 64 cosponsors. In urging support for the bill, Buchanan underscored, "Have we learned nothing from the Deepwater Horizon catastrophe? It would be a monumental mistake to lift these safeguards, which were based on recommendations by a bipartisan national commission in wake of the fatal 2010 disaster."[88] Crist added, "We should ... not hand over the fate of our coasts and oceans to the oil and gas industry. These rules are vital to protecting our coastal communities from preventable disasters and must remain in place."[88]

In contrast to Florida's representatives, who banded together with their counterparts from other coastal states to protect America's coasts and seas, Florida's Republican senators Marco Rubio and Rick Scott took a narrow, Florida-centric approach. They proposed an amendment to the 2020 senate energy bill that would extend the moratorium on drilling off eastern Florida for an additional five years (extending the end date from 2022 to 2027)[89] and sought verbal assurances from the Trump administration not to drill off Florida's coast.[90] Their not-in-my-coast approach does not protect Florida, as underscored by Floridian Rooney.

Even a potential oil spill presents an existential threat to Florida. After the Deepwater Horizon spill, the tourism industry along the Gulf Coast of Florida was crushed. Even though no oil reached the west coast of Florida, 50,000 jobs were lost. Further, if any spillage or other residual effects of a spill were to reach the Loop Current, which runs clockwise down the west coast, oil and tar balls would find their way along the coast from Tampa to Key West.[88]

Senator Lisa Murkowski from Alaska, ranking member of the Senate Committee on Energy and Natural Resources, proposed the following during the committee's hearing on the bipartisan commission report on the BP spill:

> I hope that we will at least informally agree on a threefold pledge regarding our offshore policy. That is, first, that no victim of a spill should ever go uncompensated. That [second,] taxpayers should never be on the hook for a company's damages. Third that these priorities are managed in a way that not only preserves, but also promotes a competitive, domestic offshore industry ... We absolutely need to look at ways to improve our offshore system and make those operations safer.[91]

Unfortunately, Murkowski and several members of the committee did not support legislation that could put into effect these pledges. For instance, the Senate did not implement the commission's recommendation of raising or even removing the $75 million cap on companies' liability for oil spill damages. Senator Robert Menendez, a Democrat from New Jersey, proposed a bill to remove this liability cap in almost every congress since 2010 and also in 2018,[92] but did not win support for it in the Senate. As emphasized by the commission, the $75 million cap is woefully inadequate in the era of deepwater drilling. While BP waived the cap to protect its sprawling oil empire in the United States, a smaller company with only limited operations would have little incentive to pay damages beyond the cap. In his congressional testimony, economist Michael Greenstone explained that "the removal or substantial increase of the liability cap on economic damages is the most effective way to align oil companies' incentives with the American people's interests [for safer drilling]."[93] The cap encourages companies to take excessive risks,[ii] as they reap the full benefits of taking those risks but shift the bulk of the costs to society should spills occur.[19,93,94]

Murkowski focused her legislative effort primarily on securing funding to cope with the costs of oil spills. In March 2019, Murkowski and a fellow Republican senator from Alaska, Dan Sullivan, proposed the Spill Response and Prevention Surety Act,[95] which would permanently reinstate the nine-cent tax per barrel of oil to fund the Oil Spill Trust Fund. Congress had permitted the tax to expire at the end of 2018. The trust fund provides an essential funding source to pay for oil spill response and to compensate coastal communities that bear the burden from oil spills when companies fail to do so. Recall, companies can evade payment when damages exceed the legal caps on liabilities and by declaring bankruptcy when they do not have adequate funds, whether in assets or insurance coverage, to cover damages.[96]

[ii] While raising the cap or removing it altogether would dissuade drilling in high risk areas, it is precisely in the interest of Americans not to drill in these areas [93][107].

Senate Republicans not only failed to push for safety legislation, they also exercised only anemic scrutiny of Trump's appointees to the Department of the Interior. Even those senators who had voiced alarm over Bernhardt's work in expanding offshore drilling and weakening regulations voted to confirm him in a 56–41 vote, with three Democrats breaking ranks to vote affirmatively. Florida senators Rubio and Scott and Senator Susan Collins, a Republican from Maine, claimed that they did their due diligence by securing Bernhardt's assurances that offshore drilling would not take place off the coasts of Florida and Maine, respectively.[90,97]

Concerning checks and balances over the executive branch, Congress did not abdicate its responsibilities altogether. It exercised its power of the purse to protect the Chemical Safety Board, the independent agency that investigates industrial accidents and that has earned the support of industries,[iii] trade unions and safety experts.[98] It was the only agency to successfully sue Transocean for access to the failed blowout preventer and made the critical finding about the systematic design flaws in blowout preventers, including those in the newer drilling rigs in the Gulf of Mexico[12,31] Congress blocked the Trump administration's attempt to kill off the Chemical Safety Board in its 2018, 2019 and 2020 budget proposals.[99,100] However, Congress was not able to compel the administration to nominate members to the board, leaving four out of five slots on the board unfilled as of February 2020 and jeopardizing its work.[98]

As in the case of offshore drilling expansion, congressional failure to enact offshore safety legislation left the task of battling the regulatory rollback to state attorneys general and environmental groups. A coalition of 10 state attorneys general submitted comments to BSEE, opposing the rollback of key features of the Blowout Preventer and Well Control Rule.[73] As discussed in Chapter 6, they also launched other legal challenges, for example, against seismic airgun testing in the Atlantic, a step that paves the way to exploratory drilling.

In June 2019, the Sierra Club and nine other environmental NGOs challenged the administration's partial repeal of the Well Control Rule on at least three grounds. First, by weakening the rule, the Department of the Interior failed its duty under the OCSLA to ensure safe drilling. Second, the Interior violated NEPA by failing to take into account the full environmental impact of the partial repeal. It undertook only a limited environmental analysis, rather than a comprehensive environmental impact statement. Third, the Interior violated the Administration Procedure Act by failing to provide a reasoned

---

[iii] In the face of these proposed cuts, even members of the oil industry, for instance, Stephen H. Brown, vice president of federal government affairs for Andeavor, a Texas-based refiner, underscored the important roles played by "[a] credible safety oversight entity that partners with stakeholders on accident investigations, conducts root cause analysis, and fosters learnings" [99].

explanation for its repeal of several provisions of the rule. The Interior's claim that the repeal of those provisions would not jeopardize safety contradicts the agency's own presentation of the scientific and technical evidence that each of the provisions in the original rule was necessary to reduce the risks of oil spills and workers' injuries and fatalities.

Environmental NGOs have also pursued litigation to challenge the administration's unlawful relaxation of regulations. Healthy Gulf, an environmental NGO dedicated to protecting the Gulf region, filed a lawsuit challenging the Department of the Interior's granting of about 960 waivers to oil companies from meeting existing environmental regulations. One-third of these waivers permitted companies to not abide by technical requirements on the blowout preventer, the last line of defense, which failed in the BP oil spill.[101] Healthy Gulf argued that the Department of the Interior had instituted a hidden Waiver Rule that systematically granted these waivers, and that the promulgation of the Waiver Rule violated administrative procedure. The rule was never published for notice and comment, which is required under the Administrative Procedure Act, nor subjected to environmental analysis, which is required under NEPA. In fact, the agency's granting of waivers from regulations only became known to the public when investigative journalists uncovered those actions and only then were House members able to hold hearings on these waivers (ultimately leading to Healthy Gulf filing suit).

Even the industry itself has played a role in questioning the administration's rollbacks. The administration's cuts to safety regulations, while celebrated by trade associations such as the National Ocean Industries Association and the American Petroleum Institute, were not universally supported by all members of the industry. According to media reporting, several oil and gas companies did not support the rollback of the Well Control Rule.[7] Members of the industry shared their concerns that a massive oil spill by any one company imposes financial losses and operational disruptions to all other companies and risks backlash from the public. Following the BP spill, the entire oil and gas industry listed in the US stock exchange suffered losses in their stock value and all exploratory drilling in the Gulf of Mexico was halted for six months by the Obama administration.[102,103]

Realization that regulations are in the industry's interests (by compelling all companies to adhere to a minimal baseline of safety standards) prompted some oil and gas companies to speak out against the administration's rollbacks of regulations. However, to date, they have voiced their opposition publicly only against those rollbacks that affect onshore operations. For instance, the larger oil and gas companies, such as BP, Exxon and Shell, opposed the administration's rescission of the Methane Rule.[104,105] They feared that the rescission of the rule would jeopardize their marketing of natural gas as a climate-friendly alternative to coal, and pledged to continue their methane-reduction efforts.[105]

7.6 Conclusion

When and where the next catastrophic spill might occur is up to chance, but the consequence of spills is known for certain. If, or more pessimistically when, a major spill occurs again, communities will be left to bear the financial, social and emotional costs. In the Exxon Valdez case, after an almost two-decade legal challenge by Exxon, during which time some of the plaintiffs had died, the Supreme Court reduced the punitive damages to $500 million, a pittance compared with the $2.5 billion awarded by a lower court and the initial $5 billion jury award.[82] Communities in the BP case fared no better. In 2014, the District Court in New Orleans delivered its ruling on the medical settlement for coastal residents affected by the BP spill and workers involved in the cleanup of the spill. It ruled that only those with a medical diagnosis by a hard deadline of two years after the spill would receive compensation.[2,106] As a result, the settlement excluded those whose conditions manifested only after two years, as is typical of diseases with long latency periods, and those who experienced the effects after the spill but failed to seek timely diagnosis.

The bipartisan commission's postmortem of the BP spill censured the failure of Congress and the executive branch to prioritize safety over appeasing special interests in the offshore sector. It noted,

> the federal government has never lacked the sweeping authority required to control whether, when, and how valuable oil and gas resources located on the outer continental shelf are leased, explored, or developed. The root problem has instead been that political leaders within both the Executive Branch and Congress have failed to ensure that agency regulators have had the resources necessary to exercise that authority ... and, no less important, *the political autonomy needed to overcome the powerful commercial interests that have opposed more stringent safety regulation.*[1]

Unfortunately, congressional failure to respond to the commission's recommendations in the post-BP years persisted even as the Trump administration crippled regulatory oversight. While several bipartisan members of Congress from coastal states fought intrepidly to enact safety legislation, their efforts failed in the face of opposition from many Republicans and those Democrats from oil-reliant states. The commission's warning to Congress remains prescient, "Inaction is a policy of dangerous default – of continuing to rely on chance and luck to avoid a next time."[1] The cost of a next time could well be unfathomable, as underscored by a member of the commission in the tenth anniversary of the spill. "We very well could have a disaster equal to or greater than the one ten years ago, and that would be a tragedy because we know what can happen, and yet we didn't take the steps that were necessary to address that risk ... It will probably be our epitaph."[72]

The administration's rollback of specific regulations governing drilling, both offshore and onshore, put Congress, state attorneys general and civil society to the test of safeguarding the economy, public health and the environment; and Congress's performance has been, at best, spotty and, at worst, dismal. Even more troubling and far-reaching was the administration's attack on the very core of all regulations, by undermining scientific, economic and legal analyses in the rulemaking process – an issue addressed in the next part of the book.

References

1. National Commission on the BP Deepwater Horizon Oil Spill and Offshore Drilling. *Deep Water: The Gulf Oil Disaster and the Future of Offshore Drilling*. Report to the President (Washington, DC: Government Publishing Office, 2011). www.govinfo.gov/content/pkg/GPO-OILCOMMISSION/pdf/GPO-OILCOMMISSION.pdf.
2. Associated Press. "New Wave of Suits Linked to BP Oil Spill Hitting Courts." *Associated Press*, December 6, 2019. www.businessreport.com/newsletters/new-wave-of-suits-linked-to-bp-oil-spill-hitting-courts.
3. E. Lipton. "Targeting Rules 'Written with Human Blood.'" *New York Times*, March 11, 2018.
4. J. Rowland-Shea and M. Rehmann. *The Favor Factory: President Trump's Interior Department Is Benefiting Past Political Donors and Lobbying Clients*. Center for American Progress (Washington, DC: 2018). www.americanprogress.org/issues/green/reports/2018/08/27/455150/the-favor-factory/.
5. Transportation Research Board and National Research Council. *Responding to Oil Spills in the US Arctic Marine Environment*. Washington, DC: National Academies Press, 2014. www.nap.edu/catalog/18625/responding-to-oil-spills-in-the-us-arctic-marine-environment.
6. B. Lefebvre. "Oil and Gas Allies to Trump: Slow Down." *Politico*, August 25, 2017. www.politico.com/story/2017/08/25/oil-and-gas-allies-want-trump-to-slow-down-242008.
7. A. Harder. "Big Energy's Surprise Warning: Trump, Slow Down Deregulation." *Axios*, May 15, 2017. www.axios.com/big-energys-surprise-warning-trump-slow-down-deregulation-1513302239-894581f4-3d4d-49ee-be4d-133954b2d024.html.
8. E. G. Brown, governor of California et al. *Request for Information and Comments on Preparation of 2019–2024 Outer Continental Shelf Oil and Gas Leasing Program*. Submitted to R. Zinke, secretary of the Interior. August 17, 2017.

9. L. Hogan, governor of Maryland et al. *Opposition to the Leasing, Exploration, Development and Production of Oil and Gas in the Atlantic Ocean.* Submitted to R. Zinke, secretary of the Interior. January 17, 2018.
10. R. Cooper, governor of North Carolina et al. *Opposition to Seismic Airgun Surveys and Oil and Gas Drilling Off Our Coasts.* Submitted to W. L. Ross, Jr., secretary of Commerce, and R. Zinke, secretary of the Interior. December 20, 2018.
11. O. A. Houck. "Worst Case and the Deepwater Horizon Blowout: There Ought to Be a Law." *Environmental Law Review* 40, no. 11 (2010): 11033–11040. https://marine.rutgers.edu/dmcs/ms606/2010_fall/Houck%202010.pdf.
12. US Chemical Safety and Hazard Investigation Board. *Investigation Report Executive Summary of Drilling Rig Explosion and Fire at the Macondo Well.* Report No. 2010-10-I-OS (Washington, DC: April 12, 2016). www.csb.gov/macondo-blowout-and-explosion/.
13. Office of Response and Restoration. *Largest Oil Spills Affecting US Waters since 1969.* National Oceanic and Atmospheric Administration (Silver Spring, MD: April 5, 2017). https://response.restoration.noaa.gov/oil-and-chemical-spills/oil-spills/largest-oil-spills-affecting-us-waters-1969.html.
14. S. A. Fleming, director, Physical Infrastructure, Government Accountability Office. *Oil Spills: Cost of Major Spills May Impact Viability of Oil Spill Liability Trust Fund.* Submitted to GI Testimony before the Subcommittee on Federal Financial Management, Federal Services, and International Security, Committee on Homeland Security and Governmental Affairs, US Senate GAO-10-795T. June 16, 2010.
15. Bureau of Safety and Environmental Enforcement. *Offshore Incident Statistics* (2018). www.bsee.gov/stats-facts/offshore-incident-statistics.
16. L. Muhlenbachs, M. A. Cohen and Todd Gerarden. "The Impact of Water Depth on Safety and Environmental Performance in Offshore Oil and Gas Production." *Energy Policy* 55 (2013): 699–705.
17. Bureau of Ocean Energy Management. *Deepwater Gulf of Mexico* (Washington, DC: November 9, 2017). www.boem.gov/BOEM-Deepwater-Operation-Presentation/.
18. Bureau of Safety and Environmental Enforcement. *Data Center: Production Data Online Query* (2018). www.data.bsee.gov/Production/ProductionData/Default.aspx.
19. National Academy of Engineering and National Research Council. *Macondo Well Deepwater Horizon Blowout: Lessons for Improving Offshore Drilling Safety.* Washington, DC: National Academies Press, 2012. https://doi.org/10.17226/13273.
20. S. Sneath. "After Three Workers Died in Oil Explosion, Government Struggles to Hold Offshore Contractors Accountable." *Nola.com*, September 21, 2019. www.nola.com/news/environment/article_405866cc-d672-11e9-9d72-5352ef9c14bb.html.
21. Government Accountability Office. *Climate Change: Energy Infrastructure Risks and Adaptation Efforts.* Report by F. Rusco. GAO-14-74 (January 2014). www.gao.gov/assets/670/660558.pdf.
22. Department of Energy. *Climate Change and Energy Infrastructure Exposure to Storm Surge and Sea-Level Rise.* Report by J. Bradbury, M. Allen and R. Dell. Office of Energy Policy and Systems Analysis and Oak Ridge National Laboratory (2015).
23. ABS Consulting Inc. *2016 Update of Occurrence Rates for Offshore Oil Spills.* Bureau of Ocean Energy Management, Bureau of Safety and Environmental

Enforcement (Arlington, VA: 2016). www.bsee.gov/sites/bsee.gov/files/osrr-oil-spill-response-research/1086aa.pdf.
24. F. F. Townsend. *The Federal Response to Hurricane Katrina: Lessons Learned.* Executive Office of the President, Washington, DC. 2006. https://georgewbush-whitehouse.archives.gov/reports/katrina-lessons-learned/index.html.
25. D. Fears. "After 14 Years, the US's Longest Offshore Oil Spill Is Finally Starting to Be 'Contained.'" *Washington Post*, May 17, 2019.
26. Bureau of Ocean Energy Management. *Chukchi Sea Planning Area: Draft Second Supplemental Environmental Impact Statement.* Department of the Interior (Anchorage, AK: 2014). www.boem.gov/sites/default/files/uploadedFiles/BOEM/About_BOEM/BOEM_Regions/Alaska_Region/Leasing_and_Plans/Leasing/Lease_Sales/Sale_193/Lease_Sale_193_DraftSSEIS_vol1.pdf.
27. P. Zukunft, commandant, US Coast Guard. *Keynote Address at the 7th Symposium on the Impacts of an Ice-Diminishing Arctic on Naval and Maritime Operations.* July 18, 2017.
28. A. Rowell. *US Coast Guard: We Won't Recover Oil If There's a Spill in the Arctic.* Oil Change International (Washington, DC: July 20, 2017). http://priceofoil.org/2017/07/20/us-coast-guard-we-wont-recover-oil-if-theres-a-spill-in-the-arctic.
29. Deepwater Horizon Spill Natural Resource Damage Assessment Trustees. *Gulf Spill Restoration: Gulf Resources Affected by the Deepwater Horizon Spill.* National Oceanic and Atmosphere Administration (n.d.). www.gulfspillrestoration.noaa.gov/affected-gulf-resources.
30. A. Lustgarten. *Run to Failure: BP and the Making of the Deepwater Horizon Disaster.* New York: W.W. Norton and Company, 2012.
31. W. R. Freudenburg and Robert Gramling. *Blowout in the Gulf: The BP Oil Spill and the Future of Energy in America.* Cambridge, MA: MIT Press, 2012.
32. B. Boxall and J. Tankersley. "Oil Rig Missed Inspections, Records Show." *Los Angeles Times*, June 12, 2010. http://articles.latimes.com/2010/jun/12/nation/la-na-oil-spill-20100612.
33. S. Zellmer, J. Mintz and R. L. Glicksman. "Throwing Precaution to the Wind: NEPA and the Deepwater Horizon Blowout." *Journal of Energy and Environmental Law* 62 (summer 2011): 62–70. https://digitalcommons.unl.edu/cgi/viewcontent.cgi?article=1112&context=lawfacpub.
34. D. Biello. "How Science Stopped BP's Gulf of Mexico Oil Spill." *Scientific American*, April 19, 2011. www.scientificamerican.com/article/how-science-stopped-bp-gulf-of-mexico-oil-spill.
35. R. Tillerson, CEO ExxonMobil. Testimony at the Hearing titled "Drilling Down on America's Energy Future: Safety, Security, and Clean Energy." Subcommittee on Energy and Environment, Committee on Energy and Commerce, House of Representatives, 111th Congress, 2nd Sess. June 15, 2010.
36. US Coast Guard. *Report of Investigation into the Circumstances Surrounding the Multiple Related Marine Casualties and Grounding of the MODU Kulluk* (Washington, DC: 2014). www.dco.uscg.mil/Portals/9/DCO%20Documents/5p/CG-5PC/INV/docs/documents/Kulluk.pdf.
37. Department of the Interior. *Review of Shell's 2012 Alaska Offshore Oil and Gas Exploration Program* (Washington, DC: 2013). www.doi.gov/sites/doi.gov/files/migrated/news/pressreleases/upload/Shell-report-3-8-13-Final.pdf.

38. National Transportation Safety Board. *Marine Accident Brief: Grounding of Mobile Offshore Drilling Unit Kulluk.* DCA13NM012 (Washington, DC: 2015). www.ntsb.gov/investigations/AccidentReports/Reports/MAB1510.pdf.
39. E. Markey, ranking member, Committee on Natural Resources. *Failure of Shell's Containment Dome (to Be Used in Arctic Drilling) during Testing.* Submitted to Ken Salazar, secretary, Department of the Interior. December 5, 2012.
40. R. Miller. "Shell Abandons Offshore Oil Drilling in Alaska Arctic." *Professional Mariner*, January 28, 2016.
41. US Congress. House. *Consolidated Land, Energy, and Aquatic Resources (CLEAR) Act of 2010.* H.R. 3534. 111th Congress. Introduced in House, September 8, 2010.
42. US Congress. House. *Offshore Oil and Gas Worker Whistleblower Protection Act of 2010.* H.R. 503. 112th Congress. Introduced in House, February 25, 2011.
43. S. Sturgis. "The Energy Lobbyists Linked to Trump's Offshore Drilling Plans." *Facing South*, April 15, 2017.
44. A. E. Ladd. "Pandora's Well: Hubris, Deregulation, Fossil Fuels, and the BP Oil Disaster in the Gulf." *American Behavioral Scientist* 56, no. 1 (2012): 104–127.
45. M. A. Eisner. *Regulatory Politics in an Age of Polarization and Drift: Beyond Deregulation.* New York: Routledge, 2017.
46. Transportation Research Board. *TRB Special Report 309: Evaluating the Effectiveness of Offshore Safety and Environmental Management Systems.* Washington, DC: National Academies Press, 2012.
47. D. Vaughn. *The Challenger Launch Decision: Risky Technology, Culture and Deviance at NASA.* Chicago, IL: University of Chicago Press, 1996.
48. A. Hopkins. *Disastrous Decisions: The Human and Organisational Causes of the Gulf of Mexico Blowout.* Sydney, Australia: CCH Australia Limited, 2012.
49. N. G. Leveson. *Engineering a Safer World: Systems Thinking Applied to Safety.* Cambridge, MA: MIT Press, 2012.
50. D. M. Hunter and K. McQueen-Borden. "From Santa Barbara to Macondo to SEMS." *LSU Journal of Energy Law and Resources* 4, no. 2 (2016): 233–257. https://digitalcommons.law.lsu.edu/cgi/viewcontent.cgi?article=1080&context=jelr.
51. D. Boesch. "Trump's Offshore Oil Drilling Plans Ignore the Lessons of BP Deepwater Horizon." *The Conversation*, January 5, 2018. https://theconversation.com/trumps-offshore-oil-drilling-plans-ignore-the-lessons-of-bp-deepwater-horizon-89570.
52. J. A. Watson. "A Three-Pronged Approach to Offshore Safety." *Houston Journal of International Law* 37 (May 29, 2015) 849–852. http://www.hjil.org/articles/hjil-37-3-watson.pdf.
53. *Oil and Gas and Sulfur Operations in the Outer Continental Shelf-Blowout Preventer Systems and Well Control: Final Rule.* Bureau of Safety and Environmental Enforcement. 81 Federal Register 25887–26038 (April 29, 2016). www.federalregister.gov/documents/2016/04/29/2016-08921/oil-and-gas-and-sulfur-operations-in-the-outer-continental-shelf-blowout-preventer-systems-and-well.
54. B. Salerno, director of the Bureau of Safety and Environmental Enforcement. *Testimony on the Well Control Rule and Other Regulations Related to Offshore Oil*

and Gas Production in the Hearing before the Committee on Energy and Natural Resource. US Senate, 114th Congress. 1st Sess. December 1, 2015.
55. *Oil and Gas and Sulfur Operations on the Outer Continental Shelf-Requirements for Exploratory Drilling on the Arctic Outer Continental Shelf: Final Rule.* 30 Code of Federal Regulations 250, 254 and 550. Bureau of Safety and Environmental Enforcement. 81 Federal Register 46477–46566 (July 15, 2016) www.federalregister.gov/documents/2016/07/15/2016-15699/oil-and-gas-and-sulfur-operations-on-the-outer-continental-shelf-requirements-for-exploratory.
56. A. Zibel. *Bernhardt Buddies: Conflicts of Interest Abound at Trump's Interior Department.* Public Citizen (Washington, DC: January 15, 2020).
57. American Rivers et al. *Letter to Senators: Please Oppose David Bernhardt for Deputy Secretary of the Interior.* Submitted to US Senators. May 17, 2017.
58. C. Davenport. "Trump Chooses David Bernhardt, a Former Oil Lobbyist, to Head the Interior Dept." *New York Times*, February 4, 2019.
59. C. Davenport. "Senate Confirms Bernhardt As Interior Secretary Amid Calls for Investigations into His Conduct." *New York Times*, April 11, 2019.
60. D. Van Schooten and L. Peterson. *Trump's Ethics Pledge Is Paper-Thin.* Project On Government Oversight (Washington, DC: June 2017). www.pogo.org/investigation/2017/06/trumps-ethics-pledge-is-paper-thin/.
61. C. Davenport. "Interior Chief's Lobbying Past Has Challenged the Agency's Ethics Referees." *New York Times*, November 9, 2019.
62. E. Warren and R. Blumenthal. *Concern about Potential Ethics Violations Committed by David Bernhardt, the Acting Secretary of the Department of the Interior and Request for an Inspector General Investigation of Mr. Bernhardt's Actions.* Submitted to M. L. Kendall, deputy inspector general, Department of the Interior, Office of Inspector General. February 28, 2019.
63. S. Angelle. "Energy Dominance Requires New Thinking." *Bureau of Safety and Environmental Enforcement Blog* (Washington, DC), September 22, 2017. www.bsee.gov/blog-post/energy-dominance-requires-new-thinking.
64. E. Milito, group director, Upstream and Industry Operations, American Petroleum Institute. *Testimony on the Well Control Rule and Other Regulations Related to Offshore Oil and Gas Production in the Hearing before the Committee on Energy and Natural Resource.* US Senate, 114th Congress. 1st Sess. December 1, 2015.
65. R. Davies. "To Repeal or Not to Repeal: Reviewing the US Well Control Rule." *Offshore Technology.com*, September 14, 2017. www.offshore-technology.com/features/featureto-repeal-or-not-to-repeal-reviewing-the-us-well-control-rule-5885387.
66. G. Leatherman. "NOIA Sees Trump Administration As Friend to Offshore Energy." *Ocean News and Technology*, April 7, 2017. www.oceannews.com/news/energy/noia-sees-trump-administration-as-friend-to-offshore-energy.
67. *Oil and Gas and Sulphur Operations on the Outer Continental Shelf-Oil and Gas Production Safety Systems: Final Rule.* 30 Code of Federal Regulations 250. Bureau of Safety and Environmental Enforcement. 83 Federal Register 49216–49263 (September 28, 2018) www.federalregister.gov/documents/2018/09/28/2018-21197/oil-and-gas-and-sulphur-operations-on-the-outer-continental-shelf-oil-and-gas-production-safety.
68. T. Mann. "US Warns Drillers about Bolts – Failure of Fasteners Could Result in an Oil Spill on the Scale of 2010 Disaster, Say." *Wall Street Journal*, August 30, 2016.

69. *Oil and Gas and Sulfur Operations in the Outer Continental Shelf-Blowout Preventer Systems and Well Control Revisions: Final Rule.* 30 Code of Federal Regulations 250. Bureau of Safety and Environmental Enforcement. 84 Federal Register 21908–21985 (May 15, 2019). www.federalregister.gov/documents/2019/05/15/2019-09362/oil-and-gas-and-sulfur-operations-in-the-outer-continental-shelf-blowout-preventer-systems-and-well.
70. Department of Transportation. *Blowout Prevention System Safety: 2017 Annual Report* (Washington, DC: 2017). www.safeocs.gov/2017_WCR_Annual_Report_v 4.pdf.
71. M. Voitier and D. Hoskins. *Dirty Drilling: Trump Administration Proposals Weaken Key Safety Protections and Radically Expand Offshore Drilling.* Oceana (April 2019).
72. L. Friedman. "Ten Years after Deepwater Horizon, US Is Still Vulnerable to Big Spills." *New York Times,* April 20, 2020.
73. B. E. Frosh, attorney general of Maryland. *Docket ID No. BSEE-2017-0008; RIN 1014-AA37Oil and Gas and Sulphur Operations on the Outer Continental Shelf – Oil and Gas Production Safety Systems – Revisions.* Submitted to S. A. Angelle, director, Bureau of Safety and Environmental Enforcement. January 29, 2018.
74. C. Eaton, associate attorney, Earthjustice. *Oil and Gas and Sulfur Operations in the Outer Continental Shelf – Blowout Preventer Systems and Well Control Revisions, 83 Fed. Reg. 22,128 (May 11, 2018), Docket No. BSEE-2018-0002.* Submitted to S. A. Angelle, director, Bureau of Safety and Environmental Enforcement. August 6, 2018.
75. D. Hoskins, campaign director, Climate and Energy, Oceana. *Oil and Gas Sulfur Operations in the Outer Continental Shelf – Blowout Preventer Systems and Well Control Revisions; 83 Fed. Reg. 22,128 (May 11, 2018; Dkt. No. BSEE-2018-0002; RIN 1014-AA39).* Submitted to S. A. Angelle, director, Bureau of Safety and Environmental Enforcement. August 6, 2018.
76. T. Mann. "When Safety Rules on Oil Drilling Were Changed, Some Staff Objected. Those Notes Were Cut." *Wall Street Journal,* February 26, 2020.
77. Energy Information Administration. *US Gulf of Mexico Crude Oil Production to Continue at Record Highs through 2019* (April 11, 2018). www.eia.gov/todayinenergy/detail.php?id=35732.
78. Bureau of Safety and Environmental Enforcement. *Oil and Gas and Sulphur Operations in the Outer Continental Shelf – Blowout Preventer Systems and Well Control: Regulatory Impact Analysis, Technical report.* RIN 1014-AA11, 30 CFR Part 250 (April 2016).
79. S. Mufson. "BP's Big Spill for the World's Largest Oil Spill Reaches $61.6 Billion." *Washington Post,* July 14, 2016. www.washingtonpost.com/business/economy/bps-big-bill-for-the-worlds-largest-oil-spill-now-reaches-616-billion/2016/07/14/7248cdaa-49f0-11e6-acbc-4d4870a079da_story.html?noredirect=on&utm_term=.2ddcd3bf74c6.
80. R. C. Bishop et al. "Putting a Value on Injuries to Natural Assets: The BP Oil Spill." *Science* 356, no. 6335 (2017): 253–254.
81. *Oil and Gas and Sulphur Operations on the Outer Continental Shelf-Oil and Gas Productions Safety Systems – Revisions: Proposed Rule.* 30 Code of Federal Regulations 250. Bureau of Safety and Environmental Enforcement. www

.federalregister.gov/documents/2017/12/29/2017-27309/oil-and-gas-and-sulphur-operations-on-the-outer-continental-shelf-oil-and-gas-production-safety.

82. T. P. De Sousa. "Oil over Troubled Waters: Exxon Shipping Co. v. Baker and the Supreme Court's Determination of Punitive Damages in Maritime Law." *Villanova Environmental Law Journal* 20, no. 2 (2009): 247–273.

83. R. T. Carson et al. "Contingent Valuation and Lost Passive Use: Damages from the Exxon Valdez Oil Spill." *Environmental and Resource Economics* 25, no. 3 (2003): 257–286.

84. M. Lee-Ashley. *10 Years after Deepwater Horizon, Oil Spills and Accidents Are on the Rise.* Center for American Progress (Washington, DC: March 3, 2020). www.americanprogress.org/issues/green/news/2020/03/03/481027/10-years-deepwater-horizon-oil-spills-accidents-rise/.

85. D. Fears. "US Suspends Study Aimed at Making Offshore Drilling Safer." *Washington Post*, December 26, 2017.

86. National Academies of Sciences, Engineering and Medicine. *Statement on Stop-Work Order for National Academies Study on the Department of the Interior's Offshore Oil and Gas Operations Inspection Program.* December 21, 2017.

87. US Congress. House. *Safe Coasts, Oceans, And Seaside Towns Act [the Safe COAST Act].* H.R. 1335. 116th Congress. 2nd sess. Introduced in House, February 25, 2019.

88. Office of Nanette Diaz Barragán, congresswoman of California's 44th District. "Barragán, Buchanan, Crist, Price and Rooney Announce Offshore Drilling Safety Bill." News release, May 7, 2019. https://barragan.house.gov/barragan-buchanan-crist-price-and-rooney-announce-offshore-drilling-safety-bill.

89. S. Contorno. "Marco Rubio and Rick Scott Want 10-year Ban on Drilling off Florida Coast in Senate Energy Bill." *Tampa Bay Times*, March 4, 2020. www.tampabay.com/florida-politics/buzz/2020/03/04/marco-rubio-and-rick-scott-want-10-year-ban-on-drilling-off-florida-coast-in-senate-energy-bill.

90. M. Daly. "Senate Confirms Ex-lobbyist David Bernhardt to Lead Interior." *Associated Press*, April 11, 2019. https://federalnewsnetwork.com/people/2019/04/senate-confirms-ex-lobbyist-david-bernhardt-to-lead-interior/+&cd=13&hl=en&ct=clnk&gl=us&client=firefox-b-1-d.

91. L. Murkowski. *Remarks during the Hearing on the National Commission Report on the BP Oil Spill, Committee on Energy and Natural Resources.* US Senate, 112th Congress, 1st sess. January 26, 2011.

92. US Congress. Senate. *Big Oil Bailout Prevention Unlimited Liability Act of 2018.* S. 3757. Introduced in Senate December 13, 2018.

93. M Greenstone, professor of Economics at the Massachusetts Institute of Technology. *Testimony at the Hearing on Liability and Financial Responsibility for Oil Spills under the Oil Pollution Act of 1990 and Related Statutes, Committee on Transportation and Infrastructure.* House of Representatives, 111th Congress, 2nd Sess. June 9, 2010.

94. D. Boesch. *Testimony at the Hearing on H.R. 2231, Offshore Energy and Jobs Act Part 1 and 2, Subcommittee on Energy and Mineral Resources, Committee on Natural Resources.* June 11, 2013.

95. US Congress. Senate. *Spill Response and Prevention Surety Act.* S. 865. Introduced in Senate, 116th Sess. March 25, 2019.

96. M. A. Cohen et al. "Deepwater Drilling: Law, Policy and Economics of Firm Organization and Safety." *Vanderbilt Law Review* 64 (2011): 1853–1916.
97. M. Green and R. Beitsch. "David Bernhardt Confirmed As New Interior Chief." *The Hill*, April 11, 2019. https://thehill.com/policy/energy-environment/438460-david-bernhardt-confirmed-as-new-interior-chief.
98. *Hearing on "Stakeholder Perspectives on the Importance of the US Chemical Safety and Hazard Investigation Board."* Senate Committee on Environment and Public Works. 116th Congress, 2nd Sess. January 29, 2020.
99. A. Natter. "Trump to Again Propose Eliminating Chemical Safety Board, Official Says." *Bloomberg*, July 25, 2018. www.bloomberg.com/news/articles/2018-02-01/trump-said-to-again-propose-eliminating-chemical-safety-board.
100. J. Johnson. "Chemical Safety Board Faces an Uncertain Future Despite Strong Backing. Now Down to Two Members, the Board Will Have None Come August." *Chemical and Engineering News* 98, no. 5 (January 30, 2020). https://cen.acs.org/safety/industrial-safety/Chemical-Safety-Board-faces-uncertain/98/i5+&cd=1&hl=en&ct=clnk&gl=us&client=firefox-b-1-d.
101. *Healthy Gulf v. David Bernhardt, US Department of Interior, Scott Angelle, Bureau of Safety and Environmental Enforcement*, No. 1:19-cv-02894 (D.C. Cir. 2019).
102. A. F. Herbst, J. F. Marshall and J. Wingender. "An Analysis of the Stock Market's Response to the Exxon Valdez Disaster." *Global Finance Journal* 7, no. 1 (1996): 101–114.
103. Y.-G. Lee and X. Garza-Gomez. "Total Cost of the 2010 Deepwater Horizon Oil Spill Reflected in US Stock Market." *Journal of Accounting and Finance* 12, no. 1 (2012): 73–83.
104. G. Walton, vice president, Washington Office, ExxonMobil Corporation. *Docket ID. No. EPA-HQ-OAR-2017-0483-Oil and Natural Gas Sector: Emission Standards for New, Reconstructed and Modified Sources Reconsideration: Proposed Rule (October 15, 2018)*. Submitted to Environmental Protection Agency. December 17, 2018.
105. C. Krause. "Trump's Methane Rule Rollback Divides Oil and Gas Industry." *New York Times*, August 29, 2019.
106. D. Hammer. "Thousands of Cleanup Workers That Claim BP Oil Spill Made Them Sick Haven't Had Their Day in Court." *WZZM13*, April 20, 2018. www.wzzm13.com/article/news/local/investigations/david-hammer/thousands-of-cleanup-workers-that-claim-bp-oil-spill-made-them-sick-havent-had-their-day-in-court/289-543806235.
107. A. F. Popper. "Capping Incentives, Capping Innovation, Courting Disaster: The Gulf Oil Spill and Arbitrary Limits on Civil Liability." *DePaul Law Review* 60, no. 4 (2011): 975–1006.

PART IV

America's Regulatory Process

8
Science
Undermining Facts to Understate Regulatory Benefits

Part IV of this book focuses on the US regulatory process, and more particularly on threats to its integrity and functionality posed by the Trump administration. President Trump, his congressional supporters and his political appointees demonstrated a profound hostility to regulation – and to the science that underpins it – that seems to be rooted both in ideology and in an instinctual reluctance to question the oil and gas industry. By appointing industry insiders and deregulation devotees to run key regulatory agencies, Trump put the foxes in charge of the hen house. America's long tradition of science-based regulation is under attack from within.

Since the late nineteenth century and particularly in the mid-twentieth century, Congress has enacted wide-ranging laws to protect public health and the environment. Federal administrative agencies are tasked with implementing these laws, which they contextualize into more detailed regulations that specify the best means of implementation and enforcement (alongside research, monitoring and collaboration with state governments). Public health and environmental regulations protect the quality of America's air, water and soil, setting the responsibility for companies to limit pollution from their operations and to mitigate the impacts of pollution.

Regulations serve to level the playing field for companies within a given industry, giving each company the opportunity to succeed while holding all actors to the same minimum standards of responsibility in limiting their public health and environmental impacts. However, regulations can reduce profits. Companies accustomed to ignoring the societal and environmental costs of their operations are most likely to protest this regulatory burden, but even companies that are not indifferent to their impact are still inclined to look askance at regulation that seems excessively restrictive and cuts into their profits. The oil and gas industry has thus traditionally fought against regulation, and it has found allies among politicians that are motivated not by the bottom line so much as by an ideological antipathy to the state limiting the freedom of action of private companies. This antipathy has rarely if ever been so intense and undisguised as it was under the Trump administration. As that administration recognized, while combatting individual regulations is one way to reduce the regulatory burden, another and even more powerful way is to undermine the very system within which regulations are created.

Congress previously passed environmental and administrative laws that tethered the regulatory process to scientific evidence.[i] Federal agencies were obliged to weigh scientific data, as well as dispassionate economic and legal analyses, as they developed and implemented regulations. Both Republican and Democrat administrations looked to science to inform public policy. President George H. W. Bush emphasized in 1990, "Now, more than ever, on issues ranging from climate change to AIDS research . . . government relies on the impartial perspective of science of guidance."[1]

The Trump administration sought to untether the rulemaking process from science and other forms of hard evidence and expert analysis, thereby allowing federal agencies – or at least those political appointees in charge of them – to make decisions based on personal beliefs and corporate interests.[2] The three chapters in Part IV probe the ways in which the Trump administration reshaped the rulemaking process in terms of its relationship to scientific, economic, and legal evidence and argumentation.

In this, the first chapter in Part IV, I examine the Trump administration's actions to push forward its deregulatory agenda by limiting and manipulating the role of science. I begin with a brief overview of the oil and gas industry's history of sowing doubt in the science underlying regulations it opposed and the administration's scuttling of those regulations (Section 8.1). I probe how the administration weakened the scientific integrity of rulemaking (Section 8.2), distorted the review of existing regulations (Section 8.3), published scientifically unsound analysis to repeal existing regulations (Section 8.4) and blocked valid science from informing rulemaking (Section 8.5). I then turn to the actions that Congress and civil society could have taken and what they have in fact taken to reign in the administration's attack on scientific integrity (Section 8.6). Despite strong opposition from some quarters – and with equally strong support from others, chiefly a number of politicians in Congress – the administration pushed ahead with its anti-science agenda.

---

[i] There are cases when Congress made the policy choice not to consider scientific evidence in regulating the oil and gas industry. For example, Congress deemed drilling mud and wastewater from oil and gas exploration and production as nonhazardous wastes (even if they contain hazardous chemicals such as benzene) and exempted them from the cradle-to-grave regulations for hazardous waste under the Resources Conservation and Recovery Act [2].

8.1 Sowing Doubts in Science

Regulated industries have a long history of attempting to undermine regulation by denying the consensus among scientists, anchored in extensive peer-reviewed scientific studies, that pollution adversely affects public health. This strategy undercuts health and environmental regulations at a systemic level and has proven to be far more effective than weakening one regulation at a time. The oil and gas industry's history of anti-regulatory influence (Section 8.1.1) sets the stage for the Trump administration's wholesale assault on America's regulatory institutions. The Trump administration's actions to undermine the regulatory system went far beyond those of the George W. Bush administration, even though the latter massively deregulated the oil and gas industry.

8.1.1 The Oil and Gas Industry's Historical Anti-Science Agenda

For a hundred years if not more, the oil and gas industry has opposed regulations that limit extraction and consumption by intentionally sowing doubts about the scientific data used by regulators. In the 1920s, Standard Oil, along with the Ethyl Corporation and DuPont, cast doubt on scientists' warnings about the health dangers of leaded gasoline.[3,4] In the 1950s, the oil industry (and the car industry) disparaged scientists' findings that emissions from oil fields, refineries and cars were the cause of the smog that was choking Los Angeles.[5,6] Since the 1970s, the industry has opposed the Clean Air Act, the landmark federal legislation that regulates sources of air pollutants, including oil-extraction operations, oil refineries and gasoline-powered vehicles.[7] It attempted to undermine the credibility of reports from the National Academies of Sciences, Engineering, and Medicine on the adverse public health impacts of air pollutants.[8] In the 1990s, armed with its sponsored research that dismissed the causal link between particulate matter and health,[9] the American Petroleum Institute and other industry advocates formed a group called the Air Quality Standards Coalition that challenged the scientific basis for the proposed tightening of particulate matter and ozone standards.[10,11] In the 2000s, the industry persisted in opposing the tightening of standards for worker exposure to benzene, a component of crude oil and gasoline,[12–14] even though its sponsored research concluded as early as 1948 that the only absolutely safe benzene exposure is zero exposure.[15,16]

The industry has continued to sow doubt by funding contrarian scientists and anti-regulatory think tanks such as the Citizens for a Sound Economy,[ii] the Competitive Enterprise Institute, the Heartland Institute and the Heritage

---

[ii] Citizens for a Sound Economy split into FreedomWorks and Americans for Prosperity.

Foundation.[17–21] Contrarian scientists oppose conventional scientific evidence without marshaling compelling data that would meet the standards of peer-reviewed research. These dissenting scientists generally have expertise outside the field in question, cherry pick data, peddle logical fallacies, insist on unrealistic levels of certainty and exaggerate uncertainties to mislead the public.[17,22–24] Together with the American Petroleum Institute, these scientists and think tanks have questioned the scientific consensus on the adverse public health effects from by-products of oil and gas extraction and combustion.[25–29] In the 1990s, Citizens for a Sound Economy, along with pro-oil legislators such as Representative Lamar Smith, attacked seminal epidemiological studies on human exposure to particulate matter[iii] that informed particulate matter regulations.[30–33] In response, an independent panel of scientists reanalyzed the raw confidential health data used in these studies and validated their conclusions.[34–36] Nevertheless, pro-oil legislators and conservative think tanks continue to disparage these studies as "secret science."[37,38]

The industry and think tanks have also cast doubt on the scientific evidence that shows how greenhouse gas emissions from oil and gas extraction and combustion contribute to climate change.[39] In the 1990s, they embarked on a public relations strategy to amplify doubts on climate science,[39] even though they had, as early as the 1960s and through the 1980s, acknowledged the strength of that scientific evidence.[40,41] Pro–fossil fuel legislators invited skeptics to air their views in testimony to Congress.[42,43] Support for these contrarian views grew when Philip Cooney, who had served as the American Petroleum Institute's team leader against climate action, became the chief of staff of the G. W. Bush Council on Environmental Quality. While in that position, he edited reports issued by the federal Global Change Research Group to downplay the human-caused climate crisis.[44]

The climate denial message by conservative think tanks, coupled with political pressure against climate-change conscious politicians, secured the misconception that climate science is too uncertain to justify policy action. These efforts hamstrung politicians, particularly Republicans,[iv] from supporting climate

---

[iii] The Harvard Six Cities study followed a random sample of about 8,000 white adults aged 25–74 living in six communities across the United States. The study began in the 1970s and lasted for 14 years [30]. The American Cancer Society study tracked about 550,000 adults in 154 US cities between 1982 and 1989 [31].

[iv] As documented in the research cited, climate denial groups, such as Americans for Prosperity and FreedomWorks (funded by conservative donors including the Koch Industries, a conglomerate with refineries, pipelines, petrochemicals and beyond), pressured Republican legislators to sign up to the No Climate Tax pledge. These groups threatened to finance campaigns to unseat pro-climate action Republicans and indeed successfully unseated Representative Bob Inglis and Senator Richard Lugar. These groups also convinced Tea Partiers, who were then highly influential in the Republican Party, to adopt an anti-climate position.

action. [17,18,24,44–46] A case in point is its success in compelling Senator Lindsey Graham to back out from sponsoring the Senate climate bill, the 2010 American Power Act, which he had worked on with Senator John Kerry and Senator Joseph Lieberman.[47,48] The bill died in the Senate.[v] By 2016 (amid the presidential primary race that landed Trump the Republican nomination), former Utah Governor Jon Huntsman was the only Republican candidate to acknowledge the man-made climate crisis.

After winning the election, Trump filled both his transition team and agency leadership with friends of the oil industry. Myron Ebell of the Competitive Enterprise Institute headed the EPA transition team. The Heritage Foundation boasted of its influence on the administration's deregulation agenda. Noting that "70 former Heritage employees [are] working for the Trump transition team or as part of the administration," it stated that the Trump administration "embraced nearly two-thirds of the 334 policy recommendations from the Heritage Foundation."[49] Armed with a disrespect for government oversight of the industry and an established deregulatory agenda, the stage was set for the Trump administration to orchestrate a dangerous attack on science.[50]

8.1.2 The Trump Administration's Anti-science Agenda

To reshape the regulatory process, the administration started by undermining scientific review. Political appointees at regulatory agencies sidelined scientists who worked at these agencies, referred to as career scientists, and ejected academic scientists from the science advisory bodies that oversaw agencies' work. These two groups of scientists have served to balance political pressures in agency rulemaking with expertise and fact-based legitimacy. With this bulwark removed, political appointees were free to advocate for weak air quality standards. Federal agencies then published scientifically unsound analyses justifying regulatory rollbacks. Finally, and perhaps most daunting for the future of regulatory oversight, the administration proposed rules that would prohibit regulatory agencies from even considering valid scientific studies in future rulemaking. As described in more detail below, these studies are

---

[v] The climate denial network's intimidation of Graham, along with a broader set of factors, contributed to the failure of the Senate bill. Tea Party activists in Graham's home state of South Carolina attacked Graham's support of the climate legislation, his agreement with climate science and his willingness to work with progressives on climate action [47]. Other factors were senators' fears that climate action risked further weakening the economy, which was already crippled by the Great Recession and high unemployment rates. Industry groups intensified their efforts to kill the Senate bill after the House of Representatives successfully passed Waxman-Markey's American Clean Energy and Security Act in 2010 [48].

a fundamental cog in the regulatory process, without which agencies are handicapped in promulgating effective regulations or defending existing ones.

Eager to see the further exploitation of US oil and gas resources, which he declared were being unfairly and unnecessarily obstructed by excessive regulation, President Trump ordered federal agencies to "suspend, revise, or rescind those [regulations] that unduly burden the development of domestic energy resources beyond the degree necessary to protect the public interest."[51] That belief in the undue burden of regulations aligns with his general stance against regulation or industry oversight. However, altogether less convincing was the administration's stated desire "to achieve environmental improvements" and to "employ the best available peer-reviewed science and economics."[51] Instead of doing either of those things, the administration corroded the scientific underpinning of analysis, understated the economic benefits from regulations, and narrowed legal interpretations of agencies' responsibilities to regulate polluting activities.

The Trump administration proffered two arguments to justify rollbacks. First, it argued that many regulations are needless and serve only to block the ability of the oil and gas industry to exploit the country's resources.[51] Second, it argued for a "back to basics agenda" for the EPA; in other words, it called for the agency to focus on safeguarding air, water and land, presumably because it believed the agency deviated from these goals.

Both these arguments are highly suspect,[52] as the rest of this chapter makes clear. In short, regulations imposed on the oil industry are vital to protect public health and safety. As required by law, the EPA had provided scientific and technical assessments to justify the promulgation of regulations. Indeed, regulations that restrict the industry's emissions of pollutants serve to protect air, water and land. The Trump administration, contradicting its rhetoric of safeguarding these vital resources, hacked away at the very regulations meant to ensure these protections.

8.2 Weakening Scientific Integrity at Regulatory Agencies

While the Trump administration succeeded in degrading confidence in science, agencies are still required to use evidence in rulemaking. However, the effectiveness of this safeguard depends largely on the integrity of those conducting research – in other words being honest about the findings of scientific research, rather than manipulating, distorting or suppressing the results. Scientific integrity ensures that agencies act in the public interest and not in line with special interests. Not surprisingly, this safeguard has come under attack as well.

One of the administration's major blows to scientific integrity was through agency leadership – that is, the appointment of anti-regulatory individuals to oversee agency research and analysis. Many of these appointees (from mid-level positions to the heads of agencies) had strong ties to the oil and gas sector.[53,54] To be fair, politically appointed agency heads almost inherently share at least some components of the political agenda held by their appointer. However, this selection has been balanced in many administrations, not all, with a general understanding that certain candidates simply have too many conflicts of interest to head science-based agencies. George W. Bush, excoriated for his appointees' strong ties to industry,[55] nevertheless appointed a moderate, Republican Governor Christy Whitman, as his administration's first head of the EPA. She quit after two years because she disagreed with the administration's policies.

All restraints were tossed aside under the Trump administration. Trump's first EPA administrator and his first secretary of the Interior, as well as their successors, both previously worked to oppose regulations on the oil and gas sector. Once in office, they proceeded to weaken those regulations from the inside.[56-58] Scott Pruitt, the first EPA administrator under the Trump administration, had publicly opposed the EPA's attempts to curb methane emissions from oil and gas operations – dubbed the "Methane Rule." While serving as Oklahoma attorney general, he actually sued to block that rule. Not surprisingly, he initiated a review and proposed an overhaul of the rule shortly after taking office. Andrew Wheeler, Pruitt's successor, had previously worked as a lobbyist for the coal industry. One of his former clients was Murray Energy, the largest privately owned coal company in the United States. As a lobbyist, Wheeler was paid to strongly oppose the Clean Power Plan. Not surprisingly, shortly after taking office, he finalized the rescission of the Clean Power Plan and its replacement by the Affordable Clean Energy Rule – a change that blunts efforts to reduce greenhouse gas emissions from the power sector.

Ryan Zinke, the first secretary of the Interior under the Trump administration, had a history of opposition to regulations as well. He voted against methane regulations on public lands when he was a congressman. Once in office, he promptly took steps to weaken those regulations. David Bernhardt, Zinke's successor, was even more predisposed to cut methane regulations, having worked as a lobbyist for the oil and gas sector before taking office. As expected, under his leadership, the agency finalized the weakening of regulations on oil and gas operations both on public lands and offshore.

These political appointees pressured career scientists to support the administration's positions and sidelined those who disagreed. Blocked from adhering to scientific facts, many experienced staff resigned from their positions and reported these political interferences.[58,59] For example, political appointees at the Office of Information and Regulatory Affairs, a body within the executive branch, are reported to have pressured EPA career scientists to change their analysis in support of the proposal to reduce the frequency of inspections for methane leaks in oil and gas operations.[57,60] EPA career scientists had opposed that proposal because it contradicted the EPA's responsibility under the Clean Air Act to select the best emissions-reduction systems.[57] Numerous cases of interference in the work of career scientists were documented by NGOs.[61–63]

Pruitt also ejected academic scientists from the Science Advisory Board (SAB), a body mandated by Congress to advise the EPA administrator on the science underlying regulatory policies. His directive prohibited academic scientists who hold EPA grants from serving on EPA's advisory boards.[64] Pruitt's stated justification – that these scientists face conflicts of interest – appears as hypocritical as it is inaccurate. The EPA's existing ethics rules prevent grantees from participating in discussions that could influence the outcome of their EPA grant applications.[65] Moreover, Pruitt's directive did not bar scientists who worked for industries affected by the regulations under discussion. These scientists, of course, were the very ones who might actually face a conflict of interest. The revised appointment policies raised the share of oil and gas industry affiliated scientists on the SAB from 40 percent to 68 percent, or 30 of its 44 current members[66] (though the Court of Appeals for the DC Circuit struck down the directive in April 2020).[vi] Having scrapped even the most rudimentary best practices to safeguard scientific integrity, the stage was set for a deregulation blitz.

8.3 Distorting the Review of Existing Regulations: The Case of Ambient Air Quality Standards

The Trump EPA's review of the ambient air quality standards created under the Clean Air Act illustrates the type of manipulation, suppression or distortion that can influence the regulatory process. The oil industry has long opposed this core US air pollution regulation.[67–71] However, scientific consensus on mortality and illnesses from poor air quality had kept these standards relatively unscathed. Pruitt and his successor, Wheeler, amplified the views of contrarian scientists on the scientific board that recommends air quality standards

---

[vi] In April 2020, the Court of Appeals for the DC Circuit ruled in *Physicians for Social Responsibility v. Wheeler* that the EPA had failed to provide an adequate explanation for its policy change even when that change directly affected its ability to meet its statutory scientific mandates.

(Section 8.3.1), enabling the EPA to propose weaker standards than those recommended by public health experts (Section 8.3.2).

8.3.1 Amplifying Contrarian Viewpoints on the Clean Air Science Advisory Committee

The Clean Air Science Advisory Committee (CASAC) advises the EPA administrator on setting the national ambient air quality standards (NAAQS), per its mandate under the Clean Air Act. In addition to shaping standards, the committee's recommendations are important for subsequent legal challenges, as judges generally defer to CASAC's recommendations when the reasonableness of standards is in question.[72,vii] For example, in August 2019, the US Court of Appeals for the DC Circuit upheld the EPA's ozone standard, set in 2015, noting that the EPA abided by CASAC's recommendation.[73]

Under past Republican and Democrat administrations, the EPA administrators typically selected the chair and members of CASAC from university scientists at the forefront of research on air pollution and public health. The seven-person committee, recognizing the need for in-depth expertise across disciplines, further relied on a panel of scientists for each specific pollutant they reviewed. The EPA publicized the names of nominees to CASAC and the panels so that members of the public could submit their comments.

Administrator Pruitt appointed widely contested committee members to CASAC who had close ties to industry (the American Petroleum Institute had advocated for industry representations on the committee in the past),[71] held views that contradicted the scientific consensus, and lacked direct experience in the health effects of air pollution.[66]

Pruitt appointed Dr. Anthony Cox as chair of CASAC. (Cox, who holds a PhD in Risk Analysis from MIT, had served as the director of an applied research company specializing in health, safety and environmental risk analysis. Despite these credentials, however, he had earned a reputation for being an outlier in terms of scientific consensus.) Prior to his appointment, Cox testified in 2015 against the tightening of ozone standards. He disagreed with CASAC's past reviews and cast doubt on the EPA's methods and conclusions.[74] Cox argued that the evidence did not show a causal relationship between exposure to ozone or particulate matter and mortality,[75–77] despite that

---

[vii] Courts may be less deferential to the CASAC that operated under Wheeler's oversight. As noted by Christopher Frey, who chaired CASAC during the Bush and Obama administrations, the Clean Air Act sets the criteria that the setting of these standards be based on a comprehensive scientific review. CASAC, which suffered from inadequate expertise, failed to conduct a thorough scientific review [72].

relationship being widely accepted as scientific consensus.[78–80] Critics raised the alarm that Cox's research has been funded by the oil industry and the tobacco industry, but Cox rejected claims that his work had been influenced by these industries.[81]

Pruitt's successor, Wheeler, removed several academic scientists from CASAC,[82] leaving no experts in epidemiology on the committee. Epidemiology is the main discipline that investigates the relationship between pollution and public health.[83,84] He also left only one expert in pulmonology, a discipline specializing in air pollution and public health.[82] Wheeler disbanded the 20-person particulate-matter review panel that had been established by the EPA in 2015 to assist CASAC.[81] He did so despite criticisms from at least three CASAC members who believed that CASAC needed the panel's expertise.[85] In response to the admission by CASAC as a whole of its lack of expertise and its recommendation for the panel to be reconstituted,[86] Wheeler hired a pool of consultants, several of whom had advocated weaker standards while working for the oil and gas industry.[87] The selection of consultants was far less transparent than that of the panel members, and the consultants did not deliberate freely with the CASAC nor engage in public hearings.[88]

In October 2019, CASAC member Mark Frampton, a pulmonologist and emeritus professor of medicine at the University of Rochester Medical Center, lamented that "the review process is so dysfunctional that we need to stop" and he agreed to finish CASAC deliberations "under protest."[89] Another CASAC member James Boylan, an environmental engineer with the Georgia Department of Natural Resources, shared Frampton's frustration.[89] Likewise, Christopher Frey, who had served on CASAC during the Bush and Obama administrations and in the early phase of the Trump administration, decried the review process as "border[ing] on being a total sham."[81] Frey and other members of the disbanded particulate matter panel formed the nongovernmental Independent Particulate Matter Review Panel that continued their review, held public hearings and submitted their findings to CASAC.[90]

8.3.2 Recommending Weak Public Health Protections

Under Cox's leadership, CASAC disregarded the Independent Particulate Matter Review Panel's conclusions that the particulate matter standards ($PM_{2.5}$) should be tightened.[90,viii] The panel's recommendations were based on "consistent epidemiological evidence from multiple multi-city [and] single-city studies,

---

[viii] The panel recommended the tightening of the annual $PM_{2.5}$ standard from 12 micrograms per cubic meter ($\mu g/m^3$) to a range between 8 $\mu g/m^3$ and 10 $\mu g/m^3$ and the tightening of the 24-hour standard from 35 $\mu g/m^3$ to a range between 25 $\mu g/m^3$ and 30 $\mu g/m^3$ [90].

at ... ambient concentrations ... at and below the levels of the current standards, and [which] are supported by ... experimental models in animals and humans and by accountability studies."[90] Instead, the CASAC chair submitted the committee report to the EPA in April 2019, recommending that the standards be left unchanged.[86] That report paved the way for the EPA to publish its final policy assessment in January 2020 that provided cover for the EPA administrator to leave standards unchanged. While the assessment acknowledged that the overall scientific evidence showed that the existing standards were not sufficiently protective of public health, the assessment posited that the existing standards could be seen as adequate if epidemiological evidence were given less weight and if scientific uncertainties were given more weight.[91] In December 2020, Wheeler finalized his decision to keep the particulate matter standard unchanged.[92]

The ramifications of the CASAC report go beyond giving Wheeler the greenlight not to tighten the particulate matter standards. The CASAC report – particularly its summary – gave prominence to the contrarian approach of casting doubt on the work of EPA's career scientists and of interpreting causality narrowly. The summary section of the report criticized the EPA's October 2018 draft scientific assessment as failing to "provide a sufficiently comprehensive, systematic assessment" of the science and failing to provide adequate evidence on the causal relationship between particulate matter and health impacts.[86] Some contrarians contend that only study designs employing "manipulative causality" and the use of specific statistical tests can establish a health causality.[93,94] These studies examine changes in health outcomes when policy changes cause pollution reduction or when trials randomly assign study subjects to varying levels of exposure to pollutants.

It is important to recognize that two CASAC members disagreed with the rest of the committee members[89] and took direct issue with their questioning of established science, saying the following:

> Although uncertainties remain, the evidence supporting the causal relationship between $PM_{2.5}$ exposure and mortality is robust, diverse, and convincing. ... This causality determination was first clearly promulgated in the 2009 Integrated Scientific Assessment, with full CASAC support. ... There is no credible or convincing new evidence since 2009 to question or refute this determination. ... The evidence supporting a causal relationship between $PM_{2.5}$ and mortality is even more robust now than it was in 2009.[86]

The two CASAC members' recognition of the causal relationship comports with the views of the independent panel and those of experts who had served on the previous 11 panels. The latter had undertaken the reviews of air quality standards prior to Pruitt and Wheeler taking the helm at the EPA. These experts considered the consistency of results across diverse research approaches.[78–80,95–97] They considered epidemiological studies that follow large populations over time and that control for factors other than pollution that influence health outcomes, experimental short-term exposure studies of a small number of humans, experimental studies on the effects of exposures on animals, and experimental studies of exposures at the cellular and molecular level. These experts underscored that the method preferred by some contrarians – the use of a single statistical test and reliance on the few studies with manipulative causality designs – cannot substitute for a comprehensive approach, which evaluates the overall weight of the evidence.[79,97] They warned, furthermore, that manipulative causality design studies that purposely expose populations to varying pollution levels are unethical.[79]

Sadly, the review of the ozone standards by CASAC and the EPA proceeded in a flawed manner, as did the particulate matter review. CASAC's report contradicted the view of at least one CASAC member that the standard was insufficiently protective of public health.[98,99,ix] It also contradicted the views of the independent panel for ozone, a panel set up by experts who had served on the panel that advised on EPA's 2015 ozone standard.[100,101] The CASAC report facilitated EPA's publication of its final policy assessment in May 2020 that recommended the administrator leave the ozone standards unchanged.[102]

The flawed review process, instituted by Pruitt and his successor Wheeler, has the lasting impact of catapulting the approach of contrarians – amplifying uncertainty and narrowing the interpretation of causal evidence – prominently into CASAC reports and thus providing ammunition for legal challenges against the tightening of air quality standards.

8.4 Justifying the Repeal of Regulations by Publishing Unscientific Analyses

Hindering the strengthening of existing regulations, as I have described above, weakens both public health and environmental protection in the long term. The administration dealt a further blow against these protections by repealing

---

[ix] Dr. Mark Frampton, the only pulmonologist on CASAC, wrote that "the current ozone NAAQS level of 70 ppb does not provide an adequate margin of safety for children with asthma" [99] and supported the tightening of the standard to between 60 ppb and 70 ppb, as was recommended by the 2014 CASAC. Dr. James Boylan raised the issue of why outdoor workers, who are exposed to ozone, were excluded from the assessment, and how that omission could overstate the adequacy of the ozone standard [99].

regulations altogether, using the same strategy of publishing of analyses with dubious scientific merit. One tactic was to ignore the health benefits from the reduction of particulate matter at low concentration levels and thus understate the benefits from regulations that result in the reduction of particulate matter (described in more detail in Chapter 9). This strategy was used not only to undermine those regulations that target particulate matter but also a host of other regulations that yield the ancillary benefits of reducing particulate matter.[x] The EPA applied this tactic to repeal the Clean Power Plan and to undermine the Mercury and Air Toxics Standards.

The Clean Power Plan,[103] which limited greenhouse gas emissions from power plants, had been staunchly opposed by the American Petroleum Institute.[104] If this rule had been implemented, it could have paved the way for the subsequent regulation of the oil industry's greenhouse gas emissions.[xi] The EPA's 2018 analysis assumed that reductions of particulate matter below the NAAQS do not yield public health benefits.[105] Only by making this assumption[xii] was the EPA able to make the erroneous claim that the repeal of the Clean Power Plan would yield net benefits.[105,106] With science cast aside and economic analysis manipulated, the EPA repealed the Clean Power Plan in June 2019 and, a month later, replaced it with the Affordable Clean Energy Rule.[xiii] The replacement rule, according to the EPA's own estimates, would worsen air pollution and lead to an additional 370–1,400 premature deaths and 48,000 asthma cases by 2030, relative to the case if the Clean Power Plan had been implemented.[105]

The Mercury and Air Toxics Standards Rule limits mercury emissions from coal and oil-fired power plants.[107] The EPA's 2018 reanalysis of the rule[xiv] was applauded by the American Petroleum Institute.[108,109] The EPA was able to erroneously conclude that the rule imposes net costs to society only by ignoring the public health benefits of reducing particulate matter emissions including at

---

[x] A number of regulations that target pollutants other than particulate matter yield the ancillary benefit of reducing particulate matter [121]. By ignoring these benefits, which are often substantial, the administration was able to claim that the regulations yielded little net benefits or even imposed net costs to society.

[xi] The Clean Power Plan never took effect because of a legal challenge from 24 state governments; 18 states, plus the District of Columbia, participated in that lawsuit as intervenors in support of the EPA.

[xii] It also understated benefits by choosing specific assumptions on discount rates and on global versus domestic cost of carbon (Chapter 9).

[xiii] The Clean Air Act, according to the Supreme Court case *Massachusetts v. EPA* and the EPA's endangerment finding – that greenhouse gas emissions endanger public health – mandates the EPA to regulate greenhouse gas emissions [220]. Therefore, the EPA was obliged to promulgate a replacement rule when it rescinded the Clean Power Plan.

[xiv] As discussed in Chapter 9, the rule came into effect in 2012. In response to the 2015 Supreme Court ruling in *Michigan v. EPA*, the EPA reanalyzed the rule. That reanalysis, published in 2016, had concluded that the rule was appropriate and necessary [219][221].

low concentrations. (That reduction is achieved as an ancillary benefit of efforts to reduce mercury emissions.)[110] Thus, the EPA was able to declare in May 2020 that the Mercury and Air Toxics Standards Rule is not "appropriate and necessary."[109,111] That declaration weakens the legal underpinning of the rule and opens it to legal challenges.[112,113] The weakening or repeal of that rule, which prevents about 7,000–17,000 premature deaths annually, as well as neurological, cardiovascular and pulmonary diseases,[114] would be devastating to public health.

The American Petroleum Institute, its consultants and fossil fuel–funded think tanks played a key role in undermining these rules by casting doubt on the benefits from reducing air pollutants at low concentration levels. They utilized testimonies at EPA hearings and submissions of comments to the EPA to amplify scientific uncertainties in the research on the public health effects of reducing pollution below a low threshold.[27,28,115] As I discussed in Section 8.3, there is a broad consensus among the scientific community that exposure to particulate matter, even at low concentrations and below that of the NAAQS, adversely affects public health.[116] Scientific evidence has only grown on the adverse effects of air pollutants, even at very low concentrations, on mortality and diseases.[90,117–120]

The EPA had abandoned the threshold models by the late 1970s in response to growing scientific evidence that pollutants negatively affect health even at low doses.[121] Both Republican and Democratic administrations applied the non-threshold model to set regulations, including the NAAQS for particulate matter, lead, ozone, nitrogen dioxide and sulfur dioxide.[121] The George W. Bush administration also applied the non-threshold model to regulations to limit particulates and nitrogen oxides from diesel engines.[121,122]

The rulemaking process does allow agencies to tighten, to weaken and even to repeal regulations in accordance with scientific evidence. However, the extent of the Trump administration's manipulation, distortion and suppression of scientific evidence in this process is as unethical as it is unprecedented. It simply trampled on inconvenient evidence that stood in the way of its goal to repeal regulations.[123,124]

8.5 Blocking Consideration of Valid Scientific Studies

Moving beyond publishing analyses that ignore scientific evidence, the administration proposed a far-reaching rule that would effectively prohibit the EPA from considering valid science. The proposed rule, for reasons described below, would empower an anti-regulatory administrator to block stricter regulations and even to weaken existing regulations.

In April 2018, with Pruitt still running the EPA, the administration proposed the Strengthening Transparency in Regulatory Science rule.[125] In March 2020, with Wheeler at the helm, the administration published a supplemental notice making the proposed science transparency rule far more sweeping.[126] The rule would apply to the EPA's consideration of all "influential science" and "to all data and models." Applicable studies must disclose "the information necessary for the public to understand, assess, and replicate findings." While the EPA may consider other studies, it would give greater weight to studies whose "data [is] available for independent validation," that is, those studies whose data is publicly available. It would grant the EPA administrator the powers to apply the rule on a "case-by-case basis." Additionally, the proposed rule requires the EPA to consider "various threshold models across the dose or exposure range," even when most scientists, as described above, reject the threshold model.[xv]

In defending this rule, Wheeler emphasized the value of transparency in the EPA's decision-making and the EPA's commitment not to disclose personal information. He noted,

> good science is science that can be replicated and independently validated; science that holds up to scrutiny. That is why we are moving forward to ensure that the science supporting Agency decisions is transparent and available for evaluation by the public and stakeholders ... At the same time, we will ensure that we're not disclosing confidential or personal information. ... Our proposed rule would apply prospectively to final significant regulatory actions.[127]

In December 2020, the EPA finalized the rule.[222]

At face value, Wheeler's comments seem sensible and his goals laudable. On closer scrutiny, he appears to have played a disingenuous game, pretending to support transparency while actually working to limit it and to exclude from consideration scientific evidence that does not support deregulatory actions. At least seven objections were raised against the proposed rule by members of the SAB, editors of science journals, presidents of the national academies, numerous scientists and members of Congress.

First, contrary to Wheeler's (and his predecessor's) paean to transparency, critics said that the process for developing the rule was anything but transparent. Pruitt published this draft rule without prior consultation of the SAB, prompting criticisms from even his appointees on the board.[128] The EPA continued to stall on sharing information and avoided convening meetings at which the SAB could review the rule.[129] By law, whenever the EPA administrator presents rules or standards proposed under environmental laws for

---

[xv] Several members of the SAB interpreted the proposed science transparency rule as an abandonment of the linear non-threshold models.

review at any federal agencies, the administrator must simultaneously provide to the SAB the proposal and the scientific and technical information in the EPA's possession that had informed the development of the proposal.[130,131]

Second, instead of improving science at the agency, as presidents of the national academies make clear, the rule "poses a threat to credibility of regulatory science."[132] The SAB underscored how the rule would shift the agency from "science-based decision-making" to cherry-picking information "based on non-scientific considerations." The SAB warned that "such a change could easily undercut the integrity of environmental laws, as it will allow systematic bias to be introduced with no easy remedy. Such a proposal is inconsistent with the scientific method that requires all credible data be used to understand an issue."[133,134] Even scientists working to improve transparency and reproducibility in research opposed the rule, arguing it would eliminate valid science from consideration.[135–138] The rule would scuttle the regulation not only of air pollutants but also of toxic chemicals, pesticides, herbicides, food additives and pharmaceutical drugs.[139] The rule would affect not just prospective regulations but existing regulations as well. Because the EPA is mandated to review regulations such as the ambient air quality standards periodically, it would be hampered from considering the vast numbers of studies that rely on confidential health data that had informed these standards.[140]

Third, opponents charged that the EPA's stated goal of improving knowledge of the state of science could be achieved more effectively by scientists at the EPA and serving on panels that review the cumulative evidence from a variety of studies than by the rule's insistence that studies be replicated.[141] Indeed, the EPA's established procedures (prior to their subversion by Pruitt and Wheeler) facilitate comprehensive reviews of the science pertaining to specific pollutants to guide its rulemaking. Career scientists provide an initial assessment of the state of the science. Layers of independent scientists, from panels of experts on that specific pollutant to the SAB, scrutinize that assessment. Additionally, the National Academies of Sciences, Engineering, and Medicine has weighed in on vexing issues, such as the use of non-threshold models in evaluating the effects of pollutants.

Fourth, the proposed rule was criticized for undermining transparency, contrary to its stated aims.[142] It would give the EPA administrator the discretion to exempt studies from the proposed rule;[125] in other words, it would permit the administrator to cherry-pick studies to reach a predetermined conclusion.[133,142–144] As the SAB underscored, "case-by-case exceptions may exacerbate concerns about inappropriate exclusion of scientifically important studies."[133]

Fifth, the EPA was accused of understating the difficulty in protecting individuals' confidential data, even when data is anonymized prior to release. The 1996 Health Insurance Portability and Accountability Act (HIPAA) protects patients' privacy and the Institutional Review Boards require researchers to protect confidential patient information.[142] Even when the data was anonymized and met HIPAA confidentiality requirements, researchers demonstrated that they could still identify a large number of patients.[145–147] The disclosure of confidential data, even if inadvertent, would infringe upon individuals' privacy,[148] endanger their health insurance access and their employment prospects,[149] violate privacy laws,[142] and chill individuals' willingness to participate in public health studies.[150]

Sixth, the EPA was excoriated as not having the authority to issue the rule. Critics shot down the EPA's claim of authority under the 1958 Federal Housekeeping Act, noting that the act empowered executive departments only, which does not include the EPA.[151] Moreover, critics contend that the limited powers under the act for a department to regulate its internal affairs did not authorize sweeping powers envisioned by the EPA.[151] The EPA was proposing to completely revamp how the agency approached scientific evidence, the backbone of all its operations and of all of public health and environmental regulations.

Lastly, critics – including Representative Bernice Johnson, chair of the House Science, Space, and Technology Committee and of the editorial board at the highly respected journal *Nature* – disparaged the proposed rule as a pretext for weakening regulations.[140,152] These concerns stem from strategies to block regulations, long used by the tobacco, oil, and other industries.[23,153] In 1996, lawyer Chris Horner (who later served on Trump's EPA transition team) counseled his client the tobacco company RJ Reynolds that all industrial sectors should pursue the tactic "to construct explicit procedural hurdles that the Agency must follow in issuing scientific reports ... including ... transparency, [reproducibility] and [being] not judicially reviewable." He added that they should begin with the EPA review of particulate matter and ozone.[154,155] Soon after, Lamar Smith and the oil industry attacked the landmark public health studies on particulate matter as secret science.[33] Smith attempted to pass "secret science" bills that would block federal agencies from considering those studies whose data are not publicly available.[xvi] As revealed by media investigations, Smith's staffers worked with Pruitt to resurrect Smith's failed bills as an EPA rule.[156]

[xvi] Smith sponsored H.R. 1030 Secret Science Reform Act and H.R. 1430 Honest and Open New EPA Science Treatment Act, which passed the House in 2015 and 2017, respectively, largely on partisan votes, but failed to gain traction in the Senate.

8.6 Who Can Protect Science in the Policy Process?

Since the mid-nineteenth century, Congress has recognized the value of impartial science to guide government policy from public health to national security, inaugurating the National Academy of Sciences in 1893.[157] In time, Congress established federal agencies to promulgate and implement regulations anchored in science and instituted guardrails to protect agencies' scientific analyses from politicization, manipulation and suppression.[157] Scientific integrity at agencies had enjoyed support from both Republicans and Democrats. As President George H. W. Bush underscored, "Science ... relies on freedom of inquiry; and one of the hallmarks of that freedom is objectivity ... [G]overnment relies on the impartial perspective of science for guidance." However, that system is now under tremendous strain. If it is to survive, Congress, the scientific community, NGOs and the general public must all play their part in protecting scientific integrity at federal agencies and in the rulemaking process.

8.6.1 Congress

Congress has many roles to play in both the rulemaking process and agency oversight. First and foremost, Congress has substantial sway over budgets and the nomination of agency heads. Congress has demonstrated the ability to work in a bipartisan way to protect the funding of regulatory agencies. When faced with the Trump administration's proposed budget cuts in 2017, the EPA and other federal agencies benefitted from a number of House Republicans in the Republican-controlled Congress who took the view that the United States can pursue economic progress and protect public health and the environment at the same time.[158] The chair of the House Appropriations Committee, Republican Rodney Frelinghuysen from New Jersey, underscored that "we all have the responsibility to be proper stewards of our air, land and water, and to protect the beauty of our nation for generations to come."[159] Committee member Ken Calvert, a Republican from California, added, "We all want clean air and clean water and a strong robust economy. ... It's not an 'either/or' debate."[160]

Despite such statements, Republicans in Congress have shown little readiness to oppose Trump's efforts to turn the EPA from a watchdog of polluters into an agency that prioritized the oil and gas industry's wishlist for deregulatory actions. For instance, when faced with the president's nominees for EPA administrator, only one Republican senator echoed the concerns of a large number of Democratic senators about nominees' past work in support of the oil and gas and other regulated industries. Senator Susan Collins from Maine opposed Pruitt, noting that his lawsuits against the EPA as Oklahoma's attorney general raised doubts about his commitment to the EPA's mission.[161] She also opposed Wheeler, underscoring how as assistant administrator he supported policies that were "not in the best interest of our environment and public health."[162]

Other Republicans, however, such as Senator John Barrasso of Wyoming, voiced no criticisms of Pruitt or Wheeler and instead reiterated the administration's claim that regulations harm the economy. "During the last administration," Barrasso said, "the EPA issued punishing regulations that would hurt the economy and raise costs on families. The agency is now putting forward proposals that both protect our environment and allow the country's economy to flourish."[163]

Even so, Republicans when faced with public health emergencies in their home states rejected the administration's more extreme nominees who advocate egregiously weak public health protections. Thom Tillis and Richard Burr, Republican senators from North Carolina, opposed the nomination of Michael Dourson to lead the EPA Office of Chemical Safety and Pollution Prevention.[164] They warned that Dourson would not be able to provide the necessary leadership to address public health emergencies,[164] having consistently supported far more permissive pollutant standards than those recommended by other scientists.[165,166] North Carolina, like a growing number of US states, faced drinking water contamination from per- and poly-fluoroalkyl substances (PFAS) that are hazardous to human health.[167,168] PFAS, widely utilized in the industrial world, are also used in the oil and gas industry in chemically enhanced oil extraction, in chemicals used in hydraulic fracturing to reduce corrosion, and in firefighting foam used against petroleum fires at oil refineries and storage facilities. These chemicals have been known to enter and contaminate water resources.[169]

Several Congressional representatives have also pushed back against the suppression of science reports, although they have tended to do so only when prompted by public health emergencies. Democratic Senator Joseph Machin from West Virginia, a coal-dependent state, has voted against public health regulations, and Senator Shelly Moore Caputo, his Republican counterpart, pushed the White House to release a study that recommended the tightening of standards for PFAS.[170–172]

Several members of the House have invoked Congress's investigative powers to hold agencies accountable in hearings. Representatives Bernice Johnson and Frank Pallone, the chairs of the House Committee on Energy and Commerce and the House Committee on Science, Space, and Technology, respectively, sought information from the EPA administrator on numerous actions that impaired the review process for the setting of ambient air quality standards.[173,174] Several hearings during the 116th Congress, chaired by Representative Bernice Johnson, shed light on the administration's efforts to undermine the use of valid science. These hearings contrasted with those held by other members of Congress, such as Lamar Smith, who preceded Johnson as chair of the House Committee on Science, Space, and Technology in the 115th Congress. Those hearings were used as opportunities to amplify the views of contrarians and to attack scientists working on issues such as climate change.[175]

Congressional representatives' letters to the EPA administrator, which lay out flaws in the review process or in proposed rules, serve as important documents for legal actions against the agency's deregulatory actions. Specifically, they serve as information presented to the agency, which it must adequately address before finalizing a rule. Its failure to do so would make the final rule vulnerable to challenges.[176] The proposed EPA science transparency rule sparked bipartisan opposition. No fewer than 103 representatives[xvii] signed a letter criticizing the proposed rule for "create[ing] an opaque process allowing the EPA to selectively suppress scientific evidence without accountability and in the process undermine bedrock environmental laws."[177] In the Senate, the proposal was also met with vocal opposition, with seven senators underscoring how the rule "would likely violate environmental laws that mandate the use of 'best available science.'"[178]

Violations of scientific integrity at federal agencies have prompted congressional representatives to support legislation that protects the work of career scientists from political interference. In October 2019, the House Committee on Science, Space, and Technology voted in a bipartisan manner to advance the Science Integrity Act (H.R. 1709) to the House.[179,xviii] Representative Frank Lucas of Oklahoma, who served as the ranking Republican on the committee, supported the legislation, noting, "We all agree [that] government scientists should be able to conduct their research free from suppression, intimidation, coercion, or manipulation."[180] Senator Brian Schatz introduced a companion bill in the Senate with 14 cosponsors.[181] None of these bills were enacted by the 116th Congress, sadly underscoring the unwillingness of key congressional members to stand up for science.

8.6.2 The Scientific Community

Numerous scientists and scientific organizations fought against the administration's attack on scientific integrity. As early as November 2016, in response

---

[xvii] Four Republicans, all members of the Climate Solutions Caucus, signed the letter: representatives Carlos Curbelo and Ileana Ros-Lehtinen of Florida and representatives Brian Fitzpatrick and Ryan Costello of Pennsylvania.

[xviii] Nineteen Democrats and six Republicans on the committee voted for the legislation to advance, while six Republicans voted against it.

to president-elect Trump's denial of climate science,[182] 88 prominent scientists sent an open letter to the president-elect and the 115th Congress that underscored the urgency to protect the independence of scientists, including those working in governmental agencies, to carry out their work and to communicate their findings to the public.[183] About 5,500 scientists, engineers and social scientists subsequently signed that open letter. In response to the administration's proposed regulations that ran counter to science, a spectrum of the scientific community – science editors, science organizations and individual scientists – submitted comments to set the record straight.[184] Between 2018 and 2020, editors of many leading science journals,[185,186] along with 69 public health, medical and science groups[187,188] and at least 100 other scientists[144,189,190] published letters condemning the "science transparency" rule. Editors urged their readers to take action: "Institutions and individuals must ... write to their elected representatives to call out this attempt to undermine accepted scientific practice in public-health and environmental standards."[140]

The scientists on the SAB (the board statutorily mandated to advise the EPA administrator) persisted in making their opinions known, despite the administration's attempts to sideline their work. Recall, the Trump administration had attempted to coopt the board by firing prominent academic scientists and by appointing two-thirds of its 44 members, but the board pushed back against the administration. The board proceeded with a review of the EPA's analyses and proposed rules,[191,192] even when the EPA failed to promptly provide the board with information, defying the 1978 Environmental Research, Development and Demonstration Authorization Act.[130,193] The board's numerous reports from 2018 through 2020 underscore how the EPA's analyses of proposed rules were deeply flawed.[133,143,194–196]

Just as the board did, a number of career scientists at federal agencies pushed back against the administration as well. For instance, they fought to include critical scientific facts in their reports – for example, the draft policy assessment for particulate matter, which despite opening the door for the administrator to keep the existing standards, documents how tightening the standards would save as many as 12,000 lives annually.[197,198,xix] Wheeler's failure to explain how his decision not to tighten the standards can be reconciled with these figures, along with the board's scathing reviews of the EPA's analyses, provides fodder for legal challenges against regulatory rollbacks. While the Trump administration was well within its powers to change policy directions from

---

[xix] The report states that the existing standard at 12 $\mu g/m^3$ is "associated with as many as 45,000 deaths" annually but that if the standard were tightened to 9 $\mu g/m^3$, annual deaths would fall by about 21–27 percent.

those of past administrations, courts may well push the EPA to articulate the merits of going against the advice of the board and the analyses undertaken by its career scientists.[198]

8.6.3 NGOs and the Media

For decades, a number of prominent NGOs have championed scientific integrity in federal agencies and the protection of agency scientists' work against political interference. Their tasks became even more pressing with the Trump White House seeming to see no limits in such interference and even blocking a senior analyst[xx] at the State Department from submitting written testimony on climate impacts to the House Intelligence Committee.[199–201]

The Government Accountability Project, set up in 1977 to protect whistleblowers in federal agencies, provided legal assistance to several scientists who filed whistleblower complaints and testified publicly on science erosion at agencies during the Trump administration.[202,203] Among others, Joel Clements, a Department of the Interior senior scientist, charged that his demotion to an accounting position was a ploy to silence his warnings on the adverse impacts of climate change on Alaska Native communities. Kevin Chmielewski, the deputy chief of staff of operations at the EPA, revealed Pruitt's serious ethics conflicts and his retaliation against staff who refused to condone his alleged ethics violations.[204] Chmielewski's complaint sparked federal and congressional investigations, which uncovered, among others, Pruitt's sweetheart condo-rental deal with an oil lobbyist. Pruitt eventually resigned.[205]

Another prominent NGO, founded in 1969 to examine government policies in which science is of significance, the Union of Concerned Scientists, testified to Congress and published op-eds about the Trump administration's undermining of science at federal agencies.[206,207] It submitted comments to the EPA challenging the faulty analyses underlying its deregulatory actions.[208,209] Documents uncovered by its freedom of information requests and by media investigations boosted suspicions that the EPA sidelined science and worked closely with conservative think tanks and the oil industry to pursue deregulatory actions. One set of documents revealed the Heritage Foundation's request, which foreshadowed the EPA leadership's actions, which read as follows: "Will [the Particulate Matter Review Panel] be disbanded and stopped before it rubber stamps more $PM_{2.5}$ junk science, which would ruin the repeal of the

---

[xx] The senior analyst, Rod Schoonover, warned that human-caused climate change could be "possibly catastrophic." Comments from a member of the National Security Council, who blocked the written testimony, claimed that "this is not objective testimony at all ... It includes lots of climate alarm propaganda that is not science at all" [199, 200].

Clean Power Plan?"[210] Another set of documents revealed how the EPA leadership's deliberations on the proposed science transparency rule focused not on transparency but on shielding industry's use of confidential data in the regulatory process.[156]

Likewise, media investigations uncovered how the EPA leadership appear to have ignored EPA scientists to do the bidding of the oil industry. For instance, even despite the car industry's opposition[xxi] to the Trump administration's proposal, it proceeded to freeze fuel-efficiency standards for cars at 40 miles per gallon by 2026 (instead of a modest ramp up to 55 miles per gallon per the original rule).[211,212] The administration's justification for the weakened standards – that traffic fatalities would decline – was refuted by EPA career scientists and by academics[xxii] whose research the administration had misrepresented in its analysis.[213–215] The *New York Times*, in its December 2018 and June 2019 reports, revealed how the oil industry and think tanks lobbied the administration and several members of Congress to weaken these standards.[216,217] The article reported meetings between Pruitt and the head of Marathon Oil, a major oil refiner in the United States, and an interview with Ebell, who took credit for the administration's proposed freezing of the standards. Marathon Oil had projected industry losses from greater fuel-efficiency standards, stemming from a reduction in Americans' gasoline consumption of about 350,000–400,000 barrels per day.

A number of NGOs have used the courts to challenge the Trump administration's efforts to tamper with scientific integrity in the regulatory process (discussed further in Chapter 9). For instance, the Environmental Defense Fund filed a successful lawsuit against the EPA's Science Transparency Rule. It filed the legal challenge after Wheeler finalized that rule in December 2020 and made the rule effective immediately.[218] In January 2021, the US District Court for the District of Montana ruled in *Environmental Defense Fund v. EPA* that the EPA's issuance of the rule was illegal. Judge Brian Morris ruled that the EPA could not issue substantive rules, purportedly under its housekeeping authority (i.e., an agency's powers to set its procedures), and that the EPA had to abide by the rulemaking procedure set out in the Administrative Procedure Act.[223] (Subsequently, the court vacated the rule and the Biden administration withdrew the rule in May 2021.)

[xxi] As discussed in Chapter 11, the Trump administration created unpredictability for the car industry and its supply chain by sparking a likely long-drawn out battle between California, which imposes stricter standards (and states that follow California's lead), and other states that follow the weaker proposed standards.

[xxii] William Charmley, director of the assessments and standards division of the EPA's office of transportation and air quality, wrote in a June 2018 interagency email that the "proposed standards are detrimental to safety, rather than beneficial" [214].

8.7 Conclusion

Despite the best efforts of many NGOs, some sections of the media and numerous individual scientists and scientific journals and organizations, civil society has waged an uphill battle to combat the administration's contempt for science. The political actors within America's rulemaking system are probably better placed than civil society to push back against an administration that seems to heed nothing but its fossil fuel backers. With the exception of very few members, the Republican Party, which controlled the Senate, was a willing accomplice to the Trump administration's unrelenting attack on the independence of federal agencies and on the integrity of science itself. For many Republicans, the risks of greater pollution were outweighed by what they saw as the dangers of excessive regulation, and thus they were not merely prepared but decidedly eager to let the oil and gas industry have a much greater say in how it was regulated – and, indeed, was *not* regulated.

Whether the lure of deregulation for Republicans and the voters who elect them (and Trump) will endure is difficult to judge, but one suspects that when stories start to emerge about the costs of deregulation in terms of poorer human health and a more polluted environment, then the mood among conservatives may begin to shift. At that point, however, the rulemaking system may be so degraded and the public so skeptical of scientific integrity that it may be impossible to reform and rejuvenate the system. Thus, whatever civil society can do to help preserve the institutional rulemaking architecture, faith in science and support for decisions made on the basis of evidence rather than prejudice, then the greater will be the chance of reversing the damage to America's health and environment wrought by the Trump administration in its attack on science and science-based regulation.

But it is not just science that was under attack. As the next chapter reveals, economic data and analyses, as well as their exponents, was under sustained assault if they conflicted with the political goals of the Trump administration and the financial interests of the oil and gas industry.

References

1. G. H.W. Bush. "Remarks to the National Academy of Sciences." April 23, 1990. http://bushlibrary.tamu.edu/research/papers/1990/90042301.html.
2. A. Kron. "EPA's Role in Implementing and Maintaining the Oil and Gas Industry's Environmental Exemptions: A Study in Three Statutes." *Vermont Journal of Environmental Law* 16 no. 4 (2014): 586–635.

3. W. Kovarik. "Ethyl-leaded Gasoline: How a Classic Occupational Disease Became an International Public Health Disaster." *International Journal of Occupational and Environmental Health* 11, no. 4. (2005): 384–397. https://doi.org/10.1179/oeh.2005.11.4.384.
4. D. Rosner and G. Markowitz. *Deceit and Denial: The Deadly Politics of Industrial Pollution*. Berkeley, CA: University of California Press, 2013.
5. C. Jacobs and W. Kelly. *Smogtown: The Lung-Burning History of Pollution in Los Angeles*. Woodstock, NY: Overlook Press, 2008.
6. N. Banerjee, D. Hasemyer and L. Song. "For Oil Industry, Clean Air Fight Was Dress Rehearsal for Climate Denial." *Inside Climate News*, June 6, 2016. https://insideclimatenews.org/news/05062016/oil-industry-clean-air-fight-smog-los-angeles-dress-rehearsal-climate-change-denial-exxon.
7. "Congress Faces Hard Choices on Clean Air Act." *CQ Almanac*, 31st edition (1976): 245–250. http://library.cqpress.com/cqalmanac/cqal75-1213793.
8. S. Hays. "Clean Air: From the 1970 Act to the 1977 Amendments." *Duquesne Law Review* 17, no. 1 (1977–78): 33–66.
9. J. F. Gamble. "PM2.5 and Mortality in Long-term Prospective Cohort Studies: Cause–Effect or Statistical Associations?" *Environmental Health Perspectives* 106, no. 9 (September 1998): 535–549. https://doi.org/10.1289/ehp.98106535.
10. J. Warrick. "Opponents Await Proposal to Limit Air Particulates: Industry Giants Mobilize to Block New EPA Rules." *Washington Post*, November 27, 1996.
11. J. Warrick and J. E. Yang. "Stricter Air Quality Rules May Test Hill's New Veto; Several GOP Chairmen Critical of EPA Move." *Washington Post*, November 28, 1996.
12. *Occupational Exposure to Benzene; Final Rule*. Occupational Health and Safety Administration, Department of Labor. 43 Federal Register 27962–27971 (June 27, 1978).
13. American Petroleum Institute, Benzene Task Force. *Summary of API's Benzene Research Strategy* (2000). www.documentcloud.org/documents/1373740-2000-api-summary-of-benzene-research-strategy.html#document/p2/a191357.
14. P.F. Infante. "The Past Suppression of Industry Knowledge of the Toxicity of Benzene to Humans and Potential Bias in Future Benzene Research." *International Journal of Occupational and Environmental Health* 13, no. 3 (July 19, 2013): 268–272. https://doi.org/10.1179/oeh.2006.12.3.268.
15. American Petroleum Institute. *API Toxicological Review: Benzene, Prepared under the Direction of Dr. Philip Drinker*. Harvard School of Public Health (New York: September 1948). www.documentcloud.org/documents/1373098-00010795.html.
16. K. Lombardi and J. Bennett. *A Dozen Dirty Documents: Twelve Documents That Stand Out from the Center's New Oil and Chemical Industry Archive*. Center for Public Integrity (December 2014). https://publicintegrity.org/environment/a-dozen-dirty-documents.
17. R. E. Dunlap and A. M. McCright. "Climate Change Denial: Sources, Actors, and Strategies." In *Routledge Handbook of Climate Change and Society*, edited by C. Lever-Tracy. 240–259 (London: Routledge, 2010).
18. R. J. Brulle. "Institutionalizing Delay: Foundation Funding and the Creation of US Climate Change Counter-Movement Organizations." *Climatic Change* 122, no. 4. (2014): 681–694. https://doi.org/https://doi.org/10.1007/s10584-013-1018-7.

19. R. E. Dunlap and R. J. Brulle. *Climate Change and Society: Sociological Perspectives*. New York: Oxford University Press, 2015.
20. G. Supran and N. Oreskes. "Assessing ExxonMobil's Climate Change Communications (1977–2014)." *Environmental Research Letters* 12, no. 8 (2017).
21. M. Grasso. "Oily Politics: A Critical Assessment of the Oil and Gas Industry's Contribution to Climate Change." *Energy Research & Social Science* 50 (April 2019): 106–115. https://doi.org/10.1016/j.erss.2018.11.017.
22. N. Oreskes. "The Scientific Consensus on Climate Change: How Do We Know We're Not Wrong?" In *Climate Change: What it Means for You, Your Children, and Your Grandchildren*, edited by J. F. C. DiMento and P. Doughman. 65–99. Cambridge, MA: MIT Press, 2007.
23. D. Michaels. *Doubt Is Their Product: How Industry's Assault on Science Threatens Your Health*. Oxford, UK: Oxford University Press, 2008.
24. N. Oreskes and E. M. Conway. *Merchants of Doubt: How a Handful of Scientists Obscured the Truth on Issues from Tobacco Smoke to Global Warming*. London, UK: Bloomsbury Publishing, 2011.
25. J. Schwartz. *Facts Not Fear on Air Pollution: How Regulators, Environmentalists and Scientists Exaggerate the Level and Health Risks of Air Pollution and Impose Counterproductive Regulations*. National Center for Policy Analysis Policy Report No. 294 (2006). www.heartland.org/publications-resources/publications/facts-not-fear-on-air-pollution-how-regulators-environmentalists-and-scientists-exaggerate-the-level-and-health-risks-of-air-pollution-and-impose-counterproductive-regulations.
26. W. Crews and I. Osorio. *This Liberal Congress Went to Market? A Bipartisan Policy Agenda for the 110th Congress*. Competitive Enterprise Institute. January 10, 2007. www.cei.org/sites/default/files/1-CEI%20-%20This%20Liberal%20Congress%20Went%20to%20Market.pdf.
27. H. J. Feldman, senior director, Scientific and Regulatory Affairs at American Petroleum Institute. *Testimony Regarding EPA's Proposal to Change the National Ambient Air Quality Standards for Particulate Matter*. Docket ID No. EPA–HQ–OAR–2007–0492. July 17, 2012.
28. J. E. Goodman. *Testimony on the Proposed Rule: National Ambient Air Quality Standards for Particulate Matter*. Docket ID No. EPA–HQ–OAR–2007–0492. July 17, 2012.
29. H. J. Feldman, senior director, Scientific and Regulatory Affairs at American Petroleum Institute. *Testimony of Howard J. Feldman, Public Hearings for the 2015 Ozone NAAQS Proposed Rule*. January 29, 2015.
30. D. W. Dockery et al. "An Association between Air Pollution and Mortality in Six US Cities." *New England Journal of Medicine* 329, no. 24 (December 9, 1993): 1753–1759. https://doi.org/10.1056/NEJM199312093292401.
31. A.C. Pope et al. "Particulate Air Pollution As a Predictor of Mortality in a Prospective Study of US Adults." *American Journal of Respiratory and Critical Care Medicine* 151, no. 3 (1995): 669–674.
32. FreedomWorks. "Issue Analysis 50 – The EPA's New Clean Air Standards: A Primer." News release. 1997. www.freedomworks.org/content/issue-analysis-50-epas-new-clean-air-standards-primer.

33. G. D. Thurston. "Mandating the Release of Health Research Data: Issues and Implications." *Tulane Environmental Law Journal* 11 (1998): 331–354.
34. Health Effects Institute. *Synopsis of the Particle Epidemiology Reanalysis Project* (2000). www.healtheffects.org/system/files/Reanalysis-Statement.pdf.
35. D. Krewski et al. "Overview of the Reanalysis of the Harvard Six Cities Study and American Cancer Society Study of Particulate Air Pollution and Mortality." *Journal of Toxicology and Environmental Health* 66, no. 16–19 (2003): 1507–1552. https://doi.org/10.1080/15287390306424.
36. D. Krewski. "Validation of the Harvard Six Cities Study of Particulate Air Pollution and Mortality." *New England Journal of Medicine* 350 (January 8, 2004): 198–199. https://doi.org/10.1056/NEJM200401083500225.
37. L. Smith. "The EPA's Game of Secret Science." *Wall Street Journal*, July 29, 2013. www.wsj.com/articles/SB10001424127887323829104578624562008231682.
38. A. Logomasini, senior fellow at the Competitive Enterprise Institute. "Pruitt's Rule Ending Secret Science Is Pro-Science, Pro-Consumer." *The Hill*, April 4, 2018. https://thehill.com/opinion/energy-environment/385411-pruitts-rule-ending-secret-science-is-pro-science-pro-consumer.
39. American Petroleum Institute, *Draft Global Climate Science Communications Plan.* Inside Climate News (1998).
40. C. Muffett and S. Feit. *Smoke and Fumes: The Legal and Evidentiary Basis for Holding Big Oil Accountable for the Climate Crisis.* Center for International Environmental Law (Washington, DC: November 2017). www.ciel.org/reports/smoke-and-fumes.
41. B. Franta. "Early Oil Industry Knowledge of CO2 and Global Warming." *Nature Climate Change* 8 (2018): 1024–1025. https://doi.org/10.1038/s41558-018-0349-9.
42. US House of Representatives. *Climate Science: Assumptions, Policy Implications, and the Scientific Method.* Committee on Science, Space, and Technology (2017).
43. US House of Representatives. *Paris Climate Promise: A Bad Deal for America.* Committee on Science, Space, and Technology (2016).
44. W. R. Freudenburg, R. Gramling and D. J. Davidson. "Scientific Certainty Argumentation Methods (SCAMs): Science and the Politics of Doubt." *Sociological Inquiry* 78, no. 1 (February 2008): 2–38.
45. R. E. Dunlap and A. M. McCright. "Organized Climate Change Denial." *The Oxford Handbook of Climate Change and Society*, edited by J. S. Dryzek, R. B. Norgaard and D. Schlosberg. 144–160. Oxford: Oxford University Press: 2011. https://doi.org/10.1093/oxfordhb/9780199566600.003.0010.
46. J. M. Turner and A. C. Isenberg. *The Republican Reversal: Conservatives and the Environment from Nixon to Trump.* Cambridge, MA: Harvard University Press, 2018.
47. K. P. Dineen. "Reading the Tea Leaves: The Tea Party Movement, the Conservative Establishment and the Collapse of Climate Change Legislation." Massachusetts Institute of Technology Masters Dissertation (June 2011). https://dspace.mit.edu/bitstream/handle/1721.1/66804/757149232-MIT.pdf?sequence=2&isAllowed=y.
48. T. O. McGarity. "The Disruptive Politics of Climate Disruption." *Nova Law Review* 38, no. 3 (2014): 394–472. https://nsuworks.nova.edu/cgi/viewcontent.cgi?referer=https://www.ecosia.org/&httpsredir=1&article=1002&context=nlr.

49. Heritage Foundation. "Trump Administration Embraces Heritage Foundation Policy Recommendations." January 23, 2018. www.heritage.org/impact/trump-administration-embraces-heritage-foundation-policy-recommendations.
50. P. Bharara et al. *Proposals for Reform*. National Task Force on Rule of Law & Democracy, Brennan Center for Justice at New York University School of Law. (2019). www.brennancenter.org/sites/default/files/2019-09/2019_10_TaskForce%20II_0.pdf.
51. D. J. Trump. Presidential Executive Order on Promoting Energy Independence and Economic Growth. E.O. 13783, 82 Federal Register 16093–16097. Washington, DC, March 28, 2017. www.whitehouse.gov/presidential-actions/presidential-executive-order-promoting-energy-independence-economic-growth.
52. B. Sussman. "Back to Basics or Slash and Burn? Scott Pruitt's Reign As EPA Administrator." *Environmental Law Institute* 47, no. 109 (2017): 7–26. www.eli.org/sites/default/files/docs/47.10917.pdf.
53. D. Roberts. "Donald Trump Is Handing the Federal Government over to Fossil Fuel Interests." *Vox*, June 14, 2017. www.vox.com/energy-and-environment/2017/6/13/15681498/trump-government-fossil-fuels.
54. D. Kravitz, A. Shaw and I. Arnsdorf. "What We Found in Trump's Drained Swamp: Hundreds of Ex-lobbyists and DC Insiders." *ProPublica*, March 7, 2018. www.propublica.org/article/what-we-found-in-trump-administration-drained-swamp-hundreds-of-ex-lobbyists-and-washington-dc-insiders.
55. L. Fredrickson et al. "History of US Presidential Assaults on Modern Environmental Health Protection." *American Journal of Public Health* 108, no. S2 (2018): S95–S103.
56. L. Dillon et al. "The Environmental Protection Agency in the Early Trump Administration: Prelude to Regulatory Capture." *American Journal of Public Health* 108, no. S2 (2018): S89–S94. https://ajph.aphapublications.org/doi/10.2105/AJPH.2018.304360.
57. S. Whitehouse et al. *Comments on the Environmental Protection Agency's Proposed Weakening of Rules Governing Methane Emissions from Oil and Natural Gas Facilities*. Submitted to Environmental Protection Agency. December 17, 2018.
58. C. T. Whitman. *Statement of the Honorable Christine Todd Whitman to the House Committee on Energy and Commerce*. Submitted to US House of Representatives Subcommittee on Oversight and Investigations. June 11, 2019.
59. G. Mccarthy. *Written Testimony to House Committee on Energy and Commerce*. Submitted to US House of Representatives Subcommittee on Oversight and Investigations. June 11, 2019.
60. *Oil and Natural Gas Sector: Emission Standards for New, Reconstructed, and Modified Sources Reconsideration*. 40 Code of Federal Regulations 60. Environmental Protection Agency. 83 Federal Register 52056–52107 www.federalregister.gov/documents/2018/10/15/2018-20961/oil-and-natural-gas-sector-emission-standards-for-new-reconstructed-and-modified-sources.
61. J. Carter et al. *Science Under Siege at the Department of the Interior*. Union of Concerned Scientists (December 2018). www.ucsusa.org/sites/default/files/attach/2018/12/science-under-siege-at-department-of-interior-full-report.pdf.
62. J. Carter et al. *The State of Science in the Trump Era*. Center for Science and Democracy at the Union of Concerned Scientists, Union of Concerned Scientists

(January 2019). www.ucsusa.org/sites/default/files/attach/2019/01/ucs-trump-2 yrs-report.pdf.
63. Public Employees for Environmental Responsibility. *Request for an Inquiry under the Scientific Integrity Policy into Final Rule Regarding the Definition of Water of the US*. Submitted to Acting Inspector General C. J. Sheehan, Office of Inspector General. January 18, 2020.
64. Environmental Protection Agency. Strengthening and Improving Membership on EPA Federal Advisory Committees. Report by E. S. Pruitt (Washington, DC: 2017). www.epa.gov/sites/production/files/2017-10/documents/final_draft_fac_directive-10.31.2017.pdf.
65. *Physicians for Social Responsibility v. Wheeler*, 956 F.3d 634, 638 (D.C. Cir. 2020).
66. Government Accountability Office. *EPA Advisory Committees – Improvements Needed for the Member Appointment Process*. Report by J. Alfredo Gómez. GAO-19-280 (July 2019). www.gao.gov/assets/710/700171.pdf.
67. M. Hornblower. "Businessmen Launch Drive to Soften Clean Air Rules." *Washington Post*, January 9, 1979.
68. M. Hornblower. "Major Industries Map New Attack on Clean Air Act." *Washington Post*, January 15, 1979.
69. A. C. Stern. "History of Air Pollution Legislation in the United States." *Journal of Air Pollution Control Association* 32, no. 1 (January 1982): 44–61. https://doi.org/10.1080/00022470.1982.10465369.
70. H. J. Feldman, senior director Scientific and Regulatory Affairs, American Petroleum Institute. *Regulatory Reform Task Force's Evaluation of Existing Regulations EPA-HQ-OA-2017-0190 (82 FR 17793)*. Submitted to S. K. Dravis, regulatory reform officer and associate administrator for Environmental Protection Agency Office of Policy. May 15, 2017.
71. H. J. Feldman, senior director Scientific and Regulatory Affairs, American Petroleum Institute. *Appendices to API Comments on Specific Regulations*. Submitted to S. K. Dravis, regulatory reform officer and associate administrator for Environmental Protection Agency Office of Policy. May 15, 2017.
72. "Former CASAC Chair Says Panel Dismissals Will Weaken NAAQS' Legality." *Clean Air Report* (October 18, 2018).
73. *Murray Energy Corp. v. Environmental Protection Agency*, 936 F.3d 597 (D.C. Cir. 2019).
74. L. A. Cox, Jr. *Statements on EPA'S Proposed Ozone Rule: Potential Impacts on Manufacturing*. Submitted to Subcommittee on Energy and Power and Subcommittee on Commerce, Manufacturing, and Trade. June 16, 2015.
75. L. A. Cox, Jr. "Do Causal Concentration–Response Functions Exist? A Critical Review of Associational and Causal Relations between Fine Particulate Matter and Mortality." *Critical Reviews in Toxicology* 47, no. 7 (2017): 609–637. https://doi.org/10.1080/10408444.2017.1311838.
76. US House of Representatives. *EPA's Proposed Ozone Rule: Potential Impacts on Manufacturing*. Subcommittee on Energy and Power; Subcommittee on Commerce, Manufacturing, and Trade; Committee on Energy and Commerce. 114th Cong. 1st Sess. June 16, 2015.
77. L. A. Cox, Jr., and D. A. Popken. "Has Reducing Fine Particulate Matter and Ozone Caused Reduced Mortality Rates in the United States?" *Annals of*

Epidemiology 25, no. 3 (2015): 162–173. www.ncbi.nlm.nih.gov/pubmed/25571792.
78. H. C. Frey et al. *CASAC Review of EPA's Integrated Science Assessment (ISA) for Particulate Matter (External Review Draft – October 2018)*. Submitted to Dr. L. A. Cox, Jr. and the Environmental Protection Agency. December 10, 2018.
79. G. Goldman and F. Dominici. "Don't Abandon Evidence and Process on Air Pollution Policy." *Science* 363, no. 6434 (March 29, 2019): 1398–1400. https://doi.org/10.1126/science.aaw9460.
80. J. Vanderberg. *John Vanderberg Response to Dr. Louis Anthony Cox, Jr.* Submitted to Dr. L. A. Cox, Jr. February 20, 2019.
81. S. Waldman. "Science Adviser Allowed Oil Group to Edit Research." *E&E News*, December 10, 2018. www.eenews.net/stories/1060109129.
82. J. Plautz. "Trump's Air Pollution Adviser: No Proof Cleaning Up Smog Saves Lives." *Reveal from the Center for Investigative Reporting*, October 24, 2018. www.revealnews.org/article/trumps-air-pollution-adviser-clean-air-saves-no-lives.
83. C. H. Frey, P. J. Futrell and G. E. Futrell. *Public Comment on the CASAC Review of EPA's Integrated Science Assessment for Particulate Matter (External Review Draft – October 2018)*. Submitted to Environmental Protection Agency. December 12, 2018.
84. Environmental Protection Network. *Written Comments of John Bachmann on Behalf of the Environmental Protection Network*. Submitted to Environmental Protection Agency Acting Administrator A. Wheeler and Clean Air Scientific Advisory Committee. December 9, 2018.
85. D. J. Boylan et al. *Preliminary Draft Comments from Members of the Clean Air Scientific Advisory Committee (CASAC)*. December 10, 2018.
86. L. A. Cox, Jr., Clean Air Scientific Advisory Committee chair. *CASAC Review of the EPA's Integrated Science Assessment for Particulate Matter (External Review Draft – October 2018)*. Submitted to A. R. Wheeler. April 11, 2019.
87. S. Reilly. "Documents Expose Ties among EPA's Panel Experts." *E&E News*, February 7, 2020. www.eenews.net/stories/1062289617.
88. C. H. Frey. *Context and Charge Questions for October 10–11, 2019 Meeting to Review the EPA Draft Policy Assessment for Particulate Matter*. Submitted to Members of the Independent Particulate Matter Review Panel. September 20, 2019.
89. C. Hogue. "US EPA's Science Advisers Split on Tightening Air Pollution Limit." *Chemical and Engineering News* 97, no. 44 (November 10, 2019): 20–21. https://pubs.acs.org/doi/10.1021/cen-09744-feature2.
90. Independent Particulate Matter Review Panel. *Letter on EPA's Policy Assessment for the Review of the National Ambient Air Quality Standards for Particulate Matter (External Review Draft–September 2019)*. Submitted to A. R. Wheeler, administrator of the Environmental Protection Agency. October 22, 2019.
91. Office of Air Quality Planning and Standards. Policy Assessment for the Review of the National Ambient Air Quality Standards for Particulate Matter. Environmental Protection Agency. EPA-452/R-20-002. Research Triangle Park, NC: 2020. www.epa.gov/sites/production/files/2020-01/documents/final_policy_assessment_for_the_review_of_the_pm_naaqs_01-2020.pdf.
92. *Review of the National Ambient Air Quality Standards for Particulate Matter: Final Action*. 40 Code of Federal Regulations 50. Environmental Protection Agency. 85

Federal Register 82684–82748 (December 18, 2020) www.federalregister.gov/documents/2020/12/18/2020-27125/review-of-the-national-ambient-air-quality-standards-for-particulate-matter.
93. L. A. Cox, Jr. "Re: 'Best Practices for Gauging Evidence of Causality in Air Pollution Epidemiology.'" *American Journal of Epidemiology* 187, no. 6 (March 23, 2018): 1338–1339. https://doi.org/10.1093/aje/kwy034.
94. L. A. Cox, Jr. "Modernizing the Bradford Hill Criteria for Assessing Causal Relationships in Observational Data." *Critical Reviews in Toxicology* 48, no. 8 (January 13, 2018): 682–712. https://doi.org/10.1080/10408444.2018.1518404.
95. K. M. Fedak et al. "Applying the Bradford Hill Criteria in the 21st Century: How Data Integration Has Changed Causal Inference in Molecular Epidemiology." *Emerging Themes in Epidemiology* 12, no. 1 (2015): 1–9. https://doi.org/10.1186/s12982-015-0037-4.
96. F. Dominici and C. Zigler. "Best Practices for Gauging Evidence of Causality in Air Pollution Epidemiology." *American Journal of Epidemiology* 186, no. 12 (2017): 1303–1309. www.ncbi.nlm.nih.gov/pubmed/29020141.
97. D. S. Greenbaum and R. Shaikh. "Air Quality and Human Health: The Role of Health Science in Setting National Ambient Air Quality Standards." *Magazine for Environmental Managers*, December 2018. http://pubs.awma.org/flip/EM-Dec-2018/greenbaum.pdf.
98. A. H. Saiyid. "EPA Advisers Can't Agree on Revising Ozone Limits." *Bloomberg Law*, December 6, 2019. https://news.bloomberglaw.com/environment-and-energy/epa-advisers-cant-agree-on-what-to-do-about-ozone-limits.
99. Clean Air Scientific Advisory Committee. *Consensus Responses to Charge Questions on the EPA's Policy Assessment for the Review of the Ozone National Ambient Air Quality Standards (External Review Draft – October 2019)*. Environmental Protection Agency. Washington, DC: February 19, 2020. https://yosemite.epa.gov/sab/sabproduct.nsf/264cb1227d55e02c85257402007446a4/4713D217BC07103485258515006359BA/$File/EPA-CASAC-20-003.pdf.
100. Members of the Former Clean Air Scientific Advisory Committee Ozone Review Panel (2009–14). *Letter on CASAC Advice on the EPA's Integrated Review Plan for the Ozone National Ambient Air Quality Standards (External Review Draft)*. Submitted to Dr. L. A. Cox, Jr., Chair of Clean Air Scientific Advisory Committee. November 26, 2018.
101. Members of the Former Clean Air Scientific Advisory Committee Ozone Review Panel (2009–14). *Letter on EPA's Integrated Science Assessment for Ozone and Related Photochemical Oxidants and EPA's Policy Assessment for the Review of the Ozone National Ambient Air Quality Standards*. Submitted to A. R. Wheeler, administrator of the Environmental Protection Agency. December 2, 2019.
102. Environmental Protection Agency. *Policy Assessment for the Review of the Ozone National Ambient Air Quality Standards*. Office of Air Quality Planning and Standards, Health and Environmental Impacts Division Research. Research Triangle Park, NC: May 2020. www.epa.gov/sites/production/files/2020-05/documents/o3-final_pa-05-29-20compressed.pdf.
103. *Carbon Pollution Emission Guidelines for Existing Stationary Sources: Electric Utility Generating Units; Proposed Rule*. 40 Code of Federal Regulation 60. Environmental Protection Agency. 79 Federal Regulations 34829–34958

(June 18, 2014). www.govinfo.gov/content/pkg/FR-2014-06-18/pdf/2014-13726.pdf.
104. American Petroleum Institute et al. *Comments on the Carbon Pollution Emission Guidelines for Existing Stationary Electric Utility Generation Units, Proposed Rule (published in the Federal Register 79: 34,830 (June 18, 2014))*. Submitted to Environmental Protection Agency. 2014.
105. Environmental Protection Agency. *Regulatory Impact Analysis for the Proposed Emission Guidelines for Greenhouse Gas Emissions from Existing Electric Utility Generating Units; Revisions to Emission Guideline Implementing Regulations; Revisions to New Source Review Program*. Office of Air Quality Planning and Standards, Health and Environmental Impact Division. Washington, DC: August 2018. www.epa.gov/sites/production/files/2018-08/documents/utilities_ria_proposed_ace_2018-08.pdf.
106. A. Krupnick and A. Keyes. "Hazy Treatment of Health Benefits: The Case of the Clean Power Plan." *Resources for the Future*, October 13, 2017. www.resourcesmag.org/common-resources/hazy-treatment-of-health-benefits-the-case-of-the-clean-power-plan.
107. *National Emission Standards for Hazardous Air Pollutants from Coal and Oil-Fired Electric Utility Steam Generating Units and Standards of Performance for Fossil-Fuel-Fired Electric Utility, Industrial-Commercial Institutional, and Small Industrial Commercial-Institutional Steam Generating Units*. 40 Code of Federal Regulations Parts 60 and 63. Environmental Protection Agency. 77 Federal Register 32 (February 16, 2012). www.govinfo.gov/content/pkg/FR-2012-02-16/pdf/2012-806.pdf.
108. M. Todd, chair of the Residual Risk Coalition. *Comments on "The National Emission Standards for Hazardous Air Pollutant Emissions: Coal-and Oil-Fired Electric Utility Steam Generating Units – Reconsideration of Supplemental Finding and Residual Risk and Technology Review: Proposed Rule."* Submitted to Environmental Protection Agency. Docket ID No. EPA–HQ–OAR–2018–0794. April 17, 2019.
109. *Proposed Rule: National Emission Standards for Hazardous Air Pollutants: Coal- and Oil-Fired Electric Utility Steam Generating Units – Reconsideration of Supplemental Finding and Residual Risk and Technology Review*. 40 Code of Federal Regulations 60. Environmental Protection Agency. 84 Federal Register 2670–2704 (February 7, 2019). www.govinfo.gov/content/pkg/FR-2019-02-07/pdf/2019-00936.pdf.
110. External Environmental Economics Advisory Committee. *Report on the Proposed Changes to the Federal Mercury and Air Toxics Standards* (December 2019). www.belfercenter.org/sites/default/files/files/publication/E-EEAC%20Report%20120320191330.pdf.
111. *Regulation of HAP Emissions from Coal- and Oil-fired Electrical General Units Is Not "Appropriate and Necessary": Final Rule*. 40 Code of Federal Regulations 63. Environmental Protection Agency. 85 Federal Register 31286–31320 (May 22, 2020). www.federalregister.gov/documents/2020/05/22/2020-08607/national-emission-standards-for-hazardous-air-pollutants-coal–and-oil-fired-electric-utility-steam.
112. J. Eilperin and B. Dennis. "The EPA Is about to Change a Rule Cutting Mercury Pollution. The Industry Doesn't Want It." *Washington Post*, February 17, 2020.

www.washingtonpost.com/climate-environment/the-epa-is-about-to-change-a-rule-cutting-mercury-pollution-the-industry-doesnt-want-it/2020/02/16/8ebac4e2-4470-11ea-b503-2b077c436617_story.html.
113. R. Beitsch. "EPA's Independent Science Board Says Agency Ignored Its Advice on Mercury Rule." *The Hill*, December 31, 2019. https://thehill.com/policy/energy-environment/476374-epas-independent-science-board-says-agency-ignored-their-advice-on.
114. Environmental Protection Agency. *Regulatory Impact Analysis for the Final Mercury and Air Toxics Standards* (December 2011). www3.epa.gov/ttnecas1/regdata/RIAs/matsriafinal.pdf.
115. M. Lewis, A. Logomasini and W. Yeatman. "First Steps for the Trump Administration: Champion Affordable Energy." Competitive Enterprise Institute (December 15, 2016). https://cei.org/sites/default/files/First%20Steps%20for%20the%20Trump%20Administration%20-%20Chamption%20Affordable%20Energy.pdf.
116. Office of Air Quality Planning and Standards, Health and Environmental Impact Division, and Air Benefit-Cost Group. *Summary of Expert Opinions on the Existence of a Threshold in the Concentration-Response Function for PM2.5-related Mortality*. Environmental Protection Agency. Washington, DC: 2010. www3.epa.gov/ttnecas1/regdata/Benefits/thresholdstsd.pdf.
117. L. Shi et al. "Low-Concentration PM2.5 and Mortality: Estimating Acute and Chronic Effects in a Population-Based Study." *Environmental Health Perspectives* 124, no. 1 (2015): 46–52. https://doi.org/10.1289/ehp.1409111.
118. M. Makar et al. "Estimating the Causal Effect of Fine Particulate Matter Levels on Death and Hospitalization: Are Levels below the Safety Standards Harmful?" *Epidemiology* 28, no. 5 (2017): 627–634.
119. Y. A. Awad et al. "Change in PM2. 5 Exposure and Mortality among Medicare Recipients: Combining a Semi-randomized Approach and Inverse Probability Weights in a Low Exposure Population." *Environmental Epidemiology* 3, no. 4 (2019). doi:10.1097/EE9.0000000000000054.
120. M. S. Cong Liu et al. "Ambient Particulate Air Pollution and Daily Mortality in 652 Cities." *New England Journal of Medicine* 381 (August 22, 2019): 705–715. www.nejm.org/doi/full/10.1056/NEJMoa1817364?query=featured_home#article_comments.
121. K. M. Castle and R. L. Revesz. "Environmental Standards, Thresholds, and the Next Battleground of Climate Change Regulations." *Minnesota Law Review* 103 (2019): 1349–1437. www.minnesotalawreview.org/wp-content/uploads/2019/02/4Revesz_FINAL.pdf.
122. Office of Transportation and Air Quality. *Final Regulatory Analysis: Control of Emissions from Nonroad Diesel Engines*. Environmental Protection Agency EPA420-R-04-007. Washington, DC: US Environmental Protection Agency, 2004. https://nepis.epa.gov/Exe/ZyPDF.cgi/P100K5U2.PDF?Dockey=P100K5U2.PDF.
123. Natural Resources Defense Council. *Comments of Natural Resources Defense Council on "Strengthening Transparency in Regulatory Science."* Submitted to A. R. Wheeler, acting administrator of the Environmental Protection Agency. Docket ID No. EPA-HQ-OA-2018-0259. August 15, 2018.

124. K. Bennett, Science policy director for Public Employees for Environmental Responsibility. *Comments on the US Environmental Protection Agency's Proposed "Strengthening Transparency in Regulatory Science" Rule.* Submitted to A. R. Wheeler, acting administrator of the Environmental Protection Agency. Docket ID No. EPA-HQ-OA-2018-0259-0001. August 15, 2018.
125. *Strengthening Transparency in Regulatory Science.* 40 Code of Federal Regulations 30. Environmental Protection Agency. 83 Federal Register 18768–18774 (April 30, 2018). www.federalregister.gov/documents/2018/04/30/2018-09078/strengthening-transparency-in-regulatory-science.
126. *Clarifications, Modifications and Additions to Certain Provisions in the Strengthening Transparency in Regulatory Science Proposed.* 40 Code of Federal Regulations 30. Environmental Protection Agency. 85 Federal Register 15396–15406 (March 18, 2020). www.federalregister.gov/documents/2020/03/18/2020-05012/strengthening-transparency-in-regulatory-science.
127. A. R. Wheeler, administrator of the Environmental Protection Agency. *Testimony.* Submitted to House Committee on Science, Space, and Technology. 2019.
128. M. Lavelle. "Pruitt's Own Scientist Appointees Challenge EPA Science Restrictions." *Inside Climate News,* May 17, 2018. https://insideclimatenews.org/news/17052018/scott-pruitt-epa-secret-science-health-fossil-fuel-industry.
129. M. Green and R. Beitsch. "EPA Delays Advisers' Review of 'Secret Science' Rules." *The Hill,* November 18, 2019. https://thehill.com/policy/energy-environment/470968-epa-delays-advisors-review-of-secret-science-rules.
130. C. H. Frey, former member of the Science Advisory Board, G. E. Futrell and P. J. Futrell, *EPA Has a Statutory Responsibility to Use Properly Developed and Reviewed Science.* Submitted to Science Advisory Board, Environmental Protection Agency. June 5, 2019.
131. *United States Code, Title 42*, Section 4365.
132. M. McNutt, president of the National Academy of Sciences, et al. *Strengthening Transparency in Regulatory Science* (Docket ID No. EPA-HQ-OA-2018-0259). Submitted to A. R. Wheeler, acting administrator, Environmental Protection Agency. July 16, 2018.
133. M. Honeycutt, Science Advisory Board chair. *Draft Report: Science Advisory Board (SAB) Consideration of the Scientific and Technical Basis of EPA's Proposed Rule Titled Strengthening Transparency in Regulatory Science.* Submitted to A. R. Wheeler, administrator, Environmental Protection Agency. EPA-SAB-20-xxx. October 16, 2019.
134. J. Eilperin. "EPA's Scientific Advisors Warn Its Regulatory Rollbacks Clash with Established Science." *Washington Post,* January 1, 2020. www.washingtonpost.com/climate-solutions/epas-scientific-advisers-warn-its-regulatory-rollbacks-clash-with-established-science/2019/12/31/a1994f5a-227b-11ea-a153-dce4b94e4249_story.html.
135. J. Ioannidis. "All Science Should Inform Policy and Regulation." *PLoS Medicine* 15, no. 5 (2018). https://journals.plos.org/plosmedicine/article/citation?id=10.1371/journal.pmed.1002576.
136. B. Nosek, co-founder and executive director, Center for Open Science. *Testimony at Hearing on Strengthening Transparency or Silencing Science? The Future of*

Science in EPA Rulemaking, 116th Congress. Submitted to House Committee on Science, Space, and Technology. November 13, 2019.
137. L. Teytelman, W. Gunn, and J. Kamens. "The EPA's Proposed 'Transparency Rule' Will Harm Health, Safety, and the Environment." *Stat News*, December 9, 2019. www.statnews.com/2019/12/09/epa-transparency-rule-bad-for-science-health-safety-environment.
138. W. Thomas. "Science Committee Renews Scrutiny of EPA Science Transparency Rule." *FYI Bulletin*, November 20, 2019. www.aip.org/fyi/2019/science-committee-renews-scrutiny-epa-science-transparency-rule.
139. C. Hiar. "In Battle over Pesticide Ban, Trump's EPA Aims to Undermine the Science." *E&E News*, August 23, 2018. www.sciencemag.org/news/2018/08/battle-over-pesticide-ban-trump-s-epa-aims-undermine-science.
140. "Researchers Must Unite against US Environment Agency's Attack on Scientific Evidence." *Nature* 575, no. 415 (2019). www.nature.com/articles/d41586-019-03526-z.
141. National Academies of Sciences, Engineering, and Medicine. *Reproducibility and Replicability in Science.* Washington, DC: National Academies Press, 2019. doi:10.17226/25303.
142. W. B. Jacobs, Emmett clinical professor of Environmental Law and clinic director. *Comments on Proposed Rule Strengthening Transparency in Regulatory Science.* Submitted to A. Wheeler, acting administrator, Environmental Protection Agency. 2018.
143. M. Honeycutt, Science Advisory Board chair. *Consultation on Mechanisms for Secure Access to Personally Identifying Information (PII) and Confidential Business Information (CBI) Under the Proposed Rule, Strengthening Transparency in Regulatory Science.* Submitted to A. R. Wheeler, administrator, Environmental Protection Agency. EPA-SAB-19-005. September 30, 2019.
144. D. B. Allison and H. V. Fineberg. "EPA's Proposed Transparency Rule: Factors to Consider, many; Planets to Live On, One." *Proceedings of the National Academy of Sciences of the United States of America* 117, no. 10 (March 10, 2020): 5084–5087. www.pnas.org/content/117/10/5084.
145. L. Sweeney. "Only You, Your Doctor, and Many Others May Know." *Technology Science* no. 2015092903 (2015). https://techscience.org/a/2015092903.
146. L. Sweeney et al. "Re-identification Risks in HIPAA Safe Harbor Data: A Study of Data from One Environmental Health Study." *Technology Science* no. 2017082801 (2017). https://techscience.org/a/2017082801.
147. L. Rocher, J. M. Hendrickx and Y.-A. de Montjoye. "Estimating the Success of Re-identifications in Incomplete Datasets Using Generative Models." *Nature Communications* 10, no. 1 (July 23, 2019): 1–9. www.nature.com/articles/s41467-019-10933-3#citeas.
148. National Research Council. *Improving Access to and Confidentiality of Research Data: Report of a Workshop.* Washington, DC: National Academies Press, 2000. https://doi.org/10.17226/9958.
149. L. O. Gostin, L. A. Levit, S. J. Nass. *Beyond the HIPAA Privacy Rule: Enhancing Privacy, Improving Health through Research.* Washington, DC: National Academies Press, 2009. www.ncbi.nlm.nih.gov/books/NBK9578/.

150. T. Sherer. *Testimony on Behalf of the Michael J. Fox Foundation for Parkinson's Research*. Submitted to E. B. Johnson, chairwoman, Committee on Science, Space, and Technology, and F. Lucas, ranking member, Committee on Science, Space, and Technology. November 13, 2019.
151. J. Goodwin. "The EPA's 'Censored Science' Rule Isn't Just Bad Policy, It's Also Illegal." *Union of Concerned Scientists Blog*, November 22, 2019. https://blog.ucsusa.org/guest-commentary/the-epas-censored-science-rule-isnt-just-bad-policy-its-also-illegal.
152. E. B. Johnson, chair of House of Representatives Committee on Science, Space, and Technology. *Opening Statement: Strengthening Transparency or Silencing Science? The Future of Science in EPA Rulemaking*. Submitted to House of Representatives Committee on Science, Space, and Technology. November 13, 2019.
153. D. Michaels. *The Triumph of Doubt: Dark Money and the Science of Deception*. New York: Oxford University Press, 2020.
154. C. H. Christopher, Bracewell & Patterson, LLP. *Background and Proposed Program to Address Federal Agency Science*. Submitted to T. Hyde and R. Johnson, RJ Reynolds Tobacco Company. December 23, 1996.
155. S. Lerner. "Republicans Are Using Big Tobacco's Secret Science Playbook to Gut Health Rules." *The Intercept*, February 5, 2017. https://theintercept.com/2017/02/05/republicans-want-to-make-the-epa-great-again-by-gutting-health-regulations/.
156. Y. Kothari. "Internal EPA Emails Confirm That Scott Pruitt's Secret Science Proposal Is Entirely Driven by Politics." *Union of Concerned Scientists*, April 19, 2018. https://blog.ucsusa.org/yogin-kothari/internal-epa-emails-confirm-that-scott-pruitts-secret-science-proposal-is-entirely-driven-by-politics?_ga=2.73012819.1119798374.1584912229-994550056.1584912229.
157. National Task Force on Rule of Law and Democracy. *Proposals for Reform*. Report by P. Bharara et al. (2019). www.brennancenter.org/sites/default/files/2019-09/2019_10_TaskForce%20II_0.pdf
158. D. Parkes. "Basic Science, Agricultural Research, NASA Would Finish Strong in FY 2019 Omnibus." American Association for the Advancement of Science, February 14, 2019. www.aaas.org/news/basic-science-agricultural-research-nasa-would-finish-strong-fy-2019-omnibus.
159. R. Frelinghuysen, chair, House Committee on Appropriations. *Opening Statement at the Hearing on the FY 2018 Budget for the Environmental Protection Agency*. Submitted to House Committee on Appropriations. June 15, 2017.
160. K. Calvert, chair of the Subcommittee on Interior, Environment, and Related Agencies, House Committee on Appropriations. *Opening Statement at the Hearing on the FY 2018 Budget for the Environmental Protection Agency*. Submitted to House Committee on Appropriations. June 15, 2017.
161. S. Collins. "Senator Collins to Oppose EPA Administrator Nominee's Confirmation." News release, February 15, 2017. www.collins.senate.gov/newsroom/senator-collins-oppose-epa-administrator-nominee%E2%80%99s-confirmation.
162. S. Collins. "Senator Collins to Oppose EPA Administrator Nominee's Confirmation." News release, February 27, 2019. www.collins.senate.gov/news

room/senator-collins-oppose-epa-administrator-nominee%E2%80%99s-confirmation-0.
163. M. Green. "Senate Confirms Wheeler to Lead EPA." *The Hill*, February 28, 2019. https://thehill.com/policy/energy-environment/432033-senate-confirms-wheeler-to-lead-epa.
164. T. Cama. "Two GOP Senators Oppose Trump's EPA Chemical Safety Nominee." *The Hill*, November 15, 2017. https://thehill.com/policy/energy-environment/360615-2-gop-senators-oppose-trumps-epa-chemical-safety-nominee.
165. K. Lydersen. "EPA Toxics Nominee Provided Koch-Funded Study in Chicago Petcoke Battle." *Energy News Network*, September 7, 2017. https://energynews.us/2017/09/07/midwest/epa-toxics-nominee-provided-koch-funded-study-in-chicago-petcoke-battle/.
166. Editorial Board. "Mr. Trump's Conflicted Regulators." *New York Times*, October 18, 2017.
167. Agency for Toxic Substances and Disease Registry and Division of Community Health Investigations. *Perfluoroalkyl and Polyfluoroalkyl Substances (PFAS) Frequently Asked Questions*. Center for Disease Control and Prevention (2017). www.atsdr.cdc.gov/pfas/docs/pfas_fact_sheet.pdf.
168. Z. R. Hopkins et al. "Recently Detected Drinking Water Contaminants: GenX and Other Per-and Polyfluoroalkyl Ether Acids." *American Water Works Association* 110, no. 7 (2018). https://awwa.onlinelibrary.wiley.com/doi/full/10.1002/awwa.1073.
169. T. S. Lee, I. London and J. Kindschuh. "PFAS in the Upstream Oil and Gas Industry." *Lexology*, August 2, 2019. www.lexology.com/library/detail.aspx?g=ca7bf9a5-b0f7-4089-a993-6ca387c47f3f.
170. A. Wittenberg. "After Controversy, US Releases Report Showing Elevated Health Risks from Nonstick Chemicals." *E&E News*, June 20, 2018. www.sciencemag.org/news/2018/06/after-controversy-us-releases-report-showing-elevated-health-risks-nonstick-chemicals.
171. B. Patterson. "Previously Blocked Federal Study Raises Alarm about Chemicals Like C8." *Ohio Valley Resource*, June 20, 2018. https://ohiovalleyresource.org/2018/06/20/previously-blocked-federal-study-raises-alarm-pfas-chemicals/.
172. Department of Health and Human Services. *Toxicological Profile for Perfluoroalkyls – Draft for Public Comment* (2018). www.eenews.net/assets/2018/06/20/document_gw_08.pdf.
173. House of Representatives Committee on Energy and Commerce. *Letter of Concern Regarding Rollback Measures*. Submitted to A. Wheeler. 2019.
174. House of Representatives. *Letter of Concern Regarding NAAQS and CAA*. Submitted to A. Wheeler. 2019.
175. J. Mervis and W. Cornwall. "Lamar Smith, the Departing Head of the House Science Panel, Will Leave a Controversial and Complicated Legacy." *Science*, November 5, 2017. www.sciencemag.org/news/2017/11/lamar-smith-departing-head-house-science-panel-will-leave-controversial-and-complicated.
176. A. C. Leiter. *Reversing Course: Administrative Law in a Time of Change*. American University Washington College of Law (2018). www.americanbar.org/content/dam/aba/administrative/environment_energy_resources/2018/fall/course_materials/8_Leiter.pdf.

177. US Congress. *Letter of Concern on Strengthening Transparency in Regulatory Science Rule*. Submitted to S. Pruitt. 2018.
178. Senate Committee on Environment and Public Works. *A Letter to Administrator Pruitt, EPA*. Submitted to S. Pruitt. 2018.
179. J. Mervis. "Scientific Integrity Bill Advances in US House with Bipartisan Support." *Science*, October 17, 2019. www.sciencemag.org/news/2019/10/scientific-integrity-bill-advances-us-house-bipartisan-support.
180. R. Showstack. "Scientific Integrity Act Passes House Committee." *Eos*, October 18, 2019. https://eos.org/articles/scientific-integrity-act-passes-house-committee.
181. US Congress. Senate. *Scientific Integrity Act*. S. 775, 116th Congress, 1st Sess. Introduced in Senate March 12, 2019.
182. L. M. Krauss. "Trump's Anti-Science Campaign." *New Yorker*, August 21, 2016. www.newyorker.com/news/news-desk/trumps-anti-science-campaign.
183. A. Acrivos et al. *An Open Letter to President-Elect Trump and the 115th Congress on Science and the Public Interest*. Submitted to President D. J. Trump, 115th Congress. November 30, 2016.
184. G. Goldman et al. "Ensuring Scientific Integrity in the Age of Trump." *Science* 355, no. 6326 (2017): 696–698. https://science.sciencemag.org/content/355/6326/696.
185. J. Berg, P. Campbell, V. Kiermer, N. Raikhel and D. Sweet. "Joint Statement on EPA Proposed Rule and Public Availability of Data." *Science* 360, no. 6388 (2018). https://doi.org/10.1126/science.aau0116.
186. H. H. Thorp et al. "Joint Statement on EPA Proposed Rule and Public Availability of Data." *Science* 366, no. 6470 (2019): 2. https://doi.org/10.1126/science.aba3197.
187. Academy of Integrative Health & Medicine et al. "Public Health, Medical, Academic, and Scientific Groups Oppose EPA Transparency Rule." News release, https://mcmprodaaas.s3.amazonaws.com/s3fs-public/EPA%20Transparency%20Rule%20FINAL.pdf?oNbdIjRo8Ick2LxdMeWaqWuYu4NM3unc.
188. D. G. Kirch, president and CEO, AAMC, et al. *Re: Docket Number EPA-HQ-OA-2018-0259-0025, Strengthening Transparency in Regulatory Science*. Submitted to Andrew R. Wheeler, acting administrator of the Environmental Protection Agency. July 11, 2018.
189. N. Oreskes. "Beware: Transparency Rule Is a Trojan Horse." *Nature*, May 22, 2018. www.nature.com/articles/d41586-018-05207-9.
190. A. Powell. "Letter Opposes Possible EPA Shift." *Harvard Gazette*, August 10, 2018. https://news.harvard.edu/gazette/story/2018/08/harvard-officials-hospital-leaders-call-on-epa-to-withdraw-transparency-rule-which-they-say-would-bar-use-of-best-science-data.
191. A. Cullen, chair, SAB Work Group on EPA Planned Actions for SAB Consideration of the Underlying Science. *Preparations for Chartered Science Advisory Board (SAB) Discussions of Proposed Rule: Strengthening Transparency in Regulatory Science RIN (2080-AA14)*. Submitted to Members of the Chartered SAB and SAB Liaisons. 2018.
192. Science Advisory Board. *SAB Review of the Science Supporting EPA Planned Regulatory and Deregulatory Actions*. Submitted to A. R. Wheeler. 2019.

193. J. Bachmann, former associate director for Science/Policy in EPA's Air Office. *Statement of John Bachmann for the Public Meeting of the EPA Chartered Science Advisory Board (SAB), Re: June 5–6 SAB Discussions about EPA Planned Actions and Their Supporting Science.* Submitted to Science Advisory Board. June 5, 2019.
194. M. Honeycutt, Science Advisory Board chair. *Science Advisory Board (SAB) Consideration of EPA Proposed Rule: Strengthening Transparency in Regulatory Science.* Submitted to E. S. Pruitt, administrator of the Environmental Protection Agency. EPA-SAB-18-003. June 28, 2018.
195. M. Honeycutt, Science Advisory Board chair. *Commentary on the Proposed Rule Defining the Scope of Waters Federally Regulated Under the Clean Water Act* Submitted to A. R. Wheeler. EPA-SAB-20-002. February 27, 2020.
196. M. Honeycutt, Science Advisory Board chair. *Science Advisory Board (SAB) Consideration of the Scientific and Technical Basis of the EPA's Proposed Rule titled The Safer Affordable Fuel-Efficient (SAFE) Vehicles Rule for Model Years 2021–2026 Passenger Cars and Light Trucks.* Submitted to A. R. Wheeler, administrator, Environmental Protection Agency. EPA-SAB-20-003. February 27, 2020.
197. Office of Air Quality Planning and Standards. *Policy Assessment for the Review of the National Ambient Air Quality Standards for Particulate Matter, External Review Draft.* Environmental Protection Agency EPA-452/P-19-001. Research Triangle Park, NC: 2019. www.epa.gov/sites/production/files/2019-09/documents/draft_policy_assessment_for_pm_naaqs_09-05-2019.pdf.
198. C. Davenport. "Trump's Environmental Rollbacks Find Opposition Within: Staff Scientists." *New York Times*, March 27, 2020. www.nytimes.com/2020/03/27/climate/trumps-environmental-rollbacks-staff-scientists.html.
199. J. Eilperin, J. Dawsey and B. Dennis. "White House Blocked Intelligence Agency's Written Testimony Calling Climate Change 'Possibly Catastrophic.'" *Washington Post*, June 8, 2019. www.washingtonpost.com/climate-environment/2019/06/08/white-house-blocked-intelligence-aides-written-testimony-saying-human-caused-climate-change-could-be-possibly-catastrophic.
200. L. Friedman. "White House Tried to Stop Climate Science Testimony, Documents Show." *New York Times*, June 8, 2019. www.nytimes.com/2019/06/08/climate/rod-schoonover-testimony.html.
201. R. Schoonover. "The White House Blocked My Report on Climate Change and National Security." *New York Times*, July 30, 2019. www.nytimes.com/2019/07/30/opinion/trump-climate-change.html.
202. US House of Representatives. *Scientific Integrity in Federal Agencies.* Subcommittee on Research and Technology (Committee on Science, Space, and Technology); Subcommittee on Investigations and Oversight (Committee on Science, Space, and Technology) 2019.
203. US House of Representatives. *When Science Gets Trumped: Scientific Integrity at the Department of the Interior.* Subcommittee on Oversight and Investigation, Committee on Natural Resources. July 25, 2019.
204. T. R. Carper et al. *Letter to Trump re, Request for Documents Pertaining to Kevin Chmielewski's Whistleblower Complaint.* Submitted to President D. J. Trump. April 12, 2018.
205. J. Eilperin and B. Dennis. "EPA Watchdog Closes Two Probes into Scott Pruitt's Conduct, Citing His Resignation." *Washington Post*, November 30, 2018. www

.washingtonpost.com/energy-environment/2018/11/29/epa-watchdog-closes-two-probes-into-scott-pruitts-conduct-citing-his-resignation.
206. M. Halpern, deputy director of Center for Science and Democracy, Union of Concerned Scientists. *Testimony at the Hearing on Scientific Integrity in Federal Agencies*. Submitted to the House Committee on Science, Space, and Technology, and Joint Subcommittee on Investigations and Oversight and Subcommittee on Research and Technology. July 17, 2019.
207. A. Rosenberg. "The EPA's Science Restrictions Go from Bad to Worse." *Scientific American*, November 13, 2019. https://blogs.scientificamerican.com/observations/the-epas-science-restrictions-go-from-bad-to-worse/.
208. Union of Concerned Scientists. *Comments in Disagreement with the Environmental Protection Agency's Draft Replacement for the Clean Power Plan, the Proposed Affordable Clean Energy Rule*. Submitted to A. R. Wheeler, acting administrator, Environmental Protection Agency. Docket ID No. EPA-HQ-OAR-2017-0355. October 31, 2018.
209. Union of Concerned Scientists et al. *Comments on Quantifying and Monetizing Greenhouse Gas Emissions in the Environmental Impact Statement for Model Year 2022–2025 Corporate Average Fuel Economy Standards*. Submitted to C. J. Tamm, Fuel Economy Division at the National Highway Traffic Safety Administration. Docket: NHTSA-2017-0069. September 25, 2017.
210. G. Goldman. "Here's One More Political Assault on Public Health." *Scientific American*, June 17, 2019. https://blogs.scientificamerican.com/observations/heres-one-more-political-assault-on-public-health.
211. *The Safer Affordable Fuel-Efficient (SAFE) Vehicles Rule for Model Years 2021-2026 Passenger Cars and Light Trucks: Proposed Rules*. 40 Code of Federal Regulations 85 and 86; 49 Code of Federal Regulations 523, 531, 533, 536 and 537. National Highway Traffic Safety Administration; Environmental Protection Agency. 83 Federal Register 42986–43500 (August 24, 2018). www.federalregister.gov/documents/2018/08/24/2018-16820/the-safer-affordable-fuel-efficient-safe-vehicles-rule-for-model-years-2021-2026-passenger-cars-and.
212. *The Safer Affordable Fuel-Efficient (SAFE) Vehicles Rule for Model Years 2021–2026 Passenger Cars and Light Trucks: Final Rule*. 49 Code of Federal Regulations Parts 523, 531, 533, 536 and 537. Environmental Protection Agency and National Highway Traffic Safety Administration. 85 Federal Register 84: 24174–25278 (April 30, 2020). www.govinfo.gov/content/pkg/FR-2020-04-30/pdf/2020-06967.pdf.
213. A. M. Bento et al. "Flawed Analyses of US Auto Fuel Economy Standards." *Science* 362, no. 6419 (December 7, 2018): 1119–1121. https://doi.org/10.1126/science.aav1458.
214. E. Knickmeyer. "EPA Challenged Safety of Administration Mileage Freeze." *Associated Press*, August 14, 2018. https://apnews.com/1a7551fca3294ec49029b93e994cd7f9.
215. M. Joselow. "Researchers Decry 'Misrepresented' Findings in Fuel-Efficiency Rollback Plan." *E&E News*, December 7, 2018. www.scientificamerican.com/article/researchers-decry-misrepresented-findings-in-fuel-efficiency-rollback-plan.
216. H. Tabuchi. "The Oil Industry's Covert Campaign to Rewrite American Car Emissions Rules." *New York Times*, December 13, 2018. www.nytimes.com/

2018/12/13/climate/cafe-emissions-rollback-oil-industry.html?auth=login-google.
217. H. Tabuchi. "Climate Change Denialists Dubbed Auto Makers the 'Opposition' in Fight over Trump's Emissions Rollback." *New York Times*, July 2, 2019. www.nytimes.com/2019/07/02/climate/climate-deniers-auto-emissions-rollback.html.
218. *Strengthening Transparency in Pivotal Science Underlying Significant Regulatory Actions and Influential Scientific Information: Final Rule*. 40 Code of Federal Regulations 30. Environmental Protection Agency. 86 Federal Register 469–473 (January 6, 2021). www.govinfo.gov/content/pkg/FR-2021-01-06/pdf/2020-29179.pdf.
219. *Supplemental Finding That It Is Appropriate and Necessary to Regulate Hazardous Air Pollutants from Coal- and Oil-Fired Electric Utility Steam Generating Units*. 40 Code of Federal Regulations Part 63. Environmental Protection Agency. 81 Federal Register 79 (April 25, 2016).
220. *Massachusetts v. EPA*, 549 US 497 (Supreme Court 2007).
221. *Michigan v. Environmental Protection Agency*, 576 US 743 (Supreme Court 2015).
222. *Strengthening Transparency in Pivotal Science Underlying Significant Regulatory Actions and Influential Scientific Information*. 40 Code of Federal Regulations 30. Environmental Protection Agency. 86 Federal Register 469–493 (January 6, 2020). www.federalregister.gov/documents/2021/01/06/2020-29179/strengthening-transparency-in-pivotal-science-underlying-significant-regulatory-actions-and.
223. *Environmental Defense Fund v. EPA*, 515 F. Supp. 3d 1135 (D. Mont. 2021).

9
Economics
Skewing Analyses to Justify Weaker Regulations

The economic impacts of regulations have become influential talking points in the debate over deregulation. The oil and gas industry asserts that regulations have imposed undue compliance costs on its operations and that economic benefits from health and environmental protections are uncertain at best. Conversely, public health and environmental advocates point to studies that underscore how economic benefits far exceed the compliance costs from safeguarding public health, the environment and other sectors that suffer from pollution.

In 1981, President Ronald Reagan exalted the use of formal cost–benefit analysis, that is, the explicit weighing of benefits and costs, in determining whether to promulgate regulations. Reagan issued an executive order mandating federal agencies to present the costs and benefits of proposed *major* regulations (i.e., those regulations whose estimated impacts on the economy exceed $100 million). The order also mandated regulatory agencies, *to the extent permitted by law*, to propose regulations only when their benefits exceed their costs. Federal environmental laws make clear when regulatory agencies must set regulations based only on public health considerations and when they must consider the economic impacts of proposed regulations.

Economists who support the consideration of economic benefits and costs of proposed regulations see this process as a tool to give full consideration not only to the costs of compliance for the regulated industries but also to the economic benefits of regulations, in the form of benefits to public health, the environment, the climate and economic sectors that suffer from pollution emitted by the regulated sectors.[1] To achieve comprehensive accounting of the economic benefits and costs, first, the analytical methods must capture benefits and costs as completely as possible and, second, the assumptions underlying the analyses must consider impacts on both current and future generations.

The Trump administration, departing from past practices of both Republican and Democratic administrations, applied controversial methods and assumptions to regulatory impact assessments and thus understated the economic benefits from health and environmental regulations. The administration proceeded to declare disingenuously that these regulations inflicted net costs to society and to repeal regulations, such as curbs on the industry's emissions of methane, a potent greenhouse gas. The administration's application of these methods and assumptions establishes precedents in the rulemaking process and paves the way for extensive deregulation. This approach thumbs the scales against all regulations, which is a far more potent strategy than the dismantling of regulations one at a time.

The American Petroleum Institute and industry-funded think tanks, such as the Heritage Foundation, have advocated these changes in the methods and assumptions for assessing the economic impacts of regulations. With many former members of these think tanks joining the EPA transition and management teams (discussed in Chapter 9), it may come as little surprise that the EPA implemented these strategies. These methods ignore the public health benefits from further reductions in air pollution when the ambient concentrations of these pollutants are relatively low. Another ignored category is public health benefits from the reduction of other pollutants that result from the regulation of a target pollutant. Finally, these methods ignore the public health and environment benefits that are difficult to capture in monetary terms. The industry has also advocated assumptions that result in undervaluing the benefits and costs to future generations and in ignoring the costs of climate change to non-Americans. The lack of appreciation of the well-being of future generations and of non-Americans are controversial ethical choices.

I begin with an overview of the economic analysis of regulations (Section 9.1). Then I delve into the oil and gas industry's and the administration's distortion of economic analyses and the subsequent rollback of regulations (Sections 9.2, 9.3 and 9.4) and the pushback against the administration's actions (Section 9.5). I conclude by noting how, despite the opposition, the administration pressed on with these analyses, setting the stage for legal challenges against these rollbacks.

9.1 The Economic Analysis of Regulations

US laws and their subsections specify the extent to which consideration of economic benefits and costs is required or prohibited in the promulgation of regulations. On the one hand, for example, the Clean Air Act section 109a specifies that national ambient air quality standards (NAAQS) be set to protect human health, with an adequate margin of safety benefits.[2] The weighing of benefits and costs is prohibited in setting these ambient air quality standards. The Supreme Court underscored this point in *Whitman v. American Trucking Association* in 2001.[160,i]

On the other hand, for example, the Clean Air Act section 112 (on the setting of air pollution emissions standards) mandates the consideration of compliance costs, alongside health and environmental impact and energy requirements.[2] In *Michigan v. EPA*,[3] the Supreme Court ruled that the EPA's setting of these standards must include a consideration of costs. Nevertheless, the court clarified that the EPA is not required to undertake an explicit cost–benefit analysis. Nor did that decision require the EPA to propose regulations based on weighing benefits versus costs.[4]

President Reagan's 1981 executive order, which mandated cost–benefit analyses for major rules, added another layer of complexity in rulemaking.[2] This order, while recognizing that "some costs and benefits are difficult to quantify," states that *to the extent permitted by law*, an agency should "propose or adopt a regulation only upon a reasoned determination that the benefits of the intended regulation justify its cost."[2] (President Bill Clinton's update to the executive order in 1993 emphasizes that agencies must consider benefits and costs that cannot be quantified.)[5]

Many public health and environmental advocates oppose adopting public health and environmental regulations only when their benefits exceeds costs, preferring to narrow the use of economic analysis to explore the feasibility of proposed public health protections or to explore the cost-effectiveness of alternative regulatory options.[4,6,7] Nevertheless, a number of economists have made pragmatic arguments for the use of cost–benefit analysis, alongside the consideration of the distribution of benefits and costs, as one tool to guide rulemaking.[8] They note that quantifying benefits to public health and ecosystems and assigning dollar values to these benefits places these benefits on the same monetary yardstick as compliance costs. This gives public health and environmental concerns more equal footing in regulatory debates.[8] Conversely, not assigning a dollar value would effectively equate to giving these concerns

---

[i] Congress expressly prohibits the use of formal cost–benefit analysis as a means of assessing regulations in several areas of public health. For instance, the 2016 reform to the 1976 Toxic Substances Control Act (TSCA) mandates that the EPA base its decisions on the safety of chemicals solely on the risks to human health and the environment. That revision was in part in response to the 1991 Fifth Circuit Court of Appeals decision in *Corrosion Proof Fittings v. EPA* that struck down the EPA's ban on asbestos. The court ruled that the EPA's cost–benefit analysis, which was required for its decisions to restrict chemicals under the 1976 TSCA, suffered from errors [156, 157][161].

zero value. Even so, experts caution that numerous ethical judgements[ii] lie hidden within the methods used and the assumptions applied in cost–benefit analysis.[6,7] Subsequently, economists developed techniques to place dollar values on public health benefits from regulations and on the services provided by healthy ecosystems.[9,10]

Industries largely embraced the Reagan administration's cost–benefit approach.[11] Compliance costs are more easily measured, while benefits to public health and the environment are more difficult to quantify, let alone be expressed in monetary terms, and thus they are more likely to be undercounted.[6,12] The oil and gas industry has long asserted that compliance costs to the industry are enormous, while economic benefits to public health and the environment are too uncertain. It has asserted these views in opposing air pollution regulations since the 1950s (when cities began to tackle smog and when federal air pollution regulations began to take shape)[13,14] and in fanning the anti-regulatory sentiment during the Trump administration.[14]

Contrary to the industry's assertions, a number of studies report that major public health and environmental regulations yield *net benefits* to society.[iii] As mandated by the 2000 Regulatory Right-to-Know Act, the Office of Management and Budget (OMB) (the office that serves the White House in supervising its administration of federal agencies) prepare reports summarizing the economic benefits and costs of major regulations. OMB's December 2019 report notes that for major US regulations promulgated from 2006 through 2016, their estimated benefits far exceed their estimated costs. Based on the regulatory impact assessments conducted by agencies prior to proposing and finalizing regulations, estimated benefits totaled between $287 billion and $911 billion, while estimated costs totaled between $78 billion and $115 billion.[15] The 2013 OMB report, which reviewed major regulations promulgated under the Clinton, Bush and Obama administrations from 2002 through 2012, reported similar patterns of estimated benefits far exceeding estimated costs.[16]

While companies certainly face costs in complying with regulations, studies also find the initial estimates of compliance costs often exceed the actual costs.[17–20] These ex ante estimates do not anticipate the reductions in compliance costs that arise from innovations by the regulated industry, by third parties or in regulatory strategies (such as tradable pollution permits).

[ii] Analysts conducting cost–benefit analysis make a number of ethical judgements, such as what costs and benefits are important enough to be considered and the extent to which distributional impacts across subpopulations and across generations are analyzed [7].

[iii] For instance, a peer-reviewed study on the regulatory program to reduce acid rain (caused by emissions from coal power plants) reports that annual benefits of $100 billion would dwarf the annual costs of $3 billion by the time the program was to be fully implemented in 2010 [158].

For example, in the 1970s, the oil industry opposed regulations mandating the phased removal of lead from gasoline, claiming that the high compliance costs would render gasoline unaffordable.[21] Yet, history attests to the successful phase out of lead in gasoline and no calamity to the growth of gasoline-powered cars. Studies of the lead phase out reveal that compliance costs were far less than initially estimated and that benefits outweighed costs by 10 to 1.[22] Another example is the opposition by a subset of oil and gas companies to regulations, promulgated in November 2016, mandating regular inspections of their operations to detect methane leaks. These companies warned of high compliance costs, yet subsequent analysis revealed that they overstated the costs by about four times the amounts charged by third party companies providing the equipment and services for undertaking inspections.[23]

As evidence continues to stack up on how regulations on the oil industry yield net benefits to society, the industry and its funded think tanks have advocated strategies to understate those benefits.[24,25]

9.2 How the Trump Administration Distorted the Economic Analysis of Regulations

Just as the Trump administration culled independent scientists from advisory bodies tasked with reviewing the science at regulatory agencies, it also removed independent economists from advisory boards. The George H. W. Bush administration had set up the Environmental Economics Advisory Committee to provide independent advice on economic analyses to the EPA administrator through the Science Advisory Board. Independent economists who served on the committee convened expert meetings to delve into the methods used and the assumptions applied in economic analyses and to provide scrutiny and feedback on the EPA's regulatory impact analyses. However, in June 2018, the Trump administration eliminated the Environmental Economics Advisory Committee.[26]

Academic economists set up an independent advisory body, the External Environmental Economics Advisory Committee, to provide analysis of the benefits, costs and design of environmental policies. Their findings, made available to the public, were ignored by the administration. Former advisory board members scrutinized cost–benefit analysis published by the administration and came to the following conclusion: "The pattern is clear: when environmental regulations are expected to provide substantial public benefits, assumptions are made to substantially diminish their valuations."[27]

The administration made two overt attempts to distort the norms of economic analysis. The first of these was Trump's January 2017 Executive Order 13771 that directed agencies to rescind two existing regulations of comparable costs for every new regulation issued. This unconventional approach to rulemaking focuses on costs of regulations without any acknowledgement of the benefits – departing from even Reagan's approach considering both benefits and costs of regulations. It contrasts with the standard economic analysis that would support the promulgation of regulations that yield net benefits.[28] Given the administration's focus on the weakening of regulations and its general aversion to new regulations, it was never in a position to apply this executive order.[iv] The second of these was the EPA's publication in June 2018 of an advanced notice of proposed rulemaking for a rule that would revise the agency's consideration of costs and benefits.[29] In December 2020, the EPA finalized the Cost–Benefit Rule, which would have paved the way for the administration to institute controversial methods and assumptions, detailed below, in all cost–benefit analyses of proposed regulations and proposed rescissions of regulations pertaining to the Clean Air Act.[30] (In 2021, Biden revoked the two-for-one executive order and the EPA rescinded the Cost–Benefit Rule.[31,32,v])

Even without the formal revisions discussed above, the Trump administration had considerable success in altering the norms of economic analysis on a case-by-case basis – particularly in distorting economic analyses to justify the rescission of previous analyses, the weakening of existing rules or the repeal of former rules altogether. This approach could have far-reaching consequences by setting the precedent for such approaches in future rulemaking. The consequences of weakening these regulations, a subset summarized in Table 9.1, are detailed in Chapter 8.

9.2.1 Ignoring Benefits of Reducing Air Pollutants at Low Concentrations

Economic analysis of regulations under past administrations accounted for the full economic benefits from reducing air pollution to below the NAAQS.[19,37] The Clean Air Act mandates the EPA to set these standards to protect public health with "an adequate margin of safety." While the

---

[iv] In April 2020, US District Court for the District of Columbia ruled that California, Oregon and Minnesota lacked legal standing to challenge the directive because they could not show it directly harmed them. The judge ruled that "[the states] have not shown that ... the two-for-one rule ... caused the relevant agency to act or to decline to act" [159].

[v] In January 2021, Biden issued Executive Order 13990, which revoked Trump's executive order. In May 2021, the EPA issued an interim rule that rescinded the Cost–Benefit Rule.

Table 9.1 The Trump administration's departures from standard methods and assumptions, applied by past administrations, in regulatory impact analysis

Regulatory rollback	Departures from standard methods used in economic analysis	Departures from past assumptions applied in economic analysis
Rescission of the 2016 Methane Waste Prevention Rule on public lands[33] The rule aimed to reduce methane emissions from oil and gas operations on public lands.	• Gave inadequate emphasis to the duty of federal agencies to manage public sources, i.e., to reduce the waste of natural gas on public lands and the loss of royalty revenue to the federal and state governments.	• Applied high discount rates, thus giving little weight to the interest of future generations. • Applied the domestic social cost of carbon instead of global social cost of carbon.
Rescission of the Clean Power Plan[34] The rule aimed to shift electricity generation from coal to natural gas and renewable energy; the rule caps the share of natural gas generation. Carbon dioxide is the target pollutant and particulate matter is a co-pollutant.	• Ignored benefits from reducing air pollutants at levels below the NAAQS. • Ignored benefits from reducing air pollutants at levels below the lowest detectable level.	• Applied the domestic social cost of carbon instead of global social cost of carbon. • Applied high discount rates, thus giving little weight to the interest of future generations.
Rescission of the Reanalysis of the Mercury and Air Toxics Standards[35] The rule aims to cut mercury emissions from coal-fired power plants.	• Focused on direct benefits that can be expressed in monetary terms. • Downplayed benefits that are difficult to express in monetary terms.	• Not applicable.

Mercury is the target pollutant and particulate matter is a co-pollutant. (The EPA rescinded the reanalysis, not the rule.)

- Ignored co-benefits from the reduction of particulate matter.

Sources: Environmental Protection Agency, *Repeal of the Clean Power Plan; Emission Guidelines for Greenhouse Gas Emissions from Existing Electric Utility Generating Units; Revisions to Emission Guidelines Implementing Regulations: Final Rule*, 2018;[34] Environmental Protection Agency, *National Emission Standards for Hazardous Air Pollutants*, 2020;[35] Bureau of Land Management, *Waste Prevention, Production Subject to Royalties, and Resource Conservation: Final Rule*, 2016.[36]

Notes: The American Petroleum Institute supported the rollback of the Clean Power Plan and the Mercury and Air Toxics Standards, even though these regulations target other industries, for reasons described in Chapter 8.

EPA sets the standards based on an assessment of the level that protects human health, its makes clear that the standards chosen do not eliminate all health risks.[38] For example, when the EPA chose a specific standard for particulate matter, it explicitly acknowledged that a tighter standard than that chosen would yield additional health benefits.[vi]

Past administrations, including the George W. Bush administration, acknowledged the benefits from reducing the level of particulate matter to below these standards. The EPA held this position in its analysis of regulations to control emissions from nonroad diesel vehicles[39] and in its analysis of proposed regulations to reduce emissions of sulfur dioxide and nitrogen oxide from power plants.[40] Both regulations reduced the amount of particulate matter. This approach reflects the consensus among scientific experts on the matter, discussed in Chapter 8, on the absence of a threshold below which air pollutants exert no adverse effects on human health.[41–43]

The American Petroleum Institute and industry-funded think tanks contradict this scientific consensus in their testimonies and comments on regulations.[44–46] In addition, they claim that reducing air pollution below the NAAQS cannot yield benefits because those standards were selected based on levels that would protect human health.[24,47] While compliance with existing air pollution regulations certainly does lower health risks, their argument is factually incorrect. As discussed above, even when the EPA selects a specific standard such as the NAAQS, the population still face risks from pollutants at those ambient concentrations.

The institute and allied think tanks used this argument to oppose the Obama administration's analysis of the Clean Power Plan.[47,48] Under the Trump administration, the EPA published analysis that used the incorrect logic above to justify its rescission of the Clean Power Plan in June 2019, a move supported by the American Petroleum Institute. The EPA was able to claim that the rescission of the rule yielded net benefits (and thus argue that the rescission was the correct way to proceed) only when it assumed incorrectly that reducing particulate matter at levels below the NAAQS yielded no benefits.[49] On the contrary, when the full benefits were included, the EPA's own analysis showed that the rescission of the rule would yield net costs to society.[49] Table 9.2 summarizes how the EPA flipped the conclusions of the cost–benefit analysis, from showing that the rescission would impose net costs on society to the opposite effect, by ignoring the benefits from reducing particulate matter ambient concentrations below the NAAQS.

[vi] For example the EPA chose the 15 micrograms per cubic meter ($\mu g/m^3$) standard for particulate matter, even when it noted that reducing the ambient standards from 15 $\mu g/m^3$ to 14 $\mu g/m^3$ could have prevented an additional 1,900 deaths, 3,700 heart attacks, 5,700 cases of acute bronchitis, 2,000 emergency rooms visits by asthmatic children and 200,000 lost work days [38].

Table 9.2 Repealing the Clean Power Plan: Changing conclusions by ignoring benefits

	3% discount rate	7% discount rate
(A) Full accounting of benefits from reducing levels of PM$_{2.5}$ Benefits are fully accounted (From table I-14 in EPA 2017)	Net costs of $6.3–$30.6 billion.	Net costs of $2.5–$24.6 billion.
(B) Benefits from reducing levels of PM$_{2.5}$ are understated Benefits are assumed to be zero below the NAAQS for PM$_{2.5}$ (From table I-16 in EPA 2017)	Net benefits of $7.1–$10.4 billion.	Net benefits of $9.4–$12.6 billion.

Source: Environmental Protection Agency, *Regulatory Impact Analysis for the Review of the Clean Power Plan*, 2017.[49]
Note: The conclusion of the cost–benefit analyses on repealing the Clean Power Plan flipped from (A) yielding net costs when health benefits are fully counted to (B) yielding net benefits when health benefits are ignored.

9.2.2 Ignoring Co-benefits from Reducing Non-target Pollutants

A regulation targeting one pollutant can result in the reduction of other pollutants and therefore yield additional health benefits.[50] The benefits from the reduction of these other pollutants are known as co-benefits. Likewise, a regulation that affects an industry can unintentionally raise other risks and therefore lead to indirect yet associated costs.[50]

Ignoring co-benefits directly conflicts with policy guidance issued in 2003 by the OMB. A guidance document, published during the G. W. Bush administration and still in effect, calls for the accounting of *all* benefits and *all* costs.[51] The document directs federal agencies "to look beyond the direct benefits and direct costs ... to consider any important ancillary benefits and countervailing risks."[51] Court decisions have also supported the inclusion of co-benefits and indirect costs in cost–benefit analyses.[37,52] For example, in 2016, the Court of Appeals for the DC Circuit ruled in *United States Sugar Corp. v. EPA*[53] that the EPA had acted within its powers under the Clean Air Act when it considered the co-benefits from reducing other pollutants in setting the standards for the target pollutant.

Various cost–benefit analyses of regulations under past administrations have taken into account co-benefits.[37,38,52] For example, the George H. W. Bush administration considered the co-benefits of reducing methane when setting standards on landfills. Likewise, the George W. Bush administration considered these co-benefits of particulate matter and ozone in promulgating a regulation on hazardous air pollutants from mobile sources.[37]

In opposition to the full consideration of co-benefits, the American Petroleum Institute and the industry-funded think tanks have advanced two arguments. [24,47,54,55] First, they argue that the EPA's decision on a proposed regulation should be based on the benefits from the primary pollutant being regulated and should not weigh co-benefits equally. Second, they argue that the EPA's regulation of particulate matter under section 110 of the Clean Air Act (on ambient air pollution) precludes it from pursuing indirect reductions of particulate matter under other sections of the Clean Air Act.

Neither science nor law support these arguments. First, an action can yield benefits from reducing target pollutants and additional benefits from reducing other pollutants. Obtaining two benefits for the price of one action is a testament to cost-effectiveness of that action – it does nothing to discredit the benefits created. Second, as a matter of law, the Clean Act Air does not limit the EPA to the direct targeting of pollutants to achieve pollution reductions.[38] The legislative history of the Clean Air Act indicates congressional intent for the EPA to consider benefits from reducing other pollutants when setting standards for the target pollutant.[38] Contrary to the industry's assertions, the Clean Air Act's legislative history and the OMB's instructions on the comprehensive accounting of benefits and costs support the full inclusion of co-benefits in cost–benefit analysis.

Even though ignoring co-benefits defies science, law and economic logic, the Trump administration did exactly that. As described above, its rescission of the Clean Power Plan ignored co-benefits from reducing particulate matter below the NAAQS. The administration went further to ignore the co-benefits from particulate matter *altogether* in its reanalysis of the Mercury and Air Toxics Standards.[35] By ignoring the annual co-benefits of $36–$89 billion, which dwarfed the compliance costs of $9.6 billion, the administration was able to draw the conclusion that those regulations imposed net costs to society. Table 9.3 reveals how these conclusions were flipped by ignoring co-benefits. The American Petroleum Institute supported the Trump administration's exclusion of co-benefits in its reanalysis of the mercury rules.[56]

Table 9.3 *Reanalysis of the Mercury and Air Toxics Standards Rule*

Version of analysis	The 2016 reanalysis	The 2019 reanalysis
Net benefits/costs	Benefits exceed costs	Costs exceed benefits
Benefits		
Monetized benefits from the reduction of non-target pollutants, i.e., from particulate matter	$36–$89 billion annually	Ignored and effectively set to zero
Monetized benefits from the reduction of the target pollution (hazardous air pollutants)	$4–$6 million annually	$4–$6 million annually
Non-monetized benefits	Considered alongside monetized benefits	Sidelined relative to monetized benefits from target pollutants
Costs		
Projected compliance costs	$9.6 billion annually	$9.6 billion annually

Sources: Environmental Protection Agency, *Supplemental Finding That It Is Appropriate and Necessary to Regulate Hazardous Air Pollutants from Coal- and Oil-Fired Electric Utility Steam Generating Units*, 2016;[57] Environmental Protection Agency, *National Emission Standards for Hazardous Air Pollutants*, 2019.[35]

Of note, the administration did not consistently ignore co-benefits. It cherry-picked circumstances in which to include co-benefits when doing so lent support to its replacement rule.[58] The Trump EPA promulgated the Affordable Clean Energy Rule in June 2019 to replace the Clean Power Plan.[vii] Its regulatory impact assessment of the Affordable Clean Energy Rule included co-benefits from reducing particulate matter. That inclusion was necessary to arrive at the conclusion that the rule yielded net benefits between $1.1 billion and $8.8 billion.[58,59] Had co-benefits been excluded, the rule would yield net costs between $910 million and $980 million.[58,59]

9.2.3 Ignoring Benefits That Are Not Expressed in Monetary Terms

Cost–benefit analyses are often systematically biased against regulations because of the tendency for decision-makers to focus on monetized benefits and costs. In reality, significant health and environmental benefits to ecosystems are not easily monetized. Meanwhile, compliance costs are often overstated.[6,60,61] To guard against this tendency, an EPA guidance document[62] notes that "all meaningful benefits and costs are to be included in all of the tables, even if they cannot be quantified or monetized. Not only does this provide consistency for the reader, but it also maintains important information on the context of the quantified and monetized benefits."[62] To capture non-monetized benefits, analysts have pursued a number of strategies, such as documenting health and ecosystem benefits that cannot be expressed in monetary terms.[63]

The EPA under past administrations acknowledged that significant benefits from regulations cannot be monetized and proceeded with promulgating regulations. For example, the EPA under the G. W. Bush administration underscored the urgency of limiting emissions of mercury from power plants, primarily coal-fired power plants, even when benefits were difficult to monetize. Its 2004 analysis concluded that non-monetized benefits were "large enough to justify substantial investment in emission reductions."[64] In 2016, a reanalysis of the Mercury and Air Toxics Standards rule (ordered by the Supreme Court in *Michigan v. EPA*[3] in response to the original rule promulgated in 2012) concluded that the rule's benefits exceeded its costs.[52]

In contrast to past administrations, the Trump administration focused its attention on benefits that could be monetized and gave short shrift to the benefits from mercury reduction that could not be expressed in monetary terms.[65] The

---

[vii] The Affordable Clean Energy Rule was vacated by the Court of Appeals of the DC Circuit in January 2021 (Chapter 3).

EPA was able to conclude that the Mercury Rule imposed net costs on society by limiting its focus to monetized benefits and additionally by ignoring co-benefits from particulate matter. The EPA concluded that "the total cost of compliance with MATS [Mercury and Air Toxics Standards] ($7.4 to $9.6 billion annually) vastly outweighs the *monetized* HAP [hazardous air pollutants] benefits of the rule ($4 to $6 million annually)."[66] The analysis further points to "the gross disparity between *monetized* costs and HAP benefits, which we believe to be the *primary* focus of the Administrator's determination in CAA section 112(n)(1)(A)."[66] The EPA, in contradiction to its assessment during the Bush administration, asserted that non-monetized benefits were less than the monetized compliance costs.[64] No basis for this conclusion was ever put forth.

Adding insult to injury, the EPA's focus on monetized benefits was further limited to lost earnings resulting from neurological damage among children born to a very narrow set of recreational fishers. The EPA excluded the monetary benefits of averting mercury's adverse effects on cardiovascular health, contrary to the recommendation of a scientific panel.[67,68] It also ignored the findings of a peer-reviewed study that projected that the rule would yield $43 billion in cumulative benefits (from 2016 through 2050).[69]

In April 2020, the EPA announced its rescission of the 2016 analysis of the Mercury and Air Toxics Standards but not the rule.[35] Unsurprisingly, the oil and gas industry wrote in support of that rescission.[56] The industry stands to benefit from the EPA's shift to consider primarily monetized benefits from the target pollutant and to ignore non-monetized benefits that often make up a huge share of the health and environmental benefits from regulations. Notably, several associations of power companies that were directly targeted by the rule and that had already invested in compliance measures supported keeping the rule.[70] The Edison Electric Institute, which represents a significant share of US power companies, emphasized that the repeal introduces "new uncertainty and risk for companies" that had installed the control technologies.[71] That repeal also opens the rule to legal challenges[71,72] and, importantly, institutes the precedent for ignoring non-monetized benefits in rulemaking.

9.3 How the Trump Administration Chose Assumptions That Understate Benefits from Regulations

Alongside departing from standard methods in economic analysis of regulations, the administration also departed from assumptions applied by past administrations. The departures from past assumptions can be seen most starkly in the administration's new estimates for the long-term climate costs of greenhouse gas emissions (Section 9.3.1).

The administration's estimates highlight the controversial ethical choices underlying assumptions used in economic analysis. The first ethical dilemma is the choice of discount rates, which serve to express benefits and costs that occur in the future in terms of today's dollars (Section 9.3.2). A positive discount rate places greater value on a dollar spent or received today than a dollar spent or received in the future. How should society weigh the interests of future generations relative to that of the present generation?

The second ethical dilemma is whether the United States should take into account only its own welfare or include that of other nations when its actions affect the entire world (Section 9.3.3). Should the United States, in evaluating its regulations and policies, consider only the costs of climate change that accrue domestically, or should it consider costs to the global community?

9.3.1 The Social Cost of Carbon

In 2007, the Court of Appeals for the Ninth Circuit in *Center for Biological Diversity v. National Highway Traffic and Safety Administration*[73] ruled that the federal government must account for the costs and benefits from changes in greenhouse gas emissions in its rulemaking. The court, while acknowledging the challenge in quantifying the benefit of cutting these emissions, noted that the value is "certainly not zero."

Federal agencies developed the social cost of carbon metric to capture the monetary cost of long-term climate damages resulting from the addition of one ton of carbon dioxide to the atmosphere in a given year. The social cost of methane is the analogous metric for methane emissions. Accounting for these costs is important for a range of regulations, from those directly regulating greenhouse gas emissions – such as curbs on methane emissions from oil and gas operations – to those that have a substantial but indirect effect on reducing greenhouse gas emissions – such as fuel efficiency standards for gasoline-powered cars and light trucks. Federal agencies applied these measures to 60 final rulemakings between 2008 and 2016.[74]

Accounting for these costs is also critical for projects that require federal permits and that directly affect US greenhouse gas emissions, such as the construction of interstate gas pipelines. For instance, in August 2017, the US Court of Appeals for the District of Columbia ruled in *Sierra Club v. Federal Energy Regulatory Commission*[75] that FERC had to consider the impact of greenhouse gas emissions that would result from the construction of three new interstate pipelines in the southeast.

In 2010, the Interagency Working Group, convened by the Obama administration and which comprises numerous federal agencies, used climate models and economic assessments to estimate the social cost of carbon.[76,77] Three discount rates (2.5, 3 and 5 percent) were used to express the costs of future climate damages in today's dollars.[76,77] That work was positively reviewed by the Government Accountability Office[74] and the National Academies of Sciences, Engineering, and Medicine.[1] The National Academies supported the working group's use of low discount rates and its inclusion of the global costs of climate change.[1] Nevertheless, they underscored that the climate models used are likely to understate the costs from climate disruption and recommended continued work on these assessments.[1]

The American Petroleum Institute and the oil industry-funded think tanks oppose the use of the social cost of carbon.[78–81] A number of these groups still publicly cast doubt on climate science. They claim that the process used by the Interagency Working Group was not transparent, a claim that runs counter to the report by the Government Accountability Office. That report had noted that the working group followed consensual decision-making, reviewed the academic literature and models, and updated its work based on public comments and new information.[74] The oil industry and its think tanks also criticized the working group's use of low discount rates and its inclusion of the global costs of climate change, as discussed below.

In 2017, President Trump signed Executive Order 13783,[82] which discarded the work by the Interagency Working Group as no longer representing government policy.[viii] The EPA and BLM developed interim estimates for the social costs of carbon and methane, using the same climate and economic assessment models as the Interagency Working Group. However, as seen in Table 9.4, the Trump administration drastically deflated the social cost of carbon from $62 to $1 per ton and the social cost of methane from $1,920 to $55 per ton. It did so by switching to a higher discount rate and by considering only domestic costs of climate change. These interim estimates, unlike those of the working group, were not reviewed by the National Academies.[1]

9.3.2 Discount Rates and the Welfare of Future Generations

The choice of using a higher discount rate systematically biases economic analysis against favoring regulations that typically require high upfront

---

[viii] Biden's January 2021 Executive Order 13990 paved the way for the reinstatement of the social cost of carbon as calculated by the Interagency Working Group. In March 2021, the Biden administration announced that it would use the interim value for the social cost of carbon at $42 per ton of carbon dioxide.

Table 9.4 *Comparison of the social cost of carbon and of methane under the Obama administration and the Trump administration*

Source	The Trump administration's interim estimates		The Interagency Working Group under the Obama administration	
Measure	Domestic Social cost of carbon[a] (in 2011$)	Domestic Social cost of methane[b] (in 2016$)	Global Social cost of carbon[c] (in 2007$)	Global Social cost of methane[d] (in 2007$)
Cost of 1 ton of emissions in the year 2020	$1	$55	$62	$1920
Cost of 1 ton of emissions in the year 2040	$2	$110	$84	$3119

Sources: a. Environmental Protection Agency, *Regulatory Impact Analysis for the Review of the Clean Power Plan*, 2017;[49] b. Bureau of Land Management, *Regulatory Impact Analysis for the Final Rule to Suspend or Delay Certain Requirements of the 2016 Waste Prevention Rule*, 2017.[83] c. Environmental Protection Agency, *Interagency Working Group on Social Cost of Greenhouse Gases*, 2016;[77] d. Environmental Protection Agency, *Addendum to Technical Support Document on Social Cost of Carbon for Regulatory Impact Analysis under Executive Order 12866*, 2016.[84]
Notes: Estimates from the Trump administration apply a discount rate of 7%. Estimates from the Obama administration apply a discount rate of 2.5%.

compliance costs but yield benefits later. The higher discount rate does this by deflating future benefits or costs.

The Trump administration applied a 7 percent discount rate, arguing that this figure captures private costs of investments into climate mitigation efforts. This choice departed from past Republican and Democratic administrations' use of lower rates, at most 3 percent, in analyses that raise intergenerational equity issues. The 3 percent figure is based on returns to long-term saving, as reflected in the returns to long-term government debt.[85] In fact, the OMB guidance document permits the use of discount rates lower than 3 percent when regulations yield intergenerational benefits or costs. Analysis in the 1990s had applied discount rates of 1–3 percent.[85,86]

The Interagency Working Group co-chair Michael Greenstone explained the rationale for using a lower discount rate in the analysis on climate policies.[87] He notes that climate investments that protect society's stream of consumption at a time of climate crisis are akin to purchasing insurance against catastrophic damages.[88] Research underscores the nontrivial likelihood that the climate system exceeds critical "tipping points" and society suffers catastrophic climate damage that far exceeds the average predicted levels.[89] Exceeding tipping points such as the loss of the Arctic summer sea ice and the melting of the Greenland and West Antarctic ice sheets can permanently shift the climate and cause catastrophes such as widespread drought and enormous rises in sea level.[90]

Investors are willing to accept lower returns from investments that yield benefits during periods of poor economic performance[87] because the benefits realized during the lean recession periods are more valuable than those realized during economic booms. Greenstone also noted that the returns from 10-year Treasury bills, which are the returns to holding long-term government debt, have been declining and are expected to be below 2 percent.[87,91]

The justification offered by proponents of high discount rates – that future generations tend to be wealthier than present generations – no longer holds true with the climate crisis.[92] Prior to the climate crisis, one could argue that, historically, subsequent generations have been wealthier than previous generations due to continued economic growth and technological advancement. In such a scenario, the present generation may lose more from foregoing and investing a dollar than the satisfaction gained by future generations. However, the worsening climate crisis makes future generations worse off than the present generation.[92] Indeed, future generations who live in warm regions (such as the southern United States) are projected to be absolutely poorer than current generations living in those regions.[93] Higher discount rates premised on the untrue assumption that future generations will be wealthier than present generations are therefore not defensible.

Of note, economists across the philosophical spectrum support low discount rates when the interests of future generations are at stake. Economists more supportive of government intervention to address the climate crisis have supported low discount rates.[94] Those who are generally less supportive of government interventions in the markets have also argued for a zero discount rate, pointing to how ethical considerations prohibit giving less weight to the welfare of future generations.[95]

Ethical considerations have led other scholars to advocate that society should undertake investments, including a number that are relatively modest, to protect the welfare of future generations.[96] As summarized by Clark et al., "Ultimately, this decision [to invest in climate protection] depends on human values, particularly on our valuation of intergenerational equity, food and water security, maintenance of ecosystem services, biodiversity, and the preservation of unique environments ... Are future generations entitled to the same environmental stability and biodiversity that has been afforded our generation and hundreds of generations before us?"[97]

9.3.3 The Domestic versus Global Social Cost of Carbon

The Interagency Working Group applied the global social cost of carbon that takes into account climate damages to the entire world. By contrast, the Trump administration applied the domestic social cost of carbon that narrows the consideration to damages in the United States. This choice drills down to whether Americans value a multilateral approach to addressing climate change, evidenced by the US–China Climate Accord and the Paris Climate Agreement, or choose to abandon such an approach.

Proponents of applying the global social cost of carbon point to ethical and pragmatic reasons for the United States to use the global estimate. First, the United States, which is the largest cumulative global emitter of carbon dioxide (from 1870 through 2018),[98] bears an ethical responsibility for considering the impact of its actions on the rest of the world.[99] The "do no harm" principle obligates the United States to consider the impact of its emissions on the globe.[99]

Second, it is in the self-interest of the United States to consider the global social cost of carbon. The United States can strategically encourage other countries to use the global measures, as a number of other countries do already.[100–102] The United States benefits when other countries consider their negative spillovers outside their borders while implementing policies.

Indeed, the United States coordinated with other countries to implement the social cost of carbon. The Obama administration worked with Canada and Mexico so that all three countries use common methods and values for the global social cost of carbon.[101,102] European countries apply higher figures for the global social cost of carbon than does the United States.[101] The Obama administration's promulgations of major rules to cut greenhouse gas emissions made it a more credible negotiator. This enabled it to seal the US–China Climate Accord and, months later, to achieve the important, albeit imperfect Paris Climate Accord (Chapter 11).

Opponents, including independent scholars who are not funded by the oil and gas industry, argue that global benefits can only be taken into account if foreign countries *truly* reciprocate US domestic actions to protect the climate or if US citizens, out of altruism, consider the interests of non-US citizens.[103,104] Even then, they argue that federal agencies can only consider the global social cost of carbon if Congress enacts legislation to this effect.[103,104]

However, the view that the United States should not consider global climate costs unless and until foreign countries fully reciprocate does not comport with how multilateral actions have unfolded to protect the global environment. Throughout history, leading countries have taken the first steps and subsequently steered the rest of the world toward resolving global environmental issues. The Reagan administration took early unilateral steps to address the depletion of the ozone layer[105] and its leadership culminated in a multilateral agreement.[106] Likewise, President George H. W. Bush signed on to the 1992 Framework Convention on Climate Change, with commitments in the earlier stage of his presidency to take the lead on global climate action.[107]

Moreover, the view that federal agencies are not permitted to take into account global estimates does not tally with existing case law.[101,102] In 2016, the Court of Appeals of the Seventh Circuit affirmed in *Zero Zone, Inc. v. Department of Energy*[108] that the regulatory agency acted reasonably in considering global climate impacts.[101,102] Additionally, section 115 of the Clean Air Act titled "International Air Pollution" requires the United States to address the impacts of US emissions on the health and welfare of foreign populations if foreign countries reciprocate in considering US interests.[101,102] Indeed, foreign countries including China, the largest annual global emitter of carbon dioxide since 2007,[109] have considered US interests through their commitments under the UN Framework Convention for Climate Change[110] and the Paris Climate Accord.[111]

Nevertheless, whether existing statutes permit or prohibit federal agencies to take into account the global costs of climate change is likely to be litigated in the courts. Several independent scholars,[103,112,113] the oil and gas industry[78] and its funded think tanks[79–81] challenge this statutory interpretation. This issue of whether the United States should take into account global costs is also the subject of two competing bills in Congress, discussed further in Section 9.5.

It is worth noting that the interim estimate for the domestic social cost of carbon understates the costs of climate change even to the United States. Calculating the domestic costs of climate change using the United States' share of the global economy ignores how climate impacts in other countries harm the United States. Those countries' political and economic destabilization can develop into conflicts entangling the United States and spur desperate migration flows into the United States.[1] Peer-reviewed research that explicitly models climate damages at the country level[93] put the US domestic social cost of carbon at $48 per ton.[114]

9.4 Putting the Two Together: Distorting Methods and Choosing Assumptions to Understate the Benefits from Regulations

The Trump administration's rescission of the 2016 Methane Waste Prevention Rule illustrates how the administration combined its strategies – ignoring health benefits *and* applying assumptions that lead to paltry estimates of the social cost of methane – to flip the conclusion of cost–benefit analysis to support regulatory rollback.

Table 9.5 *The waste of natural gas from oil and gas operations*

Methane loss	Site of wells	Amount of wasted gas	Number of households that could have been supplied with the wasted gas	Royalty lost
Vented or flared (cumulative methane loss from 2009 through 2015)[36]	Public lands	About 462 billion cubic feet or about 2.7% of natural gas produced on BLM-managed leases[36]	The cumulative wasted gas could supply energy needs of 6.2 million homes for one year.[36]	$83 million[a]
Methane leaks (annual estimates)	All United States	2.3% of gross US natural gas production[120]	The annual wasted gas could supply the natural gas needs for 11 million households per year.	Estimates on the royalty lost to private landholders are not available.

Sources: Bureau of Land Management, *Waste Prevention, Production Subject to Royalties, and Resource Conservation*, 2016;[36] Alvarez et al., "Assessment of Methane Emissions for the US Oil and Gas Supply Chain," 2018.[120]
Note: a. The Government Accountability Office estimated 50 billion cubic feet (bcf) of gas, or 39% of vented gas, can be captured in a cost-effective way, and the forgone royalties are $23 million. Using the same ratios, I estimate that 180.5 bcf (out of 462 bcf) of gas can be captured, and forgone royalties would amount to $83 million [116].

The Department of the Interior, through BLM, bears the statutory responsibility "to use all reasonable measures to prevent waste" on public lands.[162] In November 2016, BLM promulgated the Methane Waste Prevention Rule[36] to cut the enormous waste of natural gas from oil and gas operations on public lands.[115,116] Table 9.5 highlights the enormity of this waste. The rule limited the venting and flaring of gas, mandated the detection and repair of leaks, and required operators to pay royalty for gas that is "wasted" when operators failed to meet these performance standards. The rule was necessary as operators that faced private upfront costs regularly failed to capture the natural gas, despite the available technology to do so.[115,116]

The American Petroleum Institute and smaller oil producers opposed regulations aimed to curb methane emissions,[117–119] while the large- and medium-sized oil and gas companies supported those regulations.[119] In September 2018, the Trump administration rescinded most provisions of this regulation.[33] It returned to the previous rule, promulgated in 1979, that had failed to prevent that waste of natural gas.[116]

As summarized in Table 9.6, the Trump administration's 2019 analysis was able to conclude that the rescission of the methane rule would yield benefits to society by taking two key steps. First, the administration deflated the costs to society of failing to capture methane, a potent greenhouse gas, by employing a paltry social cost of methane of $81 or $232 per metric ton instead of the $1,729 per metric ton figure used in the 2016 analysis.

Second, the Trump administration departed from standard practice in cost–benefit analysis by ignoring the co-benefits. Specifically, it ignored the co-benefits from reducing the emissions of volatile organic compounds (VOCs), estimated at a ballpark figure of $7.2 billion.[121] This figure far exceeds BLM's estimate of the industry's compliance costs for capturing methane. This figure calls for greater effort by BLM to account for these co-benefits. The 2016 regulatory impact assessment did not consider the monetized value for these co-benefits because, in that assessment, benefits from capturing methane (using the Interagency Working Group's estimate of the social cost of methane) already far exceed compliance costs and tip the scale in favor of regulation.

The Trump administration gave short shrift to its statutory responsibility to manage natural resources for the public interest. BLM acknowledged that a repeal of the rule would lead to substantial waste of natural gas that could meet the annual needs for about 150,000 households and to forgone royalties for federal and state governments, as much as $80 million over a decade.[ix] The

---

[ix] I assume that each household uses 1.166 metric tons of methane per year.

Table 9.6 The regulatory impact assessment for the Methane Waste Prevention Rule and for the rescission of its key provisions (estimates for a 10-year period)

	Obama administration		Trump administration	
	Promulgation of the original 2016 Waste Prevention Rule		Rescission of key provisions of the 2016 rule	
Overall impact	Net benefits from promulgating the rule	$898 million to $1.2 billion	Net benefits from rescinding the rule	$581–$945 million
Benefits	Benefits		Forgone benefits	
Methane	Benefit from reducing methane losses	$814 million to $1.2 billion	Forgone benefits from methane losses	$66–$259 million
	Assumption on social cost of methane	$1,729 per metric ton	Assumption on social cost of methane	$81–$232 per metric ton
Other air pollutants	Reductions of VOC emissions		Forgone reductions of VOC emissions	Assumed to be zero but Krupnick and Echarte[121] estimate forgone benefit at $7.2 billion
Costs	Compliance costs	$110–$279 million	Forgone compliance costs	$1.36–$2.08 billion
Other issues				
Royalty for federal and state governments	Increase in royalty receipt	$65–$82 million	Forgone royalty receipt	$28.3–$79.1 million

Sources: Bureau of Land Management, *Regulatory Impact Analysis*, 2016;[122] Bureau of Land Management, *Waste Prevention, Production Subject to Royalties, and Resource Conservation*, 2016;[36] Bureau of Land Management, *Regulatory Impact Analysis for the Final Rule to Rescind or Revise Certain Requirements of the 2016 Waste Prevention Rule*, 2018;[123] Bureau of Land Management, *Waste Prevention, Production Subject to Royalties, and Resource Conservation; Rescission or Revision of Certain Requirements*, 2018.[33]

administration's revised rule creates no incentive for companies to reduce waste. It permits companies not to pay royalties on "waste gas" when their private costs of capturing that gas exceed their revenue. That operators' private costs exceed their revenue is the very condition that has led to the massive losses of natural gas.[116]

9.5 How Can the Integrity of Economic Analyses in Rulemaking Be Improved?

To understand how to safeguard economic analysis in rulemaking, it is useful to consider how rulemaking fits into the legislative process. Congress enacts laws to protect public health and the environment, and it delegates rulemaking powers to federal agencies to make more specific regulations. Agencies that respect the integrity of science and economic analysis in rulemaking are more likely to promulgate regulations that achieve the statutory goals.

For decades, Congress has recognized the need to safeguard the legitimacy of the rulemaking process and to hold the executive branch accountable for promulgating regulations that abide by statutes using fair procedures. In 1949, it enacted the Administrative Procedure Act (APA) to prevent the executive branch from taking overly broad liberties in promulgating, revising, rescinding or replacing regulations. The APA, detailed below, sets minimum standards for procedures in rulemaking and for judicial review of agencies' rulemaking. Congress has been even more specific in laws such as the Clean Air Act in instructing the EPA on when to set air quality standards based on public health criteria alone and to eschew narrow cost–benefit analysis, as well as when to take into consideration economic concerns.

Congress can choose to exercise its legislative powers to further refine its instructions to federal agencies on the proper accounting for economic benefits from regulations. For instance, significant economic benefits from a number of regulations slashed by the administration – the methane rule and the Clean Power Plan Rule – derive from mitigation of climate damages. Congress can ensure full accounting of these economic benefits, and therefore more informed rulemaking, by requiring federal agencies to apply the social cost of carbon to regulatory analyses, as recommended by the National Academies of Sciences, Engineering, and Medicine.[1]

In line with the National Academies' recommendation, Democratic Senator Michael Bennet from Colorado introduced the Carbon Pollution Transparency Act,[124,125] with 16 Democratic cosponsors, in 2018 and again in 2019. The bill mandates federal agencies to use the social costs of greenhouse gases developed by the Interagency Working Group, with adjustments for inflation. The bill would establish an interagency working group and a scientific committee to revise these estimates periodically, based on science. In requiring these estimates to account for intergenerational harm, sponsors of this bill made the ethical decision to value the welfare of future generations. Likewise, in requiring the estimates to account for the global damages from greenhouse gas emissions, they supported a multilateral approach to solve the climate crisis.

This bill echoes efforts in several states to incorporate social cost of carbon in their policy analyses. For instance, in April 2019, the state of Washington enacted legislation requiring utilities to adopt the social cost of carbon developed by the Interagency Working Group.[126] In May 2019, Colorado passed a bill requiring the Public Utility Commission to consider social costs of greenhouse gases when considering resource plans or other applications submitted by electric utilities.[127]

While only Democrats cosponsored the Senate bill, the approach of fully accounting for the costs of greenhouse gas emissions for public health and the economy has been embraced by a number of Republicans. As noted by James Baker, an elder Republican statesman, in his 2017 briefing titled "The Conservative Case for Carbon Dividends," efforts to account for the costs of carbon emissions help shape more informed decision-making.[128] Prominent Republicans who served in the Reagan, G. H. W. Bush and G. W. Bush administrations, including as chairs of the White House Council of Economic Advisors, concurred with that view.

Unfortunately, Congress failed to hold the Trump administration tethered to evidence-based economic analyses, and this failure stems in no small part from those Republicans on Capitol Hill who themselves eschewed fact-based analyses. Several Republicans pursued legislation to block federal agencies from considering the economic costs of climate change. In 2017, Representative Evan Jenkins, a Republican from West Virginia, proposed H.R. 3117[129] with 48 Republican cosponsors and Senator James Lankford, a Republican from Oklahoma, proposed the companion Senate bill S. 1512[130] with seven Republican co-sponsors. Those bills "prohibit the Department of Energy, the Environmental Protection Agency (EPA), the Department of the Interior, and the Council on Environmental Quality from considering the social cost of carbon, methane, or nitrous oxide as part of any cost benefit analysis in the rule making process, unless a federal law is enacted authorizing such consideration." Those bills also state that agencies "may" consider social costs of carbon, when calculated using specific assumptions. It is worth emphasizing that sponsors of the bill, by endorsing these assumptions, are choosing to give short shrift to the welfare of future generations.[x]

Much like its inaction on deregulation, Congress's stalemate left the battle to oppose the administration's corruption of rulemaking – through its manipulation of scientific reviews and economic analyses – to state attorneys general[xi] and to public health, environmental and community organizations. My exposition below on these legal challenges combines discussions of the administration's manipulation of both science and economic analysis in rulemaking. The legal fight – likely to take years as cases proceed from lower to high courts – will revolve around federal agencies' interpretation of their statutory duties, their compliance with procedure, their factual determinations and their discretionary decisions.[131,132] While Congress failed to enact updated laws, the APA provides an important, albeit imperfect framework that serves to hold federal agencies accountable in rulemaking.

A federal agency has the discretion to make policy shifts as long as it acts within the authorizing statute. The authorizing statute denotes the statute that delegates rulemaking powers to a federal agency to create specific regulations. That an agency has the discretion to make policy shifts means that it can choose to take a different direction from one that it has taken in the past and, accordingly, it can make changes to regulations. For instance, in response to the Trump administration's deregulatory goals,[xii] an agency may choose to reduce environmental protection and to weaken regulations to achieve that outcome. The APA and the courts, based on the recognition of the political accountability of the executive branch, underscore that agencies can make policy shifts. Even so, agencies face constraints in their discretion to make these shifts. They must follow procedures set out in the APA and in subsequent guidance and give sound reasons for their decisions.[133,134] Agencies also face checks by the courts, with the APA provisions creating a strong presumption of judicial review of agency actions. The APA section 706 sets the standard for judicial review, stating that "the reviewing court shall ... hold unlawful and set aside

---

[x] The bill states that agencies "may" consider the social costs of carbon when calculated based on domestic costs only, higher discount rates (3 percent and 7 percent) and "reasonable" time horizons. As discussed above, higher discount rates depreciate costs incurred by future generations. Limiting analysis to shorter time horizons ignores substantial climate costs that occur in the more distant future.

[xi] The diversity across state governments is worth noting. State attorneys general from states supportive of greater public health and environmental protection spearheaded challenges against deregulatory actions under the Trump administration. However, their counterparts from states opposed to regulations led to legal challenges against regulations promulgated by the Obama administration.

[xii] A federal agency may choose to pursue a different policy direction in response to new information that calls for greater public health or environmental protection or the converse.

agency action, findings, and conclusions found to be arbitrary, capricious, an abuse of discretion, or otherwise not in accordance with law." This arbitrary and capricious standard and the related principle of Chevron deference are discussed below.

State attorneys general and various groups scored a string of early wins against the Trump administration's initial efforts to tamper with agency rulemaking. Courts ruled that federal agencies contravened the APA in several ways in their delays or suspensions of rules. They failed to follow rulemaking procedures or failed to cite legal authority or to provide sound reasons for the delays or suspensions.[133,135] For instance, in July 2017, the Court of Appeals for the DC Circuit in *Clean Air Council v. Pruitt*[136] rejected the EPA's bold assertion that its delay in implementing a rule was not judicially reviewable. The EPA had granted a 90-day stay of the compliance date for the regulation curbing the emissions of methane from oil and gas facilities, pending its reconsideration of the rule. The court opined that the stay was "tantamount to amending or revoking a rule" and thus subject to review. It further ruled that the EPA violated the Clean Air Act provisions for a stay because the issues the EPA said it would reconsider in justifying the stay had already been raised by industry during the comment period and had already been deliberated on by the EPA prior to the issuance of the final rule. Another example of a win by challengers against the Trump administration is the decision by the District Court for the Northern District of California in February 2018.[137] It struck down BLM's December 2017 Suspension Rule that suspended the 2016 Waste Prevention Rule. It held the administration to reasoned and transparent decision-making by ruling that BLM had to provide "requisite good reasons and detailed justification" for its reversal. The court noted that BLM failed to address the contradictions between its new claims underlying the Suspension Rule and the evidence it had presented in promulgating the Waste Prevention Rule.

While the early legal challenges against the administration's delays or suspensions of rules were successful, the outcomes of challenges against the administration's revision and repeal of regulations were mired in uncertainty. When agencies successfully demonstrated that they had adhered to procedural requirements under the APA, these legal challenges moved to more difficult hurdles, such as the reasonableness of an agency's statutory interpretation, its claims of legitimate policy shifts and its factual determinations.

In general, courts have taken a deferential approach in agencies' statutory interpretations. The Supreme Court established the principle known as Chevron deference in the 1984 case *Chevron USA, Inc. v. Natural Resources Defense Council, Inc.*[138] In applying this principle, courts defer to an agency's action so long as the agency's action is not in conflict with the authorizing statute, and if the statute is ambiguous, courts require only that the agency's action was reasonable and in line with a permissible construction of the authorizing statute. I discuss Chevron deference in greater depth in the next chapter.

In reviewing an agency's claim of legitimate policy shift within its statutory powers, the court examines the often blurry but important line between an agency's permissible discretion in executing statutes and its non-permissible drift from a statute's original intent. In 2009, the Supreme Court in *Federal Communications Commission v. Fox Television Stations, Inc.*[139] ruled that the review standards are not more stringent simply because an agency's action alters its prior policy. Provided the agency's action is permissible under its authorizing statute, it is required only to acknowledge the change and to adhere to rulemaking guidelines under the APA in making that change. Absent more specific instructions from Congress, it is not required to show that new policies are better than old ones. Arguably, from one perspective, the Trump administration's choice of assumptions in its economic analysis – giving less weight to future generations and no weight to non-Americans – reflects a permissible policy shift from the past administration. Even so, agencies' ignorance of science and facts is not to be misconstrued as a shift in policy.

The court in *Fox* emphasized the need for an agency to provide reasons when a "new policy rests upon factual findings that contradict those which underlay its prior policy or when its prior policy has engendered serious reliance interests that must be taken into account." In 2018, the State Department's failure to address its previous findings underpinning its prior decision led to an adverse court decision.[134] Citing the *Fox* decision, the District Court for the District of Montana in *Indigenous Environmental Network v. US Department of State*[140] ruled that the State Department had failed to provide reasoned explanation in its 2018 permit approval for the Keystone XL pipeline. The pipeline would transport tar sands from Alberta to the US Gulf Coast, crossing public lands located in Montana.[xiii] While the court acknowledged that the State Department was permitted to make a policy shift in prioritizing energy security, it ruled that the department erred by simply ignoring its findings of climate impacts cited in its 2015 denial of the permit.[134]

[xiii] Federal agencies that reverse prior decisions that were premised on climate considerations may find it difficult to explain away these findings. In response to the adverse ruling against the State Department, the Trump administration issued the Keystone XL permit from the Office of the President. In the administration's view, the presidential permit would not be subject to judicial review and thus the administration could avoid addressing those past findings on climate impacts [134].

In reviewing the Trump administration's repeal and replacement of regulations, courts examined agencies' factual determinations. As documented in this chapter and in Chapter 8, federal agencies manipulated science and economic analyses in their reversal of regulations. The APA created guidelines that agencies must follow in fact finding and in the use of factual evidence. A number of scholars have taken the view that these guidelines place significant restraint on agencies in reversing regulations. They cannot make U-turns without addressing science and facts and without providing reasons for their reversals.[141–143,144] The 1983 Supreme Court decision in *Motor Vehicle Manufacturers Association v. State Farm Auto Mutual Insurance Co.* acknowledged that the court would not substitute its judgement for that of the agency.[163] Even so, it noted that courts should invalidate agency determinations that fail to "examine the relevant data and articulate a satisfactory explanation for its action including a 'rational connection between the facts found and the choice made.'"[xiv] In that case, which involved the Reagan administration's withdrawal of a rule by the Carter administration requiring automatic seatbelts, the court ruled that "the agency's explanation for rescission of the passive restraint requirement is not sufficient to enable us to conclude that the rescission was the product of reasoned decision making."[xv]

In 2009, Justice Anthony Kennedy in *Fox* further elaborated the need for agencies to address past findings. He opined, "An agency cannot simply disregard contrary or inconvenient factual determinations that it made in the past, any more than it can ignore inconvenient facts when it writes on a blank slate." Also in 2009, the court of Appeals for the Ninth Circuit in *Tucson Herpetological Society v. Salazar*[145] signaled courts should note the importance of agencies' explaining their decisions even on technical grounds. It ruled that, despite the courts' significant deference to agencies in scientific and technical issues, they would not blindly defer to scientific decisions made by federal agencies that fail to adequately explain those decisions.[141] Some scholars have argued that the courts' insistence that agencies explain the basis of their actions provides judicial checks on those decisions, even if the courts do not rule on the science.[146,147]

On the other hand, several scholars have argued that agencies enjoy extensive flexibility in judicial review, raising doubts as to the likelihood of success

---

[xiv] The court in *State Farm* gave examples of when courts would find an agency's decision arbitrary and capricious. These include "when it relies on factors that Congress did not intend for it to consider, when it fails to consider an important aspect of the problem, or when it offers an explanation running counter to the evidence or is implausible to the point where it cannot be ascribed to expertise or a difference in view."

[xv] In reversing course from its past decision, the agency had provided no empirical evidence for its assertion that people would remove the seat belts and render them ineffective.

of legal challenges. As clarified in *State Farm*, the scope of judicial review over agency decisions is "narrow," as "a court is not to substitute its judgment for that of the agency." Moreover, courts grant significant deference in scientific and technical matters, in recognition of agencies' superior expertise.[148,149] In 1983, the Supreme Court in *Baltimore Gas & Electric Co. v. Natural Resources Defense Council*[150] underscored this heightened deference, that is, "When the agency is making predictions, within its area of special expertise, at the frontiers of science ... a reviewing court must generally be at its most deferential." Scholars have also pointed out that the requirement of agencies to provide reasoned explanation is not a high bar, with courts striking down agencies' decisions only when an agency ignores relevant scientific data when making its decision.[151] Consequently, agencies are able to pursue decisions that may not be supported by the best available scientific evidence, as long as they articulate a rationale linking the data and the decision.[151] The deference of the courts toward agencies has led some commentators to point out that the key assumption justifying deference – that federal agencies act in good faith on scientific and technical matters – may not hold true in the decisions under the Trump administration,[152] thus calling for courts to conduct more searching reviews.

One important example that serves as a reminder of the executive branch's significant leeway in deregulatory actions is the December 2017 BLM repeal of a 2015 rule on hydraulic fracturing on public lands.[xvi] The state of California and a coalition of environmental groups challenged that repeal. In March 2020, the District Court for the Northern District of California ruled in favor of BLM, noting that "the record does not compel the conclusion that [BLM] arbitrarily ignored foregone benefits or arbitrarily overvalued the costs associated with the 2015 Rule ... Although BLM could have provided more detail, it did enough to clear the low bar of arbitrary and capricious review, and that is all the law requires."[153] For those who fear the decimation of public health protections, this decision serves a warning that the rollbacks of regulations, even when justified with limited information, may well survive judicial review. While judicial review provides a check against agencies acting outside their statutory powers, it does not serve as a bulwark against an administration's deregulatory actions.

When the Trump administration left office, the adjudication of many of the legal challenges against its revision or rescission of environmental regulations had not been completed.[154,xvii] The Trump administration won some of these

---

[xvi] The 2015 rule on hydraulic fracturing on public lands never came into effect because of a legal challenge.

[xvii] The Institute for Policy Integrity at the New York University Law School tracked the litigation pertaining to Trump's deregulatory agenda (see https://policyintegrity.org/trump-court-roundup).

cases and lost others in the lower courts or in the appeal courts.[xviii] While some of these lawsuits were rendered moot by the actions of the Biden administration or the 117th Congress,[xix] other important cases continued, including challenges against the administration's Navigable Waters Protection Rule that determines the reach of the Clean Water Act.[155,xx]

9.6 Conclusion

The Trump administration's strategies to weaken regulations demonstrate that the rulemaking process remains vulnerable to manipulation, despite the existence of a number of checks by Congress and the judiciary. By sidelining science and manipulating economic methods, the administration was able to undercut the rulemaking process and to repeal regulations that provide major benefits to human health, the economy and the environment.

State attorneys general and civil society fought intrepidly against the administration's scientific and economic analyses underpinning regulations that understated and even ignored the public health and environmental benefits of regulations. Even so, actions through the courts are lengthy and their outcomes are uncertain.

Ultimately, Congress holds the power and the duty to keep the administration in check. Unsurprisingly, partisan stalemate and industry influence kept productive responses to a minimum during the Trump administration. Regardless of which party controls the executive branch, the actions of the Trump administration demonstrate the need for Congress, whether through new laws or stricter oversight, to uphold the integrity of science and economic analysis in rulemaking. The assault of the Trump administration on rulemaking did not end with attacks on science or on economic analysis. As the next chapter reveals, its deregulatory actions extended to undermining values even Republicans profess to hold dear – protecting the powers of the legislature from the overreach of the executive branch and protecting the powers of states from the overreach of the federal government.

[xviii] For instance, in January 2021, the Court of Appeals of the DC Circuit vacated the Affordable Clean Energy Rule (Chapter 3). In February 2021, the District Court for the District of Montana vacated the Science Transparency Rule (Chapter 8).

[xix] For instance, in September 2020, California and 19 other states filed a legal challenge against the Trump administration's weakening of the 2016 methane rule (that aimed to curb methane leaks from oil and gas operations). In April 2021, Congress voted to reinstate the 2016 methane rule.

[xx] The litigation challenging the Navigable Waters Protection Rule continued in the District Court for the District of Colorado and in the District Court for the District of South Carolina [155].

References

1. National Academies of Sciences, Engineering, and Medicine. *Valuing Climate Damages: Updating Estimation of the Social Cost of Carbon Dioxide.* Washington, DC: National Academies Press, 2017.
2. Congressional Research Service. *Clean Air Act Issues in the 115th Congress: In Brief.* Report by J. E. McCarthy. R44744 (Washington, DC: 2018). https://fas.org/sgp/crs/misc/R44744.pdf.
3. *Michigan v. Environmental Protection Agency*, 576 US 743 (Supreme Court 2015).
4. A. Sinden. "Supreme Court Remains Skeptical of the 'Cost–Benefit State.'" *The Regulatory Review*, September 26, 2016. www.theregreview.org/2016/09/26/sinden-cost-benefit-state/.
5. W. J Clinton. *Revised Executive Order 12,886 Regulatory Planning and Review.* 58 Federal Register 190 (October 4, 1993). www.reginfo.gov/public/jsp/Utilities/EO_12866.pdf.
6. F. Ackerman and L. Heinzerling. *Priceless: On Knowing the Price of Everything and the Value of Nothing.* New York: New Press, 2004.
7. B. Fischhoff. "The Realities of Risk–Cost–Benefit Analysis." *Science* 350, no. 6260 (2015): 527–534. https://doi.org/10.1126/science.aaa6516.
8. K. J. Arrow et al. "Is There a Role for Benefit–Cost Analysis in Environmental, Health, and Safety Regulation?" *Science* 272, no. 6398 (1996): 221–222. https://doi.org/10.1126/science.aar7204.
9. R. Brent. *Cost–Benefit Analysis and Health Care Evaluations.* Cheltenham, UK: Edward Elgar, 2014.
10. Organization for Economic Cooperation and Development. *Cost Benefit Analysis and the Environment: Further Developments and Policy Use.* Organization for Economic Cooperation and Development (Paris: 2018).
11. A. Sinden. "The Cost–Benefit Boomerang." *American Prospect*, July 25, 2019. https://prospect.org/economy/cost-benefit-boomerang.
12. A. Sinden. "The Problem of Unquantified Benefits." *Environmental Law* 49 (2019): 73–129.
13. American Petroleum Institute. *API Recommends that the US Retain the NAAQS Ozone Standards: Executive Summary* (March 13, 2015). www.api.org/~/media/Files/Policy/Ozone-NAAQS/API-Recommendations-NAAQS-Ozone-Executive-Summary.pdf.
14. N. Banerjee, D. Hasmeyer and L. Song. "For Oil Industry, Clean Air Fight Was Dress Rehearsal for Climate Denial." *Inside Climate News*, June 6, 2016. https://insideclimatenews.org/news/05062016/oil-industry-clean-air-fight-smog-los-angeles-dress-rehearsal-climate-change-denial-exxon.
15. Office of Management and Budget, Office of Information and Regulatory Affairs. *2018, 2019 and 2020 Draft Report to Congress on the Benefits and Costs of Federal Regulations and Agency Compliance with the Unfunded Mandates Reform Act* (Washington, DC: 2019). www.whitehouse.gov/wp-content/uploads/2019/12/2019-CATS-5899-REV_DOC-Draft2018_2019_2020Cost_BenefitReport11_20_2019.pdf.

16. Office of Management and Budget, Office of Information and Regulatory Affairs. *2013 Report to Congress on the Benefits and Costs of Federal Regulations and Unfunded Mandates on State, Local, and Tribal Entities* (Washington, DC: 2013). www.whitehouse.gov/sites/whitehouse.gov/files/omb/inforeg/inforeg/2013_cb/ 2013_cost_benefit_report-updated.pdf.
17. M. E. Porter and C. van der Linde. "Toward a New Conception of the Environment-Competitiveness Relationship." *Journal of Economic Perspectives* 9, no. 4 (1995): 97–118.
18. W. Harrington, R. D. Morgenstern and P. Nelson. "On the Accuracy of Regulatory Cost Estimates." *Journal of Policy Analysis and Management* 19, no. 2 (2000): 297–332.
19. D. Popp. "Pollution Control Innovations and the Clean Air Act of 1990." *Journal of Policy Analysis and Management* 22, no. 4 (2003): 641–660.
20. Environmental Protection Agency. *Retrospective Study of the Costs of EPA Regulations: A Report of Four Case Studies.* National Center for Environmental Economics, Office of Policy, Environmental Protection Agency (2014). www .epa.gov/sites/production/files/2017-09/documents/ee-0575_0.pdf.
21. K. Bridbord and D. Hanson. "A Personal Perspective on the Initial Federal Health-based Regulation to Remove Lead from Gasoline." *Environmental Health Perspectives* 117, no. 8 (2009): 1195–1201.
22. R. G. Newell and K. Rogers. "Leaded Gasoline in the United States: The Breakthrough of Permit Trading." In *Choosing Environmental Policy: Comparing Instruments and Outcomes in the United States and Europe*, edited by W. Harrington, R. Morgenstern and T. Sterner. 175–191. Washington, DC: Resources for the Future, 2004. Reprint.
23. Environmental Defense Fund. "Finding, Fixing Leaks is a Cost-Effective Way to Cut Oil and Gas Methane Emissions." *Environmental Defense Fund Fact Sheet*, n.d. www .edf.org/sites/default/files/content/ldar_fact_sheet_final.pdf?utm_source=forbe s&utm_campaign=edf_methane_upd_dmt&utm_medium=cross-post&utm_id=1478 794987&utm_content=br161110.
24. H. J. Feldman, senior director of Regulatory and Scientific Affairs at the American Petroleum Institute. *Comment on the US Environmental Protection Agency's Advanced Notice of Proposed Rulemaking on "Increasing Consistency and Transparency in Considering Costs and Benefits in the Rulemaking Process" 83 Fed. Reg. 27,524 (June 13, 2018)*. Submitted to Environmental Protection Agency, National Center for Environmental Economics. Docket ID No. EPA-HQ-OA-2018-0107. April 13, 2018.
25. J. McGillis, Institute for Energy Research. *Comments on Increasing Consistency and Transparency in Considering Costs and Benefits in Rulemaking Process Advanced Notice of Proposed Rule-Making 83 Fed. Reg. 27524–27528 (June 13, 2018)*. Submitted to Environmental Protection Agency. Docket No. EPA-HQ-OA-2018-0107-1244 RIN 2010-AA12. 2018.
26. K. J. Boyle and M. J. Kotchen. "Policy Brief – The Need for More (Not Less) External Review of Economic Analysis at the US EPA." *Review of Environmental Economics and Policy* 13, no. 2 (2019): 308–316. https://doi.org/10.1093/reep/ rez006.

27. K. J. Boyle and M. Kotchen. "Retreat on Economies at the EPA." *Science* 361, no. 6404 (2018). https://doi.org/10.1126/science.aav0896.
28. K. B. Belton and J. D. Graham. "Trump's Deregulation Record: Is It Working?" *Administrative Law Review* 71, no. 4 (2019): 803–880. www.administrativelawreview.org/wp-content/uploads/2019/12/ALR_71.4-Graham-Belton.pdf.
29. *Improving Consistency and Transparency of Cost Considerations in Rulemaking: Advanced Notice of Proposed Rulemaking.* 40 Code of Federal Regulations chapter undefined. Environmental Protection Agency. 83 Federal Register 27524–27528 (June 13, 2018). www.federalregister.gov/documents/2018/06/13/2018-12707/increasing-consistency-and-transparency-in-considering-costs-and-benefits-in-the-rulemaking-process.
30. *Increasing Consistency and Transparency in Considering Benefits and Costs in the Clean Air Act Rulemaking Process: Final Rule.* 40 Code of Federal Regulations. Environmental Protection Agency. 83 Federal Register 247: 84130–84157 (December 23, 2020). www.federalregister.gov/documents/2020/12/23/2020-27368/increasing-consistency-and-transparency-in-considering-benefits-and-costs-in-the-clean-air-act.
31. *Rescinding the Rule on Increasing Consistency and Transparency in Considering Benefits and Costs in the Clean Air Act Rulemaking Process: Interim Final Rule.* 40 Code of Federal Regulations 83. Environmental Protection Agency. 86 Federal Register 26406–26419 (May 14, 2021). www.federalregister.gov/documents/2021/05/14/2021-10216/rescinding-the-rule-on-increasing-consistency-and-transparency-in-considering-benefits-and-costs-in.
32. J. R. Biden. *E.O. 13990 Protecting Public Health and the Environment and Restoring Science to Tackle the Climate Crisis.* Executive Office of the President. 86 Federal Register 7037–7043 (January 20, 2021). www.federalregister.gov/documents/2021/01/25/2021-01765/protecting-public-health-and-the-environment-and-restoring-science-to-tackle-the-climate-crisis.
33. *Waste Prevention, Production Subject to Royalties, and Resource Conservation; Rescission or Revision of Certain Requirements: Final Rule.* 43 Code of Federal Regulations Parts 3160 and 3170. Department of the Interior, Bureau of Land Management. 83 Federal Register no. 189 (September 28, 2018). www.govinfo.gov/content/pkg/FR-2018-09-28/pdf/2018-20689.pdf.
34. *Repeal of the Clean Power Plan; Emission Guidelines for Greenhouse Gas Emissions from Existing Electric Utility Generating Units; Revisions to Emission Guidelines Implementing Regulations. Final Rule.* 40 Code of Federal Regulations 60. Environmental Protection Agency. 84 Federal Register 32520–32584 (September 6, 2018). www.federalregister.gov/documents/2019/07/08/2019-13507/repeal-of-the-clean-power-plan-emission-guidelines-for-greenhouse-gas-emissions-from-existing.
35. *National Emission Standards for Hazardous Air Pollutants: Coal- and Oil-Fired Electric Utility Steam Generating Units – Reconsideration of Supplemental Finding and Residual Risk and Technology Review: Final Rule.* 40 Code of Federal Regulations Part 63. Environmental Protection Agency (22 May 2020). www.federalregister.gov/documents/2020/05/22/2020-08607/national-emission-standards-for-hazardous-air-pollutants-coal-and-oil-fired-electric-utility-steam.

36. *Waste Prevention, Production Subject to Royalties, and Resource Conservation: Final Rule.* 43 Code of Federal Regulations 3100, 3160, 3170. Bureau of Land Management, Department of the Interior. 81 Federal Register no. 223, 83008–83089 (November 18, 2016). www.govinfo.gov/content/pkg/FR-2016-11-18/pdf/2016-27637.pdf.
37. K. M. Castle and R. L. Revesz. "Environmental Standards, Thresholds, and the Next Battleground of Climate Change Regulations." *Minnesota Law Review* 103 (2019): 1349–1437.
38. Institute for Policy Integrity. *Brief by Institute for Policy Integrity at New York University School of Law As Amicus Curiae in Support of Respondents in Michigan v. EPA*, March 2015. https://policyintegrity.org/documents/SCOTUS_brief_MATS_March2015.pdf.
39. Environmental Protection Agency. *Final Regulatory Impact Analysis: Control of Emissions from Nonroad Diesel Engines.* EPA420-R-04-007 (2004).
40. Environmental Protection Agency. *Regulatory Impact Analysis for the Final Clean Air Interstate Rule* (March 2005).
41. National Research Council. *Science and Decisions: Advancing Risk Assessment.* Washington, DC: National Academies Press, 2009. doi:10.17226/12209.
42. Environmental Protection Agency. *Summary of Expert Opinions on the Existence of a Threshold in the Concentration-Response Function for PM2.5-Related Mortality* (Washington, DC: 2010). www3.epa.gov/ttnecas1/regdata/Benefits/thresholdstsd.pdf.
43. E. R. Abt, J. V. Rodricks, J. I. Levy, L. Zeise and T. A Burke. "Science and Decisions: Advancing Risk Assessment." *Risk Analysis* 30, no. 7 (2010): 1028–1036. https://doi.org/10.1111/j.1539-6924.2010.01426.x.
44. H. J. Feldman, director of Regulatory and Scientific Affairs at the American Petroleum Institute. *Testimony Regarding EPA's Proposal to Change the National Ambient Air Quality Standards for Particulate Matter.* Submitted to EPA Public hearings in Philadelphia. Docket ID No. EPA–HQ–OAR–2007–0492. July 17, 2012.
45. J. E. Goodman, toxicologist at Gradient, an environmental consulting firm. *Testimony on "National Ambient Air Quality Standards for Particulate Matter: Proposed Rule" 78 Federal Register 3085.* Submitted to Environmental Protection Agency. Public hearings in Philadelphia. Docket ID No. EPA–HQ–OAR–2007–0492. July 17, 2012.
46. M. Lewis, A. Logomasini and W. Yeatman. *First Steps for the Trump Administration: Champion Affordable Energy: Free Market Reforms to Protect the Environment and Promote Plentiful, Reliable Energy.* Competitive Enterprise Institute (December 15, 2016). https://cei.org/sites/default/files/First%20Steps%20for%20the%20Trump%20Administration%20-%20Chamption%20Affordable%20Energy.pdf.
47. D. Furchtgott-Roth, senior fellow and director, Economics21 of the Manhattan Institute for Policy Research. *The Environmental Protection Agency's Flawed Cost-Benefit Analysis Methodology.* Submitted to testimony at the hearing by the Subcommittee on Superfund Waste Management and Regulatory Oversight, Senate Committee on Environment and Public Works. October 21, 2015.

48. American Petroleum Institute et al. *Carbon Pollution Emission Guidelines for Existing Stationary Sources: Electric Utility Generation Units, Proposed Rule, 79 Fed. Reg. 34,830.* Submitted to Environmental Protection Agency. Docket ID No. EPA–HQ–OAR–2013–0602; FRL–9910-86-OAR. June 18, 2014.
49. Environmental Protection Agency, Office of Air and Radiation. *Regulatory Impact Analysis for the Review of the Clean Power Plan: Proposal* (Research Triangle Park, NC: 2017). www.epa.gov/sites/production/files/2017-10/documents/ria_proposed-cpp-repeal_2017-10.pdf.
50. J. D. Graham and J. B. Wiener. *Risk Versus Risk: Tradeoffs in Protecting Health and the Environment.* Cambridge, MA: Harvard University Press, 1995.
51. Office of Management and Budget. *Circular A-4: Guidance to Federal agencies on the Development of Regulatory Analysis As Required under Section 6(a)(3)(c) of Executive Order 12866, a Regulatory Planning and Review, the Regulatory Right-to-Know Act, and a Variety of Related Authorities* (Washington, DC: 2003). www.whitehouse.gov/sites/whitehouse.gov/files/omb/circulars/A4/a-4.pdf.
52. J. Perkins. "The Case for Co-Benefits: Regulatory Impact Analyses, Michigan v. EPA, and the Environmental Protection Agency's Mercury and Air Toxics Standards." 2015–16 Olaus and Adolph Murie Award-winning paper (co-winner), Stanford Law School (2016). https://law.stanford.edu/publications/the-case-for-co-benefits-regulatory-impact-analyses-michigan-v-epa-and-the-environmental-protection-agencys-mercury-and-air-toxics-standards.
53. *United States Sugar Corp. v. Environmental Protection Agency*, No 11-1108 (D.C. Cir. 2016).
54. D. Bakst. "Will EPA Stop Its Abuse of Costly Pollution-Control 'Co-Benefits' Assessments?" *Heritage Foundation Commentary*, October 4, 2018. www.heritage.org/agriculture/commentary/will-epa-stop-its-abuse-costly-pollution-control-co-benefits-assessments.
55. S. Beaulier and D. Sutter. "The New 'Benefits' of Environmental Regulation." *American Energy Alliance*, October 11, 2012. www.americanenergyalliance.org/2012/10/11/the-new-benefits-of-environmental-regulation.
56. M. Todd, chair of the Residual Risk Coalition (this coalition includes the American Petroleum Institute). *Comments on "The National Emission Standards for Hazardous Air Pollutant Emissions: Coal- and Oil-Fired Electric Utility Steam Generating Units – Reconsideration of Supplemental Finding and Residual Risk and Technology Review," 84 Fed. Reg. 2,670 (Feb. 7, 2019).* Submitted to Environmental Protection Agency. Docket ID No. EPA–HQ–OAR–2018–0794. April 17, 2019.
57. *Supplemental Finding That It Is Appropriate and Necessary to Regulate Hazardous Air Pollutants from Coal-and Oil-Fired Electric Utility Steam Generating Units.* 40 Code of Federal Regulations 63. Environmental Protection Agency. 81 Federal Register 24419–24452 (April 25, 2016). www.federalregister.gov/documents/2016/04/25/2016-09429/supplemental-finding-that-it-is-appropriate-and-necessary-to-regulate-hazardous-air-pollutants-from.
58. D. Burtraw and A. Keyes. "10 Big Little Flaws in EPA's Affordable Clean Energy Rule." Issue Brief 19-05. July 22, 2019. www.rff.org/publications/issue-briefs/10-big-little-flaws-in-epas-affordable-clean-energy-rule.

59. Environmental Protection Agency, Office of Air and Radiation. *Regulatory Impact Analysis for the Repeal of the Clean Power Plan, and the Emission Guidelines for Greenhouse Gas Emissions from Existing Electric Utility Generating Units* (Research Triangle Park, NC: 2019). www.epa.gov/sites/prod uction/files/2019-06/documents/utilities_ria_final_cpp_repeal_and_ace_2019-06 .pdf.
60. D. Driesen. "Is Cost–Benefit Analysis Neutral?" *University of Colorado Law Review* 77, no. 2 (2006): 339–342.
61. A. McGartland et al. "Estimating the Health Benefits of Environmental Regulations." *Science* 357, no. 6350 (2017): 457–458. https://doi.org/10.1126 /science.aam8204.
62. Environmental Protection Agency. *Guidelines for Preparing Economic Analyses*. National Center for Environmental Economics, Office of Policy, Environmental Protection Agency (December 17, 2010; updated May 2014). EPA 240-R-10-001. www.epa.gov/sites/default/files/2017-08/documents/ee-0568-50.pdf.
63. M. Mazzota et al. *Non-monetary Benefits without Apology: The Economic Theory and Practice of Ecosystem Service Benefit Indicators*. Northeast Agricultural and Resource Economics Association (Newport, RI: June 27–30, 2015).
64. Congressional Research Service. *Clean Air Act Issues in the 116th Congress*. Report by J. E. McCarthy, R. K. Lattanzio and K. C. Shouse (Washington, DC: 2019).
65. J. Goffman. "MATS, Cost–Benefit Analysis, and the Appropriate and Necessary Finding." Environmental Energy and Law Program (2018). https://eelp.law.har vard.edu/2018/12/mats-cost-benefit-analysis-and-the-appropriate-and-necessary-finding.
66. *National Emission Standards for Hazardous Air Pollutants: Coal- and Oil-Fired Electric Utility Steam Generating Units – Reconsideration of Supplemental Finding and Residual Risk and Technology Review: Proposed Rule*. 40 Code of Federal Regulations 63. Environmental Protection Agency. 84 Federal Register 2670–2704 (February 7, 2019). www.federalregister.gov/documents/2019/02/07/ 2019-00936/national-emission-standards-for-hazardous-air-pollutants-coal–and-oil-fired-electric-utility-steam.
67. H. Roman et al. "Evaluation of the Cardiovascular Effects of Methylmercury Exposures: Current Evidence Supports Development of a Dose–Response Function for Regulatory Benefits Analysis." *Environmental Health Perspectives* 119, no. 5 (2011): 607–614. https://doi.org/10.1289/ehp.1003012.
68. E. Sunderland et al. "Benefits of Regulating Hazardous Air Pollutants from Coal and Oil-Fired Utilities in the United States." *Environmental Science & Technology* 50, no. 5 (February 5, 2016): 2117–2120. https://doi.org/10.1021/acs.est.6b00239.
69. A. Giang and N. Selin. "Benefits of Mercury Controls for the United States." *Proceedings of the National Academy of Sciences of the United States of America* 113, no. 2 (2015): 286–291. https://doi.org/10.1073/pnas.1514395113.
70. Edison Electric Institute et al. *Request for Expeditious Completion of the Residual Risk and Technology Review per CAA sections 112(d)(6) and (f)(2) by April 16, 2020*. Submitted to W. L. Wehrum, assistant administrator, Environmental Protection Agency. July 10, 2018.

71. D. Brady and J. Eilperin. "EPA Overhauls Mercury Pollution Rule despite Opposition from Utilities." *Washington Post*, April 17, 2020.
72. R. Beitsch. "EPA's Independent Science Board Says Agency Ignored Its Advice on Mercury Rule." *The Hill*, December 31, 2019. https://thehill.com/policy/energy-environment/476374-epas-independent-science-board-says-agency-ignored-their-advice-on+&cd=12&hl=en&ct=clnk&gl=us&client=firefox-b-1-d.
73. *Center for Biological Diversity v. National Highway Traffic Safety Administration*, 508 F.3d 508 (9th Cir. 2007).
74. Government Accountability Office. *Regulatory Impact Analysis: Development of Social Cost of Carbon Estimates*. Report by J. A. Gómez GAO-14-663 (Washington, DC: 2014). www.gao.gov/assets/670/665016.pdf.
75. *Sierra Club v. Federal Energy Regulatory Commission*, No. 16-1329 (D.C. Cir. 2017).
76. Environmental Protection Agency. *Technical Support Document: Social Cost of Carbon for Regulatory Impact Analysis under Executive Order 12866*. Report by Interagency Working Group on Social Cost of Carbon (Washington, DC: 2010).
77. Environmental Protection Agency. *Interagency Working Group on Social Cost of Greenhouse Gases, United States Government*. Technical Support Document: Technical Update of the Social Cost of Carbon for Regulatory Impact Analysis under Executive Order 12866 (Washington, DC: 2016).
78. American Petroleum Institute et al. *Comments on "Technical Update of the Social Cost of Carbon for Regulatory Impact Analysis under Executive Order No. 12866," 78 Fed. Reg. 70,586 (Nov. 26, 2013)*. Submitted to H. Shelanski, administrator, Office of Information and Regulatory Affairs. Docket ID OMB-OMB-2013-0007. February 26, 2014.
79. M. Lewis et al., Competitive Enterprise Institute and other organizations. *Technical Update of the Social Cost of Carbon for Regulatory Impact Analysis under Executive Order No. 12866*. Submitted to H. Shelanski, administrator, Office of Information and Regulatory Affairs. Docket ID OMB-OMB-2013-0007. February 26, 2014.
80. American Energy Alliance et al. *Statement in Support of the Transparency and Honesty in Energy Regulations Act of 2016*. Submitted to E. Jenkins, Representative of West Virginia and Justice of the Supreme Court of Appeals of West Virginia. July 2016.
81. US House of Representatives. *Hearing on at What Cost? Examining the Social Cost of Carbon*. Committee on Science, Space, and Technology. 115th Congress, 1st Sess. February 28, 2017.
82. D. J. Trump. *Presidential Executive Order on Restoring the Rule of Law, Federalism, and Economic Growth by Reviewing the "Waters of the United States" Rule*. 82 Federal Register 16093. 2017. www.whitehouse.gov/presidential-actions/presidential-executive-order-restoring-rule-law-federalism-economic-growth-reviewing-waters-united-states-rule.
83. Bureau of Land Management. *Regulatory Impact Analysis for the Final Rule to Suspend or Delay Certain Requirements of the 2016 Waste Prevention Rule*. Docket ID: BLM-2017-0002 (2017).
84. Environmental Protection Agency. *Addendum to Technical Support Document on Social Cost of Carbon for Regulatory Impact Analysis under Executive Order*

12866: Application of the Methodology to Estimate the Social Cost of Methane and the Social Cost of Nitrous Oxide. Report by Interagency Working Group on Social Cost of Carbon (2016).
85. Congressional Research Service. *EPA's Proposal to Repeal the Clean Power Plan: Benefits and Costs.* Report by K. Shouse. R45119 (Washington, DC: 2018). https://fas.org/sgp/crs/misc/R45119.pdf.
86. C. R. Sunstein. "On Not Revisiting Official Discount Rates: Institutional Inertia and the Social Cost of Carbon." *American Economic Review* 104, no. 5 (2014): 547–551.
87. M. Greenstone. "Assessing Approaches to Updating the Social Cost of Carbon." Presentation at the National Academies Fifth Meeting of the Committee on Assessing Approaches to Updating the Social Cost of Carbon. May 5, 2017.
88. M. L. Weitzman. "Fat Tails and the Social Cost of Carbon." *American Economic Review* 104, no. 5 (2014): 544–546. https://doi.org/10.1257/aer.104.5.544.
89. Intergovernmental Panel on Climate Change. *The Summary for Policymakers of the Special Report on Global Warming of 1.5°C (SR15).* World Meteorological Organization (Geneva, Switzerland: 2018). www.ipcc.ch/sr15.
90. R. E. Kopp et al. "Tipping Elements and Climate-Economic Shocks: Pathways Toward Integrated Assessment." *Earth's Future* 4, no. 8 (2016): 346–372.
91. Council of Economic Advisers. *Discounting for Public Policy: Theory and Recent Evidence on the Merits of Updating the Discount Rate.* White House (Washington, DC: 2017). https://obamawhitehouse.archives.gov/sites/default/files/page/files/201701_cea_discounting_issue_brief.pdf.
92. J. Broome. *Climate Matters: Ethics in a Warming World.* New York: W.W. Norton & Company, 2012.
93. M. Burke, S. M. Hsiang and E. Miguel. "Global Non-linear Effect of Temperature on Economic Production." *Nature* 527 (2015): 235–239. https://doi.org/10.1038/nature15725.
94. N. Stern. *The Economics of Climate Change: The Stern Review.* Cambridge, UK: Cambridge University Press, 2007.
95. T. Cowen and D. Parfit. "Against the Social Discount Rate." In *Justice between Age Groups and Generations*, edited by J. S. Fishkin and P. Laslett. 144–168. New Haven, CT: Yale University Press, 1992.
96. S. M. Gardiner. *A Perfect Moral Storm: The Ethical Tragedy of Climate Change.* Oxford, UK: Oxford University Press, 2011.
97. P. U. Clark et al. "Consequences of Twenty-First-Century Policy for Multi-millennial Climate and Sea-Level Change." *Nature Climate Change* 6 (2016): 360–369. https://doi.org/10.1038/nclimate2923.
98. Global Carbon Project. "Carbon Budget and Trends 2019" (2019). www.globalcarbonproject.org/carbonbudget/index.htm.
99. D. Brown. *Climate Change Ethics: Navigating the Perfect Moral Storm.* London, UK: Routledge, 2013.
100. P. Howard and D. Sylvan. *The Economic Climate: Establishing Expert Consensus on the Economics of Climate Change.* Institute for Policy Integrity, New York University School of Law (2015). http://policyintegrity.org/files/publications/EconomicClimateConsensus.pdf.

101. P. Howard and J. Schwartz. "Think Global: International Reciprocity as Justification for a Global Social Cost of Carbon." *Columbia Journal of Environmental Law* 42, no. S (2017): 203–294. https://doi.org/10.7916/cjel.v42iS.3734.
102. R. L. Revesz et al. "The Social Cost of Carbon: A Global Imperative." *Review of Environmental Economics and Policy* 11, no. 1 (2017): 172–173.
103. T. Gayer and W. K. Viscusi. "Determining the Proper Scope of Climate Change Policy Benefits in US Regulatory Analyses: Domestic versus Global Approaches." *Review of Environmental Economics and Policy* 10, no. 2 (2016): 245–263.
104. T. Gayer and W. K. Viscusi. "Letter – The Social Cost of Carbon: Maintaining the Integrity of Economic Analysis – A Response to Revesz et al. (2017)." *Review of Environmental Economics and Policy* 11, no. 1 (2017): 174–175. https://doi.org/10.1093/reep/rew021.
105. R. Reagan. *Statement on Signing the Montreal Protocol on Ozone-Depleting Substances*. American Presidency Project, University of California, Santa Barbara. April 5, 1988. www.presidency.ucsb.edu/node/253878.
106. R. E. Benedick. *Ozone Diplomacy New Directions in Safeguarding the Planet*. Cambridge, MA: Harvard University Press, 1998.
107. G. Bush. *Statement on Signing the Instrument of Ratification for the United Nations Framework Convention on Climate Change*. American Presidency Project, University of California, Santa Barbara. October 13, 1992. www.presidency.ucsb.edu/node/266987.
108. *Zero Zone, Inc. v. United States Department of Energy*, 832 F.3d 654 (7th Cir. 2016).
109. N. Jones. "China Tops CO_2 Emissions." *Nature*, June 20, 2007. https://doi.org/10.1038/news070618-9.
110. M. Burger et al. "Legal Pathways to Reducing Greenhouse Gas Emissions under Section 115 of the Clean Air Act." *Georgetown Environmental Law Review* 28 (2016): 359–423.
111. J. Tollefson and K. R. Weiss. "Nations Adopt Historic Global Climate Accord: Agreement Commits World to Holding Warming 'Well Below' 2°C." *Nature* 582, no. 7582 (2015): 315–317.
112. S. Dudley and B. Mannix. "The Social Cost of Carbon." *Engage: Journal of the Federalist SocietyPractice Group* 15, no. 1 (June 24, 2014): 14–18.
113. A. Fraas et al. "Social Cost of Carbon: Domestic Duty." *Science* 351, no. 6273 (2016): 569. https://doi.org/10.1126/science.351.6273.569-b.
114. K. Rickie et al. "Country-level Social Cost of Carbon." *Nature Climate Change* 8 (2018): 895–900. https://doi.org/10.1038/s41558-018-0282-y.
115. Government Accountability Office. *The Federal System for Collecting Oil and Gas Revenues Needs Comprehensive Reassessment*. Report by F. Rusco. GAO-08-691 (Washington, DC: 2008). www.gao.gov/products/GAO-08-691.
116. Government Accountability Office. *Federal, Oil and Gas Leases: Opportunities Exist to Capture Vented and Flared Natural Gas, Which Would Increase Royalty Payments and Reduce Greenhouse Gases*. Report by F. Rusco. GAO-11-34 (Washington, DC: 2010). www.gao.gov/products/GAO-11-34.

117. R. Ranger, senior policy advisor, Upstream. *American Petroleum Institute's Comments on "Waste Prevention, Production Subject to Royalties, and Resource Conservation; Rescission or Revision of Certain Requirements," 83 Fed. Reg. 7924 (Feb. 22, 2018)*. Submitted to B. Steed, deputy director of Programs and Policy, Bureau of Land Management. RIN 1004-AE53. April 23, 2018.
118. M. Todd, senior policy advisor, American Petroleum Institute. *Statement for the Public Hearing on "Oil and Natural Gas Sector: Emission Standards for New, Reconstructed, and Modified Sources Reconsideration Proposed Amendments to NSPSOOOOa," 83 Fed. Reg. 52056-52107*. Submitted to public hearing. EPA-HQ-OAR-2017-0483. November 14, 2018.
119. B. Olsen and C. M. Matthews. "Trump Rollback of Methane Regulations Splits Energy Industry, Big Oil-and-Gas Companies Support Restrictions on the Powerful Greenhouse Gas, While Smaller Companies Worry about Cost." *Wall Street Journal*, August 29, 2019. www.wsj.com/articles/trump-rollback-of-methane-regulations-splits-energy-industry-11567098375.
120. R. Alvarez et al. "Assessment of Methane Emissions from the US Oil and Gas Supply Chain." *Science* 361, no. 6398 (2018): 186–188. https://science.sciencemag.org/content/361/6398/186.
121. A. Krupnick and I. Echarte. *The 2016 BLM Methane Waste Prevention Rule: Should It Stay or Should It Go?* Resources for the Future (2018). https://media.rff.org/documents/RFF-Rpt-Oil26GasRegs-BLM20methane20rule.pdf.
122. Bureau of Land Management. *Regulatory Impact Analysis for: Revisions to 43 CFR 3100 (Onshore Oil and Gas Leasing) and 43 CFR 3600 (Onshore Oil and Gas Operations) Additions of 43 CFR 3178 (Royalty-Free Use of Lease Production) and 43 CFR 3179 (Waste Prevention and Resource Conservation)*. Docket ID: BLM-2016-0001 (2016). www.regulations.gov/document?D=BLM-2016-0001-9127.
123. Bureau of Land Management. *Regulatory Impact Analysis for the Final Rule to Rescind or Revise Certain Requirements of the 2016 Waste Prevention Rule*. (Washington, DC: 2018).
124. M. Bennet, US senator for Colorado. "Bennet Introduces Bill to Lock in a Science-Based Method to Determine the Cost of Carbon Pollution." News release, 2019. www.bennet.senate.gov/public/index.cfm/2019/6/bennet-introduces-bill-to-lock-in-a-science-based-method-to-determine-the-cost-of-carbon-pollution.
125. US Congress. Senate. *Carbon Pollution and Transparency Act*. S. 1745, 116th Congress, 1st Sess. Introduced in Senate June 5, 2019.
126. D. Roberts. "A Closer Look at Washington's Superb New 100% Clean Electricity Bill." *Vox*, April 18, 2019. www.vox.com/energy-and-environment/2019/4/18/18363292/washington-clean-energy-bill.
127. J. Kohler. "'Transformative,' 'Substantial,' 'Turducken'? Colorado Lawmakers Approve a Bevy of Energy Bills in 2019 Session." *Denver Post*, May 19, 2019. www.denverpost.com/2019/05/19/colorado-clean-energy-legislature-xcel.
128. Climate Leadership Council. "The Conservative Case for Carbon Dividends." Briefing by J. A. Baker III, M. Feldstein, T. Halstead, N. G. Mankiw, H. M. Paulson, Jr., G. P. Shultz, T. Stephenson and R. Walton. 2017.
129. US Congress. House. *Transparency and Honesty in Energy Regulations Act of 2017*. H.R. 3117, 115th Congress, 1st Sess. Introduced in House June 26, 2017.

130. US Congress. Senate. *Transparency and Honesty in Energy Regulations*. S. 1512, 115th Congress, 1st Sess. Introduced in Senate June 29, 2017.
131. Congressional Research Service. *An Introduction to Judicial Review of Federal Agency Action*. Report by J. P. Cole. R44699 (Washington, DC: 2016). https://fas.org/sgp/crs/misc/R44699.pdf.
132. Congressional Research Service. *A Brief Overview of Rulemaking and Judicial Review*. Report by T. Garvey. R41546 (Washington, DC: 2017). https://fas.org/sgp/crs/misc/R41546.pdf.
133. L. Heinzerling. "Unreasonable Delays: The Legal Problems (So Far) of Trump's Deregulatory Binge." *Harvard Law and Policy Review* 12 (2018): 13–48.
134. R. Stanberry. "The APA As an Environmental Law." *Environmental Law* 49, no. 3 (2019). https://law.lclark.edu/live/files/28857-493stansberry.
135. N. Kassop. "Legal Challenges to Trump Administration Policies: The Risks of Executive Branch Lawmaking That Fails to 'Take Care'" In *Presidential Leadership and the Trump Presidency*, edited by C. M. Lamb and J. R. Neiheisel. 41–90. Cham, Switzerland: Palgrave Macmillan, 2019.
136. *Clean Air Council v. Pruitt*, No. 17-1145 (D.C. Cir. 2017).
137. *State of California, et al. v. Bureau of Land Management, et al.*, No.3:17-cv-07187-WHO (District Court for the Northern District of California 2018).
138. *Chevron USA., Inc. v. Natural Resources Defense Council, Inc.*, 467 US 837 (Supreme Court 1984).
139. *Federal Communications Commission v. Fox Television Stations, Inc.*, 556 US 502 (Supreme Court 2009).
140. *Indigenous Environmental Network v. US Department of State*, 347 F. Supp. 3d 561, 591 (District Court of the District of Montana 2018).
141. K. E. Adair and R. R. Akroyd. "Tucson Herpetological Society v. Salazar Decision and its Progeny: A Move Away from Blind Deference to Agency Decision-making." *California Water Law & Policy Reporter*, March 2012.
142. W. W. Buzbee. "The Tethered President: Consistency and Contingency in Administrative Law." *Boston University Law Review* 98 (2018): 1357–1442. https://scholarship.law.georgetown.edu/cgi/viewcontent.cgi?article=3084&context=facpub.
143. C. Cecot. "Deregulatory Cost–Benefit Analysis and Regulatory Stability." *Duke Law Journal* 68, no. 8 (2018–19): 1594–1650.
144. W. W. Buzbee. "Deregulatory Splintering the Trump Administration and Administrative Law." *Chicago-Kent Law Review* 94 (2019): 439–486.
145. *Tucson Herpetological Society v. Salazar*, 566 F.3d 870 (9th Cir. 2009).
146. E. Fisher, P. Pascual and W. Wagner. "Rethinking Judicial Review of Expert Agencies Symposium: Science Challenges for Law and Policy." *Texas Law Review* 93 (2014–15): 1681–1722.
147. W. Wagner, E. Fisher and P. Pascual. "Whose Science? A New Era in Regulatory 'Science Wars.'" *Science* 362, no. 6415 (2018): 636–639.
148. T. O. McGarity. "Judicial Review of Scientific Rulemaking." *Science, Technology, & Human Values* 9, no. 1 (1984): 97–106.
149. R. J. Kozel and J. Pojanowski. "Administrative Change." *UCLA Law Review* 59 (2011): 112–169.

150. *Baltimore Gas & Electric Co. v. Natural Resource Defense Council, Inc.*, 462 US 87 (Supreme Court 1983).
151. E. Kuhn. "Science and Deference: The 'Best Available Science' Mandate Is a Fiction in the Ninth Circuit." *Environmental Law Review Syndicate*, 2016. http://elawreview.org/environmental-law-review-syndicate/science-and-defer ence-the-best-available-science-mandate-is-a-fiction-in-the-ninth-circuit.
152. D. Dana and M. Barsa. "Judicial Review in an Age of Hyper-Polarization and Alternative Facts." *San Diego Journal of Climate and Energy Law* 9 (2017–18): 231–263. www.law.berkeley.edu/wp-content/uploads/2019/10/Judicial-Review-in-an-Age-of-Hyper-Polarization-and-Alternative-F.pdf.
153. *California v. Bureau of Land Management*, No 4:18-cv-00521 (9th Circ. 2020).
154. E. Gilmer. "Trump Environmental Record Marked by Big Losses, Undecided Cases." *Bloomberg Law*, January 11, 2021.
155. J. P. Jacobs and P. King. "Biden Races Courts for Chance to Torpedo Trump Water Rule." *E&E News*, April 28, 2021.
156. D. A. Farber. "Rethinking the Role of Cost–Benefit Analysis." *University of Chicago Law Review* 76 (2009): 1355–1380.
157. *Frank R. Lautenberg Chemical Safety for the 21st Century Act* §6, Pub L No 114–182, 130 Stat 448, 460 (2016), codified at 15 USC § 2605.
158. L. G. Chestnut and D. M. Mills. "A Fresh Look at the Benefits and Costs of the US Acid Rain Program." *Journal of Environmental Management* 77 no. 3 (2005): 252–266.
159. E. Gilmer. "Trump's 2-for-1 Regulations Order Survives States' Legal Attack." *Bloomberg Law,* April 2, 2020.
160. *Whitman v. American Trucking Association,* 531 US 457 (Supreme Court 2001).
161. *Corrosion Proof Fittings v. EPA*, 947 F.2d 1201 (5th Cir. 1991).
162. 30 U.S.C. § 225.
163. *Motor Vehicle Manufacturers Association v. State Farm Auto Mutual Insurance Co.,* 463 US 29 (Supreme Court 1983).

10

Law
Anti-regulatory Statutory Interpretations and Reshaping the Judiciary

This third and final chapter on the regulatory process turns to how the Trump administration bent the norms of statutory interpretation and judicial opinion in the administrative state in order to favor the industry and weaken regulatory protections for public health and the environment. The US Constitution vests Congress with the power to manage public lands and seas, to regulate activities that affect interstate commerce, and to make laws accordingly.[i] Congress can then delegate powers to the president and to federal agencies to carry out various functions to achieve the goals of particular statutes, that is, laws enacted by Congress. Congress has set up a number of federal administrative agencies, which are under the supervision of the executive branch, to carry out the tasks of managing the nation's resources, health and environment. Congress delegates specific powers to these agencies, detailing each agency's structure, jurisdiction, goals, regulatory authority and the procedures it must employ when exercising the power that Congress delegates through statutes.[1] These agencies that carry out various functions of governments are collectively known as the administrative state.

The executive branch is tasked with running the administrative state. In doing so, it is constrained to operate within its powers under the Constitution and within laws Congress has enacted, with limited exceptions for narrow, constitutionally endowed responsibilities. While the executive branch directs federal agencies to carry out its policies, federal agencies must operate within statutory powers Congress has delegated and abide by laws governing agencies' rulemaking, such as the Administrative Procedure Act (APA). Even so, the executive branch has ample room to maneuver by choosing how to interpret its constitutional and statutory powers and, thus, defend the legality of its

---

[i] A majority of core environmental laws are founded on Congress's authority under the Commerce Clause [146].

actions. Therefore, the presidential administration enjoys broad discretion on how it manages public lands and seas – specifically, the balance between extractive and conservation activities – and the extent to which it protects public health and the environment.

Even when all administrations are inclined to interpret the constitution and laws to justify their actions, the Trump administration's brazenness in its legal interpretations – straining the plain meaning of statutory texts and defying the Supreme Court's interpretations – was still astounding. Trump asserted powers beyond statutory texts to open to drilling public lands and seas that had been set aside for conservation. As COVID-19 engulfed the nation, Trump issued an executive order instructing federal agencies to bypass procedures under the National Environmental Policy Act (NEPA) to fast-track the permitting of oil and gas infrastructure projects, even when the law did not provide for such presidential powers.[2] Federal agencies, under the direction of Trump appointees, interpreted statutes in ways that curbed states' powers to protect their air and water resources.[ii]

The Trump administration not only diminished public trust in federal agencies, such as the Environmental Protection Agency (EPA), the Bureau of Land Management (BLM) and the Bureau of Safety and Environmental Enforcement (BSEE), he reoriented these agencies to promulgate regulations that unabashedly served the oil and gas sector. The administration also attacked the entire apparatus of the administrative state by denigrating federal agencies as an unaccountable bureaucracy that had regulated beyond its powers, even when, in reality, federal agencies face checks to ensure their accountability to Congress, to the president and to the American people.[1,3,4] That attack echoes the long-running campaign against the administrative state by think tanks that receive funding from, among others, the oil and gas sector. The demise of the administrative state would free polluters from regulations but would inexorably weaken protections for the public.

Courts often act as the final arbiter of statutory and constitutional interpretation.[iii] The judicial branch can act as an important check against the

---

[ii] For instance, its regulation that crippled states' powers to block risky oil and gas infrastructure projects contradicted the CWA, past Supreme Court interpretations of that statute, and states' past exercise of those powers. Federal agencies interpreted federal environmental statutes, whose texts speak of ambitious public health and environmental protections, in a very narrow manner and, arguably, abnegated their powers [27].

[iii] In theory, Congress can enact laws to overturn a Supreme Court decision – so long as it has the support of both chambers and of the president or enough votes to overturn a presidential veto. However, as described in previous chapters, deadlocked congresses have enacted or amended nearly no laws to protect public health or the environment since the 1990s. Congress also has no authority to overturn Supreme Court decisions based on the Constitution, e.g., when the court deems a law passed by Congress unconstitutional.

president's violations of constitutional and statutory powers and federal agencies' violation of statutory powers. Courts also play an important role in adjudicating the boundaries of power of the administrative state, as Congress's constitutional powers to create administrative agencies and to delegate powers to them are subject to the constitutional constraints of separation of powers between the legislative and the executive branches[1] as well as constitutional limits on what powers Congress can exercise in the first place. Ultimately, the Supreme Court's decisions determine the ground rules on whose rights and interests enjoy greater protection in the US legal system.

Recognizing these vital roles of the judiciary, Trump, with the support of Senate majority leader Mitch McConnell, congressional Republicans and an array of conservative and corporate interests, pursued his most potent and long-lasting legacy, that is, the reshaping of the federal judiciary. White House Counsel Don McGahn worked on assembling the list of nominees to judgeships with John Malcolm of the Heritage Foundation, a conservative think tank, and Leonard Leo, the chairman of the Federalist Society, a network of conservatives and libertarians that has worked for decades toward the nomination of judges who are antithetical to the administrative state.[5-9] McGahn underscored that "There is a coherent plan here where actually the judicial selection and the deregulatory effort are really the flip side of the same coin."[9]

This deliberate selection of judges who are inclined to read narrowly the protections offered by environmental statutes and to rein in the powers of the administrative state increases the chances of success for anti-regulatory litigants and bolsters those judges already widening their application of legal doctrines to support such interpretations.[10,11] Trump's selection of judges thumbs the scales toward judicial decisions that favor the rollback of regulations, restrictions on new regulations and the crippling of the administrative state. Trump's appointments, relative to those by previous presidents,[iv] stand out in their sheer numbers and in the shift in the balance of judges in the federal courts to a majority appointed by Republican presidents – including a 6–3 conservative-majority Supreme Court – that could last a generation.[v] These efforts will last much longer than the Trump presidency, given federal judges' lifetime appointments.

In this chapter, I describe how Trump and federal agencies interpreted laws in ways to favor the oil and gas sector (Section 10.1) but how courts served as

---

[iv] Trump appointed more Court of Appeals judges than any other president since President John F Kennedy in their first four years in office, except Carter, and almost twice the number of Court of Appeals judges than Obama did in his first term [94].

[v] By "conservative" judges, I refer to those judges who tend to take a more skeptical view of the administrative state (Section 10.3).

a check to strike down several of the administration's most brazen statutory interpretations (Section 10.2). Next, I turn to how Trump's judicial appointments may well shape future judicial decisions, by describing how judges' interpretations of statutes are shaped by their views of the administrative state (Section 10.3) and how the Federalist Society, which favors judges antithetical to the administrative state, shortlisted the judges whom Trump appointed (Section 10.4). I conclude by underscoring how the work of public health and environmental advocates has become ever more important, not only to challenge legal interpretations that narrow the protections under federal environmental statutes but also to challenge the rise of doctrines that hamper Congress's powers to enact laws and to delegate powers to federal agencies, both actions that are vital to protecting public health and the environment.

10.1 The Trump Administration's Legal Interpretations to Favor the Oil and Gas Sector

Congress enacted the Administrative Procedure Act to set out how agencies are to exercise their statutory powers. The judiciary has also developed legal rules to guide the interpretation of enabling statutes, that is, how to read the text of the statute and in case of ambiguity, examine the context of the statute. The executive branch and agencies, by choosing how they interpret laws, can favor the oil and gas sector. Trump and federal agencies directed by his appointees were remarkably brazen in their interpretations to claim expansive statutory powers to favor the oil and gas sector (Section 10.1.1), to curb states' powers to protect their health and environment (Section 10.1.2) and to weaken regulations (Section 10.1.3).

10.1.1 Claims of Expansive Executive Powers to Favor Oil and Gas Expansion

Presidents hold not only constitutional powers but also powers delegated to them by Congress through statutes. Trump distorted a number of these delegated powers in order to manipulate Congress's original intentions. For instance, in 2017, Trump shrunk two national monuments that Clinton and Obama had protected and he revoked Obama's withdrawal from offshore leases of several areas in the Arctic and Atlantic outer continental shelf. Trump's actions, as discussed in Chapters 5 and 6, pave the way for oil and gas extraction to proceed in areas formerly protected for conservation. A number of legal scholars argue that, based on the text of the two statutes – the Antiquities Act and the Outer Continental Shelf Lands Act section 12 – and their context, Congress delegated one-directional powers to presidents, that is, to impose protections but not to remove protections.[12,13]

In June 2020, Trump issued an executive order instructing federal agencies, during the COVID-19 pandemic, to bypass environmental review procedures required under NEPA.[14] Trump's actions, detailed in Chapter 5, aimed to fast-track the construction of pipelines, oil and gas drilling projects, and other infrastructure projects.[15,16] While the order claimed that NEPA granted the president "emergency authorities" to waive these procedures, several legal scholars[vi] countered that NEPA does not allow for broad waivers.[15] NEPA guidance, published by the Council for Environmental Quality, which oversees the federal agencies' implementation of NEPA, permits waivers for projects that had to proceed urgently to avoid "harm to life, property or important natural resources."[16] Fast-tracking oil and gas infrastructure does not meet these emergency needs.

10.1.2 Federal Agencies Interpreted Statutes to Curb States' Powers

Most environmental statutes are designed to address public health and environmental problems using collaborative efforts of both federal and state governments, typical of cooperative federalism.[vii] However, this arrangement is not without conflict and disagreement. Recognizing states' critical role in environmental protection, Congress reserved power for states in several federal environmental statutes and thus assured states of some constraints on federal powers. The Trump administration professed a commitment to cooperative federalism and protecting state sovereignty, but in reality it interpreted statutes in ways to curb state efforts to impose more stringent regulations than the federal baseline.[17]

For instance, under the Clean Water Act (CWA) section 401, applicants for federal permits for infrastructure projects, such as interstate gas pipelines and export or import terminals for liquid natural gas or coal, must first secure water certification permits from the state governments.[18,19] As detailed in Chapter 4,

---

[vi] Richard Lazarus, professor at Harvard Law School, and Jayni Hein, a lawyer at the Institute for Policy Integrity at New York University, were cited in the news reports.

[vii] Cooperative federalism is a statutory structure in which Congress creates a new regulatory program and establishes federal standards but leaves significant authority over implementation to the states. Most federal statutes designed to address public health and environmental protection are collaborative efforts that build on the idea of cooperative federalism.

in June 2020, the EPA finalized the 401 Certification Rule[20] that shifts the power from states to the Federal Energy Regulatory Commission (FERC) and the Army Corps of Engineers to approve energy projects. That rule enables FERC and the Corps to permit projects to proceed over a state's objection. The rule also restricts a state's review to only water quality of discharges from point sources – as opposed to states' previous reviews of the cumulative impact of the entire project on the quality and quantity of water resources and the health of the aquatic ecosystem.[21–24]

Critics underscore how the rule contradicts the CWA, in which Congress expressly recognizes states' powers to protect their water resources and grants states broad powers to review projects that affect their water resources.[21–24] The Supreme Court had affirmed these broad powers[viii] in 1994 in *PUD No. 1 v. Washington Department of Ecology*[25] and again in 2006 in *S. D. Warren Co. v. Maine Board of Environmental Protection*.[26]

The administration also interpreted other statutes in a way that curbed states' ability to protect their public health and environment. It denied the state of California a waiver to set stricter fuel efficiency standards (which would improve California's air quality but cut the demand for gasoline), even when California had exercised its powers under the Clean Air Act (CAA) to do so for decades (Chapter 11). The administration also proposed to narrow states' powers to challenge offshore drilling under the Coastal Zone Management Act (Chapter 6).

10.1.3 Federal Agencies Interpreted Statutes to Weaken Regulations

When Congress delegates powers to agencies, it provides agencies with both discretionary power and mandatory duties to help achieve statutory goals. To justify deregulation on legal grounds, federal agencies under the Trump administration interpreted statutes to claim that they did not have the discretionary powers or the mandatory duties long understood to be part of the statutory delegation.[27] By abnegating its powers, a federal agency may act against Congress's instruction, analogous to the case when an agency regulates beyond its statutory powers.

For instance, the Mineral Leasing Act of 1920 (MLA) requires BLM to ensure that lessees "use all reasonable precautions to prevent waste of oil or gas developed in the land." Reviews by the Office of the Inspector General of the

---

[viii] States may regulate the broad impacts of a project, as long as there is a discharge. States can also consider a range of conditions, not just water quality [19].

Department of the Interior and the Government Accountability Office, which report that 40 percent of gas vented or flared on BLM leases could be economically captured, had recommended that BLM update its regulations to prevent that waste.[28] In 2016, BLM promulgated the Waste Prevention Rule that restricted the venting and flaring of gas and that imposed royalty payments for gas that was not captured by oil and gas operations on public lands and was thus wasted.[28] The rule responded to BLM's statutory duty to prevent the waste of natural gas and had the additional benefit of improving air quality. The judicial review of the 2016 rule by the District Court for the District of Wyoming[29] in January 2017 concluded that BLM did not exceed its statutory authority.[ix]

However, in September 2018, under the Trump administration, BLM revised and rescinded sections of its 2016 Methane Waste Prevention Rule.[30] It chose to interpret BLM's prior action as exceeding its statutory authority. The Trump BLM claimed that "[in its previous 2016 rule] BLM exceeded its statutory authority by promulgating a rule that, rather than regulating for the prevention of 'waste,' was actually intended to regulate air quality, a matter within the regulatory jurisdiction of the EPA and the States under the Clean Air Act."

10.2 How Have Courts Ruled on the Trump Administration's Statutory Interpretations?

As of July 2020, only a few challenges against the Trump administration's interpretations of statutes to favor the oil and gas sector and deregulation have been heard in the courts.[x] I discuss three decisions that underscore how courts serve as an important check against the administration's strained legal interpretations. I also discuss a fourth decision as a reminder that the executive branch does have ample room to weaken regulations even within statutorily permissible discretion.

[ix] The court noted that "BLM has authority to promulgate and impose regulations which may have air quality benefits and even overlap with CAA regulations if such rules are independently justified as waste prevention measures pursuant to its [Mineral Leasing Act] authority." The court decided that without viewing the administrative record, it "cannot conclude that the provisions of the Rule which overlap with EPA/state air quality regulations promulgated under CAA authority lack a legitimate, independent waste prevention purpose or are otherwise so inconsistent with the CAA as to exceed BLM's authority and usurp that of the EPA, states, and tribes" [29].

[x] The Institute for Policy Integrity at New York University of Law tracks the outcomes of litigation over the Trump administration's use of agencies to deregulate and to implement its other policy priorities. See https://policyintegrity.org/trump-court-roundup.

First, the District Court for the District of Alaska in *League of Conservation Voters v. Trump*[31] ruled that Trump exceeded his statutory powers under the Outer Continental Shelf Lands Act and blocked his revocation of protections for parts of the Arctic and Atlantic seas from oil and gas drilling. In her March 2019 decision, Judge Sharon Gleason noted that the text of Section 12 (a) refers "only to the withdrawal of lands; it does not expressly authorize the President to revoke a prior withdrawal." Granting the possibility of ambiguity, she looked further into the context, and concluded that "Congress expressed one concept – withdrawal – and excluded the converse – revocation." (In April 2021, Biden issued an executive order that revoked Trump's order in its entirety.)[31,xi]

Second, the Court for the Northern District of California ruled in *California v. Bernhardt*[32] against the Trump administration's rescission of provisions in the Waste Prevention Rule and vacated those rescissions. In her July 2020 decision, Judge Yvonne Gonzalez Rogers held that the agency "ignored its statutory mandate under the Mineral Leasing Act." She noted that the statute states, "Each lease shall contain provisions for the purpose of insuring the exercise of reasonable diligence, skill, and care ... ; and for the prevention of undue waste." In addition, she held that the agency failed to abide by required procedures when making a policy shift, that is, "it repeatedly failed to justify numerous reversals in policy positions previously taken, and failed to consider scientific findings and institutions relied upon by both prior Republican and Democratic administrations." That decision was appealed to the Court of Appeals for the Ninth Circuit.

Third, the Supreme Court ruled against a narrow interpretation of the protections under the CWA in *County of Maui v. Hawaii Wildlife Fund*.[33] That April 2020 decision demonstrates how courts can serve as a bulwark against legal interpretations that paralyze federal environmental statutes. Pipeline developers, a beneficiary of weakened CWA protections, filed a brief in support of the county of Maui. The Trump administration's Department of Justice argued in defense of the county of Maui that the CWA permits are required only when a point source discharges pollution directly into navigable waters protected under the act.[xii] It claimed that permits are thus not required if pollution travelled even a short distance through groundwater before reaching navigable waters.[34]

[xi] The Court of Appeals for the Ninth Circuit heard oral arguments in June 2020, but it ruled in April 2021 that the appeal was moot following Biden's order. Parties to the lawsuit agreed that Obama's withdrawal of those areas from oil and gas leasing remained intact.

[xii] In the case in question, wastewater at a wastewater treatment plant had been stored in underground injection wells and travelled a short distance through the ground to sea.

Justice Stephen Breyer, writing for the 6–3 majority, focused on the purpose of the act to protect waters in the United States. He opined that discharges that travel within a short time and distance via groundwater to navigable waters are the "functional equivalent" to "discharges directly to navigable waters" for which the act requires permits. He noted that exempting the need for permits in these cases would create a loophole that defeats the protective purpose of the statute.[xiii] The majority decision attempts to balance, on the one hand, protections for waters and, on the other hand, a manageable permitting program that does not ensnare all small sources of pollution.[34] However, the three dissenting opinions by Justices Samuel Alito, Clarence Thomas and Neil Gorsuch, ignore the purpose of the statute and instead insist on a focus on the statute's text, which states that permits are required when pollutants are discharged into navigable waters. This case demonstrates how judges' theories about how to interpret statutes – using statutory purpose to inform interpretation or focusing only on text without regard to purpose – matter greatly for the extent of protections offered by these statutes, a point I explore in greater detail in Section 10.3.

While judicial review serves as a check on the president's and federal agencies' statutory interpretation, it does not provide a bulwark against deregulation. Federal agencies have ample room to tighten or to weaken regulations. Federal agencies' actions that are moored to reasonable statutory interpretations have survived judicial review if they meet a fairly low bar of following procedures and providing reasons for their actions. For instance, in March 2020, the District Court for the Northern District of California ruled in favor of BLM's actions in 2017 to repeal the 2015 rule on hydraulic fracturing on public lands.[35] As noted by the court, "The BLM met the requirement of providing a 'reasoned explanation' for why it is changing course after the nearly five-year long extensive rulemaking that resulted in the 2015 Rule. Although BLM could have provided more detail, it did enough to clear the low bar of arbitrary and capricious review, and that is all the law requires." (As discussed in Section 10.3, a higher bar for judicial review of agencies' actions is not necessarily a solution to protecting public health and the environment.)

These cases illustrate how judicial interpretations of statutes can serve as a check against the executive branch's brazen interpretations to justify

---

[xiii] In the past, the EPA's legal position was that permits are required for discharges of pollution that reach waters protected by the CWA through groundwater. Under the Trump administration, the EPA changed its position opining that all releases of pollutants to groundwater are categorically excluded from the CWA's permitting program. Breyer acknowledged the issue of deference to agency's interpretation but noted that the EPA's revised interpretation would open a huge loophole for clean water protection [34].

deregulation. Judicial interpretations serve as the arbiter of the extent of protections offered by landmark federal environmental laws, unless decisions are overturned by Congress through legislation. For instance, the Supreme Court's majority decision in *Massachusetts v. EPA*,[36] together with the EPA's endangerment finding, discussed in Table 10.1 (Appendix), compelled the EPA to regulate greenhouse gases. Likewise, courts' interpretation of NEPA forces agencies to take into account climate impacts in environmental assessments. This raises the question of how judges arrive at their statutory interpretations, a subject I turn to next.

10.3 Why the Judiciary Matters: Differing Views on the Constitution, Congressional Powers and the Administrative State

Judges are expected to interpret the Constitution and legislation with impartiality. Lower court judges are constrained to follow historical precedent under the doctrine of *stare decisis*, that is, to stand by things decided. The Supreme Court, in most circumstances, follows its own prior decisions as a matter of practice. Many of the decisions of the Supreme Court do not split along ideological lines. Judges jealously guard their independence and the public perception of the judiciary as an institution independent of the appointing president.[37,xiv]

Even with these caveats, scholars have documented how judges' philosophies influence their legal analysis and thus their decisions.[38,39] Empirical studies also find correlation between judges' political-party affiliation or the party of the president who appointed them and judges' decisions.[40–42] For example, Cass Sunstein and his co-authors find that in reviewing agency interpretations of law, "there is a definite 'tilt,' on the part of federal judges, in the direction of administrations of the same political party as their appointing president."[42–44] Public health and environmental policy experts have also detailed how conservative judges' statutory interpretations limit public health and environmental protection.[45–48] With a 6–3 conservative majority at the Supreme Court and new justices less committed to *stare decisis* (including Justice Gorsuch and Justice Amy Coney Barrett), these influences stand to have major impacts.[49]

[xiv] In response to Trump's criticism of a judge appointed by Obama, Chief Justice Roberts defended the judiciary's independence, saying "We do not have Obama judges or Trump judges, Bush judges or Clinton judges ... The independent judiciary is something we should all be thankful for."

Mapping judges' constitutional and statutory interpretations gives some insight into these impacts, especially in the nation's highest court (Sections 10.3.1 and 10.3.2). These interpretations shape judicial review of agencies' actions and constitutional challenges to Congress's enactment of laws and delegation of powers. The ascent of more conservative judges who view the administrative state as a threat to liberty,[17,39,50] including Trump's wave of judicial appointees who almost unanimously share these views,[51] has led to the entrenchment of doctrines[45,46] that weaken legal protections for public health and the environment (Section 10.3.3).

10.3.1 A Brief History of the Administrative State and the Supreme Court

The Constitution does not expressly establish the administrative state. It does however vest in Congress numerous powers, including the powers to enact laws that are "necessary and proper" to realize its powers,[1] and refers to "executive departments" throughout. The Supreme Court has recognized Congress's use of its Article I lawmaking powers to create federal agencies, to design agencies' structures and operations, and to specify the agencies' powers, duties and functions.[1] The Constitution's recognition of Congress's vast powers to shape the US government, coupled with the absence of direct reference in the Constitution to federal administrative agencies, has fueled the continuing debate – including but not limited to the proper contours of the administrative state, the political and legal structures to keep its exercise of powers in check, how to keep its actions accountable to the public, and the procedures to safeguard democratic and deliberative decision-making such as receiving and responding to public comments and providing reasons for decisions.[3,4,10,52–56]

The administrative state has long served as a pillar instrument of governance in the United States. Julian Mortenson and Nicholas Bagley describe how early congresses enacted dozens of broad sweeping laws, implemented through the delegation of powers to various commissions that were "vital to the establishment of a new country – to shore up its finances, regulate its industry, govern its territories, secure its revenue, and guard against internal and external threats."[57–60] William Novak details how government regulation of the economy grew in the 1800s as industrialization, while spurring economic growth, concentrated economic powers and inflicted harm on workers and communities.[61] The private sector's vast economic power posed a constitutional problem, as those powers were just as capable of "exercising social force and coercion and destroying liberty as 'public government itself.'"[61] To check those powers, cities and state governments created commissions and agencies to regulate health and safety and to regulate monopolies, while the federal government enacted protective laws, such as antitrust laws to protect consumers, and delegated powers to the Federal Trade Commission to enforce protections.[61] In the pre–New Deal era, a number of sweeping laws that offered such protections stood unchallenged, though the Supreme Court did strike down a number of protective laws.[xv]

Facing the Depression-era economic and social upheavals, President Franklin D. Roosevelt worked with Congress to enact a number of laws to promote economic recovery and to provide social protections.[62] Large businesses, while appreciative of government efforts to promote economic recovery, chafed at the protections for workers' wages, health and safety and launched attacks on the administrative state as unconstitutional.[10] The Supreme Court initially struck down a series of federal and state laws that protected workers,[xvi] ruling that those laws violated the freedom to contract. Following political conflict between the president, Congress and the courts,[xvii] by 1937 the Supreme Court, in another 5–4 decision, upheld states' economic regulations – including Washington state's minimum wage law and New Deal laws such as the National Labor Relations Act, which protected workers' right to form a union.[xviii]

The Supreme Court's split decisions along conservative and progressive lines underscore how justices' philosophies deeply shaped their interpretations

---

[xv] A number of sweeping laws were passed in this era, such as the 1887 Interstate Commerce Act, which created the Interstate Commerce Commission to oversee the conduct of the railroad industry; the 1906 Pure Food and Drug Act, which protected consumers and created the precursor to the Food and Drug Administration; and the 1914 Federal Trade Commission Act, which protected consumers by outlawing unfair practices that affected commerce.

[xvi] For instance, the court struck down New York's state minimum wage law in *Morehead v. New York ex rel. Tipaldo* [147].

[xvii] Daniel Rodriguez and Barry Weingast describe that accommodation and its importance for modern governance. The courts recognized Congress's powers to enact sweeping social and economic legislation, while Congress exercised greater care in enacting laws that meet broad markers laid out by the court to hold agencies accountable in their exercise of powers. That accommodation "enabled the federal government ... to deploy national power to solve new economic problems, to create delegations appropriate to modern needs, and to craft novel administrative instruments to carry out legislative aims" [55].

[xviii] Scholars debate on the extent to which various factors were decisive in Justice Owen Roberts shifting his support from the conservative wing to the progressive wing of the court and thus upholding the New Deal laws [55]. The later New Deal laws, which were drafted more carefully than those passed earlier in Roosevelt's presidency, addressed concerns about agencies' exercise of power. The growing public support for government interventions and the landslide victory of Roosevelt in 1936 may have persuaded Justice Roberts to shift his position. Roosevelt's proposal to increase the number of Supreme Court justices, which would weaken the influence of the anti-New Deal justices, could also have played a role in that shift.

on whether laws enacted by Congress and state governments to protect consumers and workers are constitutional. In particular, justices hold distinct views on liberty and on how government regulations help and hamper individuals' liberty. In 1905, conservative justices in the 5–4 decision *Lochner v. New York*[63] struck down the New York state law that limited working hours. They advanced a specific view of liberty, that is, the right and liberty of individuals to enter into contracts without government restrictions, particularly in the economic sphere, and argued that the freedom to contract was protected by the Due Process Clause in the Fourteenth Amendment.[xix] On the contrary, in 1937, the progressive justices, plus Justice Owen Roberts who switched to side with the progressives, in the 5–4 decision in *West Coast Hotel Co. v. Parrish*[64] upheld Washington state's minimum wage laws. Chief Justice Hughes, writing for the majority, emphasized that protecting liberty required protections for people's health, safety and welfare. He wrote,

> The Constitution does not speak of freedom of contract. It speaks of liberty and prohibits the deprivation of liberty without due process of law.... The Constitution does not recognize an absolute and uncontrollable liberty. The liberty safeguarded is liberty in a social organization which requires the protection of law against the evils which menace the health, safety, morals, and welfare of the people. Liberty under the Constitution is thus necessarily subject to the restraints of due process, and regulation which is reasonable in relation to its subject and is adopted in the interests of the community is due process.

10.3.2 The Roberts Court and Justices' Views of the Administrative State

Justices on the Roberts Court[xx] diverge in their views of liberty and in their views on the extent to which the administrative state has protected or threatened liberty.[xxi] In 2020, the Supreme Court ruled in *Seila Law v. Consumer*

[xix] It is important to note that Justice Oliver Wendell Holmes, writing for the dissenters, argued that the freedom to contract is not enshrined in the Constitution, noting long-standing laws against usury. He also argued against the majority's view of a laissez faire economy that is free of any government regulations, noting that "a constitution is not intended to embody a particular economic theory" (Justice Holmes, dissenting) [63].

[xx] Chief Justice John Roberts became chief justice of the Supreme Court, commonly referred to as the Roberts Court, in 2005. The Roberts Court is preceded by the Rehnquist Court, which was under the leadership of former Chief Justice William Rehnquist from 1986 to 2005.

[xxi] I do not assert that justices make their decisions in lock-step with other conservative-leaning or progressive-leaning judges. Kennedy was a swing vote on the court. He held strong views associated with conservatives, e.g., on protecting private property rights, but in *Murr v. Wisconsin*, he wrote in favor of regulations to protect the environment even at the costs of limiting private property rights [148]. Roberts voted with the progressives in a few cases. In

Financial Protection Bureau[65] on a constitutional challenge against the bureau,[xxii] an agency Congress established to protect consumers in the aftermath of the 2008 financial collapse.[xxiii] The case provides insights into justices' views of the administrative state and its protections for Americans, including public health and environmental protections.[xxiv]

Progressive-leaning justices underscore Congress's powers to create federal agencies and reflect on the administrative state as a positive and necessary tool of governance. Kagan, in her dissent, joined by Breyer, Ginsburg and Sotomayor, wrote, "Congress has accepted [the Framer's] invitation to experiment with administrative forms ... The deferential approach this Court has taken gives Congress the flexibility it needs to craft administrative agencies. Diverse problems of government demand diverse solutions. They call for varied measures and mixtures of democratic accountability and technical expertise, energy and efficiency." Kagan recognized how the administrative agency serves to protect the public from private exercise of power by corporations: "Congress decided that effective governance depended on shielding

---

[xxii] *Kisor v. Wilkie*, he voted on the grounds of keeping with precedent to defer to administrative agencies' reasonable interpretation of ambiguous agency regulations [149]. He also voted to uphold provisions of the Affordable Care Act deeming them constitutional in *King v. Burwell* [150]. Progressive judges have not always voted as a block, e.g., in June 2020, Breyer and Ginsburg voted with the conservatives, while Kagan and Sotomayor dissented in *US Forestry Service v. Cowpasture River Preservation Association* [151]. The 7–2 majority decided that the US Forestry Service had the authority to grant permits for pipelines to be built underneath the Appalachian Trail, even when the trail is managed by the National Park Services. The National Park Services is prohibited by law to permit energy projects to proceed on lands protected by national park designation.

[xxii] The majority decided that Congress's design of the CFPB that restricts the president to fire the CFPB's single commissioner head only for cause violated the separation of powers. The dissenters argued that the Constitution granted broad powers to Congress to shape America's government and that Congress's for-cause dismissal provision did not restrict the exercise of presidential constitutional powers and thus did not violate separation of powers. The dissenters underscored the president's constitutional duty of care to ensure agencies implemented laws and that for-cause provision served to insulate the CFPB, as other federal agencies regulating the financial sphere, from political pressure [65]. Despite the majority's assertion that the president did not have sufficient control over the agency with the for-cause firing provision, the reality demonstrated that the president wielded his powers to frustrate Congress's intent. Trump appointed Mick Mulvaney, who was the director of the Office of Management and Budget, as acting director of the CFPB and effectively put the CFPB under White House control [69].

[xxiii] Dodd-Frank Wall Street Reform and the Consumer Protection Act allowed the CFPB to enforce against "unfair, deceptive, or abusive acts and practices in the consumer financial market."

[xxiv] The *Seila* decision, which applies to agencies with singular heads whom the president can fire only for cause, does not raise constitutional issues about most agencies addressing public health and environmental issues. The administrators of the EPA, the BLM, the BSEE and the Food and Drug Administration can all be fired at will by the president. In fact, the Trump administration's interference in agencies' actions that should be informed by science have led to calls to restructure agencies to better insulate them from political pressure [152].

technical or expertise-based functions ... from political pressure (or the moneyed interests that might lie behind it)." Kagan focused on how Congress, working with the president, worked to prevent a repeat of mortgage companies' fraudulent and coercive tactics against consumers that had contributed to the 2008 financial collapse.[66]

Kagan wrote about the checks placed on agencies in general, and the Consumer Financial Protection Bureau (CFPB) in particular, to ensure they are accountable to the electorate and at the same time insulated from political pressure: "Sometimes, the arguments push toward tight presidential control of agencies ... At other times, the arguments favor greater independence from presidential involvement. Insulation from political pressure helps ensure impartial adjudications. It places technical issues in the hands of those most capable of addressing them. It promotes continuity, and prevents short-term electoral interests from distorting policy."[65] Progressive-leaning judges showed an appreciation of the role of agencies in modern governance in the 2020 Supreme Court case *Gundy v. United States*[67] on Congress's delegation of powers.[xxv] Kagan wrote, "'In our increasingly complex society, replete with ever changing and more technical problems,' this Court has understood that 'Congress simply cannot do its job absent an ability to delegate power under broad general directives.'"

On the contrary, the conservative judges see the administrative state as asserting coercive powers without sufficient checks and as threatening liberty. John Roberts, writing for the majority in *Seila*, acknowledged that "[no one] doubts Congress's power to create a vast and varied federal bureaucracy." However, he focused on the potential for agencies' unchecked powers, writing that "the CFPB Director has the authority to bring *the coercive power of the state* to bear on millions of private citizens and businesses, imposing even billion-dollar penalties through administrative adjudications and civil actions." (These penalties are against companies who commit fraud against consumers.)[xxvi] Roberts focused on the potential for insufficient checks on the CFPB's powers, despite Congress's deliberate structuring of the agency with diffuse checks on its powers while insulating it from the political pressure that

---

[xxv] Kagan was joined by Breyer, Ginsburg and Sotomayor. Alito signed up to the decision to uphold the law. However, he stated his willingness to review the nondelegation doctrine in an appropriate case. Gorsuch, Roberts and Thomas dissented. Kavanaugh was not yet seated on the court. However, in 2019 Kavanaugh wrote an opinion in *Paul v. United States*, a case the Supreme Court declined to hear, about his positive assessment of the Gorsuch opinion and that he was willing to review the nondelegation doctrine [153].

[xxvi] These penalties are imposed on corporations to protect consumers. In 2015, CFPB enforcement actions against financial corporations recovered $9.3 million per employee in refunds, redress and forgiven debts for the American consumer [154].

had undermined previous legislative attempts to protect consumers.[68,69] Roberts wrote, "the Director may unilaterally, without meaningful supervision, issue final regulations, oversee adjudications, set enforcement priorities, initiate prosecutions, and determine what penalties to impose on private parties. With no colleagues to persuade, and no boss or electorate looking over her shoulder, the Director may dictate and enforce policy for a vital segment of the economy."

Conservative justices present a view of liberty focused on freedom from government regulations.[10,51,70] Roberts, in his majority decision in *Seila*, wrote that "the Framers viewed the legislative power as a special threat to individual liberty." In an earlier case, *PHH Corp. v. Consumer Financial Protection Bureau*,[71] then Judge Kavanaugh ruled in the three judge panel decision at the Court of Appeals for the DC Circuit that the for-cause firing provision was unconstitutional. He wrote, "the independent agencies collectively constitute, in effect, a headless fourth branch of the US Government. . . . Because of their massive power . . . independent agencies pose a significant threat to individual liberty and to the constitutional system of separation of powers and checks and balances." He also wrote about how the bureau might "issu[e] a rule that affects your liberty or property."[71] Other conservative justices have expressed similar fears that the administrative state would trample on liberty.[xxvii]

The narrow view of liberty as freedom from government regulations, however, has been challenged by contemporary judges and legal scholars,[xxviii] just as it was challenged by progressive judges during the New Deal era.[54,72] Judge Cornelia Pillard, in the subsequent Court of Appeal en banc majority decision in *PHH Corp. v Consumer Financial Protection Bureau*,[71] which ruled the provision to be constitutional, presented a broader view of liberty.[54,69,73] She questioned why the adjudication of the agency's constitutionality had focused on the liberty of the finance companies that faced the agency's regulations but

---

[xxvii] Judge Gorsuch, while serving on the Tenth Circuit, warned against "permit[ting] executive bureaucracies to swallow huge amounts of core judicial and legislative power and concentrate federal power in a way that seems more than a little difficult to square with the Constitution of the framers' design." Metzger writes that "Gorsuch drew a straight line from such institutional expansion to 'governmental encroachment on the people's liberties.'" Justice Samuel Alito wrote about "an understandable concern about the aggrandizement of the power of administrative agencies as a result of the combined effect of . . . the effective delegation to agencies by Congress of huge swaths of lawmaking authority." Justice Clarence Thomas criticized the "usher[ing] in significant expansions of the administrative state" spurred by a "belief that bureaucrats might more effectively govern the country than the American people." Sharkey quotes Justice Alito's and Justice Thomas's writing in *Perez v. Mortgage Bankers Association* [70][155]. Metzger quotes Gorsuch's writing in *Gutierrez-Brizuela v. Lynch* [10][156].

[xxviii] For instance, Metzger writes, "Agencies assist with securing individual liberty, by protecting individuals against abuses of private power and ensuring access to the basic goods (safe food, a clean environment, protection against private exploitation, and so on) needed for a full and free life" [54].

"not more broadly to the liberty of individuals and families who are their customers. ... Congress understood that markets' contribution to human liberty derives from freedom of contract, and that such freedom depends on market participants' access to accurate information, and on clear and reliably enforced rules against fraud and coercion." Judge David Tatel, in his concurring opinion, invoked the Holmes dissent in *Lochner*, to question liberty being defined as the freedom of contract, without any government regulatory constraints, including those to protect against fraud and coercion.[63,72]

As summarized by Gillian Metzger, conservatives on the Roberts Court[54] have lacked "reference to the ways that the administrative state operates to constrain power, render it accountable, and advance individual liberty."[54] Instead, they have advanced doctrines that curb the administrative state and in effect weaken legal protections for public health and the environment.[45,46]

10.3.3 Doctrinal Developments against Public Health and the Environment

Ideological differences vary greatly from judge to judge. However, conservative opposition to public health and environmental protections are mostly grounded in several growing trends in judicial decisions: (1) a presumption that statutes provide narrow protections for public health and the environment, (2) a presumption that statutes delegate narrow regulatory powers to agencies, (3) limits on congressional authority to delegate powers to federal agencies and (4) limits on congressional power to enact laws.

Presumption That Statutes Provide Narrow Protections for Public Health and the Environment

Although the Constitution vests enormous powers in Congress to legislate, disputes arise on the extent of protections that Congress intended – especially concerning environmental statutes. Courts' assumptions on Congress's intentions can have major implications on how courts rule when questioning agency actions or inactions.

Progressive judges have tended to interpret statutes using the purposivist approach[xxix] that prioritizes the "policy context" in order to understand the meaning of the statutory text.[xxx] They ask how a reasonable person would

---

[xxix] For details on the similarities and differences in these interpretive approaches and for discussions on how value judgements are inherent in both these approaches, see Manning, Fallon and Krishnamurthy [74][157, 158].

[xxx] For instance, Justice Kagan wrote in her decision in *Gundy*, "It is a fundamental canon of statutory construction that the words of a statute must be read in their context and with a view

understand the problem addressed and the remedy given by the statute. The purposivist interpretation generally results in recognition of broad protections offered by these statutes that specify ambitious goals of protecting clean air, clean water and the natural environment.

Conversely, conservative judges have taken very narrow interpretations of these statutes, thus limiting the protections they can offer. Often, they apply the textualist approach that prioritizes the "semantic context," that is, how a reasonable person would use the words in the statute. The textualist approach in itself does not necessarily lead to weakened public health protections.[xxxi] However, that approach does so when specific meanings are chosen to arrive at predetermined outcomes[74] and when the ambitious purposes of statutes are downplayed. Conservative judges' narrow interpretations of these laws provided the legal scaffold for the Trump administration's deregulatory actions. Conservative judges' narrow interpretations also severely restricted protections under the CAA and curbed a more expansive interpretation in an earlier decision.

Presumption That Statutes Delegate Narrow Regulatory Powers to Agencies

Among the other powers discussed thus far, the Constitution also vests powers in Congress to shape the structure of the executive branch.[xxxii] However, the extent of Congress's powers has been another common area of dispute. Progressive judges have not only interpreted environmental statutes to grant broad protections, but they have deferred to agencies' statutory interpretations to achieve those statutory goals. Alternatively, conservative judges have interpreted agencies' powers narrowly and limited deference to agencies' statutory interpretations. They have insisted on a literal match between the agency's action and the text of the statute and that Congress spell out each and every delegation of power to agencies.[75] Because it is impossible for Congress to be clairvoyant,[51] this interpretation of statutes can hinder protective actions.

Under the Chevron doctrine, judges have deferred to federal agencies' statutory interpretation when statutes are ambiguous, as long as an agency's interpretation is reasonable.[76,77] This doctrine is named after the 1984 Supreme

---

to their place in the overall statutory scheme ... We have looked to the text in 'context' and in light of the statutory 'purpose.' ... [The] Court often looks to 'history [and] purpose' to divine the meaning of language."

[xxxi] In the 2001 Supreme Court unanimous decision in *Whitman v. American Trucking Associations*, Justice Scalia wrote that the text made clear that the CAA prohibits the EPA from considering costs in the EPA's setting of the National Ambient Air Quality Standards [136].

[xxxii] Kagan wrote in *Seila* that "Congress [has] broad authority to establish and organize the Executive Branch."

Court decision *Chevron v. NRDC*.[78] The rationale for deference, as enunciated in several decisions, is that Congress delegated powers to agencies; agencies, not courts, are politically accountable; and agencies possess the technical expertise to resolve complex problems.[xxxiii] These judges continue to undertake the judiciary's obligation to interpret the law[xxxiv] and to defer to agencies only when the statute is ambiguous and the agencies' interpretation is reasonable.[77] They have ruled against agencies that abnegate their powers, for instance in *Massachusetts v. EPA*.[36] They have also ruled against agencies that act outside their statutory bounds, for instance in *New Jersey v. EPA*,[167] when the Bush EPA shifted to regulating mercury from power plants less stringently, without following the procedure stated in the CAA to make that shift.[79,80]

Conservatives, while supportive of the Chevron doctrine under the Reagan administration, grew to be very critical of the doctrine under newer administrations.[81] For instance, in 2016, then Judge Gorsuch in the Court of Appeals for the Tenth Circuit argued that, under Chevron, "courts are not fulfilling their duty to interpret the law."[82] These judges have narrowed deference to agencies in at least two ways.[82,83] The first approach is to argue that the statute is unambiguous and to interpret the statute as delegating only narrow and clearly articulated powers to agencies. The second approach conservative judges have used to narrow deference to agencies is to apply the "major questions" presumption. That presumption requires Congress to speak clearly if it wishes to delegate powers to federal agencies on questions of special economic and political significance.

Presumption Against Congress's Delegation of Broad Powers to Federal Agencies

One of the judiciary's most important roles is to protect the constitutional separation of powers. As noted by Justice Elena Kagan in the 2019 Supreme Court decision in *Gundy v. United States*, the court has long recognized that Congress may not cede to another branch of government "powers which are strictly and exclusively legislative." Even so, progressive judges, acknowledging Congress as a co-equal branch of government, have taken a highly deferential

---

[xxxiii] For instance, Justice Elena Kagan, in her dissent in *Michigan v. EPA*, noted that courts have less insights on the technical issues underpinning environmental problems than do the EPA, and the EPA should be granted deference [159].

[xxxiv] For instance, Justice Breyer explained in his dissent in *SAS Institute v. Iancu* that courts should not treat Chevron "like a rigid, black-letter rule of law, instructing them always to allow agencies leeway to fill every gap in every statutory provision." Instead, he wrote that Chevron serves as "a rule of thumb, guiding courts in an effort to respect that leeway which Congress intended the agencies to have." He noted that in examining whether the disputed agency interpretation warranted deference under *Chevron*, judges examine the "statute's complexity, the vast number of claims that it engenders, and the consequent need for agency expertise and administrative experience" [77][166].

view, upholding a delegation of congressional power as long as Congress provides an intelligible principle that guides and constrains the administrative agency's exercise of the power that Congress has delegated.

The only time the Supreme Court has struck down laws on the basis of Congress's excessive delegation of powers was in 1935. However, the ascent of justices who advocate for the nondelegation doctrine raises the likelihood of courts striking down statutes or provisions in statutes. For instance, law professors Samuel Moyn and Aaron Belkin have raised concerns that, if Congress were to enact climate legislation that delegates broad powers to agencies, future environmental statutes could run the risk of being struck down based on the nondelegation doctrine.[84]

Constitutional Limits on Congress's Powers to Enact Laws

Conservative legal scholars have advanced constitutional analyses[xxxv] that would limit or hamper the ability of Congress to enact federal environmental laws. Environmental advocates have called attention to these arguments, which strike at the heart of Congress's legislative powers and thus at Congress's powers to protect public health and the environment.[85–89]

One constitutional analysis advanced to cripple Congress's powers to enact legislation is the narrow reading of the Commerce Clause. This provision of the Constitution grants Congress powers to regulate interstate commerce and is commonly cited as a source of authority for Congress's statutory efforts to protect human health and the environment. Congress holds only those powers vested by the Constitution and the rest of the powers are reserved for states. The narrow reading of the Commerce Clause, justified on the grounds of protecting states' rights and upholding federalism, has been used to challenge the constitutionality of federal environmental laws. These include the ESA, the CWA and the CAA, which Congress had enacted by relying on the Commerce Clause.[86]

In the 1970s, Congress's enactment of environmental laws using its Commerce Clause powers raised little controversy and enjoyed bipartisan support. The contemporary Supreme Court precedent applied a broad reading of the Commerce Clause. Judges applied a deferential review, upholding Congress's exercise of powers under the Commerce Clause as long as Congress had a rational basis for concluding that the regulated activity significantly affected interstate commerce.[86]

It is worth noting that the development of the regulatory takings doctrine and absolutist views of property rights is not simply the result of independent

---

[xxxv] Jonathan Cannon, Robert Glicksman and Erin Ryan detail how conservative legal arguments have been advanced to limit public health and environmental protections and to limit the ability of environmental advocates to bring legal challenges against federal agencies' abnegation of their powers. [17][39][50].

academic scholarship. Instead, their development since the 1980s, as documented by several studies,[90,91] has been fostered by corporations and their funded think tanks, which stood to gain from limiting environmental regulations.[xxxvi] Judge Abner Mikva, who served in the Court of Appeals for the DC Circuit, in his 2000 opinion piece in the *New York Times*, underscored that corporate influence:

> Between 1992 and 1998 ... more than 230 federal judges took one or more trips each to resort locations for legal seminars paid for by corporations and foundations that have an interest in federal litigation on environmental topics. In the seminars devoted to so-called environmental education, judges listened to speakers whose overwhelming message was that *regulation should be limited* – that the free market should be relied upon to protect the environment, for example, or that *the "takings" clause of the Constitution should be interpreted to prohibit rules against development in environmentally sensitive places.* ... Judges who attended [the seminars] wrote 10 of the most important rulings of the 1990's curbing federal environmental protections, including one that struck down habitat protection provisions of the ESA and another that invalidated regulations on soot and smog.[92,93]

Some illustrative cases concerning these doctrinal developments are summarized and discussed at the end of the chapter.

10.4 Hijacking the Courts: Trump's Wave of Judges Who Are Skeptical of the Administrative State

Bearing in mind the myriad of ways in which judicial opinion can help or hinder Congress and the administrative state in pursuing environmental and public health protections presented above, I move now to the troubling speed and focus with which the Trump administration used judicial appointments to contort the judiciary. By January 2021, President Trump had appointed three justices to the Supreme Court, 54 judges to the courts of appeals and 174 judges to the district courts.[94] The Supreme Court flipped to a 6–3 conservative majority,[xxxvii]

[xxxvi] For instance, the Reason Foundation, the Beacon Center and the Cato Institute filed briefs in support of plaintiffs in *Murr v. Wisconsin* [91]. According to Sourcewatch.org, these think tanks have received funding from the Koch Foundation, the Donors Trust or the Donors Capital Fund. Charles and David Koch advocate the libertarian position of limited government and limited regulations, from which their business conglomerate stands to benefit.

[xxxvii] The term "conservative" in this chapter refers to judges who view the administrative state skeptically or antithetically. As discussed in Section 10.3.3, the opinions of Justices John Roberts, Neil Gorsuch, Brett Kavanagh, Clarence Thomas and Samuel Alito reflect skepticism against the administrative state. Some commentators have claimed that Roberts and Gorsuch abandoned conservatism in their decisions in *Bostock v. Clayton County* in support of LGBTQ rights under the Equal Protection Act. That distinct definition of conservative thought is not the one examined in this chapter [160].

with conservative Justice Brett Kavanaugh replacing Justice Anthony Kennedy, a Reagan appointee, who was often the swing vote on the court[38] and conservative Justice Amy Coney Barrett[xxxviii] replacing Justice Ruth Bader Ginsburg, a progressive who had penned opinions supportive of the administrative state's protection of public health and the environment.[47,48,95] Three out of the eleven judges who serve on the Court of Appeals for the DC Circuit, which reviews many of the cases on the environment and on federal agencies, are Trump appointees.[94] In the federal appeals courts, which hear hundreds of cases annually, far more than the Supreme Court, Trump appointed more than one-quarter of the judges. Judges who had been appointed by Republican presidents became the majority in three circuits and an even larger majority in other circuits.[96]

The series of 5–4 decisions in the Supreme Court on environmental cases, prior to the passing of Ginsburg, demonstrate how closely contested judicial constitutional and statutory interpretations have become – including on Congress's powers to enact laws and to delegate powers to agencies, on the protections offered by landmark environmental statutes and on agencies' discretionary powers. A growing number of judges view the administrative state as a foe of personal liberty and freedom from government restrictions; and these judges are supporting and advancing doctrines that rein in the administrative state. In advancing their interpretations of separation of powers, federalism and property rights, these judges almost inevitably narrow public health and environmental protections. The appointments of even more federal judges who share views antithetical to the administrative state can cripple these protections. Indeed, this reshaping of the judiciary is precisely the strategy that has been long in the making by conservative megadonors – a plan which was jolted into action with alarming precision and haste by Trump with the backing of Senate leadership like Mitch McConnell's.

White House Counsel Don McGahn laid bare the ideology behind Trump's judicial nominations in his November 2016 speech to the Federalist Society. McGahn stated "the greatest threat to the rule of law in our modern society is the ever-expanding regulatory state and the most effective bulwark against that threat is a strong judiciary ... Regulatory reform and judicial selection are so deeply connected."[97] Trump outsourced the judicial nomination to the Federalist Society, having vowed as a presidential candidate, "We're going to have great judges, conservative, all picked by Federalist Society."[97] Promising to appoint conservative judges, whose names he released, was Trump's key election strategy, and appointing them as promised was his key reelection

---

[xxxviii] Nevitt and Freeman suggest that, based on Barrett's academic writing and her decisions in the Court of Appeals, that Barrett is likely to read agencies' statutory powers narrowly and to hold a high bar on standing against pro-environment plaintiffs [48][95].

strategy.^xxxix McGahn confirmed the role of the Federalist Society in the judicial nominations: "Our opponents of judicial nominees frequently claim the president has outsourced his selection of judges ... I've been a member of the Federalist Society since law school ... So, frankly, it seems like it's been insourced."[98]

Indeed, a case in point on the intertwining of judicial selection and deregulation is the appointment of Judge Neomi Rao, whose prior record indicated her support for deregulation, to the Court of Appeals for the DC Circuit in 2019. Rao, a tenured professor at George Mason Law School and the founding director of the Center for Administrative State at the law school, has argued in favor of a strict reading of doctrines that hobble the functioning of the administrative state, such as the nondelegation doctrine.[99] She served in the Trump administration from 2017 through 2019 as the administrator of the Office of Information and Regulatory Affairs, overseeing deregulatory actions of federal agencies. Rao's opinion piece in the *Washington Post* in 2018 touted the costs savings from the Trump administration's deregulatory actions but did not pay much attention to the forgone societal benefits from the retrenchment of protective regulations.[100,101] In April 2020, Rao dissented in the 2–1 ruling in the court of appeals against the Trump EPA's suspension of an entire rule on hydroflurocarbons (HFCs), a potent greenhouse gas, without abiding by the notice and comment procedure mandated by the Administrative Procedure Act.[102]

While all presidents strive to select judges who share their philosophy, Trump's nomination process stood out for its large number of judicial appointments and its laser focus on the nomination and appointment of judges who view the administrative state skeptically or even antithetically. McConnell and Republican allies paved the way for a record number of 103 unfilled judicial positions left vacant at the start of Trump's presidency, made

---

xxxix Trump released the shortlist of conservative judges whom he would consider appointing to fill the Scalia seat, put together by John G. Malcolm of the Heritage Foundation and Leonard Leo of the Federal society. On November 6, 2019, when Trump celebrated with Republican senators the confirmation of his 158th judge, Mitch McConnell remarked that "So we got a chance to set the agenda ... What's the most important thing? Clearly, it was the Supreme Court. You had been helped enormously by a decision that I made ... not to let President Obama fill that Scalia vacancy on the way out the door." According to a poll, one-fifth of voters in the 2016 presidential elections said the Supreme Court was the most important issue in their decision, and among them, 57 percent voted for Trump [109]. Journalist Bill Sher wrote in his opinion piece for the *Washington Post* that "President Trump has retained support from many Republicans and conservatives thanks to a Faustian bargain: So long as he stacks the judiciary with friendly judges, they'll look the other way when he [takes] positions out of step with recent conservative orthodoxy." Not all conservatives were aligned with Trump or remained aligned with Trump. In July 2020, Steven Calabrese, a co-founder of the Federalist Society, wrote in a *New York Times* opinion piece that he had voted for Trump, but Trump's tweet seeking to postpone the 2020 election was "fascistic" and "grounds for the president's immediate impeachment" [161, 162].

possible by the Senate's relentless blocking of proceedings for Obama's judicial nominees.[xl] They took the unprecedented action of blocking hearings for Judge Merrick Garland, Obama's nominee to the open Scalia seat.[103] The hyperpartisanship in the Senate and the end of norms in the Senate to accommodate the minority parties[xli] enabled Trump, with a narrow Republican Senate majority to confirm judges who are more conservative than those judges selected by prior Republican presidents.[38,103,104,xlii]

Trump's outsourcing of the nomination of federal judges to the Federalist Society almost inevitably stacks the deck in the courts against public health and environmental protections. The Federalist Society describes itself as "a group of conservatives[xliii] and libertarians," "founded on the principles that the state exists to preserve freedom, that the separation of governmental powers is central to our Constitution, and that it is emphatically the ... duty of the judiciary to say what the law is, not what it should be."[105]

The society promotes legal interpretations of the separation of powers in ways that debilitate the administrative state and cripple its powers to regulate powerful economic interests. The society defines freedom as individuals' freedom from government regulations. However, as Judge Pillard in *Seila* noted, this view of freedom ignores the need to protect individuals' freedoms from imbalances of power in the marketplace, for which government regulations of corporations are needed. The society undergirds these legal interpretations as the application of textualism that abides loyally by the Constitution and the laws.[xliv] However, as

---

[xl] Only President George H. W. Bush had a comparable number of vacancies, which was as a result of the expansion of the federal bench under his administration [104]. McConnell and Republican allies blocked proceedings for nominees to the federal courts, including those who had cleared the senate judiciary committee [103].

[xli] Republican senators voted to replace the filibuster, i.e., the 60-vote threshold for confirmation of a Supreme Court judge, with a requirement for only a simple majority vote. They also ended the blue slip norm, under which the senators from the majority party respect the decision of senators from a judge's home state to oppose a nomination. Trump, unlike previous presidents, did not need to moderate his judicial choice to gain the support of Democratic senators [103, 104]. In November 2013, the Democrats had ended the filibuster federal judicial appointments but not for the Supreme Court judges, justifying the action as a response to Republicans blocking Obama's judicial appointments.

[xlii] The *New York Times* analysis of decisions in the appellate courts through December 2020 finds that the Trump appointees are more likely than judges appointed by other Republican presidents to disagree with Democratic appointed judges [104]. Many lifetime appointees to the appellate courts were more openly engaged in causes important to Republicans, such as opposition to gay marriage and to government funding for abortion [104].

[xliii] Conservative legal thinkers are not a monolith. The society serves as a "mediating" organization, which brings diverse viewpoints together under the conservative tent and fosters communication and cooperation with its members [5].

[xliv] The false distinction that textualists apply the law while purposivists make the law has been excoriated in the literature. Critics of this false distinction point to how the imprecision of language leads judges, both textualists and purposivists, to make judgement calls that in turn reflect their philosophies. The more ardent critics argue that textualism is a fig leaf to justify

discussed above, the application of textualism, through the lens that is antithetical to the administrative state, results in narrow interpretations of the protections offered by existing environmental regulations and, in the more extreme case, blocks Congress's powers to enact protective legislation.

While the society prides itself as an organization that promotes debates on legal doctrines and that does not take policy positions, its activities belie this distinction.[xlv] A number of studies have documented how the society served to nurture and amplify legal doctrines in ways that limit the functions of the administrative state.[7,98,106] Political scientist Amanda Hollis-Brusky's study of 12 federalism and separation of powers cases decided by the Supreme Court between 1983 and 2001 reveals how network members were involved in key aspects of these cases: "crafting the arguments, arguing the cases, clerking for the judges or issuing the rulings."[106] The society's network of members serve to develop, vet and legitimize ideas, moving more fringe ideas into the mainstream, and thus facilitating judges to incorporate these ideas into their decisions. (McGahn himself boasted "We have seen our views go from the fringe, views that in years past would inhibit someone's chances to be considered for the federal bench to being the center of the conversation."[104])

As documented by numerous studies, the society has worked for decades toward nomination of judges who are antithetical to the administrative state.[5–8] It played a pivotal role in the appointments of Supreme Court judges in the George W Bush administrations.[7,xlvi] Hollis-Brusky writes how the society's gatherings and functions provide a venue for aspiring judges to audition for judgeships by speaking in favor of conservative legal scholarship. The gatherings also serve as a disciplinary device for judges, who not only learn conservative legal thinking but also criticism of those decisions that drift away from those principles.[7] Analysis of appellate court decisions reveal that decisions of

---

judges' decisions that in reality are colored by their philosophy. Critics of textualism include Judge Richard Posner, who is no liberal [163]. Critics have also excoriated the claims of originalism made by some conservative judges, i.e., that their interpretations abide by the views of the Founders. For instance, Eric Segal documents how originalism has been used by conservative judges to justify ahistorical interpretations that align with conservative values [164].

[xlv] Fostering debates and sponsoring scholarship of legal conservative thought are legitimate exercises of freedom of speech and freedom of association. The society separates itself from the actions of its network members. For instance, the society does not take direct legal or political positions and does not participate in litigation, filing briefs or supporting political candidates, though its network members do [106]. However, the society's ties to megadonors and anonymous donors while influencing judicial decisions and appointments have raised concerns.

[xlvi] During the George W. Bush administration, Leo began working as an outside adviser for the White House on initiatives related to judicial nominations and worked alongside Brett Kavanaugh, then White House associate counsel. In 2005 and 2006, Leo led the campaigns supporting the nomination of Roberts and Alito to the Supreme Court [110].

judges who are Federalist Society members are significantly more conservative than non-members.[107]

The prominent role of the society in the judicial selection process, intertwined with undue influence by the web of powerful donors and nonprofits linked to Chairman Leo, strikes at the heart of US democracy.[108–110] A sacred tenet at the core of American democracy is that all persons are equal under the law and that judges are impartial in their decisions. As Justice Kennedy underscored in 1999, "There must be both the perception and the reality that in defending these values,[xlvii] the judge is not affected by improper influences or improper restraints. That's neutrality."[93]

Media investigations reveal that the Federalist Society and the network of non-profits advocating conservative causes that are linked to Leo have received large amounts of funding from conservative donors who support retrenchment of the administrative state. The society has received funds from the Koch Foundation, whose corporate conglomerate would benefit from retrenchment in regulations, the Mercer Foundation, which supported the Trump presidential campaign and funds climate denial,[xlviii] and from undisclosed donors, such as the Donors Trust.[108–110,xlix]

The *Washington Post* reports that from 2014 through 2017 alone, $250 million flowed to non-profits linked to Leo.[110] That report's quartet of observations – including Leo's disclosure of BH Group as his employer, BH Group's contribution of $1 million to Trump's inauguration, the anonymity of funder(s) of the BH Group, which appears to be a shell company, and Leo's prominent role in the judicial nomination process – create, at the minimum, an appearance of a potential link between a political donation and access to the selection of judges.

Senator Shelton Whitehouse, Democrat on Senate Judiciary Committee, warned that the possibility of these funds coming from anonymous donors who have interests before the federal courts raises red flags for donor influence on judicial appointments.[111] Whitehouse's statement is not without basis. Research that tracks funding, albeit of judicial elections in state courts, reveals that "undisclosed money often comes from donors with strong economic interests in issues coming up in state court – some of whom have cases pending at the time of an election."[112]

[xlvii] Kennedy spoke of three parts to the rule of law: "the government is bound by the law; ... all people are treated equally and ... there are certain enduring human rights that must be protected."

[xlviii] ConservativeTransparency.org, which tracks funding using Forms 990 filed with the Internal Revenue Service, provides more details on the funding sources for the society, including the Bradley Foundation. According to Sourcewatch.org, the Bradley Foundation has funded opposition to renewable energy.

[xlix] According to Sourcewatch.org, the Donors Trust receives funds from donors, who can choose legally to remain anonymous, and then distributes funds to conservative and libertarian causes. The Koch brothers have used the Donors Trust to cloak their funding.

Media investigations also revealed links between the Federalist Society and political activity by the Koch Seminar Network[1] against regulations and in favor of conservative judges.[113] The Koch Seminar Network, in writing about its "judicial confirmation strategy," explains that Network organizations, such as Americans for Prosperity and Freedom Partners Chamber of Commerce, have been working with the Federalist Society. It affirmed its goals "to ensure [that] a strict constructionist is confirmed to the court" whenever a seat on the Supreme Court becomes vacant.[113,168] The network took credit for launching a pressure campaign on three Democratic senators who eventually voted for Gorsuch. The Federalist Society has also worked with the Charles Koch Foundation on "educating judges" against the administrative state.

According to a report by Allison Pienta, an alumna of the George Mason Law School, the society and the foundation have been involved in a "Federal Judges Initiative" at that law school "designed to train 'newly confirmed judges' in, among other things, the 'tricks agencies [use] to portray their proposed regulations as having substantial benefits.'"[114] Pienta also reports that the $30 million Grant Agreement from the foundation and an anonymous donor to rename that law school "requires support for the School's Center that is dedicated to 'dismantling the Administrative State.'"[115,li]

10.5 The Uphill Battle for Laws and Regulations to Protect Public Health and the Environment

The court decisions that struck down the Trump administration's strained legal interpretations underscore the importance of the courts to serve as a check against the executive branch – in the case of the Trump administration, against unconstrained favoring of oil and gas expansion and deregulatory actions that

---

[1] The network describes itself as "a network of conservative groups assembled by billionaire donors Charles and David Koch." Lee Fang and Nick Surgey from the *Intercept* reported on the network's activities based on its memos to its donors [113].

[li] Based on documents obtained under Virginia's Freedom of Information Act request to George Mason University, Pienta writes, "In March 2016, the [anonymous] donor, now believed to be Barre Seid, entered into the $20 million Grant Agreement ($4 million per year for five years) with another $10 million coming from the Charles Koch Foundation that renamed George Mason's Law School after Justice Scalia. The Agreement also requires support for the School's Center that is dedicated to 'dismantling the Administrative State' and a new 'Center for Liberty & Law.' The Grant Agreement installed the Federalist Society's Leonard Leo in an oversight role over the Law School through a newly created entity called the 'BH Fund' ... The $4 million annual pledges under the Grant Agreement are conditioned on 'annual written proposals' by the University Foundation that the donor must 'approve' in his 'sole and absolute discretion' before each contribution is due." Neomi Rao, the founding director of the Center for Administrative State, was subsequently appointed by Trump to the Court of Appeals.

violate federal statutes. At the same time, Trump's actions in reshaping the judiciary with the complicity of Senate Republicans demonstrates the fragility of this vital guardrail. Bill Kristol, a neoconservative turned critic of Republicans who support Trump, lamented, "Under McConnell's leadership, the Senate, far from providing a check on the executive branch, has acted as an accelerant."[116]

Trump's outsourcing of judicial appointments to Leonard Leo, the Federalist Society and its anonymous donors almost certainly promises to reshape the judiciary to lean in favor of more conservative legal doctrines. This Pyrrhic victory for conservatives undermines public trust in the judiciary's independence from moneyed interests. And many Americans are alarmed at how Trump accelerated the erosion of America's democratic institutions.[117–119] Americans' trust in the judiciary to act in the public interest, while still higher than that for politicians, is far from universal. A poll by the Annenberg Center at the University of Pennsylvania in August 2019 revealed that only "68 percent of the public trusts the Supreme Court to act in the public's best interest" and "only 49 percent of the respondents hold the view that Supreme Court justices set aside their personal and political views and make rulings based on the Constitution, the law, and the facts of the case."[120]

The Senate Democrats took some preliminary steps to shine light on the judicial nomination process. In March 2020, Whitehouse and four other Democrats on the Senate Judiciary Committee launched an investigation on potential conflicts of interest in "Mr. Leo's prominent role in the Trump Administration's judicial selection and nominations process while maintaining a financial interest in advocacy efforts related to this process."[121] They requested Attorney General Barr to provide communications to the Department of Justice on matters pertaining to Leo's potential, actual or suggested judicial nominations.[111] Leo, who took leave from the Federalist society while advising on Trump's judicial appointments, denied any link between his work on judicial appointments and his other work, including fundraising for non-profits that advocate for conservative issues.[110] In July 2020, Senators Dianne Feinstein and Sheldon Whitehouse, Democrat members of the Senate Judiciary Committee, proposed the Judicial Ads Act. The bill, cosponsored by eight other Democrats, would require groups that spend $50,000 in a calendar year on advertisements linked to federal judicial nominations to release the names of their donors who give over $5,000.[122]

Congress's responsibility in enacting legislation has become even more important with the narrow readings of existing environmental statutes. Congress urgently needs to enact laws that address existential threats such as climate change, while taking care not to trip over doctrines that have been advanced to invalidate laws. At the same time, Congress would need to amend laws in response to judicial interpretations that narrow protections. With

Congress's failure to enact major environmental legislations since the 1990s, these legislative duties might appear to be a tall order for Congress. Even so, rare windows of opportunities for bipartisan support of environmental legislation, albeit less ambitious ones, provide some reason for hope. In June 2020, in the face of competitive Senate races in November, Republican Senators Cory Gardner of Colorado and Steve Daines of Montana worked to secure the support of Republicans for the American Outdoors Act, which provides $900 million in annual funding for the Land and Water Conservation Fund.[lii] The Senate voted 73–25 in favor of the bill, over the objection of some Republicans that the federal government would misuse the fund to expand public lands.[123] The House also passed the bill, which Trump subsequently signed into law.

While Congress holds legislative powers to protect agencies' abilities to regulate, it is also congressional representatives who have spearheaded the attack on the administrative state and agencies' regulatory powers. Ever since the 111th Congress, anti-regulatory Republicans in Congress have proposed the Regulations from the Executive in Need of Scrutiny (REINS) Act, which requires Congress and the president, within 70 days, to approve regulations that impose costs of $100 million or more per year.[124] Under that proposed bill, if Congress failed to act, the agency would also be prohibited from promulgating another related rule for the duration of that congressional session. This bill, which would end protections including those under long-standing landmark environmental laws, has failed to win enough votes.[liii] While the regulatory process provides checks and balances and some insulation for agencies to enact regulations based on scientific evidence, the REINS Act provides a major opening for regulated industries to lobby politicians to block regulations. Far from encouraging democratic deliberations in Congress on regulations, in 2017 Congress exercised its powers to review regulations under the Congressional Review Act to strike down regulations along party lines without much debate.[125] Given the reality of Congress's operations, such as Mitch McConnell's blocking of hearings in the Senate for bills passed with bipartisan support in the House, many proposed regulations are not likely to see a vote within 70 days, and Americans' health and environment would be left unprotected.

Turning to federal agencies, there is some good news. As noted by Ann Carlson "one of the reasons [for the EPA's legal] wins, even with conservative courts, is that they're very careful in really examining the science and building an

---

[lii] The fund, which had been chronically underfunded since Congress established it in 1965, provides federal, state and local government with funds to acquire water and land and easements to protect national parks and forests and wildlife areas.

[liii] In 2017, the Republican-controlled House voted 237–187, with two Democrats crossing the aisle, in favor of the bill. It secured 39 Republican cosponsors in the Senate.

administrative record that demonstrates expertise, and care, and thoughtfulness."[126] Federal agencies will need to spell out their statutory powers to promulgate those regulations, abide by procedures under the enabling statute and the APA, and undertake thorough scientific and technical analyses to build a robust administrative record. Because of the EPA's careful approach in promulgating rules under the Obama administration, the Trump administration faced legal defeats when it made U-turns on regulations without addressing the agency's prior legal interpretations and factual findings.

For lawyers, the battlefield has widened beyond mounting challenges in individual public health and environmental cases. Faced with litigants that advanced doctrines limiting health and environmental protections, lawyers, with the assistance of legal scholars, have written legal briefs challenging these doctrines and laying out flaws in the historical and empirical assumptions of these doctrines. Legal scholars have also pointed out how doctrines to block regulatory agencies, justified on the basis of separation of powers, in reality jeopardize separation of powers by shifting powers from Congress to the judiciary.

Legal scholars have also assembled historical documents to question the assumptions that these legal interpretations are based on the founding fathers' vision and the history of the early Republic. As Gillian Metzger writes, "This historical scholarship holds important lessons for current debates over administrative law and the administrative state more broadly. The extensive history of administrative agencies operating from the nation's beginnings to today undercuts efforts to paint contemporary administrative government as a fundamental deviation from the Constitution."[54] For instance, Bagley and Mortenson challenge Gorsuch's assertion that the Framers opposed Congress's delegation of powers.[liv] They argue that while the Constitution states that Congress holds the legislative power, it does not state that Congress cannot delegate its powers. They also detail how early congresses enacted expansive laws fundamental to administrative government (Section 10.3.1).[57–59]

These three chapters on rulemaking underscore how the administration has reshaped science, economic and legal analysis to thumb the scales for oil and gas extraction and away from public health and environmental protection. The consequences of these actions in jeopardizing human lives have been most evident in the Trump administration's obstruction of Americans' efforts to address the climate crisis, the issue I turn to next.

[liv] Keith Whittington and Jason Iuliano assembled a dataset of every federal and state case that involved a nondelegation challenge between 1789 and 1940. They conclude that the courts did not apply the nondelegation doctrine to constrain expansive delegations of power [165].

Appendix

Table 10.1 *A brief discussion of pertinent doctrinal developments at the Supreme Court*

Case	Synopsis
Presumptions on the breadth of statutory protections	
Mass. v. EPA	The 5–4 majority decision in the 2005 Supreme Court case *Massachusetts v. EPA*[36] presumed broad statutory goals under the CAA, granting powers to the EPA to address emerging problems.[51,127] Justice Stevens, writing for the majority, reasoned that "While the Congresses that drafted [section 202(a)] might not have appreciated the possibility that burning fossil fuels could lead to global warming, they did understand that without regulatory flexibility, changing circumstances and scientific developments would soon render the CAA obsolete."
Rapanos v. US	The court's 2006 decision in *Rapanos v. United States*[128] demonstrates how conservative judges' interpretations can severely limit protections offered under the CWA. At issue was whether the CWA protected seasonal wetlands that are linked to waters traditionally protected under the act. Five Justices recognized that the CWA offers expansive protections for clean water. Justice John Paul Stephens, joined by Justices Stephen Breyer, Ruth Ginsburg and David Souter, wrote that "Congress' intent in enacting the [act] was clearly to establish an all-encompassing program of water pollution regulation." Conversely, Justice Scalia, whose opinion was joined by Justices Roberts, Thomas and Alito, focused on the dictionary meaning of the word "waters" in the statute. This meaning divorced from scientific assessment on the need to protect wetlands in order to protect downstream waters. Textualists' selection of the meaning of words in statutes is not value-free. Scalia used the definition from the older 1932 Webster dictionary. But the 1961 Webster dictionary, already published when the CWA was drafted and enacted, defined waters as "the water *occupying* or flowing in a particular bed." That 1961 definition, arguably, would include the seasonal wetlands.[129]
UARG v. EPA	In 2014, Justice Scalia's opinion in *Utility Air Regulatory Group v. EPA*,[130] joined by three conservative judges and Kennedy, a swing vote, undercut the court's previous presumption of EPA's broad statutory powers to regulate greenhouse gas emissions under the CAA. Scalia narrowed the broad presumption in

Table 10.1 *(cont.)*

Case	Synopsis
	Massachusetts v. EPA, opining instead that the EPA's powers under the CAA are to be assessed on a section-by-section basis.
Presumptions for and against the delegation of powers	
EPA v. EME Homer City	The court's decision in the 2014 case *EPA v. EME Homer City*,[131] written by Justice Ginsburg, demonstrated the court's presumption that a statute delegates to an agency the authority to fill in gaps to achieve statutory goals. The CAA's "good neighbor" provision, which mandates that upwind states must not emit pollutants in "amounts which will ... contribute significantly to nonattainment" in downwind states, is one such example of this phenomenon. Ginsburg acknowledged the complexity of the interstate air pollution problem and ruled that the statute did not dictate how to allocate the emissions reductions among upwind states[132,133] – thus delegating to the EPA the authority to decide how to do so.
UARG v. EPA	Justice Scalia's opinion in *Utility Air Regulatory Group v. EPA*,[130] in addition to its impacts discussed above, also demonstrates the court shifting away from its past presumption for the delegation of broad regulatory agency power, emphasizing "[the court would] expect Congress to speak clearly if it wishes to assign to an agency decisions of vast 'economic and political significance.'" This "major questions" presumption is not neutral in its values or its policy impacts, as noted by law professor Lisa Heinzerling.[51] By reading statutes in ways that take away agencies' powers to promulgate regulations, this presumption gives less consideration to the economic and political impacts of not regulating. It also gives less consideration to the interests of beneficiaries of regulations than to the interests of regulated entities.
Presumptions for and against congressional authority to delegate	
Panama Refining Co. v. Ryan	The court has not struck down a law via the nondelegation doctrine, i.e., overruling a law based on Congress's excessive delegation of power to another branch, since 1935 in *Panama Refining Co. v. Ryan*.[134] The court struck down a provision of the National Industrial Recovery Act, in which Congress attempted to give the president broad powers to set production quotas for individual oil producers. Justice Hughes ruled that the provision "established no standard, [and] laid down no rule" for the executive branch. The court also struck down another provision of the same statute in the 1935 case of *ALA Schechter Poultry Corp. v. United States*.[135]
Gundy v. US	In June 2019, the Supreme Court came two votes short of bringing the nondelegation doctrine back into action in *Gundy v. United States*.[67] The case centered on Congress's delegation of powers

Table 10.1 *(cont.)*

Case	Synopsis
	attorney general to set up a registry for sex-offenders who had been convicted prior to the act. The majority ruled 5–3 to uphold the statute as not violating the nondelegation doctrine. Kagan, joined by three other justices, followed the court's highly deferential approach of upholding a delegation as long as Congress provides an intelligible principle for the delegee. Kagan pointed to examples of broad delegations upheld even by Scalia, including the delegation to the EPA "to issue whatever air quality standards are 'requisite to protect the public health.'"[136] Kagan underscored how delegation is fundamental to Congress's obligations to govern. However, Gorsuch, writing for the two dissenters, Thomas and Roberts, took the view that the statute should be struck down for Congress's excessive delegation of powers. Emphasizing the separation of powers, Gorsuch argued the Framers insisted that Congress should not delegate its legislative powers because "They believed the new federal government's most dangerous power was the power to enact laws restricting the people's liberty." A number of legal scholars point out how Gorsuch's opinion, while centered on the Framers' vision, does not accord with history.[57] In effect, this stance gives judges tremendous discretion to curtail Congress's legislative powers – masking a lack of approval for particular legislative actions with concern over separation of powers.[59,137–141]
Constitutional limits on Congress's powers to enact laws	
US v. Lopez	In 1995, the Supreme Court's 5–4 decision in *United States v. Lopez*[142] narrowed its interpretation of Congress's power to enact law by ruling that Congress went beyond its power under the Commerce Clause in passing the Gun Free School Zones Act of 1990. The majority expressed their skepticism of the federal regulation of primarily noncommercial activity in areas that were historically reserved to state governments, especially when there was only a weak link between the regulated activity – gun violence – and interstate commerce.[86,87]
Rancho Viejo LLC v. Norton	Subsequently, a number of judicial opinions have questioned congressional power to enact certain laws under the Commerce Act – including environmental laws. In 2003, prior to joining the Supreme Court, Judge John Roberts' suggested in a dissent for *Rancho Viejo LLC v. Norton*[143] for the Court of Appeals for the DC Circuit that it would be unconstitutional for the ESA to protect an imperiled species that exists only in one state.[89]

Table 10.1 *(cont.)*

Case	Synopsis
PETPO v. USFWS	In 2014, the US District Court for the District of Utah in *People for Ethical Treatment of Property Owners v. US Fish and Wildlife Service*[144] held that regulation under the ESA of a prairie dog species found only in Utah is beyond Congress's power under the Commerce Clause.[145] The ruling was overruled however by the Tenth Circuit in 2017, and on appeal the Supreme Court declined to hear the case. Congressional power to enact statutory environmental protections under the Commerce Clause remains a target for anti-regulatory skeptics.

References

1. Congressional Research Service. *Congress's Authority to Influence and Control Executive Branch Agencies*. Report by T. Garvey, legislative attorney and D. Sheffner, legislative attorney. R45442 (December 19, 2018). https://fas.org/sgp/crs/misc/R45442.pdf.
2. M. B. Gerrard. "Emergency Exemptions from Environmental Laws." In *Law in the Time of COVID-19*, edited by K. Pistor. New York: Columbia Law School, 2020.
3. J. D. Michaels. *Constitutional Coup: Privatization's Threat to the American Republic*. Cambridge, MA: Harvard University Press, 2017.
4. B. Emerson. *The Public's Law*. New York: Oxford University Press, 2019.
5. A. Southworth. *Lawyers of the Right: Professionalizing the Conservative Coalition*. Chicago, IL: University of Chicago Press, 2008.
6. S. M. Teles. *The Rise of the Conservative Legal Movement: The Battle for Control of the Law*. Princeton, NJ: Princeton University Press, 2010.
7. A. Hollis-Brusky. *Ideas with Consequences: The Federalist Society and the Conservative Counterrevolution*. New York: Oxford University Press, 2015.
8. M. Avery and D. McLaughlin. *The Federalist Society: How Conservatives Took the Law Back from Liberals*. Nashville, TN: Vanderbilt University Press, 2013.
9. R. Barnes and S. Mufson. "White House Counts on Kavanaugh in Battle against 'Administrative State.'" *Washington Post*, August 12, 2018. www.washingtonpost.com/politics/courts_law/brett-kavanaugh-and-the-end-of-the-regulatory-state-as-we-know-it/2018/08/12/22649a04-9bdc-11e8-8d5e-c6c594024954_story.html.
10. G. E. Metzger. "1930s Redux: The Administrative State under Siege." *Harvard Law Review* 131, no. 1 (November 10, 2017): 1–95. https://harvardlawreview.org/2017/11/1930s-redux-the-administrative-state-under-siege/.
11. A. Southworth. "Lawyers and the Conservative Counterrevolution." *Law & Social Inquiry* 43, no. 4 (September 24, 2018): 1698–1728. https://doi.org/https://doi.org/10.1111/lsi.12363.
12. M. Squillace et al. "Presidents Lack the Authority to Abolish or Diminish National Monuments." *Virginia Law Review Online* 103 (June 9, 2017): 55–71. www.virginialawreview.org/sites/virginialawreview.org/files/Hecht%20PDF.pdf.
13. J. Hein. "Monumental Decisions: One-Way Levers Towards Preservation in the Antiquities Act and Outer Continental Shelf Lands Act." *Environmental Law* 48, no. 125 (April 11, 2018): 126–166.
14. D. J. Trump. *Presidential Executive Order on Accelerating the Nation's Economic Recovery From the COVID-19 Emergency by Expediting Infrastructure Investments and Other Activities*. Executive Order 13927, 85 Federal Register 35165–35170 (2020). www.federalregister.gov/documents/2020/06/09/2020-12584/accelerating-the-nations-economic-recovery-from-the-covid-19-emergency-by-expediting-infrastructure.
15. C. Davenport and L. Friedman. "Trump, Citing Pandemic, Moves to Weaken Two Key Environmental Protections." *New York Times*, June 4, 2020. www.nytimes.com/2020/06/04/climate/trump-environment-coronavirus.html.

16. D. Reeves. "Trump Suspends Environmental Rules for Infrastructure, Citing Pandemic." *Energy Washington Week*, June 4, 2020. https://insideepa.com/daily-news/trump-suspends-environmental-rules-infrastructure-citing-pandemic.
17. R. L. Glicksman. "The Firm Constitutional Foundation and Shaky Political Future of Environmental Cooperative Federalism." In *Controversies in American Federalism and Public Policy*, edited by C. P. Banks. 132–150. London, UK: Routledge, 2018.
18. D. Duncan and C. Ellis. "Clean Water Act Section 401: Balancing States' Rights and the Nation's Need for Energy Infrastructure." *Hastings Environmental Law Journal* 25, no. 2 (2019): 235–262. https://repository.uchastings.edu/cgi/viewcontent.cgi?article=1568&context=hastings_environmental_law_journal.
19. Congressional Research Service. *Clean Water Act Section 401: Background and Issues*. Report by C. Copeland, specialist in Resources and Environmental Policy. 97-488 (Washington, DC: 2015). https://fas.org/sgp/crs/misc/97-488.pdf.
20. *Clean Water Act Section 401 Certification Rule: Final Rule*. 40 Code of Federal Regulations Part 121. Environmental Protection Agency. 85 Federal Register 42210–42287 (July 13, 2020). www.epa.gov/sites/production/files/2020-06/documents/pre-publication_version_of_the_clean_water_act_section_401_certification_rule_508.pdf.
21. Attorneys General of California, Connecticut, Maryland, Maine, Massachusetts, Minnesota, New Jersey, New Mexico, New York, Oregon, Pennsylvania, Rhode Island, Vermont, Washington and Pennsylvania Department of Environmental Protection. *Objection to "Clean Water Act Section 401 Guidance for Federal Agencies, States and Authorized Tribes" Issued by the US Environmental Protection Agency*. Submitted to A. Wheeler, administrator, Environmental Protection Agency. July 25, 2019.
22. T. Carper, ranking member of the Committee on Environment and Public Works, T. Duckworth, ranking member of the Subcommittee on Fisheries, Water, and Wildlife, and C. A. Booker, ranking member of the Subcommittee on Superfund, Waste, Management, and Regulatory Oversight. *Comments on Proposed Clean Water Act Section 401 Certification Rule*. Submitted to A. Wheeler, administrator, Environmental Protection Agency. October 21, 2019.
23. P. Parenteau. "EPA's Latest Power Grab Is Aimed at States' Rights." *The Hill*, August 14, 2019. https://thehill.com/opinion/energy-environment/457426-epas-latest-power-grab-is-aimed-at-states-rights.
24. M. Nasmith, staff attorney of Earthjustice, et al. *Comments on EPA Proposed Rule Updating Regulations on Water Quality Certification*. Submitted to A. Wheeler, administrator, Environmental Protection Agency. October 21, 2019.
25. *PUD No. 1 of Jefferson County v. Washington Department of Ecology*, 511 US 700 (Supreme Court 1994).
26. *S. D. Warren Co. v. Maine Board of Environmental Protection*, 547 US 370 (Supreme Court 2006).
27. W. W. Buzbee. "Agency Statutory Abnegation in the Deregulatory Playbook." *Duke Law Journal* 68, no. 8 (2019): 1509–1591. https://scholarship.law.duke.edu/dlj/vol68/iss8/1.

28. *Waste Prevention, Production Subject to Royalties, and Resource Conservation: Final Rule*. 43 Code of Federal Regulations Parts 3100, 3160 and 3170. I. Bureau of Land Management. 81 Federal Register 83008–83089 www.govinfo.gov/content/pkg/FR-2016-11-18/pdf/2016-27637.pdf.
29. *Wyoming v. US Department of Interior*, No. 2:16-CV-0280-SWS, 2017 WL 161428 (D. Wyo. Jan. 16, 2017).
30. *Waste Prevention, Production Subject to Royalties, and Resource Conservation; Rescission or Revision of Certain Requirements: Final Rule*. 43 Code of Federal Regulations Parts 3160 and 3170. Bureau of Land Management. 83 Federal Register 49184–49214. www.regulations.gov/document?D=BLM-2018-0001-223600.
31. *League of Conservation Voters v. Trump*, 363 F. Supp. 3d 1013 (D. Alaska 2019), vacated and remanded sub nom. *League of Conservation Voters v. Biden*, 843 F. Appendix 937 (9th Cir. 2021).
32. *California v. Bernhardt*, 472 F. Supp. 3d 573 (N.D. Cal 2020).
33. *County of Maui, Hawaii v. Hawaii Wildlife Fund*, 140 S. Ct. 1462 (Supreme Court 2020).
34. L. Heinzerling. "Opinion Analysis: The Justices' Purpose-full Reading of the Clean Water Act." *SCOTUSblog*. 2020. www.scotusblog.com/2020/04/opinion-analysis-opinion-analysis-the-justices-purpose-full-reading-of-the-clean-water-act.
35. *California v. US Bureau of Land Management*, No. 18-CV-00521-HSG, 2018 WL 3439453 (N.D. Cal. July 17, 2018).
36. *Massachusetts v. Environmental Protection Agency* 549 US 497 (Supreme Court 2007).
37. M. Sherman. "Roberts, Trump Spar in Extraordinary Scrap over Judges." *Associated Press News*, November 21, 2018. https://apnews.com/c4b34f9639e141069c08cf1e3deb6b84.
38. L. Baum and N. Devins. *The Company They Keep: How Partisan Divisions Came to the Supreme Court*. New York: Oxford University Press, 2019.
39. J. Z. Cannon. *Environment in the Balance: The Green Movement and the Supreme Court*. Cambridge, MA: Harvard University Press, 2015.
40. R. L. Revesz. "Environmental Regulation, Ideology, and the DC Circuit." *Virginia Law Review Online* 83, no. 8 (November 1997): 1717–1772. https://doi.org/10.2307/1073657.
41. L. Epstein, W. M. Landes and R. A. Posner. *The Behavior of Federal Judges: A Theoretical and Empirical Study of Rational Choice*. Cambridge, MA: Harvard University Press, 2013.
42. C. R. Sunstein et al. *Are Judges Political? An Empirical Analysis of the Federal Judiciary*. Washington, DC: Brookings Institution Press, 2006.
43. C. R. Sunstein and T. J. Miles. "Do Judges Make Regulatory Policy? An Empirical Investigation of Chevron." *University of Chicago Law Review* 73 (June 2006): 1–55.
44. C. Sunstein. "Beyond Marbury: The Executive's Power to Say What the Law Is." *Yale Law Journal* 115, no. 371 (2004): 1–34. https://doi.org/10.2307/20455706.
45. O. Lawrence, J. D. Gostin and J. G. Hodge, Jr. "Substantial Shifts in Supreme Court Health Law Jurisprudence." *JAMA* 320, no. 14 (October 9, 2018): 1431–1432. https://doi.org/10.1001/jama.2018.12331.

46. J. G. Hodge, Jr. et al. "Public Health Law and Policy Implications: Justice Kavanaugh." *Journal of Law, Medicine & Ethics* 47, no. 2 suppl. (July 12, 2019): 59–62. https://doi.org/10.1177/1073110519857319.
47. L. O. Gostin, W. E. Parmet and S. Rosenbaum. "Health Policy in the Supreme Court and a New Conservative Majority." *JAMA* 324, no. 21 (2020): 2157–2158. https://doi.org/10.1001/jama.2020.21987.
48. J. Freeman. "What Amy Coney Barrett's Confirmation Will Mean for Joe Biden's Climate Plan." *Vox*, October 26, 2020. www.vox.com/energy-and-environment/21526207/amy-coney-barrett-senate-vote-environmental-law-biden-climate-plan.
49. T. B. Edsall. "The Right's Relentless Supreme Court Justice Picking Machine." *New York Times*, October 1, 2020. www.nytimes.com/2020/10/01/opinion/amy-coney-barrett-supreme-court.html.
50. E. Ryan. "Environmentalists: Brace for Preemption, Propertization, and Problems of Political Scale." In *Environmental Law, Disrupted*, edited by J. Owley and K. Hirokawa. Washington, DC: Environmental Law Institute, 2019.
51. L. Heinzerling. "The Power Cannons." *William & Mary Law Review* 58, no. 6 (October 7, 2017): 1932–2004. https://papers.ssrn.com/sol3/papers.cfm?abstract_id=2757770.
52. J. M. Beerman. "The Never-Ending Assault on the Administrative State." *Notre Dame Law Review* 93, no. 4 (July 2018): 1599–1652.
53. S. G. Calabresi and G. Lawson. "The Depravity of the 1930s and the Modern Administrative State." *Notre Dame Law Review* 94, no. 2 (January 2019): 821–866. https://scholarship.law.nd.edu/cgi/viewcontent.cgi?article=4825&context=ndlr.
54. G. E. Metzger. "The Roberts Court and Administrative Law." *Supreme Court Review*, no. 1 (2020): 1–71.
55. D. B. Rodriguez and B. R. Weingast. "Engineering the Modern Administrative State, Part I: Political Accommodation and Legal Strategy in the New Deal Era." *Northwestern Public Law Research Paper*, no. 19-03 (February 15, 2019): 1–65.
56. Galperin, J. "The Death of Administrative Democracy." *University of Pittsburgh Law Review* 82, no. 1 (2020).
57. J. D. Mortenson and N. Bagley. "Delegation at the Founding." University of Michigan Public Law Research Paper, no. 658 (December 31, 2019). https://papers.ssrn.com/sol3/papers.cfm?abstract_id=3512154.
58. J. D. Mortenson and N. Bagley. "There's No Historical Justification for One of the Most Dangerous Ideas in American Law." *The Atlantic*, May 26, 2020. www.theatlantic.com/ideas/archive/2020/05/nondelegation-doctrine-orliginalism/612013.
59. E. Bazelon. "How Will Trump's Supreme Court Remake America?" *New York Times*, February 27, 2020. www.nytimes.com/2020/02/27/magazine/how-will-trumps-supreme-court-remake-america.html.
60. J. L. Mashaw. *Creating the Administrative Constitution: The Lost One Hundred Years of American Administrative Law*. New Haven, CT: Yale University Press, 2012.
61. W. Novak. "Law and the Social Control of American Capitalism." *Emory Law Journal* 60 (2010): 377–405. https://law.emory.edu/elj/_documents/volumes/60/2/symposium/novak.pdf.

62. J. S. Hacker and P. Pierson. *American Amnesia: How the War on Government Led Us to Forget What Made America Prosper.* New York: Simon & Schuster, 2017.
63. *Lochner v. New York*, 198 US 45 (Supreme Court 1905).
64. *West Coast Hotel Co. v. Parrish*, 300 US 379 (Supreme Court 1937).
65. *Seila Law LLC v. Consumer Financial Protection Bureau*, 140 S. Ct. 2183 (Supreme Court 2020).
66. N. Fligstein and A. F. Roehrkasse. "The Causes of Fraud in the Financial Crisis of 2007 to 2009: Evidence from the Mortgage-Backed Securities Industry." *American Sociological Review* 81, no. 4 (June 23, 2016): 617–643. https://doi.org/10.1177/0003122416645594.
67. *Gundy v. United States*, 139 S. Ct. 2116 (Supreme Court 2019).
68. L. Kennedy, P. A. McCoy and E. Bernstein. "The Consumer Financial Protection Bureau: Financial Regulation for the 21st Century." *Cornell Law Review* 98, no. 5 (2012): 1141–1176.
69. P. A. McCoy. "Inside Job: The Assault on the Structure of the Consumer Financial Protection Bureau." *Minnesota Law Review* 103, no. 6 (2018): 2543–2615.
70. C. M. Sharkey. "The Administrative State and the Common Law: Regulatory Substitutes or Complements." *Emory Law Journal* 65, no. 6 (2016): 1705–1740. https://law.emory.edu/elj/content/volume-65/issue-6/articles-essays/administrative-state-common-law-substitutes-complements.html.
71. *PHH Corporation, et al., v. Consumer Financial Protection Bureau*, No 15-1177 (D.C. Cir. 2016).
72. S. Sarkar and J. A. Rosenthal. "PHH Corporation v. Consumer Financial Protection Bureau: Financial Fairness and Administrative Anxiety." *University of Pennsylvania Law Review Online* 166, no. 14 (2018): 265–272. https://scholarship.law.upenn.edu/penn_law_review_online/vol166/iss1/14.
73. S. Harrington. "Kavanaugh on the Executive Branch: PHH Corp. v. Consumer Financial Protection Bureau." *SCOTUSblog*, 2018. www.scotusblog.com/2018/08/kavanaugh-on-the-executive-branch-phh-corp-v-consumer-financial-protection-bureau.
74. R. H. Fallon. "Three Symmetries between Textualist and Purposivist Theories of Statutory Interpretation – and the Irreducible Roles of Values and Judgment within Both." *Cornell Law Review* 99, no. 4 (May 2014): 685–734. https://scholarship.law.cornell.edu/cgi/viewcontent.cgi?article=4628&context=clr.
75. A. Kaswan. "Our New Pro-Liberty Justice–and What That Means for Environmental Law." *Trends* 50, no. 3 (January/February 2019): 4–7. www.americanbar.org/groups/environment_energy_resources/publications/trends/2018-2019/january-february-2019/our-new-pro-liberty.
76. Congressional Research Service. *Chevron Deference: A Primer.* Report by V. C. Brannon, legislative attorney and J. P. Cole, legislative attorney. R44954 (Washington, DC: September 19, 2017). https://fas.org/sgp/crs/misc/R44954.pdf.
77. Congressional Research Service. *Deference and its Discontents: Will the Supreme Court Overrule Chevron?* Report by V. C. Brannon, legislative attorney and J. P. Cole, legislative attorney. LSB10204 (Washington, DC: October 11, 2018). https://fas.org/sgp/crs/misc/LSB10204.pdf.
78. *Chevron v. Natural Resources Defense Council.* 467 US 837 (Supreme Court 1984).

79. Congressional Research Service. *The DC Circuit Rejects EPA's Mercury Rules: New Jersey v. EPA*. RS22817 (Washington, DC: April 9, 2008). www.everycrsreport.com/reports/RS22817.html.
80. N. Morales. "New Jersey v. Environmental Protection Agency." *Harvard Environmental Law Review* 33 (2009): 263–282. https://harvardelr.com/wp-content/uploads/sites/12/2019/07/33.1-Morales.pdf.
81. J. A. Pojanowski. "Without Deference." *Missouri Law Review* 81 (February 24, 2017): 1076–1094.
82. J. R. Siegel. "The Constitutional Case for Chevron Deference." *Vanderbilt Law Review* 71, no. 3 (2018): 937–993.
83. C. R. Sunstein. "Chevron As Law." *Georgetown Law Journal* (January 9, 2019): 1–61.
84. Take Back the Court. *The Roberts Court Would Likely Strike Down Climate Change Legislation*. Report by S. Moyn and A. Belkin (September 2019). https://static1.squarespace.com/static/5ce33e8da6bbec0001ea9543/t/5d7d429025734e4ae9c92070/1568490130130/Supreme+Court+Will+Overturn+Climate+Legislation+FINAL.pdf.
85. Environmental Law Institute. "Challenges to Environmental Protection in the Courts Continued." www.eli.org/constitution-courts-and-legislation/challenges-environmental-courts-continued#commerce.
86. E. Biber and E. O'Dea. "Is the Endangered Species Act Constitutional? How the Utah Prairie Dog Case May Impact California." *State Bar of California Environmental Law News* 24, no. 1 (Summer 2015).
87. E. Biber. "The ESA and the Commerce Clause." *Legal Planet*, 2014. https://legal-planet.org/2014/11/18/the-esa-and-the-commerce-clause.
88. J. L. Dunec. "Book Review: Global Chemical Control Handbook: A Guide to Chemical Management Programs, Lynn L. Bergeson, ed." *Review of Natural Resources & Environment* (Spring 2016): 61–62.
89. B. Parker, executive director at Earthjustice. *The Commerce Clause and the Environment*. Analysis for Judging the Environment (2015). www.judgingtheenvironment.org/library/reports_analysis/the-commerce-clause-and-the-environment.pdf.
90. D. T. Kendall and C. P. Lord. "The Takings Project: A Critical Analysis and Assessment of the Progress So Far." *Boston College Environmental Affairs Law Review* 25, no. 3 (May 1, 1998): 509–587. https://lawdigitalcommons.bc.edu/cgi/viewcontent.cgi?article=1280&context=ealr.
91. J. Pollack. "The Takings Project Revisited: A Critical Analysis of This Expanding Threat to Environmental Law." *Harvard Environmental Law Review* 44 (2020): 235–278.
92. A. Mikva. "The Wooing of Our Judges." *New York Times*, August 28, 2000. www.nytimes.com/2000/08/28/opinion/the-wooing-of-our-judges.html.
93. D. Barnhizer. "On the Make: Campaign Funding and the Corrupting of the American Judiciary." *Catholic University Law Review* 50, no. 2 (Winter 2001): 361–428.
94. R. Wheeler. "How Close Is President Trump to His Goal of Record-Setting Judicial Appointments?" *Brookings Institution Blog*, May 5, 2020. www.brookings.edu/blog/fixgov/2020/05/05/how-close-is-president-trump-to-his-goal-of-record-setting-judicial-appointments.

95. M. Nevitt. "The Remaking of the Supreme Court Implications for Climate Change Litigation and Regulation." *Cardozo Law Review* 42, no. N (2020): 101–115.
96. C. Hulse. "Protégé Confirmed, McConnell Is One Judge Closer to His Goal." *Washington Post*, June 19, 2020. www.nytimes.com/2020/06/18/us/mcconnell-courts-justin-walker.html.
97. J. Zengerle. "How the Trump Administration Is Remaking the Courts." *New York Times*, August 22, 2018. www.nytimes.com/2018/08/22/magazine/trump-remaking-courts-judiciary.html.
98. D. Montgomery. "Conquerors of the Courts." *Washington Post Magazine*, January 2, 2019. www.washingtonpost.com/news/magazine/wp/2019/01/02/feature/conquerors-of-the-courts.
99. N. Rao. "Administrative Collusion: How Delegation Diminishes the Collective Congress." *New York University Law Review* 90, no. 5 (October 16, 2015): 1463–1526.
100. N. Rao. "The Trump Administration's Deregulation Efforts Are Saving Billions of Dollars." *Washington Post*, October 17, 2018. www.washingtonpost.com/opinions/the-trump-administration-is-deregulating-at-breakneck-speed/2018/10/17/09bd0b4c-d194-11e8-83d6-291fcead2ab1_story.html.
101. R. L. Revesz. "Destabilizing Environmental Regulation: The Trump Administration's Concerted Attack on Regulatory Analysis." *Ecology Law Quarterly* 47 (2020): 887–956.
102. R. Frazin. "Court Strikes Down EPA Suspension of Obama-Era Greenhouse Gas Rule." *The Hill*, April 7, 2020. https://thehill.com/policy/energy-environment/491568-court-strikes-down-epa-suspension-of-obama-era-hfc-rule.
103. M. J. Gerhardt and R. W. Painter. "Majority Rule and the Future of Judicial Selection." *Wisconsin Law Review* 2017, no. 2 (2017): 263–284.
104. R. R. Ruiz et al. "Trump Stamps GOP Imprint on the Courts." *New York Times*, March 15, 2020. https://static01.nyt.com/images/2020/03/15/nytfrontpage/scan.pdf.
105. Federalist Society. "About Us." 2020. https://fedsoc.org/about-us.
106. A. Hollis-Brusky. "'It's the Network': The Federalist Society As a Supplier of Intellectual Capital for the Supreme Court." In *Studies in Law, Politics, and Society*, edited by A. Sarat. 137–178. Bingley: Emerald Group Publishing, 2013.
107. N. Scherer and B. Miller. "The Federalist Society's Influence on the Federal Judiciary." *Political Research Quarterly* 62, no. 2 (May 1, 2009): 366–378. https://doi.org/10.1177/1065912908317030.
108. J. González, A. Goodman and E. Lipton. "Inside How the Federalist Society & Koch Brothers Are Pushing for Trump to Reshape Federal Judiciary." *Democracy Now!*, March 21, 2017. www.democracynow.org/2017/3/21/inside_how_the_federalist_society_koch.
109. E. Lipton and J. Peters. "Conservatives Press Overhaul in the Judiciary." *New York Times*, March 19, 2017.
110. R. O'Harrow, Jr. and S. Boburg. "The Activist behind the Push to Reshape US Courts." *Washington Post*, May 21, 2019. www.washingtonpost.com/graphics/2019/investigations/leonard-leo-federalists-society-courts.
111. S. Whitehouse, senator, and R. J. Durbin, senator. *Request for Communications at the Department of Justice Pertaining to Mr. Leo Involvement in Potential, Actual,*

or Suggested Judicial Nominations. Submitted to Attorney General W. Barr. March 4, 2020.
112. Brennan Center for Justice. *Who Pays for Judicial Races? The Politics of Judicial Elections 2015–16*. Report by A. Bannon (December 14, 2017). www.brennancenter.org/our-work/research-reports/who-pays-judicial-races-politics-judicial-elections-2015-16.
113. L. Fang and N. Surgey. "Koch Document Reveals Laundry List of Policy Victories Extracted from the Trump Administration." *The Intercept*, February 25, 2018. https://theintercept.com/2018/02/25/koch-brothers-trump-administration/.
114. A. Pienta. *Update to Report on "The Federalist Society's Takeover of George Mason University's Public Law School"* (December 2018). https://static1.squarespace.com/static/5400da69e4b0cb1fd47c9077/t/5c1d302b4fa51a38153ac07e/1545416751164/Update+to+Report+on+Federalist+Society+Takeover+of+GMU+Law+12+17+2018.pdf.
115. A. Pienta. *New Evidence Suggests Chicago Billionaire "Closely Allied" with the Koch Brothers and Implicated in Funding Climate-Change Denial and Islamophobia Is Anonymous "Dark Money" Donor behind Renaming of George Mason University's Law School* (December 16, 2019). https://medium.com/@acaalim/new-evidence-suggests-chicago-billionaire-closely-allied-with-the-koch-brothers-and-implicated-2abc9bcbd102.
116. J. Mayer. "How Mitch McConnell Became Trump's Enabler-in-Chief." *New Yorker Magazine*, April 12, 2020. www.newyorker.com/magazine/2020/04/20/how-mitch-mcconnell-became-trumps-enabler-in-chief.
117. D. Ziblatt and S. Levitsky. *How Democracies Die*. New York: Crown, 2018.
118. D. Frum. *Trumpocracy: The Corruption of the American Republican*. New York: Harper Collins, 2018.
119. S. Stevens. *It Was All a Lie: How the Republican Party Became Donald Trump*. New York: Knopf, 2020.
120. Annenberg Public Policy Center. "Most Americans Trust the Supreme Court, But Think It Is 'Too Mixed Up in Politics.'" News release, 2019, www.annenbergpublicpolicycenter.org/most-americans-trust-the-supreme-court-but-think-it-is-too-mixed-up-in-politics.
121. Staff to Sheldon Whitehouse. "Senate Democrats Request Documents Related to Leonard Leo's Work Leading Trump Administration Judicial Selections." News release, March 5, 2020, www.whitehouse.senate.gov/news/release/senate-dems-request-docs-related-to-leonard-leos-work-leading-trump-admin-judicial-selections-say-leo-hand-picked-trump-judges-while-raking-in-undisclosed-sums-from-related-advocacy-groupsa-potential-conflict-of-interest-and-violation-of-federal-ethics-law.
122. S. Whitehouse. "Dark Money and US Courts: The Problem and Solutions." *Harvard Journal on Legislation* 57, no. 2 (2020): 273–301. https://harvardjol.com/wp-content/uploads/sites/17/2020/05/Sen.-Whitehouse_Dark-Money.pdf.
123. C. Foran and T. Barrett. "Senate Passes Sweeping Conservation Legislation in Bipartisan Vote." *CNN Politics*, June 17, 2020. www.cnn.com/2020/06/17/politics/conservation-legislation-senate/index.html.
124. US Congress. House. *Regulations from the Executive in Need of Scrutiny Act of 2019*. H.R. 3972, 116th Congress, 1st Sess. Introduced in House August 15, 2019.

125. C. Coglianese and G. Scheffler. "What Congress's Repeal Efforts Can Teach Us about Regulatory Reform." *Administrative Law Review Accord* 3 (November 29, 2017): 43–58.
126. R. Meyer. "The EPA Needs Lots of Money to Gut Itself." *The Atlantic*, March 20, 2017. www.theatlantic.com/science/archive/2017/03/the-paradox-of-defunding-the-epa/520002.
127. J. Freeman and A. Vermeule. "Massachusetts v. EPA: From Politics to Expertise." *Supreme Court Review* 2007 (2007): 51–110. https://doi.org/https://doi.org/10.1086/655170.
128. *Rapanos v. United States*, 547 US 715 (Supreme Court 2006).
129. A. Wittenberg. "Clean Water Rule: Will Scalia's Dictionary Haunt Trump's WOTUS Overhaul?" *E&E News*, May 15, 2017. www.eenews.net/stories/1060054554+&cd=1&hl=en&ct=clnk&gl=ca.
130. *Utility Air Regulatory Group v. Environmental Protection Agency*, 573 US 302 (Supreme Court 2014).
131. *Environmental Protection Agency v. EME Homer City Generation*, 572 US 489 (Supreme Court 2014).
132. L. Denniston. "Opinion Analysis: Paying for Blocking Ill Winds." *SCOTUSblog*, 2014.
133. A. Serfess. "EPA v. EME Homer City Generation, LP: Supreme Court Upholds Transport Rule – Third Time's a Charm for Good Neighbor Provision Enforcement." *Tulane Environmental Law Journal* 28, no. 1 (2014): 115–126.
134. *Panama Refining Company v. Ryan*, 293 US 388 (Supreme Court 1935).
135. *ALA Schechter Poultry Corporation v. United States*, 295 US 495 (Supreme Court 1935).
136. *Whitman v. American Trucking Associations, Inc.*, 531 US 457 (Supreme Court 2001).
137. J. Hall. "The Gorsuch Test: Gundy v. United States, Limiting the Administrative State, and the Future of Nondelegation." *Duke Law Journal* 70, no. 1 (March 9, 2020): 175–215. http://dx.doi.org/10.2139/ssrn.3550906.
138. M. Sohoni. "Opinion Analysis: Court Refuses to Resurrect Nondelegation Doctrine." *SCOTUSblog*, 2019. www.scotusblog.com/2019/06/opinion-analysis-court-refuses-to-resurrect-nondelegation-doctrine.
139. D. Farber. "Just in From the Supreme Court." *Legal Planet*, 2019. https://legal-planet.org/2019/11/25/just-in-from-the-supreme-court.
140. D. Farber. "Justice Gorsuch Versus the Administrative State." *Center for Progressive Reform Blog*, 2019. http://progressivereform.org/cpr-blog/justice-gorsuch-versus-the-administrative-state.
141. M. D. Tortorice. "Nondelegation and the Major Questions Doctrine: Displacing Interpretive Power." *Buffalo Law Review* 67, no. 4 (August 1, 2019): 1075–1131.
142. *United States v. Alfonso D. Lopez, Jr.*, 514 US 549 (Supreme Court 1995).
143. *Rancho Viejo LLC v. Norton*, 334 F.3d 1158 (D.C. Cir. 2003).
144. *People for Ethical Treatment of Property Owners v. United States Fish & Wildlife*, 852 F.3d 990 (10th Cir. 2017).
145. M. C. Blumm. "Defending the Constitutionality of the Endangered Species Act: The Case of the Utah Prairie Dog." *On the Merits* (Washington Legal Foundation), June 5, 2015. https://papers.ssrn.com/sol3/papers.cfm?abstract_id=2615357.

146. J. R. May. "Healthcare, Environmental Law, and the Supreme Court: An Analysis under the Commerce, Necessary and Proper, and Tax and Spending Clauses." *Environmental Law* 43, no. 2 (2013): 233–254.
147. *Morehead v. New York ex rel. Tipaldo*, 298 US 587 (Supreme Court 1936).
148. *Murr v. Wisconsin*, 137 S. Ct. 1933 (Supreme Court 2017).
149. *Kisor v. Wilkie*, 139 S. Ct. 2400 (Supreme Court 2019).
150. *King v. Burwell*, 576 US 473 (Supreme Court 2015).
151. *United States Forest Service v. Cowpasture River Preservation Association*, 140 S. Ct. 1837 (Supreme Court 2020).
152. H. Kitrosser. "Accountability in the Deep State." *UCLA Law Review* 65 (2018): 1532–1550.
153. *Paul v. United States*, 140 S. Ct. 342 (Supreme Court 2019).
154. C. L. Peterson "Consumer Financial Protection Bureau Law Enforcement: An Empirical Review." *Tulane Law Review* 90 (2015): 1057–1112.
155. *Perez v. Mortgage Bankers Association*, 135 S. Ct. 1199 (Supreme Court 2015).
156. *Gutierrez-Brizuela v. Lynch*, 834 F.3d 1142, 1149 (10th Cir. 2016).
157. J. F. Manning. "What Divides Textualists from Purposivists?" *Columbia Law Review* 106 (2006): 70–111.
158. A. S. Krishnakumar. "Backdoor Purposivism." *Duke Law Journal* 69 (2020): 1275–1352.
159. *Michigan v. EPA*, 576 US 743 (Supreme Court 2015).
160. *Bostock v. Clayton County, Georgia*, 140 S. Ct. 1731 (Supreme Court 2020).
161. S. Calabrese. "Trump Might Try to Postpone the Election. That's Unconstitutional." *New York Times*, July 30, 2020.
162. B. Scher. "The GOP Traded Its Principles for Conservative Judges." *Washington Post*, July 5, 2020.
163. R. A. Posner. "The Incoherence of Antonin Scalia." *New Republic*, August 24, 2012.
164. E. Segall. *Originalism as Faith*. Cambridge, UK: Cambridge University Press, 2018.
165. K. E. Whittington and J. Iuliano. "The Myth of the Nondelegation Doctrine." *University of Pennsylvania Law Review* 165, no. 2 (2017): 379–431.
166. *SAS Institute Inc. v. Iancu*,138 S. Ct. 1348, 1358–59 (Supreme Court 2018).
167. *New Jersey v. EPA*, 517 F.3d 574 (D.C. Cir. 2008).
168. Koch Seminar Network. *Efforts in Government: Advancing Principled Public Policy*. n.d. www.documentcloud.org/documents/4364737-Koch-Seminar-Network.html.

PART V

The Global Climate

11

Endangering the Climate
Attacking Global Cooperation, State Governments' Leadership and the Private Sector's Economic Restructuring

Addressing the climate crisis requires the United States to take bold and comprehensive actions to shift away from oil and gas dependency. The federal government possesses key tools to reshape the rules of the economic system within which private actors operate and thus shepherd a more orderly energy and economic transformation. These include correcting economic signals to better capture the costs of greenhouse gas emissions and climate disruptions, imposing well-designed regulations to prompt cuts in those emissions, investing in research, development and deployment to accelerate innovations and adoption of low carbon technologies, and directing its procurement powers to strengthen the early demand for low carbon technologies. Equally importantly, it can assist fossil-fuel-dependent communities to undertake the challenging tasks of diversifying their economies and to bridge the energy transition.

Addressing the climate crisis drew support early on from both Republicans and Democrats. Republican Senator Lincoln Chafee, chair of the Environment and Public Works Subcommittee on Environmental Pollution, urged climate action, warning that "the scientific evidence is telling us . . . we have a serious problem . . . this generation may be committing all of us to severe economic and environmental disruption."[1] Republican George H. W. Bush raised hopes for climate action, pledging to the use the White House to address the greenhouse effect and convening a summit on climate change science and economics at the White House in 1990. Bush underscored that "we must leave this Earth in better condition than we found it" when signing the 1992 global agreement establishing the UN Framework Convention for Climate Change. Unfortunately, faced with divergence among Republicans on climate action,[i] Bush reneged on his

---

[i] The G. H. W. Bush administration enjoyed strong Republican support for climate action, but it also faced opposition from Republicans such as John Sununu, White House chief of staff, and

support for limiting greenhouse gas emissions.² He was, however, more successful in more politically feasible environmental policy initiatives, for example, working with Republican Representative Sherwood Boehlert and Democrat Representative Henry Waxman to successfully address the US acid rain problem.³

In the early 2000s, Republican Senator John McCain and Democrat Senator Joe Lieberman worked together to advocate for climate legislation but failed to gain traction. The 2008 Republican platform underscored that "Republicans support technology-driven, market-based solutions that will decrease emissions" and "mitigate the impact of climate change where it occurs."[4],ii In 2009, Congress came closest to enacting a cap-and-trade program[iii] on greenhouse gas emissions through the American Clean Energy and Security Act. The House passed the bill, but the Senate did not take up the bill.

Nevertheless, Congress has moved the dial slightly forward on energy transition. Congress's provision of tax credits (albeit without reliable continuity) for wind projects, first authored by Republican Senator Chuck Grassley in 1992, and for solar projects, first signed into law by G. W. Bush in 2005, helped expand their adoption.[5–7] As part of the 2009 America Reinvestment and Recovery Act, Congress provided modest but noteworthy clean energy funding that created a modest number of jobs.[8] In 2015, Congress allocated (limited) funding through the Obama administration's POWER initiative, which expanded assistance to Appalachian coal communities to diversify their local economies, to sustain social services and to remediate environmental damages.[9]

---

from think tanks, such as the Cato Institute, that downplayed the dangers of climate change [2]. James Hansen, climate scientist at the National Aeronautics and Space Administration, complained that the G. H.W. Bush administration's Office of Management and Budget toned down the severity of his conclusions on the climate crisis before they were delivered to a Senate committee headed by then Senator Al Gore [2].

ii James Morton Turner and Andrew Isenberg's *Republican Reversal* traced the transformation of the Republican party to one in which a majority of its members peddled climate denial. Naomi Oreskes and Erik Conway's *Merchants of Doubt* detail the campaign of disinformation undertaken by oil companies, while Robert Brulle's journal articles track how think tanks, funded by oil companies and dark money, entrenched climate denial and hampered climate action [4][228, 229, 230].

iii In a cap-and-trade program, the government regulator sets the total amount of pollution that regulated entities are permitted to emit. The regulator auctions emission permits (or chooses other methods to allocate those permits). The price of the permit is set by the market, i.e., the relative demand and supply for those permits. A regulated entity must purchase enough permits to cover its total emissions. An entity that is efficient at reducing its emissions has the option of selling its permit to those entities that are less efficient at reducing their emissions. The cap-and-trade system, by encouraging more efficient entities to reduce their emissions, can reduce the overall costs of achieving emissions reductions.

Unfortunately, the federal government has failed to correct many misleading economic signals in the US economy (e.g., failing to reform the tax code to eliminate existing tax preferences for oil and gas and failing to ensure polluters pay the price for emitting carbon dioxide). Corrections to these signals are necessary to grow jobs and investments in a low-carbon economy. The failed 2009 cap-and-trade proposal would have established a price for greenhouse gas emissions from the power sector, a strong first step to signaling the shift of economic activities and investment away from carbon-intensive activities. G. W. Bush, who on the one hand peddled climate denial within his administration, nevertheless remarked that oil companies could drill profitably in the United States, even in the absence of the industry's numerous tax preferences.[10] However, Bush made no formal effort to pare back those preferences and Obama's attempt to do so was rejected by Congress.[11] Obama successfully negotiated the Paris Climate Agreement, but America's voluntary commitments to cut greenhouse emissions remained vulnerable without congressionally enacted climate legislation.[iv]

Trump and his administration undid the Obama administration's actions and more. Trump's claims that oil and gas expansion would secure well-paying jobs for Americans (alongside his false promises to bring back coal jobs) cemented the misleading narrative of jobs versus climate. As detailed in Chapter 2, these promises, which were already questionable when Trump made them on the campaign trail, had evaporated by 2019 with the poor financial performance and dismal outlook for both the oil and gas sector and the coal sector. While the Trump administration peddled climate denial and entrenched many more advantages for the oil and gas sector, many state governments and companies that are operating in sectors other than the fossil fuel sector (hereafter, the nonfossil-fuel private sector) continued to take prudent actions to respond to the market and climate realities.

Americans can still take bold actions to help stabilize the climate.[12–14] The 2021 National Academies of Sciences, Engineering, and Medicine report outlines feasible pathways to decarbonize the US electricity sector, achieving 75 percent clean energy by 2035 and carbon neutrality by 2050 to help mitigate the worse impacts of climate change.[15] Legislative solutions can win public support if climate action is framed as a strategy to create jobs,

---

[iv] The Senate rejected the ratification of the Kyoto Protocol, negotiated by Bill Clinton and Al Gore in 1997, arguing that it imposed unequal burdens on the United States versus China and India. The protocol exempted China and India from mandatory cuts on the basis that China and India contributed a far smaller share of the cumulative anthropogenic greenhouse gas emissions than did industrialized countries. However, China's and India's annual greenhouse gas emissions were rising rapidly.

economic opportunities and investments in fossil-fuel-reliant communities and coupled with the acknowledgement that the clean energy sector alone cannot provide the panacea to resolve Americans' job losses that preceded the energy transition. The provisions enacted by Congress in December 2020 to fund clean energy innovations, to finance low-income households' adoption of renewable energy and to cut one potent class of greenhouse gases provide some optimism that climate action can win sufficient support in Congress.

I begin by framing the climate crisis in economic terms – explaining how climate action is cheaper than climate inaction (Section 11.1). Next, I turn to how the Trump administration attempted to derail the prudent efforts by state governments and the nonfossil-fuel private sector to respond to the climate crisis (Section 11.2). I discuss legislative efforts in Congress to correct economic signals in the economy (Section 11.3) and to create programs to address job creation and economic diversification in fossil-fuel-reliant communities (Section 11.4). Next, I review Congress's December 2020 "climate legislation" (Section 11.5). I conclude by noting how provisions in the 2020 "climate legislation" is only a down payment to the extensive work ahead for Congress to reorient the US economy toward a lower-carbon future while helping to grow jobs and economic opportunities and to assist communities in the energy transition.

11.1 Costs of Climate Disruptions to the US Economy

Every region in the United States has suffered climate change induced droughts, wildfires, heatwaves, water scarcity, flooding, crop failures or smog (Table 11.1).[16,17] The American South and the lower Midwest are projected to bear the brunt of the economic costs associated with climate change.[18] The poor would suffer most, with the poorest third of US counties losing 2–20 percent of their county income by 2100.[18,v]

The costs of climate action are dwarfed by the costs of inaction. The total costs of reducing US emissions in line with a global pathway to achieving the 2°Celsius target in the Paris Climate Agreement are estimated at $1–$4 trillion.[19,vi] These investments compare favorably with US benefits from

---

[v] These estimates are based on projections that greenhouse gas emissions will continue largely unabated, with the temperature rise reaching 4.9°Celsius by 2100. This scenario is known as the representative concentration pathway (RCP) 8.5 [18].

[vi] The reductions amount to 80 percent of 2005 US emissions. These reductions would be achieved through changes in the energy system undertaken from 2015 through 2050 [17].

Table 11.1 *Adverse climate impacts in the United States (selected events in 2000–18)*

Droughts	The record-breaking drought in California from 2012 to 2014 affected the agriculturally important southern Central Valley and the populated coastal areas.[26] *Climate change contributed to an estimated 8%–27% of the observed 2012–14 drought and 5%–18% of that in 2014.*[26] *Climate change, by causing very hot and very dry years, makes droughts in California more frequent, longer and more intense.*[27] Montana, North Dakota and South Dakota, the top US wheat producers, endured severe drought in the summer of 2017 and suffered $2.5 billion in agriculture losses.[28] *Climate change made such intense droughts up to 1.5 times more likely, by increasing evaporation from the land surface and transpiration from plants.*
Wildfires	4.2 million hectares of forest burned in the American West from 1984 to 2015. *This is nearly double the area that would be expected to suffer fires in the absence of climate change.*[29]
Water loss	The snowpack in the American West suffered 10%–20% water losses between the 1980s and the 2000s, leading to decreased streamflow, reducing water availability for human use and affecting the health of ecosystems.[30] *These losses are attributed to a combination of climate change and natural variability but not natural variability alone.*[30]
Heavy rainfall	Louisiana suffered heavy rainfall and subsequent flooding in 2016 that cost $8.7 billion.[31,32] *Climate change made this event 1.4 times more likely.*[31]
Flooding	High tide flooding and days of flooding increased in the Southeast Atlantic Coast, Northeast Atlantic and the Eastern and Western Gulf Coast between 2000 and 2018.[33] *Climate-change-induced sea-level rise causes flooding.*[34] More than one-third of the $199 billion in US flood damage from 1988 to 2017 was related to trends in intensifying precipitation tied to climate change.[35]
More intense hurricanes	Hurricanes have become more supercharged, i.e., they bring larger amounts of rainfall. Examples of this phenomena include Hurricane Harvey, which hit Texas and Louisiana in 2017, and Hurricane Florence, which hit the Carolinas in 2018.[36,37] *Rising sea-surface temperatures cause more evaporation and more water in the air, which turns into heavy rainfall.*[36]
Ocean food chain	Heatwaves in ocean waters around Alaska in 2016 stunted the growth of zooplanktons and subsequently led to the deaths of birds and sea mammals up the food chain.[38–40] *Human-caused climate change made the 2016 marine heat waves 50 times more likely.*[38]
Displaced communities	By 2003, flooding and erosion affected 184 of the 213 Alaska Native villages, requiring many communities to relocate.[41,42]

Notes: Advances in the science of attribution have enabled scientists to link climate change to specific events.[43,44]

averting *annual* climate damages – an estimated $170–$206 billion by 2050.[19,20,vii] Already, costs to the United States have mounted. Between 2006 through 2016, disaster response alone for crop failures, flooding and wildfires cost taxpayers more than $350 billion and these costs are predicted to rise.[21,22,viii] In 2020, damages from hurricanes, wildfires and severe storms amounted to $95 billion, almost double that in 2019.[23,24] Political support for climate investments can dwindle when they are framed as sacrifices imposed on current generations for benefits that accrue to future generations. However, a number of climate actions, such as the shift to wind and solar power and the shift away from gasoline powered vehicles, yield immediate benefits. Reducing fossil fuel combustion can also improve local air quality and avert excess mortality from exposure to particulate matter. The eastern United States experiences excess deaths from exposure to particulate matter from fossil fuel combustion.[25]

11.2 Obstruction of Climate Action by the Trump Administration

Although Congress holds much of the authority to enact major government action against climate change, the executive branch can play critical leadership roles in climate action by facilitating international cooperation and coordination among the federal and state governments and the private sector. The Trump administration went far to obstruct the US response to climate disruptions by blocking international cooperation (Section 11.2.1), scuttling actions by state governments and nonfossil-fuel private sectors (Section 11.2.2) and manipulating the financial sector's lending decisions (Section 11.2.3). The administration also undermined a climate action program intended to incentivize carbon sequestration by easing oil and gas companies' ability to secure tax credits without achieving verified sequestration (Section 11.2.4).

[vii] These are the annual averted costs in 2050 under the RCP 4.5 scenario as compared to the RCP 8.5 scenario. The RCP 4.5 scenario assumes that emissions peak in 2040, leading to average temperature rise of 2°C–3°C by 2100. The averted losses include those related to human health, water scarcity, agriculture losses, damages to ecosystems and infrastructure, and avoided rise in electricity demand.

[viii] Costs to taxpayers are expected to triple to $35 billion per year by the middle of the twenty-first century and to quadruple to $112 billion per year by late century [21]. These costs – disaster responses, crop and flood insurance payout, wildfire management and federal infrastructure repairs – are only a fraction of the climate costs faced by Americans [22].

11.2.1 Hindering International Cooperation to Reduce Global Emissions

Under the Paris Climate Agreement that came into force in 2016, countries pledged to make voluntary cuts to their greenhouse gas emissions to keep the global average temperature rise at below 2°Celsius above the preindustrial period and to pursue efforts to further limit the rise to below 1.5°Celsius.[45] The Paris Agreement is only a starting point for countries to make further commitments to emissions reductions. If countries were to achieve their commitments under the Paris Agreement, temperatures are projected to rise by 3°Celsius by 2100.[46] US participation in global efforts matters as it is the largest contributor to cumulative greenhouse gases emissions,[ix] the second largest global carbon dioxide emitter in 2019 (at 15 percent of global emissions) and the largest emitter of carbon dioxide per person in 2019.[47] The United States had pledged to cut its emissions by 26–28 percent below its 2005 level by 2025 and set an aspirational target of an 80 percent reduction by 2050.

In June 2017, Trump announced that he would withdraw the United States from the Paris Climate Agreement.[x] Even prior to the formal US withdrawal in November 2020, the Trump administration took steps that undermined international efforts to address the climate crisis. Its actions to block negotiations post-Paris and to rescind US funding commitments[48] damaged US credibility as a partner in the global efforts to reduce emissions.[49] In 2018, 200 countries called for strong support of the Intergovernmental Panel on Climate Change report, which underscored that keeping the average global temperature rise below 1.5°Celsius could mitigate the more disastrous climate disruptions that would occur in the 2°Celsius scenario (Table 11.2).[50] Based on current emissions patterns, the average rise in global temperatures is projected to break the 1.5°Celsius mark by around 2040.[46] Strong consensus support for the report would have highlighted the urgency to cut global emissions to half the level of the 2017 emissions by 2030 and to achieve net zero emissions by 2050.[50]

[ix] In terms of cumulative carbon dioxide emissions between 1850 and 2019, the top contributors are the United States (25 percent), the European Union (27 countries; 17 percent), China (13 percent), Russia (7 percent), United Kingdom (5 percent), Japan (4 percent) and India (3 percent) [47]. Among the top emitters of carbon dioxide, US per capita carbon dioxide emissions of 16.1 tons per person annually is more than double that of China (7.1 tons per person) and the European Union (6.6 tons per person), and eight times that of India (1.9 tons per person) [47].

[x] On the first day of his presidency, Biden signed an executive order to return the United States to the Paris Climate Agreement. The United States rejoined that agreement in February 2021.

Table 11.2 *Adverse global impacts with average global temperature rise at 1.5° Celsius and 2°Celsius*

	1.5° Celsius	2° Celsius
Global population exposed to heatwaves at least once every five years (%)	14	37
Global population exposed to water scarcity (million people)	270	390
Global population exposed to droughts (million people)	130	190
Decline in wheat yields relative to 1995–2014 yields (%)	4	5
Decline in maize yields relative to 1995–2014 yields (%)	6	9
Decline in marine fisheries (million tons)	1.5	3
Decline in coral reefs (%)	70–90	99

Sources: Intergovernmental Panel on Climate Change[46] and Carbon Brief.[61]
Notes: To put these figures in perspective, the global population in 2018 was 7.7 billion and the world total marine catch in 2016 was 79.3 million tons.[62] Already in 2016, two-thirds of the global population lived under conditions of severe water scarcity for at least one month of the year.[63]

However, the United States joined Russia, Saudi Arabia and Kuwait in pushing for a watered-down statement of merely "taking note" of the landmark report by the Intergovernmental Panel on Climate Change.[51] US intransigence also emboldened Australia, Brazil and Saudi Arabia to oppose the transparent accounting of emissions that was necessary for the success of carbon markets in the Paris Agreement.[52,53]

The National Academies of Sciences, Engineering, and Medicine[54] alongside several Republicans and Democrats (Section 11.5.1), had underscored that enhancing US leadership in clean energy innovation can be an effective strategy to address the twin challenges of the climate crisis and the increasingly competitive global manufacturing sector.[54,55] China and several countries within the European Union, particularly Germany and Denmark, pursuing that very strategy, have deepened their strengths in wind and solar technologies and manufacturing, serving both domestic and export markets.[56–58] Unsurprisingly, that economic bonus has motivated these countries to adopt more aggressive climate targets. In October 2020, the European Parliament voted to cut emissions by 60 percent by 2030, with the goal of reaching carbon neutrality by 2050.[59] China, which is on track for exceeding its Paris commitments,[59] pledged to become carbon neutral by 2060, while South Korea and Japan pledged to achieve carbon neutrality by 2050.[60]

11.2.2 Obstructing the Progress of States and Private Industry

The Trump administration did not stop at climate inaction but took intentional steps to obstruct the proactive efforts of state governments and private actors. Most notably, the Trump administration blocked California's fuel-efficient and zero-emissions vehicle programs. By doing so, it scuttled the auto manufacturers' and the parts manufacturers' restructuring of their operations toward meeting the rising global demand for such vehicles.[64] General Motors' decision to manufacture only electric cars by 2035 is a testament to this market shift.[65]

California has implemented fuel-efficient vehicle programs since the 1950s and zero-emissions vehicle programs subsequently.[66,67] Its standards influence one-third of the US auto market, as 13 states and the District of Columbia follow California's stricter fuel-efficiency standards, and nine states modeled their programs on California's zero-emissions vehicle program.[68] The Clean Air Act section 209(b) empowers the EPA to grant California a waiver to set stricter vehicle emissions standards[66,67,69] and permits other states to follow California's stricter standards. While fuel taxes are a more economically efficient strategy to reduce gasoline consumption, implementing and strengthening fuel-efficiency standards have been far more politically feasible than attempts to increase fuel taxes.[69,70,xi]

During the 2008 Great Recession, the Obama administration negotiated with the US auto manufacturing industry, as part of their bailout package, to ramp up fuel-efficiency standards in a predictable manner. That strategy was meant to help the US automakers become more competitive to meet the rising global demand for fuel-efficient vehicles.[71,72] At the same time, these standards were projected to reduce greenhouse gas emissions from the auto sector by six billion metric tons, thus halving the auto sector's emissions by 2025.[73,74] The negotiations between the administration, California and the automakers resulted in one uniform set of fuel-efficiency standards and greenhouse gas standards for the entire United States.[xii] The 2016 midterm review conducted by the Obama administration concluded that the fuel-efficiency standard of 54.5 miles per gallon by 2025 was technologically and economically feasible.[73]

[xi] In 2008, Secretary of Energy Steven Chu explained how higher gasoline prices in the United States could incentivize consumers to reduce their gasoline consumption. That strategy could be coupled with tax credits or rebates to minimize burdens on lower-income households. During a congressional hearing held by the Senate Committee on Environment and Public Works, several members of Congress excoriated Chu for the Obama administration's "war on affordable energy."

[xii] The EPA and NHTSA jointly promulgated Corporate Average Fuel Economy and greenhouse gas standards for 2017–25 model vehicles.

When Trump took office, the auto industry requested that the Trump administration ease the ramp up of the standards. Instead, the administration delivered a proposal with far weaker standards than those envisioned by automakers, eliciting pushback from automakers.[71] The administration's proposal of 37 miles per gallon by 2025,[75] which would freeze standards to those originally expected to be attained in 2020,[76] prompted the California Air Resources Board to vote in March 2017 to pursue stricter standards.[77] The battle between the administration and California created major uncertainties in the automobile markets, with two different standards, subject to litigation.

The Trump administration implemented its deregulatory actions in two parts. In September 2019, the EPA announced its revocation of California's waiver, while the National Highway Traffic Safety Administration (NHTSA), under the Department of Transportation,[78] simultaneously announced its preemption of California's fuel-efficient and zero-emissions vehicle programs.[75] In April 2020, the EPA and NHTSA finalized the weakened standards[79] via the Safer Affordable Fuel-Efficient (SAFE) Vehicles Rule.

Automakers lobbied *against* extensive deregulation and advocated instead for moderate ramp up of standards that "promote advanced technology for the sake of long-term environmental gains and US global competitiveness."[71,80,81] Trump raged against some automakers' refusal to support his deregulatory actions.[82] In July 2019, four automakers signed a deal with California, pledging to comply with California's standards even if the Trump administration chose to finalize severely weakened standards.[80] About a year later, in August 2020, five automakers – BMW, Ford, Honda, Volkswagen and Volvo – finalized agreements with California to achieve average fuel economy of 51 miles per gallon by 2026.[83]

In November 2019, prior to the finalization of those agreements, the Department of Justice launched an antitrust investigation against the automakers that negotiated with California but dropped the investigation three months later. Representative Jerry Nadler, chairman of the House Judiciary Committee and Representative David Cicilline, chair of the Subcommittee on Antitrust Issues, launched a hearing into that investigation, noting "there is virtually no antitrust theory that the Justice Department can use to prove that this agreement will unreasonably restrain trade or otherwise violate the antitrust laws." John W. Elias, who served as acting chief of staff for the Department of Justice's Antitrust Division for the first half of the Trump administration, testified at the hearing that the investigation served as "evidence that our nation's antitrust laws were being misused . . . [Under US laws] companies are free to collectively lobby the government for regulation . . . The events here constituted an abuse of authority."[84]

The administration went far to assist the oil sector at a cost to other economic sectors.[78,85] According to a *New York Times* investigation, "Marathon Petroleum, the country's largest refiner, worked with powerful oil-industry groups and a conservative policy network financed by the billionaire industrialist Charles G. Koch to run a stealth campaign to roll back car emissions standards."[78] That investigation described the campaign's exchanges with the Trump administration and quoted Marathon informing its shareholders that the Trump administration's rollback would increase the gasoline demand to between 350,000 and 400,000 barrels per day.[78,xiii] The administration's actions forfeited $1.7 trillion in fuels savings for American households, with an average saving of $8,000 for a 2025 model vehicle over its lifetime.[76]

11.2.3 Attempting to Dictate the US Finance Sector's Lending Decisions

The Trump administration's actions to block the nonfossil-fuel private sectors from adapting to market realities in a carbon-constrained world extended to the US financial sector. It finalized a rule to impede the decisions of US banks and financial institutions not to invest in the riskiest oil and gas operations. As early as 2017, financial analysts warned that investments in Arctic oil and gas were not financially sound, noting "the idea that we have to go into the Arctic to find new resources ... has been dispelled by the enormous cheap, easier to produce and quicker time-to-market resources in the Permian onshore."[86] Beginning in December 2019, a number of US banks, including Goldman Sachs, JP Morgan Chase, Wells Fargo, Morgan Stanley, Citigroup and Bank of America[87–89] announced their decisions not to invest in Arctic oil and gas projects.

The banks' announcements incensed the Alaskan congressional delegation, a number of other Republicans and Trump,[90] who staunchly supported drilling in the Arctic Sea and Arctic National Wildlife Refuge (ANWR). In 2018, the Trump administration approved the permit for Hilcorp's Liberty project in federal Arctic waters.[91,xiv] Senator Lisa Murkowski of Alaska was the architect of Congress's decision to open ANWR to drilling (Chapter 6). In January 2021, the Trump administration sold oil and gas leases on 575,000 acres in

---

[xiii] That investigation also revealed the campaign's actions to block state legislatures' support for fuel-efficient vehicles.

[xiv] In December 2020, the Court of Appeals for the Ninth Circuit ruled that the permit approval violated environmental laws and remanded the case back to the Bureau of Oceans and Energy Management [91].

ANWR.[92,xv] That same month, it finalized plans to allow oil and gas development in 18 million acres of the National Petroleum Reserve in Alaska that spans 23.6 million acres.[92]

In May 2020, 44 Republicans wrote to Trump asking his administration to "*use every administrative and regulatory tool* at your disposal to prevent America's financial institutions from discriminating against America's energy sector while they simultaneously enjoy the benefits of federal government programs."[93] Trump denounced the banks' decisions as discrimination against energy companies and as cowering to the "radical left."[94] In response to Republican lawmakers' complaints, in November 2020, the Office of the Comptroller of the Currency (OCC) proposed the Fair Access to Financial Services Rule that would apply to large banks with at least $100 billion in assets and to federal savings associations.[95] That rule prohibited these institutions from denying financing to any "politically controversial but lawful businesses" without providing an objective, quantifiable risk-based analysis established by the bank in advance.[96] The proposed rule would impose obligations on banks to extend financing even if there is adequate qualitative evidence on risky lending. In justifying its rule, the OCC claimed that banks took decisions not to lend sectors "based on criteria unrelated to safe and sound banking practices, including personal beliefs and opinions." It also posited climate change as one of the risks unrelated to financial exposure that banks are not equipped to evaluate and asserted that that issue "is the purview of Congress and Federal energy and environmental regulators."[95]

The Trump administration's OCC rule reflects its broad commitment to block the financial sector from considering environmental, social and governance factors in making its investment decisions. The financial sector has increasingly recognized how these factors influence the financial returns to investments.[97] In November 2020, the Department of Labor finalized a rule that hindered investment fund managers who managed employee retirement benefits plans from considering "non-pecuniary" factors in their investment choices.[98,99] That rule, promulgated under the Employee Retirement Income Security Act of 1974, which governs retirement plans in the private sector, covers a large amount of investment funds. Fear of running afoul of that rule may discourage fund managers from considering environmental, social and governance factors that can influence the financial returns from investments.

[xv] Biden's executive order on January 20, 2021 placed a temporary moratorium on any oil- and gas-related activities in ANWR.

11.2.4 Co-opting Support for Carbon Sequestration Projects to Subsidize Oil Extraction

The Trump administration took steps to co-opt the tax credit scheme aimed at supporting carbon capture and sequestration (CCS) projects into primarily subsidizing the expansion of oil extraction. In 2008, Congress granted Section 45Q tax credits[xvi] to CCS projects. These projects, which are intended to capture and sequester carbon dioxide for the long term, are seen by some but not all energy analysts as one option in the technology toolbox to reduce net greenhouse gases emissions. (The debate on the CCS technology is detailed in Section 11.5.3.) Congress provided the 45Q tax credits to two distinct types of projects under the CCS umbrella term: first, projects that capture carbon dioxide for long-term sequestration and, second, projects that inject carbon dioxide into oil wells to enhance oil recovery.[100] Even prior to the Trump administration's deregulatory actions, energy analysts and researchers had warned that under present regulations and companies' practices,[xvii] the injection of carbon dioxide into oil wells serves as a technology for recovering oil and not as a technology for long-term carbon dioxide sequestration.[101–104,xviii]

Most CCS projects in the United States focus on capturing carbon dioxide for enhanced oil recovery.[101,104] In the past, companies had been required to submit monitoring, reporting and verification plans to the EPA on their carbon dioxide sequestration. That requirement provided an important check so that the 45Q credits are not misused to simply expand oil extraction without any sequestration. However, thanks to the lobbying efforts of the oil industry,[104] in January 2021, the Department of the Treasury finalized the Rule on Carbon Oxide Sequestration Credit.[105] The rule permits companies to choose to comply with standards on carbon sequestration set by the International Organization for Standardization (ISO) and to hire an engineer or geologist to verify the claims. Critics contend that the EPA's requirements are more

---

[xvi] As the result of 2018 legislation and the December 2020 spending bill, projects that commence by 2025 can receive the 45Q tax credit.

[xvii] Wells that inject carbon dioxide for enhanced oil recovery are not required to meet conditions for well integrity that make it more likely for carbon dioxide to be stored for the long-term [101]. Moreover, the numerous abandoned wells in oil fields provide pathways for the carbon dioxide to escape to the surface or to pollute groundwater [101]. The International Energy Agency report notes "At present, no CO_2-EOR (carbon dioxide injection for enhanced oil recovery) site is pursuing this dual objective. Today, EOR operations are carried out with the aim of maximizing oil output" [102]. The Intergovernmental Panel on Climate Change report also emphasized that while CO_2-EOR is a mature technology, combining CO_2-EOR with CO_2 storage is financially feasible only under limited conditions [46].

[xviii] Oil companies' ability to extract more oil per well does not translate necessarily to their decision to drill fewer wells. Thus, the technology can contribute to overall expansion in oil extraction activities.

stringent than that of the ISO and external verification by a truly independent third-party assessor is essential.[106] The weakening of verification exacerbates the existing problems in the tax credit program, in which companies' claims for tax benefits for sequestered carbon dioxide exceeded that amount for which they had submitted verification plans.[106,107]

11.3 Congress's Legislative Efforts: Correcting Economic Signals to Incorporate the Consideration of Climate Costs

Transforming the economy with job growth and economic opportunities in the low-carbon sectors requires Congress to correct economic signals so that economic activities and investment flows take into account the costs from climate disruptions. Legislation efforts in the 116th Congress to correct these signals include proposals for a carbon tax (Section 11.3.1) and for federal agencies to apply the social cost of carbon in their decision-making (Section 11.3.2). Legislative efforts to correct the misdirection of investments include proposals for correcting tax preferences that favor the oil and gas sector (Section 11.3.3) and for mandating greater examination and disclosure of the financial sector's exposure to climate risks (Section 11.3.4).

11.3.1 Implementing a Carbon Tax

Imposing a carbon tax on fossil fuels and other sources of greenhouse gas emissions is one of the policy instruments to reduce emissions across the economy.[108] The tax, by setting a price on carbon emissions, provides the incentive for emissions reductions and the signal for economic activities and investments to shift away from carbon-intensive sectors.[108]

One of the carbon tax bills introduced in the 116th Congress is the American Opportunity Carbon Fee Act, sponsored by Senator Sheldon Whitehouse with three other Democrat Senators.[109,xix] That bill would impose a tax on carbon dioxide and other greenhouse gas emissions at $52 per ton of carbon dioxide equivalent,[xx] rising annually at 6 percent to account for inflation. The tax would be imposed on fossil fuel products (i.e., petroleum products, natural gas and coal), facilities that emit greenhouse gases, and other activities that emit greenhouse gases (e.g., venting, flaring and leakage of methane across the

---

[xix] During the 116th Congress, eight carbon tax bills were introduced, including bills that had Republican cosponsors, and one cap-and-trade bill was also introduced.
[xx] This $52 figure is tied to the social cost of carbon (Section 11.3.3).

supply chain). To mitigate potential negative effects on vulnerable populations, the bill directs revenue from the carbon tax to state governments to finance programs to assist low-income and rural households to reduce their energy expenses and to provide job training for workers to transition to a low-carbon economy (Section 11.4). To safeguard the competitiveness of US industries, the bill also imposes taxes on the imports of energy-intensive goods, known as border carbon adjustments.

A well-designed carbon tax can help shift economic activities toward less carbon-intensive sectors, thereby changing the composition of jobs and economic activities within the economy but without causing overall economic contraction.[110,111] Studies of the carbon tax introduced in British Columbia in 2008 find no significant impact on the gross domestic product[111] and no significant impact on overall employment (though employment fell in carbon-intensive sectors).[112] Studies of the cap-and-trade program in the European Union likewise found that carbon pricing did not reduce gross domestic product even as it reduced emissions.[113,114,xxi]

Tax rates need to adequately reflect the true economic costs of carbon emissions to substantially reduce emissions. The social cost of carbon (discussed in Chapter 9 and in Section 11.3.2) captures the monetary cost of long-term climate damages resulting from the addition of one ton of carbon dioxide to the atmosphere in a given year. Implementation of carbon pricing in various jurisdictions at levels lower than the true costs of carbon emissions (in order to overcome political opposition) falls short of bringing deep cuts in emissions.[115] Economists have proposed a refinement in the design of the carbon tax to ensure specific emissions-reduction milestones are met, that is, the tax rate would rise more rapidly if emissions-reduction targets are not being met but would rise more slowly otherwise.[116]

A number of Republicans have supported carbon taxes, indicating their acknowledgement of the economic rationale for a carbon tax.[117] Oil companies and the American Petroleum Institute, which have long opposed carbon taxes, have also came out in support for low carbon taxes but with preconditions of retrenching regulations and waiving any liability for climate damages. The argument that congressional representatives should support a low carbon tax with preconditions that bring the oil and gas industry to the table, rather than to hold out for higher carbon taxes with no preconditions, deserves a closer look.

[xxi] Metcalf and Stock caution that their study is likely to capture the lower bound effects of carbon prices on the economy, in light of the low carbon prices in the European Union's cap-and-trade program [113].

Such a "compromise" can weaken the overall climate response, for reasons discussed below.

Exxon, BP and Shell pledged $1 million each to Americans for Carbon Dividends, a group that lobbies for a $40 carbon tax coupled with the repeal of the EPA's powers to regulate greenhouse gases and the waiver of liability from climate damages for oil companies.[118] One of the carbon tax bills, the proposed Energy Innovation and Carbon Dividend Act,[xxii] contains a provision of "regulatory simplification," which would suspend the EPA's powers to regulate carbon dioxide emissions from stationary sources but reinstate those powers if emissions-reductions targets are not met by 2030. Another carbon tax plan, proposed by the Climate Leadership Council and originally authored by former Republican secretaries of state James Baker and George Shultz, takes a similar tack of proposing a carbon tax combined with "the removal of regulations on current and future federal stationary source carbon regulations."[119] The council's 2017 plan also included a waiver of liability for oil companies from climate damages, a provision subsequently dropped in its 2019 proposal.[120,121]

Oil companies continue to insist on immunity from climate change lawsuits in exchange for support of a limited carbon tax.[122] They argue that climate liability is not appropriate for litigation in the courts but is a matter for the legislative and executive branches.[123] As of December 2020, oil companies face lawsuits from 24 local and state governments seeking to recover damages related to climate disruptions.[xxiii] Such litigation, if successful, would be costly for the industry.[124,125] Tobacco companies, which had similarly engaged in disinformation campaigns and whose products adversely affected human health, paid at least $200 billion in legal settlements.[124,125]

Congress should be very wary of preconditions such as waiving climate liability or suspending EPA's regulatory authority in exchange for a carbon tax for at least five reasons. First, a mere $40 per ton tax, without supplemental EPA regulation, would not achieve reductions in greenhouse gas emissions adequate to stabilize the climate.[117] Simulations run by Katherine Jordan at Carnegie Mellon University find that a $40 per ton tax alone would not be sufficient to incentivize consumers to shift out of gasoline-powered cars. Instead, fuel-efficiency regulations would be needed to incentivize such a shift.[117] Other

---

[xxii] The Energy Innovation and Carbon Dividend Act's proposed provision of returning tax revenue to taxpayers may help win popular support, but the policy choice to return all revenue to taxpayers would preclude using the funds to assist fossil-fuel-reliant communities or low-income households.

[xxiii] In May 2021, the Supreme Court in *BP v. Baltimore* ruled that these lawsuits should be heard in federal courts. State courts are perceived as more favorable to local and state governments [123].

studies suggest carbon taxes that are far higher than $40 are needed to incentivize economic actors to transition away from fossil fuels.[126]

Second, regulations to cut emissions can be a more cost-effective approach for some sectors, even in the presence of a carbon tax.[127] In fact, regulations under the Montreal Protocol to cut ozone-depleting chemicals (which turned out to be potent greenhouse gases) have played an important role in addressing the climate crisis.[128] Third, in the absence of supplemental regulations, a carbon tax would need to be much higher to achieve major cuts in emissions. Such a high tax is not likely to be politically feasible.[129]

Fourth, if the EPA's regulatory powers to cut emissions are temporarily suspended (even with the caveat of potential reinstatement should emissions cuts not be met – as envisioned by the Energy Innovation and Carbon Dividend Act), the oil industry and their political supporters would likely block the reinstatement.[130] The Trump administration's promulgation of the Pollutant-Specific Significant Contribution Finding for Greenhouse Gas Emissions Rule,[xxiv] which blocked the EPA's regulation of greenhouse gas emissions from stationary sources (except power plants) and which effectively nullified the Clean Air Act provisions used by the Obama administration to cut emissions, showcases the industry's extraordinary ability to push for deregulation with the support of some members of Congress.[131]

Fifth, pressure from regulations and from potential liability for damages to public health and the environment has driven the oil sector's inching toward greater willingness to accept a carbon tax. Unsurprisingly, the House Democrat's Climate Crisis Report warned that "Congress should not offer liability relief or nullify Clean Air Act authorities or other existing statutory duties to cut pollution in exchange for a carbon price."[132]

11.3.2 Mandating Federal Agencies to Apply the Social Cost of Carbon

While the carbon tax in the marketplace helps economic actors take into account the cost of emissions, the social cost of greenhouse gas emissions is a metric that government agencies can use in making their decisions on

---

[xxiv] Under that rule, any stationary source whose industrywide greenhouse gas emissions make up less than 3 percent of US greenhouse gas pollution will be deemed "necessarily insignificant without consideration of any other factors." The rule would therefore effectively block the EPA from using its powers under section 111(b) of the Clean Air Act, which the Obama administration used to regulate methane emissions from oil and gas operations. In March 2021, the Biden administration requested the DC Circuit Court to vacate the rule, which had been promulgated without the notice and comment procedure, and the court vacated the rule in April 2021.

projects that either raise or reduce these emissions. Under the Obama administration, the Interagency Working Group proposed the social cost of carbon at $42–$62 per ton for the year 2020, whereas the Trump administration's application of a miniscule rate of $1 per ton of carbon dioxide downplayed the climate impacts.[133]

In 2019, Democratic Senator Michael Bennett and 16 Democratic cosponsors introduced the Carbon Pollution Transparency Act,[134] which would mandate federal agencies to use the social cost of greenhouse gases developed by the Interagency Working Group under the Obama administration, with adjustments for inflation. In September 2020, Democratic Rep. Donald McEachi introduced a companion bill in the House.[135] The bill requires the social cost of greenhouse gases to incorporate damages from climate change, including increases in energy use, losses in agricultural productivity, adverse impacts on human health, property damage from increased flood risk and sea-level rise, and impairments to ecosystem services.

The bill takes at least three steps forward in more comprehensive accounting of the costs of greenhouse gas emissions. First, the bill mandates consideration of costs to society that cannot be easily monetized. It specifically emphasizes costs to communities that suffer from environmental injustice. Second, to account for the intergenerational harm, the bill supports the use of the interest rate for consumption for weighting future consumption relative to present consumption.[xxv] That choice of pegging the interest rate to the lower, consumption-related discount rates, as opposed to the higher, investment-related discount rates, helps mitigate the understatement of losses faced by future generations when economists apply the discounting method (Chapter 9). Third, by requiring estimates to account for *global* damages from greenhouse gas emissions, the bill recognizes a multilateral approach to solving the climate crisis.

The bill establishes an interagency working group to revise these estimates and a scientific review committee to provide guidance on revisions, with experts in climate science, climate economics and decision analysis. The bill also sought to ensure deliberation, transparency and responsiveness to evidence by requiring the interagency working group to respond to public comments.

A number of economists have supported updates to these metrics. In January 2021, Michael Greenstone, the chief economist of the Council of

---

[xxv] The discount rate captures the relative value placed on the consumption of present generations versus future generations in assessing the damages from climate change. The bill states that the choice of discount rates should reflect "the intergenerational nature of the harm caused by climate change or be consistent with the interest rate of consumption used by Federal agencies to reflect climate risk" [134].

Economic Advisors, who co-led the working group in the Obama administration, and his co-author, Tamma Carleton, proposed that a social cost of carbon of $125 per ton would more appropriately capture the costs of climate damages,[136] arguing that the rates of the 10-year US treasury bonds, which are used as a reference point for discount rates, are far lower than the historic rates used by the working group.[xxvi] The figure of $125 per ton has been adopted in some jurisdictions and agencies, for instance, the New York Department of Environmental Conservation. Other economists have called for these metrics to account for the distributional impacts of climate losses, recognizing that a dollar lost by poorer households causes greater welfare loss than a dollar lost by richer households. Accounting for these inequities would likely push up the social cost of carbon.[137]

In contrast to these proposals, others have advocated for a lower social cost of carbon, estimated using higher discount rates (Chapter 9). The justification advanced for higher discount rates is that through history, successive generations are wealthier than previous generations as a result of technological progress. Under such a scenario, current generations would bear a greater burden from forgoing consumption today to invest for climate protections than the benefits gained by future generations from those climate investments. This argument fails to pass muster because climate disruptions that threaten the survivability of the planet mean that future generations may well be poorer than the current generation[138] and, indeed, future generations who live in warm regions (such as the southern United States) are projected to be absolutely poorer than current generations living in those regions.[139]

11.3.3 Removing Tax Preferences for the Oil and Gas Sector

The tax preferences enjoyed by the oil and gas industry inflate returns to investments into that sector, thus artificially drawing more investments relative to the case had tax structures been sector neutral. The removal of these tax preferences should garner the support of fiscal conservatives as these tax preferences cost the Department of the Treasury an estimated $41.4 billion in forgone revenue over a 10-year period.[140,141,xxvii]

[xxvi] Carleton and Greenstone argued in favor of a lower discount rate, pointing to the returns from the inflation-adjusted 10-year US treasury bond that hovered at 1 percent, while the Interagency Working Group had referenced the historical rates of 3 percent [136].

[xxvii] In 2017, these tax preferences cost the federal government $3.2 billion from the expensing of intangible drilling costs, $1.7 billion from taking the depreciation as a percentage of revenue and $1.1 billion from the domestic manufacturing deduction [145].

In 2017, as well as in earlier sessions of Congress, Senator Robert Menendez and 22 cosponsors introduced the Close Big Oil Tax Loopholes bill that would remove tax preferences for major integrated oil companies.[142] In September 2020, Rep. Earl Blumenthal introduced the End Oil and Gas Tax Subsidies Act,[143] which would remove these tax preferences for all oil and gas companies.

Oil and gas companies enjoy three main tax preferences. The first is the treatment of intangible drilling costs as expenses, as opposed to taken as depreciation over the lifetime of the well.[xxviii] On the contrary, capital investments in other sectors are typically depreciated over the useful lifetime of the capital. As a result, income from investments into oil and gas wells face lower tax rates than investments in other sectors.[144,145] Second, since 1926, oil and gas companies have been able to choose to take the depreciation of their capital investments as a percentage of their revenue, rather than as depreciation of the actual costs of investments over the productive lifetime of a given well.[144,145] Third, oil and gas companies enjoy a section-199 domestic-manufacturing tax deduction, which provides a 6 percent reduction to taxable income. The manufacturing tax reduction, enacted by Congress in 2004, was aimed at disincentivizing factories in the United States from relocating abroad. Extending this tax preference to oil and gas extraction activities is inappropriate as these activities must remain at the oil and gas reservoir.[144,145]

Opponents of the removal of these tax preferences argue that their removal would result in reduced oil and gas production and job losses. However, several studies show that oil and gas production expands primarily in response to the projected global oil prices. In other words, these tax preferences, while costly to the US Treasury, do little to expand production or jobs. As documented by Metcalf,[145] the removal of these tax preferences would have fairly limited impacts on production, resulting in a short-run oil production decline of 9 percent in domestic oil production and 11 percent in domestic gas production and a collective long-run decline of 4–5 percent.

11.3.4 Improving the US Financial Sector's Management of Climate Risks

The economic transformation toward a low-carbon economy requires investment flows to account for climate risks. The financial sector, as discussed in

---

[xxviii] Independent oil companies can treat all their intangible drilling costs as expenses, while integrated oil companies can deduct 70 percent of these costs and must depreciate the balance over five years. The percentage depletion tax provision also disproportionately benefits smaller firms [144].

Section 11.2.3, is aware of the increasing risks of investing in the oil and gas sector, as well as the fossil fuel sector more generally. However, major US banks and financial institutions still hold significant investments in fossil fuels. From 2016 through 2019 alone, four US banks – JP Morgan Chase, Wells Fargo, Citibank and Bank of America – accounted for 30 percent of new fossil fuel financing.[89,146]

The Federal Reserve's November 2020 report drew attention to the US financial sector's exposure to climate risks.[147] These "climate risks" stem from the adverse impacts of climate change on oil and gas infrastructure and other assets, from oil and gas investments falling far below their initial valuation as governments worldwide tighten regulations on greenhouse gas emissions, and from economies transitioning away from carbon-intensive activities.[148,149,xxix] A number of central banks and financial regulators have warned that losses from climate disruptions and from the precipitous fall in the value of oil and gas assets, termed stranded assets, could destabilize the entire financial sector.[148,150]

To safeguard the US financial sector and to disincentivize banks and financial institutions from making loans that expose them to excessive climate risks, Rep. Sean Casten and Senator Brian Schatz proposed the Climate Change Financial Risk Act of 2019.[xxx] That bill would require the Federal Reserve to test every two years if banks and financial institutions can withstand financial losses under a variety of climate scenarios. It would also set up a climate risk scenario committee to develop possible scenarios and to determine the financial and economic risks in these scenarios.[151,152]

Climate disruptions impose not only systemic risks to the entire financial sector but also challenges to individual investors. Investors can better direct investments if they have information on companies' exposure to climate risks. In October 2020, Rep. Sean Casten and Senator Elizabeth Warren introduced the Climate Risk Disclosure Act of 2019.[153,154] The bill would mandate the Securities and Exchange Commission (SEC) to require an issuer of securities – that is, a company that issues financial instruments used to raise capital – to disclose information on its holdings of oil and gas and other fossil fuel assets, its vulnerability from climate risks and its mitigation actions taken to address climate risks.

[xxix] At least one study suggests that significant oil and gas assets would become stranded even under the current technological trajectory, even if no new climate policies are adopted, with significant loses to fossil fuels producers and exporters such as the United States [149].

[xxx] External regulatory action is needed as individual banks and financial institutions focus only on their individual portfolio of oil and gas investments and may give inadequate attention to the systemic risks to the entire financial sector. Moreover, executives' incentives may lead them to focus on short-run returns, rather than the long-run health and stability of their institution.

While these bills would compel the Federal Reserve and the SEC to take action, the US Commodities Trading and Financial Commission, an independent federal agency that regulates the derivatives market, underscores that existing legislation empowers US financial regulators to address financial climate-related risk.[155] According to its November 2020 report, the Dodd-Frank Wall Street Reform and Consumer Protection Act of 2010, which establishes mechanisms for federal regulators to address risks that threaten the financial system, empowers regulators to conduct stress tests on banks' exposure to climate risks.[155,156] The report also emphasized the SEC's powers to require publicly traded companies and other companies to disclose climate risks.[155]

11.3.5 Federal Agencies' Direct Procurement of Renewable Energy

The federal government is the largest energy consumer in the United States, spending $15.6 billion in energy expenditures in 2017.[157] Congress has deployed the federal government procurement powers as one strategy to strengthen the demand for renewable energy and thus incentivize the private sector's investments in renewable energy generation. Specifically, in the Energy Policy Act of 2005, signed into law by G. W. Bush, Congress mandated federal agencies to meet 7.5 percent of its total renewable electricity consumption by 2013.

To enhance federal agencies' ability to purchase electricity from renewable energy developers, Congress would need to modify the 10-year limits on federal agencies' procurement contracts. Renewable energy developers offer power-purchase agreements to large corporate clients that sell electricity at cheaper than retail prices but that require a contractual commitment in the range of 15–20 years. Longer contracts provide the certainty in demand for renewable energy developers to invest in new projects. Without Congress's authorization for longer contracts, federal agencies are limited to purchases of renewable energy from the retail electricity suppliers in restructured electricity markets (Chapter 3).

The Department of Defense, which consumes three-quarters of the energy used by federal agencies, has enjoyed greater flexibility to shift to a greater share of renewable energy on military bases because of its ability to sign 30-year contracts.[157] In 2017, the Department of Defense met 6 percent of its total electricity consumption from renewable energy. The department has committed to more ambitious renewable energy targets, recognizing the advantages of energy diversification to enhance resiliency on military bases.[157,158]

11.4 Congress's Legislative Efforts: Addressing the Job Transition and Equity

Congress's corrections of the economic signals alone are insufficient to sustain the transition to a low carbon economy. Jobs are understandably the focus of attention in the debate on energy transition in Congress and among the general public.[159] Indeed, the loss of jobs in the fossil fuel sector have become the focal narrative of the opposition against climate action.[160] I discuss Congress's efforts to address the jobs transition (Section 11.4.1) and review programs aimed at generating clean energy jobs (Section 11.4.2) and supporting economic diversification in fossil-fuel-reliant communities (Section 11.4.3).

11.4.1 Job Transition in a Changing Energy Economy

The framing of the energy transition debate as an opportunity to create jobs and economic development is motivated by the very real need for a new economy that supports livelihoods. Nevertheless, it is useful to maintain a realistic expectation about the magnitude of jobs that renewable energy and energy efficiency sectors can generate. These sectors alone cannot shoulder the burden of addressing the job losses in the United States that long preceded the changing energy landscape. Job losses, including the hollowing out of manufacturing sector, are due to a multiplicity of factors, including automation, product and skill obsolescence, the mismatch of skills to emerging technologies, competition from innovators and producers in the United States and abroad, and more.[161,162] Even in the oil and gas sector, automation had led to a decline in higher-paying extractive jobs for workers with a high school education, despite growth in production (Chapter 2).

Noteworthy job growth in the renewable energy and energy efficiency sectors has cemented support for fostering continued growth of those sectors. Numbers vary according to how jobs are classified.[163] Using data from 2016, a study from the Brookings Institution estimates that, at the time, the United States had about 1.3 million workers in the clean energy sector and about 4.4 million workers in the energy efficiency sector.[164,xxxi] The organization E2 estimate that there were three times as many jobs in clean energy as there were in the fossil fuel sector in 2019.[165,xxxii] The Climate Alliance of 25 states,

---

[xxxi] The study estimates an additional 877,000 workers in the environmental management sector.
[xxxii] According to the E2 report, in 2019, clean energy jobs (3.36 million) outnumbered total fossil fuel employment (1.19 million). Clean energy jobs include those in energy efficiency (2.4 million), renewable energy (520,000), clean vehicles (270,000), grid and storage (150,000), and fuels (39,000) [165].

headed by Republican and Democrat governors, reports 2.1 million clean energy jobs in 2019.[166]

As of 2019, the manufacturing sector hosts 530 factories across 43 states that build parts for wind turbines.[167] The relatively high domestic content in wind turbine projects and the difficult of transportation of large wind turbines can help sustain domestic manufacturing.[168] While the domestic manufacturing of solar panels faced headwinds from the availability of cheaper imports, the installation segment of the solar industry grew, thanks to these lower-cost panels.[169] The Bureau of Labor Statistics reports that wind turbine technicians and solar installers are among the fastest-growing jobs. While starting from a small base, this growth does provide some sense of the budding opportunities in the sector.[170] Even so, these jobs are concentrated in the installation of projects and fewer workers are needed to maintain installed systems.

While jobs in the clean energy sector were on an upward trajectory prior to the COVID-19 pandemic, that economic shock also hit the clean energy sector hard. By December 2020, the sector lost 238,000 jobs in energy efficiency, 53,000 jobs in renewable energy, 25,000 jobs in clean vehicles and 22,000 jobs in clean fuels, transmission, storage and distribution.[171] Congress's economic recovery package, enacted in March 2020 (Chapter 2), and its December 2020 "climate legislation" (Section 11.5.1) offered some support for the renewable energy sector to bridge the pandemic.

While there is bipartisan support on growing renewable energy jobs, views diverge on how to address potential job losses in the oil and gas sector in the energy transition. While oil and gas jobs make up a limited share of jobs in the United States, they make up significant shares of jobs in localities that rely primarily on that industry, including in parts of the Permian Basin, the Bakken Shale and the Marcellus Shale.[172] While oil and gas jobs numbered 1.9 million in the United States in 2019, they can make up 30–50 percent of jobs in some counties in West Texas, Oklahoma, Wyoming, North Dakota and West Virginia.[172] Moreover, job losses in oil and gas sectors exacerbated earlier waves of job losses in manufacturing in many of the Rust Belt communities. The industry often serves as a major source of revenue for local governments, schools and a host of social services.[173,174]

One approach has coupled targeted investments into renewable energy, storage and grid modernization with investments in oil and gas reliant communities to ease the transition away from fossil fuels. For instance, Democrat Rep. Matt Cartwright of Pennsylvania sponsored the Consortia-Led Energy and Advanced Manufacturing Networks Act, which targets investments in renewable energy, energy efficiency and smart grid technologies. At the same time, Cartwright sponsored bills to assist local communities in fossil-fuel-reliant states, such as the Revitalizing the Economy of Coal Communities by Leveraging Local Activities and Investing More "RECLAIM" Act.[175] The RECLAIM Act would make funds from the Abandoned Mine Land Reclamation Fund available to state governments and tribes to promote economic revitalization, diversification and development in economically distressed mining communities. The funding would be available for reclamation and restoration of land and water resources adversely affected by coal mining. Analogous proposals have emerged on federal funding for combining efforts to remediate the lands pockmarked with abandoned oil and gas wells with local economic development strategies.[176]

A contrasting approach to address job losses in the fossil fuel sector has been to combine some support for renewable energy with significant investments into fossil fuels, including investments into developing technologies such as carbon capture and sequestration that would permit the continued combustion of oil and gas and even coal in a carbon-constrained world. For instance, in September 2020, the House passed the Clean Energy Jobs and Innovation Act (H.R. 4447) that provided a mix of support for fossil fuels and renewable energy, skewed in favor of fossil fuels.[177,178,xxxiii] Nevertheless, the alternative approach of funding innovation in fossil fuels, which promises to lengthen the longevity of fossil fuels, does not address fossil fuel communities' immediate and ongoing losses of jobs and tax revenue.

These divergent approaches reflect the disagreement on how best to assist fossil-fuel-reliant communities. Fossil fuel innovations, although short-sighted and unsustainable, have the benefit of both local trust in fossil fuel jobs and the ability at least in theory to directly benefit the communities most reliant on fossil fuels. Strategic investments in renewable energy coupled with direct community support, by contrast, takes a long-term and far more realistic approach to community assistance. However, direct assistance to these communities is essential because new clean energy jobs are not created in the same geographical locations nor do they rely on the same set of skills as the jobs that have been lost in the fossil fuel sector. At the same time, funding for supporting the transition of fossil fuel communities is limited (Section 11.4.3).

[xxxiii] Among provisions that raised concerns among environmental groups were funding for a new nuclear fuel, high-assay low-enriched uranium, which poses high risks of accidents and proliferation, as well as funding for the use of carbon dioxide for enhanced oil recovery [177].

11.4.2 Past Successes and Failures in Reshaping Jobs

Congress's past actions funding clean energy as a strategy to create jobs during the Great Recession provide insights on the strengths and limitations of such actions. In 2008, Congress enacted the American Reinvestment and Recovery Act (ARRA). About 17 percent of ARRA funding was directed into clean energy funding. Provisions included Department of Energy grants for renewable energy and energy efficiency retrofit, EPA grants for brownfield redevelopment, as well as tax breaks and loan guarantees for renewable energy.[179,180] Additionally, the Department of Energy and the Department of Labor together spent about $700 million on green job training programs.[180]

Studies on the ARRA and on green jobs creation provide a number of insights. First, these investments generated green jobs only in the long run. On average, about $1 million of green ARRA investments created about 15 jobs.[8] Notably, the creation of green jobs, while durable, proceeded at a slower pace than government stimulus spending on construction and highway infrastructure.[181,182] Projects in renewable energy and the remediation of brownfields require design, planning and permitting and are often not "shovel-ready."

Second, job creation was more successful in communities with existing skill sets to support green jobs.[8] This finding implies that investments would need to be preceded by preliminary work to build the prerequisite skill sets or investments would need to be targeted to communities that already possess those skill sets. Third, there is a strong overlap of skills between technically oriented jobs in the oil and gas sector and those in the clean energy sector.[180] This finding suggests that at least one subset of workers in the oil and gas sector can move with some ease into the clean energy sector and the training needed to fill gaps in skills is likely to be more successful.[180]

11.4.3 Addressing Equity through Economic Diversification and Revitalization

While the energy transition has focused attention on job losses in the oil and gas sector,[183] in reality, many regions have already experienced or are likely to experience job losses as a result of the inherent boom–bust nature of the industry. The economic and social costs from dependency on the oil and gas sector have prompted some efforts to diversify local economies.[184,185]

To its credit, Congress has supported several programs to assist communities in diversifying their economies, including federal–state partnerships such as the

Appalachian Regional Commission and the Economic Development Administration programs run by the Department of Commerce.[9,186] A number of these programs responded to distressed Appalachian coal communities that were displaced by automation in mining and outcompeted by cheaper coal from western United States and then cheaper natural gas.[187] Congress has continued to support these programs and rejected the Trump administration's proposed cuts of $340 million to the Appalachian Regional Commission and the Economic Development Administration programs in its 2018 budget proposal.[188]

Studies on economic diversification emphasize the variation across fossil-fuel-reliant communities and caution against a one-size-fits-all approach.[189] Support for locally grown entrepreneurship and self-employment, customized job training, apprenticeship and earn and learn programs, some of which are provided by extension services from local colleges, universities and state agencies, have proven to be effective in promoting local economic development.[172,190,191] Likewise, investment in schools and healthcare, as well as direct federal or state assistance to offset the declining tax base in these communities (which have relied on extractive taxes or property taxes), are essential for the economic turnaround of these communities.[192–194] A number of studies emphasize that collaborative engagement and technical assistance from extension services can help communities explore economic diversification strategies.[195–197] The task of spurring new economic ventures is all the more challenging for many areas that need to undertake costly remediation of abandoned mines and wells prior to repurposing the lands.[198]

Some efforts at diversification can unintentionally perpetuate reliance on fossil fuel extraction.[199] For instance, the Appalachian Regional Commission's sponsorship of several programs that focused on transitioning workers from the coal sector to the natural gas sector runs the risk of replicating the region's history of reliance on a narrow set of extractive industries.[200] Conversely, Reimagine Appalachia, an umbrella group of NGOs and community and grassroots organizations, proposed a broader range of economic diversification strategies that emphasize local community benefits, such as promoting decentralized renewable energy generation, sustainable farming and restoring and protecting Appalachia's natural resources.[201]

These programs require funding, careful planning, oversight and accountability. Funding could potentially come from implementing a nationwide carbon tax (Section 11.3.1) and an orderly drawdown of the tax preferences for the oil and gas sector (Section 11.3.2). The prioritization of targeted worker and community programs over the poorly designed bailouts of oil and gas companies that were provided in the 2020 COVID-19 rescue package

(Chapter 2) can better assist oil and gas reliant communities to bridge the energy transition.

11.5 Congress's December 2020 "Climate Legislation"

In earlier sections, I focused on the important legislation efforts in Congress that had yet to become law. Next, I turn to the most significant "climate legislation" successfully passed into law during the Trump era. The 2021 Consolidated Appropriations Act (H.R. 133) is the omnibus spending bill signed into law by Trump in December 2020.[202] Provisions in that legislation implement two key strategies for Congress to address the climate crisis, that is, funding clean energy innovations and deployment (Section 11.5.1) and mandating cuts to greenhouse gas emissions (Section 11.5.2).

11.5.1 Clean Energy Innovation and Deployment

The 2021 Consolidated Appropriations Act includes provisions from the Senate's American Energy Innovation Act (S. 2657), sponsored by Senator Joe Manchin of West Virginia and Senator Lisa Murkowski of Alaska, and from the House's Clean Economy Jobs and Innovation Act (H.R. 4447), sponsored, among others, by Republican Brian Fitzpatrick and Democrat Conor Lamb, both from Pennsylvania. These provisions focus on research, development and deployment (RD&D) of clean energy and of less carbon intensive technologies. Federal funding of RD&D has reduced the costs of renewable energy technologies,[xxxiv] and, in turn, the cost-competitiveness of these technologies has spurred even greater adoption.[203–205]

To its credit, Congress targeted assistance for renewable energy deployment to low-income and rural communities, thus ensuring that the benefits from the energy transition are more widely shared. It allocated $1.7 billion for lower-income communities to install renewable energy sources and to weatherize in their homes and earmarked $10 million for the secretary of agriculture to carry out a pilot program for rural communities to adopt renewable energy. In addition, it established a Department of Energy grant program to assist utilities and electric cooperatives with energy storage and microgrid projects that rely on renewable energy. A subset of these

---

[xxxiv] Popp lays out why government funding in energy innovation is justified and how government funding can be targeted to avoid crowding out private investments [204].

cooperatives, which powered America's rural electrification, continue to serve rural communities.

Nevertheless, the legislation's allocation of funding is skewed in favor of nuclear and fossil, continuing the historical trends.[206] The legislation includes RD&D funding of $4 billion for solar and wind, hydropower and geothermal; $6 billion for carbon capture, utilization and storage; and $11 billion for nuclear.[207] Research support to decarbonize the grid was directed to developing technologies such as battery recycling, electricity storage at various timescales and smart grid technologies.[208] Congress provided offshore wind tax credits through 2025 and extended the Investment Tax Credit for solar projects for two years and the Production Tax Credit for wind projects for one year[209] but left unreformed the tax preferences for the oil and gas sector (Section 11.3.3).

Taxpayer-funded programs focused on funding projects with high risks but potentially high returns. Despite some high-profile failures, overall the Department of Energy's past loan program received more repayments with interest than losses. Investments in renewable energy and electric cars have seen mixed success.[210,211,xxxv] Investments in fossil-fuel-related technologies such as CCS have seen a mixed record as well.[104,212–214,xxxvi]

The scarcity of taxpayer funding calls for more careful prioritization of RD&D funding, not only in the allocation between renewable energy and fossil fuels (Section 11.5.3) but also among renewable energy pursuits (e.g., prioritizing storage and smart grid technologies, for which technological improvements and cost reductions are urgently needed,[205] versus wind and solar more generally that has seen significant cost reductions). Meeting the goals of significant emissions reductions in the electricity sector (e.g., by 2035, a target set by President Biden) requires channeling funds into efforts that have a greater likelihood of successfully reducing significant shares of

---

[xxxv] The program made a $535 million loan to Tesla Inc. to open its first factory in Silicon Valley and Tesla became one of the world's leading makers of electric cars and battery storage. Other projects, such as the solar manufacturer Solyndra, which received $539 million in loan guarantees, went bankrupt, as it was unable to compete with cheaper solar panels from China [215]. Fisker Automotive received $192 million in loan guarantees to produce plug-in hybrid vehicles but declared bankruptcy in 2013 [210].

[xxxvi] In 2017, the International Energy Agency identified two commercial power plants in North America equipped with CCS technology. Petro Nova, located in Texas, was built in 2017 but mothballed in 2020. The plant cost $1billion, generated power at $4,200 per kilowatt and captured 33 percent of emissions from one power unit. Boundary Dam, located in Saskatchewan, installed a retrofit for only one power unit for Can$1.3 billion and operates at 50 percent of capacity [213, 214]. An earlier project, the Kemper CCS plant in Mississippi, cost $7.5 million and ended its CCS operations. From 2010 through 2017, the Department of Energy spent $1.1 billion on nine major CCS demonstration projects. All but three were abandoned by 2017 [212].

emissions in a cost-effective way within a shorter time frame. Researchers suggest that one prudent strategy, given uncertainties in these investments, is to allocate some RD&D funds to nearly competitive technologies and some to high-risk, high-reward options, based on expected benefits.[215]

11.5.2 Emission Cuts to Drive Innovation

The 2021 Consolidated Appropriations Act also enacted ambitious cuts to a class of potent greenhouse gas, used in refrigerants and air conditioning. It mandated cuts to hydrofluorocarbons (HFCs) and chlorofluorocarbons (CFCs) of 40 percent by 2024 and 85 percent by 2036 and empowered the EPA to implement these cuts. The phase out of HFCs mirrors the Kigali Amendment to the Montreal Protocol, which is projected to avert at least as 0.5°Celsius of warming by the end of the century.[207,xxxvii]

Republican Senator John Kennedy of Louisiana, who sponsored this provision under the American Innovation and Manufacturing Act with Democrat Senator Tom Carper of Delaware, framed this provision as a strategy to support American innovation and jobs.[216] US companies' position as innovators and producers of substitute products motivated Republican senators to request (unsuccessfully) that the Trump administration present the Kigali Amendment, signed by Obama in 2016, for Senate ratification.[217] Senators emphasized the amendment would create a projected 33,000 jobs, increase exports by $4.8 billion and enable the United States to outcompete China, which is still producing HFCs.[217] With the regulatory certainty of the phased drawdown, US innovators and producers of the substitute technologies can more confidently direct their investment.[218] (Notably, US companies' leadership in the manufacturer of replacement products had motivated the United States to negotiate and ratify the Montreal Protocol.[219])

11.5.3 Shortcomings in the Legislation

While the 2020 "climate legislation" gives reason for optimism, the legislation reflects the tensions between competing views of how to achieve the energy transition (Section 11.4.1) and the compromises made to garner broad enough support among members of Congress.

[xxxvii] The Montreal Protocol is an international agreement that came into force in 1989 in which countries cut their emissions of ozone-depleting chemicals. Unfortunately, one popular substitute to these chemicals is HFCs which are a potent greenhouse gas. The Kigali Amendment focuses on cuts to HFCs.

As noted earlier, some members of Congress favor focusing RD&D funding on renewable energy, storage and smart grid technologies and pursuing separate efforts to assist fossil-fuel-reliant communities to diversify and to rebuild their local economies. That strategy confronts the reality that jobs in oil, gas and coal declined even before the United States put in place policies for a climate-motivated energy transition. That strategy is also compatible with the view of some energy experts that the United States can achieve significant emissions cuts in the electricity sector by focusing on accelerating renewable energy adoption.[12,220,221,xxxviii]

Conversely, other members of Congress, predominantly from states reliant on fossil fuel extraction, have championed RD&D in CCS technologies and research into developing new products from coal, arguing that these technologies would permit fossil fuel communities to continue their reliance on extractive jobs. While CCS can play a narrower role for addressing emissions in hard to decarbonize industrial sectors, such as cement,[222] promises that CCS can stem the decline in jobs in fossil-fuel-reliant communities run the risk of creating false hopes for these communities. The failure of highly subsidized CCS projects underscores that these technologies are far from commercially feasible.[104,212–214] These failures also caution against assumptions that CCS permits large-scale oil and gas or coal combustion to continue unabated, when major cuts in emissions are needed to keep the average rise in global temperatures below the 1.5°Celsius or even the 2°Celsius target. The amount of carbon captured equates to a miniscule share of the emissions from oil and gas extraction and combustion. For instance, Exxon Mobil reported that its carbon capture in 2019 amounted to 1.2 percent of the total emissions from the combustion of oil and gas the company extracted.[223,xxxix]

Environmental NGOs voiced concerns about provisions in the legislation that, on closer inspection, do not support clean energy innovation but appear to be a giveaway to special interest groups.[224] For instance, critics excoriated the allocation of $75 million to create a federal uranium reserve through the

---

[xxxviii] Jacobson finds that spending on wind, solar and storage to replace fossil fuel combustion is more cost-effective than spending on carbon capture and storage or synthetic direct air capture. The latter additionally contributes to local air pollution and other public health and environmental costs during the extraction of fossil fuels [220]. Lenzi warns of the potential that CCS is used simply to justify prolonging fossil fuel emissions and the risk that the technology fails to deliver on the large-scale sequestration needed to stabilize the climate [231]. On the contrary, Cunliff notes CCS remains an important option in the technology toolbox for hard-to-decarbonize industries, such as the cement industry [222].

[xxxix] Exxon Mobil estimated that the combustion of natural gas and oil that it produced in 2019 would contribute to 190 million tons of carbon dioxide (CO_2) equivalent and 380 million tons of CO_2 equivalent, respectively. Estimates of the emissions based on final product sales (after refining and other processing) is 730 million tons of CO_2 equivalent. Exxon Mobil reported it had captured 6.8 million tons of CO_2 equivalent in 2019.

purchase of domestic uranium at above market price, an action that does not advance innovation in nuclear energy technologies.[224,225] Such a reserve is not necessary as US nuclear plants have never faced any uranium shortage and US uranium supply comes from allies such as Australia and Canada.[226] The report by the Government Accountability Office emphasized that the Nuclear Fuel Working Group, convened by the Trump administration, did not articulate a clear rationale for its recommendation to boost the US domestic uranium industry.[225,227] Waste of taxpayers' scarce funds can jeopardize public support for RD&D to accelerate the clean energy transition.

11.6 Conclusion

Congress holds the powers to implement a variety of tools to support an orderly transition away from oil and gas dependency. The legislative efforts in Congress to correct signals in the economy on the costs of climate disruption and to assist the fossil-fuel-reliant communities to diversify their economies, if enacted into law, can do much to help move the energy transition forward. Provisions in the December 2020 legislation to support clean energy innovation and to cut potent greenhouse gas emissions underscore that Congress can harness bipartisan cooperation to enact legislation to address climate change. The framing of climate legislation to emphasis support for jobs, economic opportunities and innovation focuses on the priorities of Americans, albeit clean energy is one among the multiplicity of strategies needed to revitalize the US economy. The 116th Congress bookended the Trump era with a down payment on climate action, offering some hope for Congress's collaboration with the incoming Biden administration, which I turn to next in the concluding chapter.

References

1. US Senate. "Climate Change: 114th Cong., 2nd Sess." *Congressional Record* 162, no. 94 (June 14, 2016). www.congress.gov/congressional-record/2016/6/14/senate-section/article/S3871-3.
2. S. Waldman and B. Hulac. "This Is When the GOP Turned Away from Climate Policy." *E&E News*, December 5, 2018.
3. L. R. Ember. "The Last of a Breed." *Chemical & Engineering News*, August 21, 2006. https://cen.acs.org/articles/84/i34/Last-Breed.html.
4. R. J. Brulle. "Institutionalizing Delay: Foundation Funding and the Creation of US Climate Change Counter-Movement Organizations." *Climatic Change* 122, no. 4 (2014): 681–694.

5. National Renewable Energy Laboratory. *Utility-Scale Solar 2013: An Empirical Analysis of Project Cost, Performance, and Pricing Trends in the United States*. Report by M. Bolinger and S. Weaver (September 17, 2014).
6. C. Hitaj. "Wind Power Development in the United States." *Journal of Environmental Economics and Management* 65, no. 3 (2013): 394–410.
7. Congressional Research Service. *The Renewable Electricity Production Tax Credit: In Brief*. Report by M. F. Sherlock. R43453 (Washington, DC: April 29, 2020). https://fas.org/sgp/crs/misc/R43453.pdf.
8. D. Popp et al. "The Employment Impact of Green Fiscal Push: Evidence from the American Recovery Act." National Bureau of Economic Research working paper 27321 (June 2020).
9. Congressional Research Service. *The POWER Initiative: Energy Transition as Economic Development*. Report by M. H. Cecire. R46015 (Washington, DC: November 20, 2019). https://fas.org/sgp/crs/misc/R46015.pdf.
10. M. O'Brien. "Boehner Opens Door to Ending Tax Breaks for Big Oil Companies." *The Hill*, April 25, 2011. https://thehill.com/blogs/blog-briefing-room/news/157643-boehner-opens-door-to-ending-tax-breaks-for-big-oil-companies.
11. Congressional Research Service. *Oil and Natural Gas Industry Tax Issues in the FY2014 Budget Proposal*. Report by R. Pirog. R42374 (Washington, DC: October 30, 2013). https://fas.org/sgp/crs/misc/R42374.pdf.
12. M. E. Mann. *The New Climate War: The Fight to Take Back Our Planet*. New York: Public Affairs, 2021.
13. C. Zhou et al. "Greater Committed Warming after Accounting for the Pattern Effect." *Nature Climate Change* 11, no. 2 (January 4, 2021): 132–136. https://doi.org/10.1038/s41558-020-00955-x.
14. Princeton University, Andlinger Center for Energy and the Environment, and High Meadows Environmental Institute. *Net-Zero America: Potential Pathways, Infrastructure, and Impacts*. Report by E. Larson et al (December 15, 2020). https://environmenthalfcentury.princeton.edu/sites/g/files/toruqf331/files/2020-12/Princeton_NZA_ Interim_Report_15_Dec_2020_FINAL.pdf.
15. National Academies of Sciences, Engineering, and Medicine. *Accelerating Decarbonization of the US Energy System*. Washington, DC: National Academies Press, 2021. www.nationalacademies.org/our-work/accelerating-decarbonization-in-the-united-states-technology-policy-and-societal-dimensions.
16. US Global Change Research Program. *2017: Climate Science Special Report: Fourth National Climate Assessment*. Report by D. J. Wuebbles et al. (Washington, DC: 2017).
17. US Global Change Research Program. *2018: Impacts, Risks, and Adaptation in the United States: Fourth National Climate Assessment*. Report by D. R. Reidmiller (Washington, DC: 2018).
18. S. Hsiang et al. "Estimating Economic Damage from Climate Change in the United States." *Science* 356, no. 6345 (June 1, 2017): 1362–1369. https://doi.org/10.1126/science.aal4369.
19. R. Birdsey et al. "Executive Summary." In *Second State of the Carbon Cycle Report* edited by N. Cavallaro et al. US Global Change Research Program

(November 2018). https://carbon2018.globalchange.gov/downloads/SOCCR2_Executive_Summary.pdf.
20. Environmental Protection Agency. *Multi-model Framework for Quantitative Sectoral Impact Analysis: A Technical Report for the Fourth National Climate Assessment* (May 11, 2017). https://cfpub.epa.gov/si/si_public_record_report.cfm?Lab=OAP&dirEntryId=335095.
21. Office of Management and Budget. *Climate Change: The Fiscal Risks Facing the Federal Government* (Washington, DC: November 2016).
22. Government Accountability Office. *Climate Change: Information on Potential Economic Effects Could Help Guide Federal Efforts to Reduce Fiscal Exposure.* Report by J. Alfredo Gómez (September 28, 2017). www.gao.gov/products/gao-17-720.
23. C. Flavelle. "US Disaster Costs Doubled in 2020, Reflecting Costs of Climate Change." *New York Times*, January 7, 2021.
24. NOAA National Centers for Environmental Information. *Billion-Dollar Weather and Climate Disasters: Overview.* www.ncdc.noaa.gov/billions.
25. K. Vohra et al. "Global Mortality from Outdoor Fine Particle Pollution Generated by Fossil Fuel Combustion: Results from GEOS-Chem." *Environmental Research* 195 (2021). www.sciencedirect.com/science/article/abs/pii/S0013935121000487.
26. A. P. Williams et al. "Contribution of Anthropogenic Warming to California Drought during 2012–2014." *Geophysical Research Letters* 42, no. 16 (August 20, 2015): 6819–6828. https://doi.org/10.1002/2015GL064924.
27. N. S. Diffenbaugh, D. L. Swain and D. Touma. "Anthropogenic Warming Has Increased Drought Risk in California." *Proceedings of the National Academy of Sciences of the United States of America* 112, no. 13 (March 2, 2015): 3931–3936. https://doi.org/10.1073/pnas.1422385112.
28. A. Hoell et al. *Anthropogenic Contributions to the Intensity of the 2017 United States Northern Great Plains Drought.* American Meteorological Society (December 2018). www.ametsoc.net/eee/2017a/ch6_EEEof2017_Hoell.pdf.
29. J. T. Abatzoglou and A. P. Williams. "*Impact of Anthropogenic Climate Change on Wildfire across Western US Forests.*" 113, no. 42 (October 10, 2016): 11770–11775. https://doi.org/10.1073/pnas.1607171113.
30. J. C. Fyfe et al. "Large Near-Term Projected Snowpack Loss over the Western United States." *Nature Communications* 8, no. 14996 (April 18, 2017). https://doi.org/10.1038/ncomms14996.
31. K. van der Wiel et al. *Rapid Attribution of the August 2016 Flood-Inducing Extreme Precipitation in South Louisiana to Climate Change.* Hydrology and Earth System Sciences (2017). www.hydrol-earth-syst-sci.net/21/897/2017/hess-21-897-2017.pdf.
32. Lewis Terrell and Associates, LLC. *The Economic Impact of the August 2016 Floods on the State of Louisiana.* Report by D. Terrell, Louisiana Economic Development (2016). http://gov.louisiana.gov/assets/docs/RestoreLA/SupportingDocs/Meeting-9-28-16/2016-August-Flood-Economic-Impact-Report_09-01-16.pdf.
33. W. V. Sweet et al. *2017 State of US High Tide Flooding with a 2018 Outlook.* National Oceanic and Atmospheric Administration (NOAA) Center for Operational

Oceanographic Products and Services, NOAA Office of Coastal Management, NOAA National Centers for Environmental Information and the Baldwin Group, Inc. (June 6, 2018). www.ncdc.noaa.gov/monitoring-content/sotc/national/2018/may/ 2017_State_of_US_High_Tide_Flooding.pdf.
34. *Patterns and Projections of High Tide Flooding along the US Coastline Using a Common Impact Threshold*. National Ocean Service Center for Operational Oceanographic Products and Services, Department of Commerce (Silver Spring, MD: February 2018). https://tidesandcurrents.noaa.gov/publications/ techrpt86_PaP_of_HTFlooding.pdf.
35. F. V. Davenport, M. Burke and N. S. Diffenbaugh. "Contribution of Historical Precipitation Change to US Flood Damages." *Proceedings of the National Academy of Sciences of the United States of America* 118, no. 4 (January 26, 2021). https://doi.org/10.1073/pnas.2017524118.
36. K. E. Trenberth et al. "Hurricane Harvey Links to Ocean Heat Content and Climate Change Adaptation." *Earth's Future* 6, no. 5 (May 9, 2018). https://doi .org/10.1029/2018EF000825.
37. K. A. Reed et al. *The Human Influence on Hurricane Florence* (September 11, 2018). https://crd.lbl.gov/assets/Uploads/Wehner/climate-change-Florence -0911201800Z-final.pdf.
38. E. C. J. Oliver et al. *Anthropogenic and Natural Influences on Record 2016 Marine Heat Waves*. American Meteorological Society (January 2018). www .ametsoc.net/eee/2016/ch9.pdf.
39. C. Welch. "Huge Puffin Die-Off May Be Linked to Hotter Seas." *National Geographic*, November 8, 2016. https://news.nationalgeographic.com/2016/11/ tufted-puffins-die-off-bering-sea-alaska-starvation-warm-water-climate-change.
40. J. E. Walsh et al. "The High Latitude Marine Heat Wave of 2016 and Its Impacts on Alaska." *Bulletin of the American Meteorological Society* 99, no. 1 (January 1, 2018): S39–S43. https://doi.org/10.1175/BAMS-D-17-0105.1.
41. Government Accountability Office. *Alaska Native Villages: Most Are Affected by Flooding and Erosion, But Few Qualify for Federal Assistance*. Report by A. Mittal and J. D. Malcolm (Washington, DC: December 12, 2003). www .gao.gov/assets/gao-04-142.pdf.
42. Government Accountability Office. *Alaska Native Villages: Limited Progress Has Been Made on Relocating Villages Threatened by Flooding and Erosion*. Report by A. Mittal (Washington, DC: June 3, 2009). www.gao.gov/assets/gao-09-551.pdf.
43. Q. Schiermeier. "Droughts, Heatwaves and Floods: How to Tell When Climate Change Is to Blame." *Nature* 560 (July 30, 2018): 20–23. https://doi.org/10.1038 /d41586-018-05849-9.
44. National Academies of Sciences, Engineering, and Medicine. *Attribution of Extreme Weather Events in the Context of Climate Change*. Washington, DC: National Academies Press, 2016.
45. *Paris Agreement to the United Nations Framework Convention on Climate Change*. T.I.A.S. No. 16-1104. December 12, 2015. https://unfccc.int/files/meet ings/paris_nov_2015/application/pdf/paris_agreement_english_.pdf.
46. Intergovernmental Panel for Climate Change. *Special Report: Global Warming of 1.5°C. Summary for Policymakers*. Report by V. Masson-Delmotte et al. (October 2018). https://report.ipcc.ch/sr15/pdf/sr15_spm_final.pdf.

47. Global Carbon Project. *Global Carbon Budget 2020*. Report by P. Friedlingstein et al. (December 11, 2020). www.globalcarbonproject.org/carbonbudget/20/files/GCP_CarbonBudget_2020.pdf.
48. N. Aizenman. "A Little-Known Climate Fund Is Suddenly in the Spotlight." *National Public Radio*, June 6, 2017. www.npr.org/sections/goatsandsoda/2017/06/09/532106567/a-little-known-climate-fund-is-suddenly-in-the-spotlight.
49. Q. Schiermeier. "The US Has Left the Paris Climate Deal – What's Next?" *Nature*, November 4, 2020. www.nature.com/articles/d41586-020-03066-x.
50. Intergovernmental Panel for Climate Change. *Special Report: Global Warming of 1.5°C. Summary for Policymakers*. Report by M. Allen et al. (May 2019). www.ipcc.ch/sr15/chapter/spm.
51. J. Chemnick. "US Stands with Russia and Saudi Arabia against Climate Science." *E&E News*, December 10, 2018.
52. C. Stam. "Australia, Brazil and Saudi Arabia Blocking Climate Talks, Says Green MEPs." *EuroActiv*, December 11, 2019. www.climatechangenews.com/2019/12/11/australia-brazil-saudi-arabia-blocking-climate-talks-says-green-mep/+&cd=23&hl=en&ct=clnk&gl=ca.
53. S. Evans and J. Gabbatiss. "In-depth Q&A: How 'Article 6' Carbon Markets Could 'Make or Break' the Paris Agreement." *CarbonBrief*, November 29, 2019. www.carbonbrief.org/in-depth-q-and-a-how-article-6-carbon-markets-could-make-or-break-the-paris-agreement.
54. National Academies of Sciences, Engineering, and Medicine. *Rising above the Gathering Storm: Energizing and Employing America for a Brighter Economic Future*. Washington, DC: National Academies Press, 2007. https://doi.org/10.17226/11463.
55. R. Matthews. "A Parting Letter from Energy Secretary Steven Chu." *Green Market Oracle*, February 4, 2013. https://thegreenmarketoracle.com/2013/02/04/a-parting-letter-from-energy-secretary.
56. C. von Hirschhausen et al. *Energiewende "Made in Germany": Low Carbon Electricity Sector Reform in the European Context*. Cham: Springer, 2018. www.springer.com/gp/book/9783319951256.
57. M.M. Jackson, J. I. Lewis and X. Zhang. "A Green Expansion: China's Role in the Global Deployment and Transfer of Solar Photovoltaic Technology." *Journal of Energy for Sustainable Development* 60 (2021): 90–101.
58. World Bank Group Trade & Competitiveness. *Accelerating Innovation in China's Solar, Wind and Energy Storage Sectors*. Report by S. Kuriakose et al. (2017). https://openknowledge.worldbank.org/bitstream/handle/10986/28573/120374-WP-PUBLIC-11-10-2017-15-10-11-ChinaGreenInnovationFINALSEP.pdf?sequence=1.
59. J. Chen et al. *EU Climate Mitigation Policy*. International Monetary Fund (September 16, 2020). www.imf.org/en/Publications/Departmental-Papers-Policy-Papers/Issues/2020/09/16/EU-Climate-Mitigation-Policy-49639.
60. L. Friedman. "US Quits Paris Climate Agreement: Questions and Answers." *New York Times*, November 4, 2020.
61. Carbon Brief. *The Impacts of Climate Change at 1.5, 2°C and Beyond*. Report by R. McSweeney, R. Pearce and T. Prater (October 4, 2018). https://interactive.carbonbrief.org/impacts-climate-change-one-point-five-degrees-two-degrees/?utm_source=web&utm_campaign=Redirect.

62. Food and Agriculture Organization of the United Nations. *The State of World Fisheries and Aquaculture: Meeting the Sustainable Development Goals* (2018). www.fao.org/3/i9540en/I9540EN.pdf.
63. M. M. Mekonnen and A. Y. Hoekstra. "Four Billion People Facing Severe Water Scarcity." *Science Advances* 2, no. 2 (2016). https://doi.org/10.1126/sciadv.1500323.
64. S. Helper, J. S. Miller and M. Muro. "Why Undermining Fuel Efficiency Standards Would Harm the US Auto Industry." *Brookings*, July 2, 2018. www.brookings.edu/blog/the-avenue/2018/07/02/why-undermining-fuel-efficiency-standards-would-harm-the-us-auto-industry.
65. S. Mufson. "GM Plans Ambitious Pivot from Gasoline." *Washington Post*, January 29, 2021.
66. J. E. Krier and E. Ursin. *Pollution and Policy: A Case Essay on California and Federal Experience with Motor Vehicle Air Pollution, 1940–1975.* Berkeley: University of California Press, 1977.
67. G. Collantes and D. Sperling. "The Origin of California's Zero Emission Vehicle Mandate." *Transportation Research Part A: Policy and Practice* 42, no. 10 (2008): 1302–1313.
68. M. Honeycutt, Science Advisory Board chair. *Draft Report: Science Advisory Board (SAB) Consideration of the Scientific and Technical Basis of EPA's Proposed Rule Titled Strengthening Transparency in Regulatory Science.* Submitted to A. R. Wheeler, administrator, Environmental Protection Agency. EPA-SAB-20-xxx. October 16, 2019.
69. S. T. Anderson et al. "Automobile Fuel Economy Standards: Impacts, Efficiency, and Alternatives." *Review of Environmental Economics and Policy* 14, no. 2 (2020). https://nature.berkeley.edu/~sallee/apsf-reep.pdf.
70. M. Dempsey and K. Brown. "Sec. Chu Refuses to Retract Statement That the Goal Is to Boost Price of Gas to Levels in Europe." US Senate Committee on Environment and Public Works. News release, 2012. www.epw.senate.gov/public/index.cfm/2012/3/post-d06a83f9-802a-23ad-421c-d407d1d706d2.
71. Aston Martin Lagonda, Ltd. et al. *Letter to Donald J. Trump on the Review of the CAFE and GHG Rule for Automobiles.* Submitted to President D. J. Trump. June 6, 2019.
72. Energy Innovation: Policy and Technology, LLC. *Economic, Emissions Impact of Trump Administration Fuel Economy and GHG Emissions Standards Freeze; Implications for US, California, "Section 177" States, Canada.* Report by M. Mahajan and R. Orvis (August 2019). https://energyinnovation.org/wp-content/uploads/2019/08/Impacts-of-Trump-Fuel-Economy-Standard-Rollback-on-US-Section-177-States-Canada-8.7.19.pdf.
73. White House Office of the Press Secretary. "Obama Administration Finalizes Historic 54.5 MPG Fuel Efficiency Standards." News release, August 28, 2012, https://obamawhitehouse.archives.gov/the-press-office/2012/08/28/obama-administration-finalizes-historic-545-mpg-fuel-efficiency-standard.
74. *2017 and Later Model Year Light-Duty Vehicle Greenhouse Gas Emissions and Corporate Average Fuel Economy Standards.* 49 Code of Federal Regulations Parts 523, 531, 533, 536 and 537. National Highway Traffic Safety Administration. 77 Federal Register 62623–63200 (October 15, 2012). www.govinfo.gov/content/pkg/FR-2012-10-15/pdf/2012-21972.pdf.

75. *The Safer Affordable Fuel-Efficient (SAFE) Vehicles Rule Part One: One National Program*. 40 Code of Federal Regulations 85 and 86; 49 Code of Federal Regulations 531 and 533. Environmental Protection Agency and National Highway Traffic Safety Administration, Department of Transportation. 84 Federal Register 188. www.federalregister.gov/documents/2019/09/27/2019-20672/the-safer-affordable-fuel-efficient-safe-vehicles-rule-part-one-one-national-program.
76. J. Freeman. "The Auto Rule Rollback Only Trump Seems to Want." *New York Times*, September 11, 2019.
77. H. Tabuchi. "California Upholds Auto Emissions Standards, Setting Up Face-Off with Trump." *New York Times*, March 24, 2017.
78. H. Tabuchi. "Big Oil Angles, Quietly, to Ease Emissions Cuts." *New York Times*, December 14, 2018.
79. *The Safer Affordable Fuel-Efficient (SAFE) Vehicles Rule for Model Years 2021–2026 Passenger Cars and Light Trucks*. 40 Code of Federal Regulations 86 and 600; 49 Code of Federal Regulations 523, 531, 533, 536, and 537. National Highway Traffic Safety Administration, Environmental Protection Agency. 85 Federal Register 24174–25278 (April 30, 2020) www.govinfo.gov/content/pkg/FR-2020-04-30/pdf/2020-06967.pdf.
80. C. Davenport. "Trump to Revoke California's Authority to Set Stricter Auto Emissions Rules." *New York Times*, September 17, 2019.
81. C. Davenport. "Automakers Tell Trump His Pollution Rules Could Mean 'Untenable' Instability and Lower Profits." *New York Times*, June 6, 2019.
82. Editorial Board. "A Cruel Parody of Antitrust Enforcement." *New York Times*, September 6, 2019.
83. R. Beitsch and R. Frazin. "California Finalizes Fuel Efficiency Deal with Five Automakers, Undercutting Trump." *The Hill*, August 17, 2020. https://thehill.com/policy/energy-environment/512414-california-finalizes-fuel-efficiency-deal-with-five-automakers.
84. R. Beitsch. "DOJ Whistleblower: California Emissions Probe Was 'Abuse of Authority'." *The Hill*, June 24, 2020. https://thehill.com/policy/energy-environment/504384-doj-whistleblower-california-emissions-probe-was-abuse-of-authority.
85. A. Durkee. "Trump Is Angry That Automakers Don't Want His Anti-Climate Change Policy." *Vanity Fair*, August 22, 2019. www.vanityfair.com/news/2019/08/trump-auto-industry-fuel-emission-standards-california.
86. G. Acton. "There's Almost Zero Rationale for Arctic Oil Exploration, Says Goldman Sachs Analyst." *CNBC News*, March 23, 2017. www.cnbc.com/2017/03/23/theres-almost-zero-rationale-for-arctic-oil-exploration-says-goldman-sachs.html.
87. J. A. Dluohy. "Wall Street Is Feeling the Pressure to Stop Arctic Oil Funding." *World Oil*, April 27, 2020.
88. L. Nguyen. "Bank of America Says It Won't Finance Oil and Gas Exploration in the Arctic." *Bloomberg*, November 30, 2020. www.bloomberg.com/news/articles/2020-11-30/bofa-says-it-won-t-finance-oil-and-gas-exploration-in-the-arctic.
89. Rainforest Action Network et al. *Banking on Climate Chaos: Fossil Fuel Finance Report*. Report by A. Kirsch et al. (March 24, 2021). www.ran.org/bankingonclimatechaos2021.

90. R. Frazin. "Republicans Say Trump Should Act against Financial Institutions That Are Unwilling to Fund Certain Fossil Fuel Projects." *The Hill*, May 8, 2020. https://thehill.com/policy/energy-environment/496881-republicans-say-trump-should-act-against-financial-institutions.
91. A. DeMarban. "Federal Appeals Court Rejects Trump Administration Permit for Offshore Oil Project in Arctic Alaska." *Anchorage Daily News*, December 7, 2020. www.adn.com/business-economy/energy/2020/12/07/federal-appeals-court-rejects-trump-administration-permit-for-offshore-oil-project-in-arctic-alaska.
92. H. Fountain. "Sale of Drilling Leases in Arctic Refuge Fails to Yield a Windfall." *New York Times*, January 6, 2021. www.nytimes.com/2021/01/06/climate/arctic-refuge-drilling-lease-sales.html.
93. D. Sullivan et al. *Letter to POTUS Regarding Energy Financial Institutions*. Submitted to President D. J. Trump. May 7, 2020.
94. R. Frazin. "Trump Criticizes Banks Withholding Funds from Certain Fossil Fuel Projects." *The Hill*, April 24, 2020. https://thehill.com/policy/energy-environment/494568-trump-criticizes-banks-withholding-funds-from-certain-fossil-fuel.
95. *Fair Access to Financial Services: Notice of Proposed Rulemaking*. 12 Code of Federal Regulations 55. Office of the Comptroller of the Currency, Department of Treasury. 85 Federal Register 75261–75266 (November 25, 2020). www.federalregister.gov/documents/2020/11/25/2020-26067/fair-access-to-financial-services+&cd=7&hl=en&ct=clnk&gl=ca.
96. J. C. Kirn. "OCC Proposes 'Fair Access' Rule: Potential Implications for ESG Analysis." *National Law Review* 10, no. 328 (November 23, 2020).
97. M. M. Schanzenbach and R. H. Sitkoff. "ESG Investing: Theory, Evidence, and Fiduciary Principles." *Journal of Financial Planning* (October 1, 2020).
98. *Financial Factors in Selecting Plan Investments: Proposed Rule*. 29 Code of Federal Regulations 2550. Employee Benefits Security Administration, Department of Labor. 85 Federal Register 39113–39128 (June 30, 2020). www.federalregister.gov/documents/2020/06/30/2020-13705/financial-factors-in-selecting-plan-investments.
99. *Financial Factors in Selecting Plan Investments: Final Rule*. 29 Code of Federal Regulations 2509 and 2550. Employee Benefits Security Administration, Department of Labor. 85 Federal Register 72846–72885 (November 13, 2020). www.federalregister.gov/documents/2020/11/13/2020-24515/financial-factors-in-selecting-plan-investments.
100. Congressional Research Service. *Injection and Geologic Sequestration of Carbon Dioxide: Federal Role and Issues for Congress*. Report by A. C. Jones. R46192 (Washington, DC: January 24, 2020). https://fas.org/sgp/crs/misc/R46192.pdf.
101. Natural Resources Defense Council. *Strengthening the Regulation of Enhanced Oil Recovery to Align It with the Objectives of Geologic Carbon Dioxide Sequestration*. Report by B. Mordick and G. Peridas. 16-09-B (November 2017). www.nrdc.org/sites/default/files/regulation-eor-carbon-dioxide-sequestration-report.pdf.
102. International Energy Agency Carbon Capture and Storage Unit. *Storing CO_2 through Enhanced Oil Recovery: Combining EOR with CO_2 Storage (EOR+)*

for Profit. Report by Dr. W. Heidug et al. (2015). https://nachhaltigwirtschaften.at/resources/iea_pdf/reports/iea_ghg_storing_co2_trough_enhanced_oil_recovery.pdf.
103. Working Group III of the Intergovernmental Panel on Climate Change. *IPCC Special Report on Carbon Dioxide Capture and Storage*, edited by B. Metz et al. New York: Cambridge University Press, 2005. www.ipcc.ch/site/assets/uploads/2018/03/srccs_wholereport-1.pdf.
104. N. Kusnetz. "Exxon Touts Carbon Capture As a Climate Fix, But Uses It to Maximize Profit and Keep Oil Flowing." *Inside Climate News*, September 27, 2020. https://insideclimatenews.org/news/27092020/exxon-carbon-capture.
105. *Credit for Carbon Oxide Sequestration: Final Regulations*. 26 Code of Federal Regulations Part 1. Internal Revenue Service, Department of Treasury (December 31, 2020). www.irs.gov/pub/irs-drop/td-9944.pdf.
106. B. Mordick, Natural Resources Defense Council, and J. Noël, Greenpeace. *Comments on Credit for Carbon Oxide Sequestration*. Submitted to Internal Revenue Service. IRS REG-112339-19. 2020.
107. A. Grinberg, Clean Water Action. *Comments on Proposed Regulations Credit for Carbon Oxide Sequestration*. Submitted to Internal Revenue Service. IRS REG-112339-19. July 31, 2020.
108. G. E. Metcalf. *Paying for Pollution: Why a Carbon Tax Is Good for America*. New York: Oxford University Press, 2019.
109. US Congress. Senate. *American Opportunity Carbon Fee Act of 2019*. S. 1128, 116th Congress, 1st Sess. Introduced in Senate April 10, 2019.
110. L. H. Goulder et al. "Impacts of a Carbon Tax across US Household Income Groups: What Are the Equity-Efficiency Trade Offs?" *Journal of Public Economics* 175 (2019): 44–64. https://doi.org/10.1016/j.jpubeco.2019.04.002.
111. G. E. Metcalf. "On the Economics of a Carbon Tax for the United States." *Brookings Papers on Economic Activity* (February 24, 2019).
112. A. Yamazaki. "Jobs and Climate Policy: Evidence from British Columbia's Revenue-Neutral Carbon Tax." *Journal of Environmental Economics and Management* 83 (May 2017): 197–216.
113. G. E. Metcalf and J. H. Stock. "The Macroeconomic Impact of Europe's Carbon Taxes." National Bureau of Economic Research working paper 27488 (July 2020). www.nber.org/papers/w27488.
114. P. Bayer and M. Aklin. "The European Union Emissions Trading System Reduced CO_2 Emissions despite Low Prices." *Proceedings of the National Academy of Sciences of the United States of America* 117, no. 16 (April 6, 2020): 8804–8812. www.pnas.org/content/117/16/8804.
115. J. F. Green. "Does Carbon Pricing Reduce Emissions? A Review of Ex-post Analyses." *Environmental Research Letters* 16 (2021). https://iopscience.iop.org/article/10.1088/1748-9326/abdae9/pdf.
116. G. Metcalf. "An Emissions Assurance Mechanism: Adding Environmental Certainty to a Carbon Tax." *Review of Environmental Economics and Policy* 14, no. 1 (2020): 114–130.
117. G. Wagner. "The Numbers behind Exxon's Support for a Carbon Tax." *Bloomberg*, October 9, 2020. www.bloomberg.com/news/articles/2020-10-09/the-numbers-behind-exxon-s-support-for-a-carbon-tax.

118. J. A. Dlouhy. "Oil Companies Join Corporate Lobbying Push for US Carbon Tax." *Bloomberg*, May 20, 2019. www.bloomberg.com/news/articles/2019-05-20/oil-companies-join-corporate-lobbying-push-for-u-s-carbon-tax.
119. Climate Leadership Council. *The Four Pillars of Our Carbon Dividends Plan* (September 2019). https://clcouncil.org/our-plan.
120. Climate Leadership Council. *The Conservative Case for Carbon Dividends*. Report by J. A. Baker III et al. (February 2017). www.clcouncil.org/media/2017/03/The-Conservative-Case-for-Carbon-Dividends.pdf.
121. K. Savage. "Climate Liability Waiver Dropped from Major Carbon Tax Proposal." *Climate Docket*, September 12, 2019. www.climatedocket.com/2019/09/12/climate-liability-waiver-carbon-tax-baker-schultz/.
122. D. Roberts. "Energy Lobbyists Have a New PAC to Push for a Carbon Tax. Wait, What?" *Vox*, June 23, 2018. www.vox.com/energy-and-environment/2018/6/22/17487488/carbon-tax-dividend-trent-lott-john-breaux.
123. Brief of Amici Curiae Senators Whitehouse, Cardin, Blumenthal, Warren, Markey, and Van Hollen in Support of Respondent. *BP PLC, et al. v. Mayor and City Council of Baltimore*. US Supreme Court Case No. 19-1189, 2020.
124. G. E. Kelder, Jr., and R. A. Daynard. "The Role of Litigation in the Effective Control of the Sale and Use of Tobacco." *Stanford Law and Policy Review* 8 (1997): 63–98.
125. B. Meier. "Cigarette Makers and States Draft a $206 Billion Deal." *New York Times*, November 14, 1998. www.nytimes.com/1998/11/14/us/cigarette-makers-and-states-draft-a-206-billion-deal.html.
126. G. Heal. "Economic Aspects of the Energy Transition." National Bureau of Economic Research working paper 27766 (September 2020). www.nber.org/papers/w27766.
127. G. E. Metcalf. *Harnessing the Power of Markets to Solve the Climate Problem*. Aspen Institute Economic Strategy Group (December 2020). https://works.bepress.com/gilbert_metcalf/129/.
128. R. Goyal et al. "Reduction in Surface Climate Change Achieved by the 1987 Montreal Protocol." *Environmental Research Letters* 14, no. 124041 (December 6, 2019). https://iopscience.iop.org/article/10.1088/1748-9326/ab4874.
129. Senate Democrats' Special Committee on the Climate Crisis. *The Case for Climate Action: Building a Clean Economy for the American People*. Report by Senator B. Schatz et al. (August 25, 2020). www.schatz.senate.gov/imo/media/doc/SCCC_Climate_Crisis_Report.pdf.
130. M. Levy. "President-Elect Biden Supports a 'Carbon Enforcement Mechanism' – Could that Mean a Price on Carbon?" *Harvard Environmental & Energy Law Program*, November 14, 2020. https://eelp.law.harvard.edu/2020/11/president-elect-biden-supports-a-carbon-enforcement-mechanism-could-that-mean-a-price-on-carbon/#_ftn43.
131. *Pollutant-Specific Significant Contribution Finding for Greenhouse Gas Emissions from New, Modified, and Reconstructed Stationary Sources: Electric Utility Generating Units, and Process for Determining Significance of Other New Source Performance Standards Source Categories: Final Rule*. 40 Code of Federal Regulations 60. Environmental Protection Agency. 86 Federal Register

2542–2558 (January 13, 2021) www.federalregister.gov/documents/2021/01/13/2021-00389/pollutant-specific-significant-contribution-finding-for-greenhouse-gas-emissions-from-new-modified.
132. House Select Committee on the Climate Crisis. *Solving the Climate Crisis: The Congressional Action Plan for a Clean Energy Economy and a Healthy, Resilient, and Just America*. Report by Majority Committee Staff (June 2020). https://climatecrisis.house.gov/sites/climatecrisis.house.gov/files/Climate%20Crisis%20Action%20Plan.pdf.
133. Government Accountability Office. *Social Cost of Carbon: Identifying a Federal Entity to Address the National Academies' Recommendations Could Strengthen Regulatory Analysis*. Report by J. A. Gómez (Washington, DC: June 2020). www.gao.gov/assets/710/707776.pdf.
134. US Congress. Senate. *Carbon Pollution Transparency Act*. S., 116th Congress, 1st Sess. Introduced in Senate June 5, 2019.
135. US Congress. House. *Carbon Pollution Transparency Act*. H.R. 8174, 116th Congress, 2nd Sess. Introduced in House September 4, 2020.
136. T. Carleton and M. Greenstone. "Updating the United States Government's Social Cost of Carbon." University of Chicago, Becker Friedman Institute for Economics working paper 2021-04 (January 14, 2021). https://papers.ssrn.com/sol3/papers.cfm?abstract_id=3764255.
137. M. Fleurbaey et al. "The Social Cost of Carbon: Valuing Inequality, Risk, and Population for Climate Policy." *The Monist* 102, no. 1 (December 20, 2018): 84–109. https://doi.org/10.1093/monist/ony023.
138. J. Broome. *Climate Matters: Ethics in a Warming World*. New York: W. W. Norton & Company, 2012.
139. M. Burke, S. Hsiang and E. Miguel. "Global Non-linear Effect of Temperature on Economic Production." *Nature* 527 (October 21, 2015): 235–239. https://doi.org/10.1038/nature15725.
140. Office of Management and Budget. *FY 2013 Administration Budget*. White House (Washington, DC: 2012).
141. Congressional Budget Office. *Options for Reducing the Deficit: 2014 to 2023* (Washington, DC: 2013).
142. US Congress. Senate. *Close Big Oil Tax Loopholes Act*. S. 1710, 115th Congress, 1st Sess. Introduced in Senate August 2, 2017.
143. US Congress. House. *End Oil and Gas Tax Subsidies Act of 2020*. H.R. 8411, 116th Congress, 2nd Sess. Introduced in House September 29, 2020.
144. J. E. Aldy and J. Patashnik. "Eliminating Fossil Fuel Subsidies." In *15 Ways to Rethink the Federal Budget*, edited by M. Greenstone et al. 31–35. Washington, DC: Hamilton Project, Brookings Institution, 2013.
145. G. Metcalf. "The Impact of Removing Tax Preferences for US Oil and Gas Production." *Journal of the Association of Environmental and Resource Economists* 5, no. 1 (2018): 1–37.
146. Rainforest Action Network et al. *Banking on Climate Change: Fossil Fuel Finance Report*. Report by A. Kirsch et al. (March 18, 2020). http://priceofoil.org/content/uploads/2020/03/Banking_on_Climate_Change_2020.pdf.

147. Board of Governors of the Federal Reserve System. *Financial Stability Report* (November 2020). www.federalreserve.gov/publications/files/financial-stability-report-20201109.pdf.
148. E. Campiglio et al. "Climate Change Challenges for Central Banks and Financial Regulators." *Nature Climate Change* 8, no. 6 (2018): 462–468.
149. J.-F. Mercure et al. "Macroeconomic Impact of Stranded Fossil Fuel Assets." *Nature Climate Change* 8 (June 4, 2018): 588–593. www.nature.com/articles/s41558-018-0182-1.
150. M. Arnold. "ECB Stress Test Reveals Economic Impact of Climate Change." *Financial Times*, March 18, 2021.
151. US Congress. House. *Climate Change Financial Risk Act of 2019*. H.R., 5194, 116th Congress, 2nd Sess. Introduced in House November 20, 2019.
152. US Congress. Senate. *Climate Change Financial Risk Act of 2019*. S., 116th Congress, 1st Sess. Introduced in Senate November 20, 2019.
153. US Congress. House. *Climate Risk Disclosure Act of 2019*. H.R. 3623, 116th Congress, 2nd Sess. Introduced in House July 5, 2019.
154. US Congress. Senate. *Climate Risk Disclosure Act of 2019*. S., 2075, 116th Congress, 1st Sess. Introduced in Senate July 10, 2019.
155. Climate-Related Market Risk Subcommittee. *Managing Climate Risk in the US Financial System*. Report by D. Gillers et al. US Commodity Futures Trading Commission (Washington, DC: 2020). www.cftc.gov/sites/default/files/2020-09/9-9-20%20Report%20of%20the%20Subcommittee%20on%20Climate-Related%20Market%20Risk%20-%20Managing%20Climate%20Risk%20in%20the%20U.S.%20Financial%20System%20for%20posting.pdf.
156. Great Democracy Initiative. *A Regulatory Green Light: How Dodd-Frank Can Address Wall Street's Role in the Climate Crisis*. Report by G. Steele (January 2020). https://greatdemocracyinitiative.org/wp-content/uploads/2020/01/Final_Greenlight_Steele.pdf.
157. Congressional Research Service. *Department of Defense Energy Management: Background and Issues for Congress*. Report by H. L. Greenley. R45832 (Washington, DC: July 25, 2019). https://fas.org/sgp/crs/natsec/R45832.pdf.
158. C. Samaras, W. J. Nuttalla and M. Bazilian. "Energy and the Military: Convergence of Security, Economic, and Environmental Decision-Making." *Energy Strategy Reviews* 26 (November 2019): 1004009–100414.
159. D. J. Hess. *Good Green Jobs in a Global Economy: Making and Keeping New Industries in the United States*. Cambridge, MA: MIT Press, 2012.
160. W. F. Lamb et al. "Discourses of Climate Delay." *Global Sustainability* 3, no. e17 (July 1, 2020). https://doi.org/https://doi.org/10.1017/sus.2020.13.
161. P. Osterman et al. *Working in America: A Blueprint for the New Labor Market*. Cambridge, MA: MIT Press, 2002.
162. D. M. West. *The Future of Work: Robots, AI, and Automation*. Washington, DC: Brookings Institution Press, 2019.
163. R. J. Lambert and P. P. Silva. "The Challenges of Determining the Employment Effects of Renewable Energy." *Renewable and Sustainable Energy Reviews* 16, no. 7 (2012): 4667–4674.

164. Metropolitan Policy Program. *Advancing Inclusion through Clean Energy Jobs*. Report by M. Muro, A. Tomer, R. Shivaram and J. Kane. Brookings Institution (April 2019). www.brookings.edu/wp-content/uploads/2019/04/2019.04_metro_Clean-Energy-Jobs_Report_Muro-Tomer-Shivaran-Kane.pdf.
165. E2. *Clean Jobs America 2020: Repowering America's Economy in the Wake of Covid-19* (April 2020). https://e2.org/wp-content/uploads/2020/04/E2-Clean-Jobs-America-2020.pdf.
166. US Climate Alliance. *2020 Clean Energy Employment Report*. Report by BW Research (2020). www.usclimatealliance.org/jobsreport.
167. American Wind Energy Association. *Made-in-the-USA Wind Power Jobs* (2020). www.awea.org/wind-101/benefits-of-wind/powering-job-growth.
168. Berkeley Lab. *Wind Technologies Market Report*. Report by M. Bolinger and R. Wiser (2020). https://emp.lbl.gov/wind-technologies-market-report.
169. L. Hughes and J. Meckling. "The Politics of Renewable Energy Trade: The US–China Solar Dispute." *Energy Policy* 105 (June 2017): 256–262.
170. Bureau of Labor Statistics. "Fastest Growing Occupations." *Occupational Outlook Handbook*. 2020. www.bls.gov/ooh/fastest-growing.htm.
171. P. Jordan. *Memorandum: Clean Energy Employment Initial Impacts from the COVID-19 Economic Crisis, December 2020, Revised*. BW Research Partnership. Submitted to E2, E4TheFuture and ACORE. February 8, 2021.
172. A. Tomer, J. W. Kane and C. George. *How Renewable Energy Jobs Can Uplift Fossil Fuel Communities and Remake Climate Politics*. Brookings Institution (February 23, 2021). www.brookings.edu/research/how-renewable-energy-jobs-can-uplift-fossil-fuel-communities-and-remake-climate-politics/?preview_id=1414272.
173. R. G. Newell and D. Raimi. "US State and Local Oil and Gas Revenue Sources and Uses." *Energy Policy* 112 (January 2018): 12–18. https://doi.org/10.1016/j.enpol.2017.10.002.
174. J. Yahn. "Power and Powerlessness in the Shale Valley Schools: Fracking for Funding." *West Virginia Law Review* 120, no. 3 (2018): 943–971. https://researchrepository.wvu.edu/wvlr/vol120/iss3/11.
175. US Congress. House. *Revitalizing the Economy of Coal Communities by Leveraging Local Activities and Investing More Act of 2019*. H.R., 116th Congress, 1st Sess. Introduced in House April 9, 2019.
176. Columbia Center on Global Energy Policy and Resources for the Future. *Green Stimulus for Oil and Gas Workers: Considering a Major Federal Effort to Plug Orphaned and Abandoned Wells*. Report by D. Raimi et al. (New York: July 2020). www.energypolicy.columbia.edu/sites/default/files/file-uploads/OrphanWells_CGEP-Report_071620.pdf.
177. M. Boom. "House Bill Would Deliver Needed Steps toward a Clean Economy." *Expert Blog*. National Resource Defense Council. 2020. www.nrdc.org/experts/marc-boom/house-bill-would-deliver-needed-steps-toward-clean-economy.
178. J. Dillon and N. Sobczyk. "Clean Energy Push Caught in Congressional Chaos." *E&E News*, September 21, 2020.
179. J. E. Aldy. "Policy Monitor – A Preliminary Assessment of the American Recovery and Reinvestment Act's Clean Energy Package." *Review of Environmental Economics and Policy* 7, no. 1 (2013): 136–155.

180. Z. Chen et al. "Green Stimulus in a Post-pandemic Recovery: The Role of Skills for a Resilient Recovery." *Environmental & Resource Economics*, special issue *Economics of the Environment in the Shadow of Coronavirus* 76, no. 4 (2020): 901–911.
181. D. J. Wilson. "Fiscal Spending Jobs Multipliers: Evidence from the 2009 American Recovery and Reinvestment Act." *American Economic Journal: Economic Policy* 4, no. 3 (2012): 251–282.
182. A. Garin. "Putting America to Work, Where? Evidence on the Effectiveness of Infrastructure Construction As a Locally Targeted Employment Policy." *Journal of Urban Economics* 111, no. C (2019): 108–131.
183. F. Vona. "Job Losses and Political Acceptability of Climate Policies: Why the 'Job-Killing' Argument Is So Persistent and How to Overturn it." *Climate Policy* 19, no. 4 (2019): 524–532.
184. J. Haggerty et al. "Long-Term Effects of Income Specialization in Oil and Gas Extraction: The US West, 1980–2011." *Energy Economics* 45 (2014): 186–195. https://doi.org/10.1016/j.eneco.2014.06.020.
185. G. D. Jacobsen and D. P. Parker. "The Economic Aftermath of Resource Booms: Evidence from Boomtowns in the American West." *Economic Journal* 126, no. 593 (2016): 1092–1128.
186. J. Ford. "Equality of Opportunity in Appalachia." McCourt School of Public Policy, 2018. http://hdl.handle.net/10822/1050855.
187. M. R. Betz et al. "Coal Mining, Economic Development, and the Natural Resources Curse." *Energy Economics* 50 (2015): 105–116.
188. V. Volcovici. "Trump Seeks to Ax Appalachia Economic Programs, Causing Worry in Coal Country." *Reuters*, March 16, 2017. www.reuters.com/article/us-usa-trump-budget-appalachia-idUSKBN16N2VF.
189. S. J. Goetz, M. D. Partridge and H. M. Stephens. "The Economic Status of Rural America in the President Trump Era and Beyond." *Applied Economic Perspectives and Policy* 40, no. 1 (2018): 97–118.
190. H. M. Stephens, M. D. Partridge and A. Faggian. "Innovation, Entrepreneurship and Economic Growth in Lagging Regions." *Journal of Regional Science* 53, no. 5 (2013): 778–812.
191. T. J. Bartik. "Using Place-Based Jobs Policies to Help Distressed Communities." *Journal of Economic Perspectives* 34, no. 3 (2020): 99–127.
192. B. G. Kaufman et al. "The Rising Rate of Rural Hospital Closures." *Journal of Rural Health* 32, no. 1 (2016): 35–43.
193. P. J. Kannapel and M. A. Flory. "Postsecondary Transitions for Youth in Appalachia's Central Subregions: A Review of Education Research, 1995–2015." *Journal of Research in Rural Education* 32, no. 6 (2017): 1–17.
194. H. M. Stephens and J. Deskins. "Economic Distress and Labor Market Participation." *American Journal of Agricultural Economics* 100, no. 5 (2018): 1336–1356.
195. B. Taylor, M. Hufford and K. Bilbrey. "A Green New Deal for Appalachia: Economic Transition, Coal Reclamation Costs, Bottom-Up Policymaking (Part 1)." *Journal of Appalachian Studies* 23, no. 1 (2017): 8–28.
196. L. Tarus, M. Hufford and B. Taylor. "A Green New Deal for Appalachia: Economic Transition, Coal Reclamation Costs, Bottom-Up Policymaking (Part 2)." *Journal of Appalachian Studies* 23, no. 1 (2017): 151–169.

197. J. H. Haggerty et al. "Planning for the Local Impacts of Coal Facility Closure: Emerging Strategies in the US West." *Resources Policy* 57 (2018): 69–80.
198. C. Zipper and J. Skousen (eds.). *Appalachia's Coal-Mined Landscapes: Resources and Communities in a New Energy Era.* Cham: Springer, 2021.
199. J. M. Cha. "A Just Transition for Whom? Politics, Contestation, and Social Identity in the Disruption of Coal in the Powder River Basin." *Energy Research & Social Science* 69 (2020): 101657.
200. Department of Energy et al. *The Appalachian Energy and Petrochemical Renaissance: An Examination of Economic Progress and Opportunities* (June 2020). www.energy.gov/sites/prod/files/2020/06/f76/Appalachian%20Energy%20and%20Petrochemical%20Report_063020_v3.pdf.
201. Appalachia, R. *Reimagine Appalachia Blueprint: A New Deal That Works for Us.* (September 2020). https://reimagineappalachia.org/wp-content/uploads/2020/09/ReImagineAppalachia_Blueprint_092020.pdf.
202. US Congress. House. *Consolidated Appropriations Act, 2021.* H.R. 133, 116th Congress, 2nd Sess. Introduced in House January 3, 2019.
203. National Academies of Sciences, Engineering, and Medicine. *An Assessment of ARPA-E.* Washington, DC: National Academies Press, 2017. https://doi.org/10.17226/24778.
204. Hamilton Project. *Promoting Innovation for Low-Carbon Technologies.* Report by D. Popp (October 23, 2019). www.hamiltonproject.org/papers/promoting_innovation_for_low_carbon_technologies.
205. N. Kittner, F. Lill and D. M. Kammen. "Energy Storage Deployment and Innovation For the Clean Energy Transition." *Nature Energy* 2, no. 17125 (2017): 1–6. www.nature.com/articles/nenergy2017125.
206. Congressional Research Service. *Renewable Energy R&D Funding History: A Comparison with Funding for Nuclear Energy, Fossil Energy, Energy Efficiency, and Electric Systems R&D.* Report by C. E. Clark. RS22858 (Washington, DC: June 18, 2018). https://fas.org/sgp/crs/misc/RS22858.pdf.
207. S. Kaplan and D. Grandoni. "Stimulus Deal Includes Raft of Provisions to Fight Climate Change." *Washington Post*, December 21, 2020. www.washingtonpost.com/climate-solutions/2020/12/21/congress-climate-spending.
208. S. K. Johnson. "The Omnibus Bill Was Packed with Energy and Environment Policy." *Ars Technica*, December 23, 2020. https://arstechnica.com/science/2020/12/heres-the-energy-and-environment-policy-passed-with-the-relief-bill.
209. J. St. John. "Congress Passes Spending Bill with Solar, Wind Tax Credit Extensions and Energy R&D Package." *Greentech Media*, December 22, 2020. www.greentechmedia.com/articles/read/solar-and-wind-tax-credit-extensions-energy-rd-package-in-spending-bill-before-congress.
210. C. R. Wise and E. M. Witesman. "Direct Government Investment: Perverse Privatization or New Tool of Government?" *Public Administration Review* 79, no. 2 (2018): 168–179. https://onlinelibrary.wiley.com/doi/full/10.1111/puar.12987.
211. D. Iaconangelo. "DOE Program May Save – or Thwart – Biden's Energy Plan." *E&E News*, January 26, 2021.
212. Government Accountability Office. *Advanced Fossil Energy: Information on DOE Provided Funding for Research and Development Projects Started from*

Fiscal Years 2010 through 2017. Report by F. Rusco et al. (September 2018). www.gao.gov/assets/gao-18-619.pdf.
213. Institute for Energy Economics and Financial Analysis. *Carbon Capture and Storage Is about Reputation, Not Economics*. Report by C. Butler (July 2020). https://ieefa.org/wp-content/uploads/2020/07/CCS-Is-About-Reputation-Not-Economics_July-2020.pdf.
214. Institute for Energy Economics and Financial Analysis. *Petra Nova Mothballing Post-Mortem: Closure of Texas Carbon Capture Plant Is a Warning Sign*. Report by D. Wamsted and D. Schlissel (August 2020). https://ieefa.org/wp-content/uploads/2020/08/Petra-Nova-Mothballing-Post-Mortem_August-2020.pdf.
215. M. Webster et al. "Should We Give Up after Solyndra? Optimal Technology R&D Portfolios under Uncertainty." *Journal of the Association of Environmental and Resource Economists* 4, no. S1 (2017): S123–S151.
216. N. Sobczyk and G. Koss. "Congress Passes Major Climate Legislation in Year-End Omnibus." *E&E News*, December 22, 2020.
217. J. Kennedy, S. M. Collins et al. *Letter in Support of Sending the Kigali Amendment to the Montreal Protocol to the Senate for Its Advice and Consent*. Submitted to President D. J. Trump. June 4, 2018.
218. C. Mooney. "Nearly 200 Nations to Reduce Use of Super-Polluting Hydrofluorocarbons." *Washington Post*, October 15, 2016.
219. R. E. Benedick. *Ozone Diplomacy: New Directions in Safeguarding the Planet*. Cambridge, MA: Harvard University Press, 1998.
220. M. Z. Jacobson. "The Health and Climate Impacts of Carbon Capture and Direct Air Capture." *Energy & Environmental Science* 12, no. 12 (2019): 3567–3574.
221. S. Sgouridis et al. "Comparative Net Energy Analysis of Renewable Electricity and Carbon Capture and Storage." *Nature Energy* 4 (2019): 456–465. https://doi.org/10.1038/s41560-019-0365-7.
222. C. Cunliff. "An Innovation Agenda for Hard-to-Decarbonize Energy Sectors." *Issues in Science and Technology* 36, no. 1 (2019): 74–79.
223. Exxon Mobil. *Energy and Carbon Summary: Scope 3 Emissions* (January 5, 2021). https://corporate.exxonmobil.com/Sustainability/Energy-and-Carbon-Summary/Scope-3-emissions.
224. J. Axelrod and G. H. Fettus, "Fossils and Nukes: The Downsides to Congress' Latest Actions." *Expert Blog*, National Resource Defense Council, 2020.
225. A. Barich and J. Holzman. "Biden Victory Could Be a Boon for Uranium, But Obstacles Remain." *S&P Global Market Intelligence*, November 20, 2020. www.spglobal.com/marketintelligence/en/news-insights/latest-news-headlines/biden-victory-could-be-a-boon-for-uranium-but-obstacles-remain-61317068.
226. J. Krzyzaniak. "Trump's $1.5 Billion Uranium Stockpile: A Solution in Search of a Problem." *Bulletin of the Atomic Scientists*, February 24, 2020. https://thebulletin.org/2020/02/trumps-1-5-billion-uranium-stockpile-a-solution-in-search-of-a-problem.
227. Government Accountability Office. *Uranium Management: Actions to Mitigate Risks to Domestic Supply Chain Could Be Better Planned and Coordinated*. Report by A. Bawden (Washington, DC: December 2020). www.gao.gov/assets/gao-21-28.pdf.

228. J. M. Turner and A. C. Isenberg. *The Republican Reversal: Conservatives and the Environment from Nixon to Trump*. Cambridge, MA: Harvard University Press, 2018.
229. N. Oreskes and E. M. Conway. *Merchants of Doubt: How a Handful of Scientists Obscured the Truth on Issues from Tobacco Smoke to Climate Change*. New York: Bloomsbury, 2011.
230. R. J. Brulle, G. Hall, L. Loy and K. Schell-Smith. "Obstructing Action: Foundation Funding and US Climate Change Counter-Movement Organizations." *Climatic Change* 166, no. 1 (2021): 1–7.
231. D. Lenzi. "The Ethics of Negative Emissions." *Global Sustainability* 1, no. e7 (2018): 1–8. https://doi.org/10.1017/sus.2018.5.

12
America at a Crossroads

Trump's tumultuous one-term presidency came to an end on January 20, 2021, bookended by misinformation and public deception and concluded in violence. Joe Biden won the presidency decisively, with 306 electoral college votes and 81.3 million popular votes – 7.1 million more than went to Trump. However, for weeks after the election results were finalized and confirmed, 12 Republican senators and 140 Republican House members pressed for the rejection of Biden's electoral college win. Their peddling of misinformation on election fraud, coupled with Trump inciting his supporters with calls to action, culminated in the violent storming of Capitol Hill on January 6, 2021.[1-6] For some, President Biden's win serves as a testament to the enduring power of democracy: the majority of voters rejected Trump's divisive rhetoric, damaging policies and disregard for human life. For others, the Biden win fails to assuage concerns over a deeply fractured America.

The Biden presidency commenced amid a multiplicity of crises: a raging pandemic, a tattered economy, civil unrest and climate crisis. The Trump years had laid bare America's many vulnerabilities, not least the damage it could suffer at the hands of an ardently pro–fossil fuel, anti-science administration indifferent to the fate of the country's environment. If America is to reduce that vulnerability, it must understand not only what the Trump administration did and how it was able to achieve its goals but also what reforms are needed to avert a repeat of the Trump era or something even worse.

An equally destructive but savvier and more efficient administration, building on the Trump administration's playbook, could move more quickly to finalize far-reaching rules that would make government a servant of the oil and gas industry. Such an administration could capitalize on Trump's legacy: a weakened administrative state (i.e., the federal agencies tasked by Congress with overseeing the implementation of legislation); a strengthened conservative judicial philosophy further undermining the functioning of that state; a grave mistrust of government; and a widespread readiness to repudiate facts.

12.1 Lessons from the Trump Years

Over many decades, Republicans worked with Democrats to build economic opportunities while protecting public health and the environment, even though the two parties disagreed on how to strike the balance among competing goals.[7] A speech by George W. Bush in 1989 reflected that bipartisan philosophy: "A sound ecology and a strong economy are not mutually exclusive," declared Bush. "They go hand in hand."[8] Despite conservative values of limited government, Bush used the levers of government to impose restrictions on polluting economic activities in order to protect public health and the environment, noting that "we bear a sacred trust in our tenancy here and a covenant with those most precious to us – our children and theirs."[9] He worked with Congress to address air-quality problems, using a mixture of market-based and more traditional regulatory approaches under the 1990 Clean Air Act Amendments.

In the first decade and a half of the twenty-first century, a number of states had made significant strides toward renewable energy (as discussed in Chapter 3). These states – some headed by Republican governors, some by Democrats – recognized how the energy transition can secure affordable energy, build jobs and protect public health, the environment and the climate. A number of states pursued community-based solar and wind energy because, in part at least, they were committed to revitalizing communities through more equitable economic development strategies. They were not blind to the fact that making an energy transition from fossil fuels to renewables is a formidable task, but they were also not ignorant of the urgent need to make such a shift. Even legislators in oil- and gas-reliant states came to realize the need to transition quickly, having experienced the economically painful shale gas bust as early as 2012 in some regions, the shale oil bust in mid-2014 through 2016 and the further decline of the oil and gas sector in 2019 (Chapter 2).

When Trump took office in 2017, he dismissed the value of renewable energy as an energy source, despite renewable energy supporters such as Governor Sam Brownback of Kansas and Senator Chuck Grassley of Iowa pointing out that wind and solar were critical contributors to the economies of rural regions of Republican-voting states.[10] Instead, Trump's America First Energy Strategy insisted that promoting oil and gas extraction was the only way to drive America's economic progress. The Trump administration peddled the jobs-versus-the-environment message. As the preceding chapters of this book explain, the Trump administration capitalized on the deep-seated and long-running political, economic and legal support for the US oil and gas sector. It took far-ranging actions within just four years to open public lands and seas to oil and gas drilling and to slash regulations. Political appointees with ties to the oil and gas sector took charge of federal agencies and reoriented these agencies to serve oil and gas interests. The administration accelerated lease sales, locking the federal government into contractual agreements that will make future restoration and conservation efforts difficult and expensive.

Regulations in a market economy force polluters to incorporate the costs of pollution into their decision-making processes, rather than leaving them free to dump those costs onto the rest of society. The Trump administration misleadingly framed regulations as job-killing bureaucratic red tape. It manufactured doubt about evidence-based science and manipulated economic analyses to rescind specific regulations. It bulldozed or reshaped much of the regulatory landscape, enacting rules that blocked agencies from considering valid science and taking account of the economic benefits of regulations, thereby systematically understating the benefits from health and environmental regulations and the costs from oil and gas extraction and infrastructure. It interpreted statutes narrowly to strip federal agencies of their duties to protect America's air, water and land. It also interpreted presidential powers broadly in order to legitimize its removal of protections for public lands. Contrary to its rhetoric about states' rights, the administration offered disingenuous statutory interpretations to curb states' powers to protect their air and waters. Behind its limited government rhetoric, the administration wielded the power of the federal government in order to force the nonfossil-fuel private sector to serve the interests of the oil and gas sector; for instance, the administration issued rules to curb US banks' reluctance to fund Arctic drilling.[11]

For those Americans who want to return to the days of a bipartisan consensus that the economy and the environment are *both* important, and especially for those who believe the time has come for America to take bold steps toward renewable energy, the Trump years offer three important lessons. The first is that executive leadership has the power to inflict long-term damage when left largely unchecked by Congress (as discussed in Section 12.1.1). The second is that Congress can enable such damage not just by what actions it takes but also by what it fails to do. During the Trump era, and even before it, Congress failed to put in place legislation to protect America's air, waters, lands and seas and to strengthen governance of those resources (Section 12.1.2). The third lesson is that when ideology trumps science and government is cast by politicians as the enemy, then America is vulnerable to the influence of powerful actors such as the oil and gas sector (Section 12.1.3).

12.1.1 The Long-Term Damage Caused by the Trump Administration

While most of Trump's personal power fell away with his presidency, many aspects of his legacy will last long into the future. Most notably, perhaps, his attack on the administrative state will continue through his appointments of three Supreme Court justices, 54 court of appeals judges and 174 federal district court judges. Trump's judicial selections have exacerbated the tendency within the US judiciary, or at least among its conservative members,[i] to put personal freedom and private property rights above the responsibilities of government to protect individuals from more powerful economic actors. Faced with a choice between, on the one hand, defending Congress's powers to enact legislation to protect public health and the environment, as well as the administrative state's statutory powers to implement these protections, and, on the other hand, championing the rights of private entities that would be affected by such regulations, judges appointed by Trump are likely to choose the latter.

The Trump administration increased the longevity of a pro–fossil fuel and anti-regulatory ideology that is long overdue for retirement. Despite evidence to the contrary, many Republicans and some Democrats, particularly from oil- and gas-reliant regions, continue to see only a narrow path for America's energy future, one that will continue in much the same direction it has gone before, with fossil fuels providing affordable energy, secure jobs and a strong American economy. Trump's administration certainly favored the oil and gas sector; indeed, the administration did not tilt the playing field in that sector's direction so much as abandon any pretense of a level playing field. The administration's economic approach privatized profits and socialized costs, transferring the wealth in public lands and seas to private enterprises and shifting the health and safety risks from poorly regulated drilling to the rest of the economy. The administration accelerated fire sales of oil and gas leases on public lands and seas and cut royalty payments (purportedly to incentivize drilling) at a time when the world was awash in oil and gas. At the same time, it rolled back regulations governing drilling, including those put in place in response to the catastrophic BP oil spill that inflicted economic losses on so

---

[i] By "conservative" judges, I refer to those judges who tend to take a more skeptical view of the administrative state. I discuss how judicial philosophies can shape judges' decisions in Chapter 10.

many coastal communities, the very communities who will bear the costs of the next oil spill. Its false "job-killing regulations" narrative obscured the reality that regulations, by reducing the risk of catastrophic disasters, in fact, protect livelihoods that rely on healthy ecosystems. The Trump administration's deregulatory zeal even provoked concern among some oil and gas companies, which urged the administration not to eliminate regulations to curb methane leaks so as not to undermine their marketing of natural gas as a clean fuel.

Contrary to its free market rhetoric, the Trump administration wielded government power against the nonfossil-fuel private companies when they acted to protect their profitability (in the face of the climate crisis) in a way that reduced the profits of the oil industry. For instance, Trump's Department of Justice launched a probe into those auto manufacturers that agreed to abide by California's stricter fuel-efficiency standards, a prudent move for companies competing in global auto markets with increasingly strict standards (Chapter 11).

The Trump administration's readiness to jeopardize the long-term financial health of the US economy for the sake of the oil and gas sector even extended to directing funds from the March 2020 Coronavirus Aid, Relief, and Economic Security Act (CARES) to the bailout of oil and gas companies, including those that had performed poorly even before the pandemic hit in February 2020 (Chapter 2).[12–14] The Treasury's provision of $828 million of loans to the oil and gas sector and purchases of $475 million of corporate bonds from that sector are expected to result in significant losses for taxpayers.[15,16] Sadly, the administration's bailout strategy failed to deliver on its promises to protect workers and communities reliant on oil and gas, and instead enriched oil and gas executives.[14,17,18] At the same time, the administration rejected funding programs geared directly toward workers. For instance, it declined to support a request by 31 state governments to employ oil and gas workers to cap abandoned oil and gas wells.[19]

12.1.2 Congress's Role in Enabling a Destructive Administration

The conduct of the Trump administration demonstrates how any administration that prioritizes oil and gas interests can pursue policies that harm the US economy *if* Congress fails to fulfil its responsibilities as a steward of America's air, water, public lands and seas. Congress enabled the Trump administration as much by what it did not do as by what it did. The Trump administration was able to open the Atlantic, Pacific and Arctic coasts and large tracts of public lands to drilling because Congress let lapse its moratoria on drilling in these areas. Congress also made a choice not to designate conservation protections for these lands and seas. Had Congress enacted legislation on climate change and updated legislation governing oil and gas extraction onshore and offshore to keep up with advancements in extraction technologies, the Trump administration would not have been able to weaken as easily the regulatory restraints against oil and gas companies' polluting activities.

The Republican-controlled 115th Congress also actively abetted the Trump administration's actions. Congress, through a budget reconciliation process, opened the Arctic National Wildlife Refuge to oil and gas drilling (Chapter 7). (A handful of Republicans did voice their opposition to sacrificing this unique section of Alaskan Arctic coastline, which had not before been opened to drilling and which was now being sacrificed for oil reserves amounting to at most two years of US consumption.) Congress passed the Congressional Review Act to rescind 14 Obama-era regulations, including a Bureau of Land Management regulation that would improve public input into the management of public lands. Those rescissions were then signed into law by President Trump. The Trump administration's anti-regulatory agenda built on the work of Republican congressional representatives with a history of favoring the oil and gas industry. For instance, on January 5, 2021, Trump's EPA promulgated the deceptively named Science Transparency Rule, which replicated Representative Lamar Smith's equally deceptively named HONEST Act. Smith had worked on similar bills over the years before the HONEST Act passed the House in 2017 but not the Senate.

In a blatant case of putting the foxes in charge of the henhouses, the Trump administration appointed to federal agencies' top leadership and science advisory positions individuals who had served the oil and gas industry as lawyers, lobbyists and consultants or as analysts at think tanks bankrolled by the oil and gas industry. As detailed in Chapter 8, the Republican-controlled Senate, along with some Democrats from oil- and gas-producing states, chose to support these nominees, eschewing the norm of blocking appointees with real or apparent conflicts of interest. Think tanks such as the Heritage Foundation boasted of how their analysts filled various ranks of political appointments at agencies and how those agencies then quickly implemented the foundation's priorities.[20] Many of those priorities are in line with a deregulatory wish-list published in 2017 by the American Petroleum Institute.[21,22] Ed Feulner, a founder and former president of the Heritage Foundation, in endorsing the Trump administration's policies, noted that Trump "has an incredible advantage, even over Ronald Reagan. Because Ronald Reagan knew there were certain things government couldn't do."[20]

The majority of Republicans in Congress did not protest against the attack on science waged by federal agencies that were now in the grip of political appointees with ties to the oil and gas industry. At the helm of federal agencies' reviews of pollution standards were scientists with a long history of manufacturing doubt on scientific consensus, of amplifying scientific uncertainties and of holding fringe viewpoints that were rejected by epidemiologists (the central discipline in the study of pollution and public health). These approaches and fringe viewpoints commanded center stage in agencies' science reports, despite objections from career scientists at federal agencies and academic scientists. A number of influential Republicans on Capitol Hill stacked congressional hearings with scientists and analysts notorious for voicing fringe views.[23]

Notwithstanding the economic and environmental arguments for renewable energy, Congress continued to support massive subsidies and tax breaks for the oil and gas industry, which amounted to $4 billion per year,[24] while providing only limited financial support for the transition toward renewable energy. As detailed in Chapter 2, by 2019, before COVID-19 struck, Congress was well aware of the poor fiscal outlook for the oil and gas industry that was being widely reported by consulting firms and the media.[25] Many banks and investors had already shut the funding spigot for the shale sector. But Congress did not specify rules in its March 2020 economic rescue package that would have prohibited the Treasury from providing loans to or purchasing corporate bonds from oil and gas companies that had performed poorly prior to the pandemic.[12,13,26]

Congressional representatives were fully cognizant of the Trump administration's economic approach of privatizing profits and socializing risks, as well as, in the case of some of those representatives, their constituencies' opposition to that arrangement. For instance, the Trump administration's decision to open the Atlantic coast to drilling was opposed by many voters in the Republican-majority states of North Carolina, South Carolina, Georgia and Florida. Fears that offshore drilling might doom their election chances motivated congressional representatives to work to protect their individual states. The vast majority of Republicans in Congress declined to support legislation proposed in the House and in the Senate to block drilling expansion in the Atlantic coast that would have protected all states located on the Atlantic coast. Instead, they focused their efforts on petitioning Trump to exempt federal waters abutting their states from the drilling expansion, even though that approach still exposed their states' economies to spills from drilling in federal waters abutting other states.[27] Trump's speech granting the moratoria on offshore drilling for these five states two months before the 2020 elections underscored that the representatives had to rely on Trump's "benevolent" powers, rather than good economic sense, to spare their states: "If you want to have oil rigs out there, just let me know – we'll take it off. I can understand that, too."

Congress, not the president, has the ultimate authority to protect public health and the environment. However, in the absence of congressional self-assertion, a president can wreak havoc. With control of the House shifting from the Republicans to the Democrats in the 116th Congress, Democrat legislators, with support from very few Republicans, passed bills that could have reined in the Trump administration. However, without sufficient support from the Republicans who controlled the Senate, bills stalled. The bills that passed in the House included bans on new offshore oil and gas drilling off the Atlantic and Pacific coasts and Florida and in Alaska's Arctic National Wildlife Refuge. The House also passed a scientific integrity bill to shield government scientists. Other proposed bills would have protected public lands against drilling (e.g., the Bears Ears and the Grand Escalante national monuments on federal lands in the state of Utah); would have blocked deregulatory actions; and would have improved agencies' evidence-based decision-making (e.g., by requiring agencies to incorporate the costs of climate change in their decision-making process).

Why did these efforts fail, even with widespread public support? A large part of the answer lies in the fact that many congressional representatives, not to mention members of the Trump administration, have been far more attentive to the interests of the oil and gas sector than to voters' interests – a sad state of affairs, but one confirmed by numerous studies on the influence of donors in US elections.[28–30] For instance, one study found that members of Congress with a strong anti-environmental voting record in one election cycle receive more political contributions from oil and gas companies in the next election cycle.[31] Investigations by journalists into the passage of the CARES Act revealed how several members of Congress who receive substantial support from the oil and gas sector worked to reshape the rules of Federal Reserve programs to benefit the oil and gas sector (Chapter 2).[32–34]

Congress had and still has the authority to solve many of the issues described in this book. It can protect public and private lands, conserve the seas, correct injustices inflicted on Native Americans whose lands are coveted by the oil and gas industry and on low-income and minority communities that bear the brunt of pollution emitted by the oil and gas sector, assist fossil-fuel-reliant communities, mandate scientific integrity in scientific and economic analyses, and much more. However, Congress also has tremendous leeway to ignore these issues if the members' constituencies allow it.

12.1.3 The Republicans' Long Road from Environmental Stewards to Science Deniers

Congressional inaction on public health and environmental protection is rooted in decades of political dissonance within both the Republican and the

Democratic parties. In earlier decades, Republicans in Congress had championed conservation, public health, environmental protection, science-based policies and innovative pollution-reduction strategies such as "cap and trade." However, as detailed by James Morton Turner and Andrew Isenberg in *The Republican Reversal*, the voices of moderate Republicans who viewed stewardship of public lands and natural resources as part of the conservative philosophy and who believed science should always inform policymaking have been sidelined.[159] Scholars have tracked several aspects of this shift: the administrative state, a mechanism for implementing policy, came to be vilified as "big government" and attacked for infringing on personal freedoms and property rights;[35] staff at federal agencies, whose congressionally mandated tasks can only be performed efficiently if personnel are insulated from political interference, were cast as unaccountable, unelected bureaucrats;[36] and science linking pollution to public health and climate change became delegitimized[37] while fringe positions were fervently embraced.[38]

In the early 1970s, amid environmental catastrophes such as the Santa Barbara oil spill, President Richard Nixon responded to a growing popular demand for government action to combat pollution. He created the EPA and signed a number of landmark environmental acts that enjoyed bipartisan support, including the National Environmental Policy Act and the Clean Air Act. When Nixon vetoed the Clean Water Act, arguing the $24 billion bill was too expensive, Congress overrode that vote. The Senate voted 52–12 to override the veto, with 17 Republicans in favor; the House voted 247–23, with 96 Republicans in favor.

In the 1980s, however, the Republican Party began to embrace a different ideology. President Ronald Reagan portrayed "big government" as a danger to America and professed support for the "Sagebrush rebels" who opposed the federal government's ownership of public lands. He appointed an avowed opponent of public lands, James Watt, as secretary of the Interior and he installed the anti-regulatory Anne Gorsuch as administrator of the EPA. Reagan's administration implemented the anti-science playbook long before Trump – pushing misinformation in science; dismissing academic scientists from advisory boards; and elevating the views of the polluters over those of career scientists at regulatory agencies.[39,40] However, opposition from coalitions of moderate Republicans and Democrats forced Reagan to switch to rebuilding the EPA and Department of the Interior.[39,40] Reagan went on to sign legislation, which won the support of the Republican-dominated Senate, to protect 10 million acres of public lands as wilderness, placing him second only to Jimmy Carter in terms of the amount of wilderness protected by presidential action.[41]

In the late 1980s, Republican Senator Lincoln Chafee convened a number of hearings on environmental challenges. NASA scientist James Hansen's warning that humans had caused climate change that was already adversely affecting the United States motivated both Republicans and Democrats to commit to taking action. President George H. W. Bush vowed to use the "White House effect" to combat the "greenhouse effect." The House voted by voice-vote and the Senate voted unanimously to mandate federal agencies to conduct a series of quadrennial National Climate Assessments documenting climate change's impacts on the United States. However, by 1992 Bush no longer supported national targets for reductions in greenhouse gas emissions, contending that "the American way of life is not up for negotiation. Period."[42,43] Bush nevertheless worked with Congress to implement environmental protections, including the first ever cap-and-trade program that successfully combated acid rain.[44]

In the mid-1990s, Newt Gingrich became speaker of the House, and he led fellow Republicans on a crusade against "big government," rebranding government protection of public lands and seas and the environment as an intolerable infringement on Americans' personal freedom that paralyzed economic growth.[45] In 1995, the Republican-controlled Congress eliminated the science-based Office of Technology Assessment, portending the forthcoming victory of politics over science. The Clinton administration was able to push through some protective actions through executive orders, including the landmark executive order on environmental justice highlighting the plight of minorities and low-income communities, who face the highest level of exposure to pollution in the United States. However, on climate issues, the administration ran into a brick wall, with the Senate voting 95–0 to block the ratification of the Kyoto Protocol, the international agreement for developed countries to limit their greenhouse gas emissions.

During the presidency of George W. Bush, a majority in Congress, including many Democrats, supported the 2005 Energy Policy Act. That act exempted hydraulic fracturing from certain provisions of the Safe Drinking Water Act, benefiting, among others, Halliburton, a major oil and gas service company in which Vice President Cheney had served as CEO. Lawmakers accepted reduced regulatory oversight as a necessary trade-off to secure increased domestic production of oil and gas. Republican Sherwood Boehlert, chair of the House Science Committee, was willing to step out of the party line and voted against the Energy Policy bill, arguing that by emphasizing drilling over energy efficiency, the bill failed to address the energy crisis. The Bush administration made frequent use of the anti-science playbook.[39,40] For instance, Philip Cooney, a former lobbyist for the American Petroleum Institute who was working at the White House, altered an EPA draft report to downplay global warming risks,[46,47] and NASA's public affairs officials warned scientist James Hansen of dire consequences if he continued to speak out against global warming.[47,48]

As protection of the environment and the climate was gradually erased from the Republican Party platform, the oil and gas sector continued to enjoy support from both parties. The Obama administration pursued an "all of the above" energy strategy, continuing its support for oil and gas, albeit while also introducing regulations to reduce the industry's adverse impacts and providing some support for the development and use of renewable energy (Chapter 2).

The administration promoted America's shale expertise across the world, even in countries where local communities resisted shale operation.[49,50] It put a positive spin on the emerging evidence of the adverse public health impacts of shale operations and downplayed concerns raised in an EPA study (Chapter 2). It touted natural gas as a bridge fuel that would enable the electricity sector to shift from coal to reduce its greenhouse gas emissions and to subsequently shift to renewable sources of energy when they become cheaper. It did so even as scientific evidence was mounting about the scale of methane leaks from natural gas extraction, transportation and distribution.[51-53] (Cheap natural gas did assist the initial stages of the US transition away from coal, but the growing cost-competitiveness of renewable energy made it financially feasible to shift directly to renewable energy, as discussed in Chapter 3.)

The administration proposed to open drilling in the Atlantic coast but was forced to halt its plans when the BP oil spill paralyzed the economies of the Gulf Coast states, including their fishing and tourism industries. Democratic congressional representatives from these states resisted legislation to improve safety in offshore drilling even though the bipartisan investigation into the spill revealed deep-seated problems and warned that another devastating spill was simply a matter of time. These Democrats cited their states' reliance on the oil and gas industry, even as they acknowledged the double-edged sword of that continued dependency. In the face of Congress's inaction on legislation, the administration pushed forward with regulations to mitigate the public health and environmental impacts of the oil and gas sector (Chapters 4 and 7).

On the renewable energy front, the administration was able to chart some progress. It secured $90 billion in funds to support clean energy from the 2009 American Recovery and Reinvestment Act that had allocated $840 billion in new public spending to respond to the crash in financial markets.[54] Studies reveal that the public funding contributed to stimulating complementary private investments, creating clean energy jobs, and spurring clean energy innovations.[55-58]

By 2016, the marginalization of moderate voices within the Republican Party meant that denial of climate change had become a litmus test for Republican values. The long-running misinformation campaign conducted by think tanks funded by the oil and gas sector and conservative donors and amplified by pro-oil politicians had entrenched science denial.[59–61] A lone candidate for the 2016 Republican presidential nomination, former Utah governor Jon Huntsman, acknowledged the scientific fact that humans caused climate change.

By the time Trump took office, most Republicans in Congress were unlikely to object to an administration that routinely denied science, denigrated the work of civil servants and normalized misinformation. And, indeed, as the new administration made misinformation its stock in trade, from climate denial to COVID-19, the vast majority of Republicans in Congress either loudly supported the Trump administration's actions or stayed silent. Others, such as Trump's Supreme Court appointee Amy Comey Barrett, took the subtler science-denial strategy and claimed that they lacked the scientific expertise to judge whether human activity was causing climate change.[62]

12.2 America at a Crossroads

Today, pro–fossil fuel and anti-regulatory voices not only dominate the Republican Party and dictate its energy and environmental platform but also remain influential in the Democratic Party. Their views, however, are increasingly being challenged by reality in the form of evolving technology, shifting public opinion and changing economic calculations.

As discussed in Chapter 3, renewable energy has become more price competitive while technologies to support renewable energy, such as storage, continue to advance. The share of renewables in the US energy mix has increased without sacrificing energy affordability or grid reliability (Chapter 3). Americans' demand for clean air, clean water and unpolluted lands has grown stronger and louder. But economic activities that rely on healthy lands and oceans still vie with oil and gas extraction for scarce private lands, tribal lands and public lands and seas. At the same time, advances in scientific understanding have led to higher estimates of the public health, environmental and climate costs of oil and gas extraction and combustion.[63]

Trade-offs made in favor of the oil and gas sector are no longer warranted. It is folly for the executive branch, Congress and state governments to permit the oil and gas industry to externalize major public health and environmental costs and to impose them on society at large. Their rationale for allowing this – that sacrifices must be made to secure reliable supplies of oil and gas, which fuel America's economic growth – is no longer valid. To be sure, extractive activities played a key role in America's early economic growth. In response, common law evolved to prioritize the rights of mineral owners over surface owners (Chapter 2). State and federal legislation in many ways prioritized extractive activities over public health and the environment (Chapter 2). However, today, renewable energy offers affordable, price-competitive energy options with a smaller public health and environmental footprint. Indeed, as detailed in Chapter 3, states have made tremendous strides in shifting to renewable energy and generating jobs, innovation and revenue (even some states with significant oil and gas extraction, such as Colorado). Likewise, the nonfossil-fuel private sector has made prudent shifts in response to market and climate realities (Chapters 3 and 11).

12.2.1 The Twenty-First-Century Transition to Renewable Energy

Between February 2020 and December 2020, the clean energy sector lost about 400,000 jobs, about 12 percent of its pre-COVID employment.[64],[ii] One bright spot during the pandemic was the corporate sector's continued commitment to shift to renewable energy, thereby providing sustained demand for renewable energy projects.[65] The January 2021 report by the market research company S&P Global Market Intelligence forecast that the renewable energy sector would rebound in 2021 and continue to grow thereafter, especially in light of decisions by numerous state governments to accelerate the clean energy transition and falling costs for wind and solar.[66]

In its 2021 *Annual Energy Outlook* report, the US Energy Information Administration forecast (based on policies in place in 2020) that renewable energy would grow from 20 percent of US electricity generation in 2020 to 39 percent in 2035. In contrast, the shares from other sources are expected to decline.[67],[iii] It also projected that 60 percent of new capacity added between 2021 and 2050 would come from wind and solar, and the rest from natural gas. Administrator Linda Capuano had noted a year earlier, "We see renewables as the fastest-growing source of electricity generation through 2050 as cost

---

[ii] E2's report, *Clean Energy & COVID-19 Crisis*, defined the clean energy sector to include a variety of sectors: renewable energy, energy efficiency, clean/electric vehicles, grid and storage, and clean fuels [64].

[iii] The report projected that natural gas would decline to 34 percent (down from 40 percent in 2020), coal to 14 percent (down from 19 percent in 2020) and nuclear to 13 percent (down from 20 percent in 2020).

declines make them economically competitive beyond the expiration of existing federal and state policy supports."[68] The outlook has also been positive in the electric car sector. In January 2021, General Motors announced plans to shift to an all-electric fleet of cars and light trucks by 2025, while other US automakers doubled down on investments in electric vehicles.[69,70]

In contrast, the outlook for the oil and gas sector is uncertain at best and dismal at worst. In the first quarter of 2020, the top 100 US oil producers collectively wrote down $90 billion worth of assets; in the second quarter, the amount was $66 billion. Those write-downs amount to almost 9.2 percent and 6.5 percent, respectively, of the net present value of the companies' assets in the previous year.[71,72] Energy analysts, even those working for oil and gas companies, forecast declines in the global demand for oil within two decades, if not earlier.[73,iv] In November 2019, the International Energy Agency (IEA), whose reports influence the decisions of investors, energy companies and governments, published its prediction that global oil demand will plateau around 2030.[74] A year later, the agency's executive director, Fatih Birol, declared that "the era of global oil demand growth will come to an end in the next decade."[75] In May 2021, the IEA published its starkest warning yet on fossil fuel investments, emphasizing that investors would need to end investments into new coal, oil and gas projects in 2021 to mitigate the climate crisis.[76,77,v]

Oil and gas companies also face headwinds from lawsuits filed in courts across the world that seek to limit their greenhouse gas emissions or seek to hold them liable for climate damages.[78,79] In May 2021, the District Court of the Hague ruled that Royal Dutch Shell, headquartered in the Netherlands, must reduce the company's greenhouse gas emissions by 45 percent by 2030, compared to its 2019 levels and that Royal Dutch Shell has a responsibility to undertake its best efforts to reduce the emissions of its customers and suppliers.[80,81] While the ruling will no doubt be appealed, that decision may well be a harbinger of other similar lawsuits in Europe, where courts have ordered governments to take affirmative actions to protect the climate.[82,vi]

[iv] A number of projections expect oil demand to peak in the next two decades, before leveling or even declining. These include energy research company Rystad (by 2028), consulting firm McKinsey (by 2033), clean energy research group Bloomberg New Energy Finance (by 2035) and energy industry advisors Wood Mackenzie (by 2035). Analysis by oil and gas companies and organizations report similar projections, including the Norwegian energy company Equinor (by 2027–28), French oil major Total (by 2030) and OPEC (by 2040) [73].

[v] The IEA's acknowledgment of the necessity for a rapid transition away from oil and gas departs from its traditional support of the oil and gas sector and for only incremental changes in the global energy markets. Even as recently as 2016, the IEA took the position that "fossil fuels, in particular, natural gas and oil will continue to be the foundation of the world's energy system for decades to come" [76].

[vi] For instance, In December 2019, the Supreme Court of the Netherlands in *Urgenda Foundation v. Netherlands* ordered the Dutch government to do more to cut greenhouse gas emissions, i.e., to

Not surprisingly, prominent investors globally are abandoning the oil and gas sector. In March 2019, Norway's sovereign wealth fund, the largest sovereign wealth fund in the world, announced that it would divest from companies that explore for oil and gas but will still hold stakes in conglomerates such as BP and Shell that have renewable energy divisions. In January 2020, Larry Fink, CEO of BlackRock, which manages about $7 trillion out of the $80 trillion in global investments, announced that BlackRock would begin shifting away from oil and gas investments.[83] In December 2020, New York State's $226 billion pension fund announced its move away from the riskiest oil and gas companies by 2024 and its plans to decarbonize by 2040.[84] That same month, Lloyds, the world's largest insurance company, announced it would end stop-loss insurance coverage for oil sands projects and Arctic oil and gas projects by 2030.[85]

12.2.2 Biden's Clean Energy Transition and Environmental Justice Agenda

This premonitory shift away from fossil fuels by investors wary of economic losses and by governments committed to addressing the climate risks warns against the United States' continued reliance on oil and gas extraction as an economic development strategy. American policymakers who are prepared to heed these warning signs need to rebuild the economy toward lower carbon intensity, while acknowledging the past contributions and future needs of fossil fuel workers and communities and redressing the environmental injustice in communities that bore the brunt of pollution. To achieve all this will require not only reversing Trump's policies but also going beyond restoring the status quo ante Trump and charting a course for America's energy policy compatible with a low-carbon future.

Candidate Biden signaled a pivotal shift in the US energy trajectory by proposing the Clean Energy Transition and Environmental Justice Plan in July 2020.[86],[vii] Biden's plan posited that "the economy and the environment are completely and totally connected."[8] It envisions a clean energy revolution that will focus on job creation, build a more equitable economy and protect public health, the environment and the climate.

The plan makes the energy transition and correcting environmental injustices a central part of US economic recovery efforts and of the US energy

---

reduce its greenhouse gas emissions by at least 25 percent compared with 1990 levels by the end of 2020. In April 2021, the German Constitutional Court ordered the German government to tighten its climate targets to protect future generations.

[vii] Biden's policy focuses on "clean energy" as opposed to "renewable energy." The Biden plan foresees the use of existing nuclear plants to contribute to a carbon-free electricity grid.

strategy moving forward; 40 percent of the federal government's climate investments are meant to help "environmental justice communities" (i.e., low-income, minority and Native American communities who have borne the brunt of pollution and climate changes.[87,88]) The plan also provides bridge funding from the federal government to help local communities whose revenues have already been falling because of the decline of the coal sector and the shale bust. To pay for the energy transition, the plan envisions a series of tax reforms. (Biden's legislative proposal to implement the clean energy transition, released in March 2021, is discussed in Section 12.3.2.)

In a presidential debate in October 2020, candidate Biden said he "would transition away from the oil industry. . . . The oil industry pollutes, significantly. It has to be replaced by renewable energy over time."[89] Biden envisioned an end to the federal government granting new permits for extraction of oil and natural gas on public lands, where companies are already not producing on 50 percent of the public lands they have leased (Chapter 4). Biden did not commit to banning fracking.[viii]

Biden's plan to transition away from oil echoes the vision in the 2008 Democrat and Republican party platforms during the presidential contest between John McCain and Barack Obama.[90,91] The 2008 Republican platform noted that "by decreasing the long term demand for oil, we will be well positioned to address the challenge of climate change and continue our long-standing responsibility for stewardship over the environment ... Alternate power sources must enter the mainstream ... Technological developments ... will increase their economic viability."[92,ix]

Nevertheless, Biden's plan emphasizes the urgency for immediate action. It marks a clear shift from the Obama administration's strategy of expanding oil and gas extraction while promoting renewable energy but with overall government policy continuing to favor fossil fuels.[93] Obama supported the expansion of oil and gas, despite studies revealing that limiting global warming below 1.5 or 2°Celsius requires leaving substantial oil and gas reserves in the ground.[94,95] Obama framed the expansion of the clean energy sector as one other important source of jobs in the economy, not as a central strategy to restructure the economy. Nevertheless, his support for renewable energy and energy efficiency contributed to its declining costs

---

[viii] The federal government does not have the power to block fracking on private lands, where most shale oil and shale gas production takes place. However, the federal government can regulate that activity.

[ix] The platform was a compromise between Republicans who diverged in their views of the US energy strategy and on climate change [91]. While the platform urged the long-term shift from oil, it emphasized the near-term expansion of oil and gas extraction.

and thus paved the way for the Biden administration to embrace the clean energy transition.[54]

Biden's push for a shift to clean energy and away from oil extraction as his central strategy to rebuild the economy responds to the enormity of the COVID-19 economic crisis and the urgency of the climate crisis; reflects the strong pressure from a coalition of economic justice, environmental justice and climate movements; and takes advantage of cheap wind and solar. He framed the US economic recovery post-COVID-19 as building economic prosperity "from the bottom up and the middle out."[96]

12.3 Implementing the New Agenda

Implementation of the clean energy transition and environmental justice agenda will require long-term support from Congress and sustained action by the executive branch. As discussed in earlier chapters, the federal government, by shaping the rules of the economic system within which private actors operate, can incentivize economic actors to shift their activities and investments away from fossil fuel and into the clean energy sector. A number of strategies to support the transition, detailed by various studies and summarized in earlier chapters, have been adopted by the administration. Much can be accomplished during Biden's first term through federal agencies operating under existing federal laws (Section 12.3.1), but major legislative initiatives and the rebalancing of the courts with judges who are not skeptical of or opposed to the administrative state will require longer-term efforts (Sections 12.3.2 and 12.3.3). Assistance to local communities to restructure their economies away from reliance on fossil fuel extraction will necessitate even more persistent efforts across several administrations and congresses (Section 12.3.4).

12.3.1 Action by Federal Agencies under Existing Legislation

Under the Biden administration, leaders with experience in promoting renewable energy and protecting public health and the environment have taken the helm of federal agencies and have reorientated these agencies' actions to help support the clean energy transition. The transition from fossil fuels to renewable energy relies on private sector investments to flow into the renewable energy sector and out of the fossil fuel sector. Federal agencies have the ability to encourage this redirection of funds (Chapter 11). For instance, a September 2020 report by the Commodity Futures Trading Commission, an independent federal agency, noted that, under existing regulations, the Securities and Exchange Commission can require a vast number of publicly traded companies and other firms to disclose climate risks to investors[97,98,x] building on the commission's existing requirements for companies to disclose information pertinent to their environmental liabilities.[99] The report also noted that the Federal Reserve can use its authority to assess the performance of financial institutions under various climate scenarios, just as it assesses their performance under various economic scenarios.[97]

Another way that federal agencies can influence the flow of private investments into the oil and gas sector is through their decisions to approve or reject applications for federal permits (e.g., for oil and gas extraction on public lands and for the construction of interstate gas pipelines). The Biden administration has reinstated decision-making processes at federal agencies to appropriately account for the costs of climate damages from greenhouse gas emissions.[100,xi] It set the "social cost of carbon" (i.e., the metric for the monetary cost of climate damages from one ton of carbon dioxide) at the value established by the Interagency Working Group on the Social Costs of Carbon during the Obama administration and established a process to update the metric based on the latest scientific evidence (Chapters 9 and 11). (Predictably, the Trump administration had adopted a value that understated climate costs.)

Meanwhile, the EPA, the Bureau of Land Management and the Bureau of Ocean and Energy Management have begun steps to issue regulations on oil and gas extraction, using their authority under existing laws (Chapters 4 and 7). These include regulations to curb methane leaks from oil and gas operations and regulations on oil and gas operations on public lands and offshore.[101,102] The EPA has also begun work to repeal several rules introduced by the Trump administration that had systematically understated adverse public health and environmental impacts (Chapters 8 and 9).[xii] Court rulings that have vacated some of the controversial Trump-era rules and remanded them to the EPA have made it easier for the agency to replace these rules or to repeal them altogether.[103,xiii] The Department of

---

[x] These include direct risks (e.g., offshore oil rigs may be damaged by intense hurricanes and require expensive remediation) and indirect risks (e.g., an oil extraction project may become unprofitable because of the added costs from government's adoption of climate policies such as carbon taxes).

[xi] In 2007, the Court of Appeals in the Ninth Circuit in *Center for Biological Diversity v. National Highway Traffic and Safety Administration* ruled that the federal government must account for the costs and benefits from changes in greenhouse gas emissions in its rulemaking (Chapter 9).

[xii] The Biden administration has not always asked courts to vacate Trump-era rules. For instance, in April 2021, it asked the District Court for the Western District of Virginia not to vacate the Trump era rule pertaining to the National Environmental Policy Act (Chapter 4) while the Biden administration was reviewing that rule.

[xiii] For instance, in January 2021, the Court of Appeals for the DC Circuit vacated the Trump-era Affordable Clean Energy Rule (Chapter 3) in *American Lung Association v. EPA* and remanded

the Interior has paused federal leasing on public lands and seas, in order to undertake a review of these programs so that future decisions on leasing properly account for the benefits from conservation activities versus extractive activities (Chapters 4 and 6).[xiv]

The Biden administration has also directed the federal government, the largest energy consumer in the United States,[104] to purchase renewable energy.[101] Those purchases, by enhancing demand, incentivize investments into renewable energy generation (Chapter 11). Several federal agencies, such as the Department of Defense, have the authorization to sign contracts beyond the 10-year limit set by Congress,[105] and longer-term contracts accord with the business model of renewable energy suppliers. Such procurement agreements enable federal agencies to pay competitive prices for their electricity, thereby saving money for taxpayers while increasing the proportion of the nation's energy consumption that is from renewable sources.[104]

The Biden administration has begun steps to promote the use of offshore wind in federal waters along the New England and mid-Atlantic coasts and thus support the commitments of several state governments in those regions to purchase electricity generated from offshore wind.[106] The administration has also begun to work with states to facilitate the construction of transmission lines along public roads and railways, in part so as to link regions with high wind and solar resources in the Midwest and Southwest to population centers on the East and West coasts (Chapter 3).[107]

12.3.2 Working with Congress on Major Legislation

In March 2021, Biden announced in Pittsburgh the first detailed legislative proposal to implement his clean energy transition and environmental justice vision.[xv] The American Jobs Plan proposes to spend $2.25 trillion (i.e., 1 percent of US GDP), beginning in 2022, over eight years, on clean energy and

---

it back to the EPA. In February 2021, the District Court for the District of Montana in *Environmental Defense Fund v. EPA* vacated the Science Transparency Rule (Chapter 8) and remanded it back to the EPA.

[xiv] The Biden administration's actions have not all been in favor of reining in leasing on public lands. In May 2021, the Department of Justice defended the Trump administration's 2020 Record of Decision for the Willow Project in the National Petroleum Reserve-Alaska, despite environmental groups' criticisms of the Trump's administration's environmental review of that project. Nevertheless, that same month, the Biden administration took a different action in another part of Alaska. It suspended oil and gas leases in the Arctic National Wildlife Refuge that the Trump administration had sold in January 2021 (Chapter 6).

[xv] Biden's first major legislation, the American Rescue Plan enacted in March 2021, provided some funding for infrastructure, including $30 billion for mass transit systems (whose ridership fell during the pandemic) and funds for state and local governments to invest in water and sewage systems.

other infrastructure investments that, Biden argues, can transform the energy system, modernize infrastructure, enhance the competitiveness of industries and spur the growth of "good paying and union jobs."[108–110] It also provides $400 billion in tax credits to the clean energy sector. While the plan's fate was undecided at the time of writing (May 2021), it demonstrates the multifaceted strategies needed to accelerate the clean energy transition.[111]

The plan sets out a clean electricity standard. By 2035, it envisages a shift to carbon-free electricity generation, inclusive of solar, wind, nuclear and hydro, with natural gas providing a small share of the generation.[xvi] It proposes government funding to build out the transmission lines, tax credits to incentivize wind, solar, storage and other renewable energy projects and government funding for RD&D to spur energy innovations.[110]

The plan frames the clean energy transition as a job creation strategy, growing jobs in the clean energy sector (e.g., in the construction or installation of solar, wind and storage projects and in building transmission lines); in the manufacturing sector (e.g., in electric vehicles and along the manufacturing supply chain to support the clean energy and electric vehicle sectors); and in the housing and building sector (e.g., in weatherizing buildings for energy efficiency). The plan invests in workforce development and in RD&D in the clean energy and manufacturing sectors. It provides funds to employ workers to plug abandoned oil and gas wells and to remediate abandoned coal and uranium mines.

To pay for these investments, the plan proposes a series of tax reforms to improve overall corporate tax collection and redirects incentives away from the oil and gas sector. It ends various subsidies and tax preferences enjoyed by the fossil fuel sector that are estimated to cost the Treasury $4 billion per year[24] and removes tax loopholes for larger corporations.[112–114] It raises the corporate tax rate from 21 percent to 28 percent (but this is still below the 35 percent prior to the 2017 Trump tax cut).

The goal of job creation through investment in renewable energy is certainly achievable but it may not be entirely straightforward to accomplish, given the experience with similar investment (at a far smaller scale) in the 2009 American Reinvestment and Recovery Act (Chapter 11).[58,115] First, renewable energy and remediation, which require project planning, will be slower to yield jobs than other infrastructure investments that are more "shovel ready." Second, while there is some overlap in skills between a subset of jobs in the

---

[xvi] Expanding renewable energy to provide 75–90 percent of electricity needs by 2035, with existing nuclear and natural gas generation supplying the remainder, is financially and technically feasible, according to studies from the National Academies of Sciences, Engineering, and Medicine and the University of California at Berkeley (Chapter 3).

clean energy sector and a subset of jobs in the fossil fuel sector, significant training is needed to prepare workers for the clean energy sector and to boost their productivity. Close collaboration between training programs and businesses is necessary to prepare workers for the rapidly evolving labor markets. Third, jobs created through the clean energy transition may not be located in the same places where people have been employed in the oil and gas industry (Chapters 3 and 11). Hence, at least some of the employment programs that target fossil fuel workers should put them to work in their home communities (e.g., capping oil and gas wells and remediating abandoned mines) so as to maintain economic activity in these communities and support the efforts to diversify the economy by returning scarred lands to productive use.

An important legislative step on the path to a clean energy transition is to establish a carbon tax. A carbon tax would provide a price signal for economic activities and investments to move out of carbon-intensive sectors (Chapter 11). Such a tax can be designed with complementary policies to ensure that low-income households are not faced with high energy costs. The administration will need to negotiate for a carbon tax that more closely reflects the full "social cost of carbon" without surrendering to demands from oil and gas companies for waivers against liabilities for climate damages and for the suspension of greenhouse gas regulations in exchange for a carbon tax that is far below the social cost of carbon (Chapter 11).[xvii]

The Biden administration will also need to work with legislators to protect local communities. The administration can build on legislative initiatives taken in the 116th Congress (Chapters 4 and 7). For instance, reinstating congressional moratoria on drilling in the Atlantic and the Pacific coasts and extending the moratorium on drilling in the eastern Gulf of Mexico beyond 2022 will help protect the livelihoods of communities that rely on healthy oceans. The administration may be able to win the support of some Republicans in Congress by drawing attention to the opposition to drilling in Republican-leaning states such as North Carolina, South Carolina, Georgia and Florida (Chapter 6).

The administration will also need to guard against the danger that its efforts to protect public health and the environment will not be unraveled by a future pro-oil, anti-regulatory administration. Steps to govern oil and gas extraction (e.g., to curb methane emissions and to improve safety measures in drilling on public lands and offshore) will be more durable if they are enacted into law (Chapters 4 and 7) rather than left as regulations. Likewise, new legislation should specify the analytical procedures agencies must use, thereby making it

---

[xvii] Exxon, BP and Shell support a $40 per ton carbon tax with preconditions of retrenchment of regulations and waiver for climate liabilities.

harder for future administrations to emulate the Trump administration and jettison scientific integrity in regulatory and permitting decisions (Chapters 8 and 9). For instance, legislation should be introduced that requires federal agencies to apply the social cost of carbon and that sets out a transparent and evidence-based process for updating that cost, so as to ensure that agencies consider climate costs in their decisions and shepherd activities away from carbon-intensive sectors, such as oil and gas extraction.

12.3.3 Judicial Nominations to the Federal Courts

Biden is not likely to be able to significantly change the composition of the courts in the near future. As of March 2021, Biden has fewer vacancies to fill than Trump encountered, with only seven vacancies in the courts of appeals, 61 in the district courts and none in the Supreme Court.[116,117] Trump entered office with one vacancy in the Supreme Court, as a result of Senate Majority leader McConnell's blocking of Obama's nominee Merrick Garland, 19 vacancies in the courts of appeals and 97 in the district courts.[116] In March 2021, Russell Wheeler of the Brookings Institution analyzed the 13 circuits of the courts of appeals and concluded that Biden's nominations are not likely to tip the balance in the fifth through the eighth circuits from a majority of Republican appointees to a majority of Democratic appointees. Biden's nominations could potentially tip the balance in the second circuit, which has a narrow Republican-appointed majority and in the tenth circuit, which has equal numbers of Republican and Democrat appointees.[116,xviii] (See Section 12.4.2 for a fuller discussion of the Trump-appointed judges' decisions that lean against environmental protection.)

Congress, of course, has the power to enact new legislation that can overcome the hurdles of relying on existing laws to address complex issues such as the climate crisis and guard against narrow interpretations of existing laws imposed by the Supreme Court. However, in dealing with complex public health and environmental issues, Congress often has to give discretion to federal agencies to tailor regulations to particular circumstances and deal with ambiguities. To prevent these laws from being struck down by conservative justices on the grounds that they are instances of an unconstitutional congressional delegation of powers, Congress will need to articulate laws that clearly lay out the powers and duties of federal agencies (Chapter 10).

[xviii] Democrat-appointed judges are in a majority in the federal, district of Columbia, first, second, fourth and ninth circuits of the courts of appeals.

The executive branch's promulgation of regulations will need to pay close attention to the language of the statutes that empower agencies' actions. Conservative judges are apt to interpret statutes narrowly, granting only limited powers to federal agencies and giving them very little discretion to read between the lines (Chapter 10). Moreover, the level of specificity conservative judges require of statutes (including those enacted decades before scientists came to understand processes such as climate change) gives agencies limited room in which to maneuver. Nevertheless, the success of the Obama administration's regulations in withstanding legal challenges (because they were anchored within agencies' powers as laid down in federal environmental laws and because they abided by rulemaking procedures) provide some reason for optimism.

12.3.4 The Long-Term Process of Helping Fossil Fuel Workers and Their Communities

The transformation of the US economy from one anchored in fossil fuels to one reliant on clean energy will require long and sustained action by Congress and the executive branch to help fossil-fuel-reliant workers and communities diversify their local economies.[118–120] Congress will need to enact laws to give these workers and communities the funds to navigate the energy transition. Federal agencies will have to work collaboratively with state and local governments, grassroots organizations and unions (Chapters 3 and 11).

The wave of closures of coal mines and coal power plants and the contraction of the shale industry have left many communities struggling with high levels of unemployment and reduced revenues to local governments that finance services such as K-12 schools.[118–120] Coal miners have lost not only their jobs but also pension benefits, while their communities are dotted with abandoned mines that contaminate land and water.[121–123] Biden's establishment of an Interagency Working Group on Coal and Power Plant Communities and Economic Revitalization is a first step in charting the long-term collaborative strategies to rebuild and diversify these coal-reliant economies.[124]

As difficult as the coal transition has been, the transition from oil and gas that is on the horizon will be far more challenging given the larger number of communities reliant on that sector. The long-term planning for that transition will need to begin sooner rather than later if oil- and gas-reliant communities are to avoid the same fate as coal-mining communities.[118,119,125,xix]

[xix] Losses of retirement benefits and the costs for reclamation of mines were estimated at $2.3 billion and $3 billion, respectively [118].

Congress will need to examine and mitigate oil and gas companies' use of corporate reorganization and bankruptcy proceedings to shed their financial and environmental liabilities.[121,125] Coal companies' use of these strategies in the waning days of their operations warn of potential copycat actions by oil and gas companies. Coal companies pushed their obligations under federal laws to remediate lands and to pay for healthcare and retirement benefits for miners to the backend of their operations. Several coal companies shielded themselves from these liabilities by creating underfunded subsidiaries to assume the regulatory liabilities; these subsidiaries were later sold off and many of these went into bankruptcy and did not pay off the liabilities.[121] From 2012 through 2017, four major coal companies used chapter-11 bankruptcies to discharge $5.2 billion of their liabilities to provide environmental and retirement benefits.[121,xx]

Congress will also need to enact laws that ensure companies set aside funds upfront that cover the full cost of remediation of oil and gas wells on federal lands and offshore.[126] (State governments will need to do the same for oil and gas wells under their jurisdiction.) As of 2016, between 2.1 million and 2.4 million abandoned wells in the United States had not been capped.[127,128,xxi] The cost of plugging and remediating a mere 56,000 of those wells is estimated to range from $1.4 billion to $2.7 billion.[128] The number of these wells will increase because presently producing wells will dry up and will need to be capped and because shale companies, driven by the need to pay their debtors, will continue to drill new wells (Chapter 2). Mitigating this problem requires new laws that require companies to assign funds upfront to regulatory bodies, prior to drilling new wells, that pay for the entire costs of capping those wells when they are no longer productive.

Paying for the efforts to rebuild these communities will be a challenge. Among the proposals put forward for raising the required revenue is one for Congress to enact a carbon tax at the value of the "social cost of carbon." Such a tax could achieve the twin goals of signaling to economic actors the advantages of shifting away from oil and gas and of raising revenue, a segment of which could assist oil- and gas-reliant communities. Congress will also need to revisit the system of public finance that currently requires local communities to shoulder a significant share of the costs of providing local services – a burden

---

[xx] A chapter-11 bankruptcy proceeding allows a company to restructure its debts and remain in operation.

[xxi] The federal government and state governments have required companies to post performance bonds prior to drilling wells, but they have set these bonds at values far below the cost of capping the wells. Companies can thus choose the cheaper option of forgoing their bonds and not cap the wells.

that may be unbearable for communities facing economic disruption. Transfers from the federal government and from state governments could help communities in maintaining high-quality public schools, community colleges, apprentice opportunities and entrepreneurship training, as well as other forms of social support.

12.4 Potential Roadblocks in the Path of the Energy Transition

Accomplishing many of the tasks listed above will require concerted government action that extends well beyond the first (and any second) term of the Biden administration. The length of the path ahead means that there will be many opportunities for opponents of the energy transition to try to block or derail its progress. Republican state attorneys general and the oil and gas sector are likely to oppose the administration's decisions and regulations. Indeed, state attorneys general filed lawsuits against the administration as early as March 2021.[129,xxii] The most formidable opponents are likely to be found within Congress (Section 12.4.1) and in the courts (Section 12.4.2). The obstacles they can put in America's path away from its reliance on oil and gas should not be underestimated – but neither should they be exaggerated, for they are not insurmountable.

12.4.1 Congressional Opposition

A number of examples from the first months of the Biden administration illustrate the kinds of challenges Biden is likely to face throughout his term from both Democrats and Republicans in advancing his clean energy agenda.

Within the Democratic Party, there are tensions between those who advocate for a rapid restructuring of the economy toward clean energy and those who are apprehensive about the impact of the shift on oil- and gas-reliant communities. Some Democrats from states that depend on fossil fuel revenue have been willing to support the overall energy transition agenda and requested for accommodations to be made for their states during the transition; other

---

[xxii] Hours after Biden's inauguration, Texas's attorney general, Republican Ken Paxton, vowed, "I promise my fellow Texans and Americans that I will fight against the many unconstitutional and illegal actions that the new administration will take" [129]. In March 2021, 14 states sued the Department of the Interior for halting the approval of new oil and gas leases on public lands. Also in March 2021, Missouri with 11 other states filed a lawsuit challenging the Biden administration's use of the social cost of carbon. In April 2021, Louisiana with nine other states filed a similar lawsuit.

Democrats, however, have opposed restraints on the natural gas sector.[130,xxiii] Programs that help fossil fuel workers find secure jobs and enable local communities to rebuild their economies are essential to assuage Democrats' anxieties.

Several Republicans have cited economic reasons for opposing government investments in the clean energy transition, claiming that a growing national debt, corporate tax hikes and reductions in corporate subsidies will cause the economy to contract. These claims. however, are contentious. On the issue of government investments, economists, both progressive and conservative-leaning, support government spending and investments that can jumpstart the economy following the pandemic.[131–134] Running up the national debt in the short run to undertake productive investments to set workers and US companies on a trajectory of greater productivity can raise the national income and pay off the national debt in the longer run.[111,135] On the issue of corporate subsidies, fiscal conservatists would support drawing down taxpayer-funded subsidies to the oil and gas sectors that had performed poorly prior to the pandemic. Reversing, even partially, the Trump corporate tax cuts – which are estimated to raise the national debt by \$1.9 trillion within a decade[136,137] – would dampen the adverse impacts from those tax cuts. The position of Republicans in Congress departs from that of the majority of voters, including many Republicans, who support government investments into clean energy funded by corporate tax hikes.[138,xxiv]

Other tactics used by some Republicans depart from the democratic norm of debating competing economic viewpoints. They have labelled the clean energy transition as a "far-left" agenda, turning the nonpartisan issue of renewable energy, which has been championed by Republicans and Democrats (Chapters 3 and 11), into a partisan one. Some have resorted to fear-mongering that Biden will increase taxes on households making less than \$400,000 to finance government spending. Still others have amplified bizarre claims, such as the notion that Biden's climate protection goals will curb Americans' meat consumption.[139]

[xxiii] For instance, senators Michelle Lujan Grisham and Martin Heinrich from New Mexico proposed that the Biden administration permit oil and gas extraction on public lands situated in New Mexico to continue, in recognition of the state's overall efforts to reduce greenhouse gas emissions across its economy [130]. Congressional members from natural-gas-reliant states, such as Representative Conor Lamb from Pennsylvania, have strongly supported the expansion of the natural gas sector.

[xxiv] An April 2021 poll of some 2,000 registered voters found that 65 percent of respondents supported the American Jobs Plan funded with the corporate tax hike, with 21 percent opposed. Republicans were split, with 42 percent supporting and 47 percent opposing. A majority of independents, 60 percent, and an overwhelming majority of Democrats, 85 percent, were supportive [138].

Sadly, as the retired Republican politician John Boehner observes in his book *On the House: A Washington Memoir*, by the time he became speaker in 2011, Republicans who were willing to work with Democrats on legislation had become a rarity in Congress.[38,140,xxv] Until then, many Republicans who were committed to both economic progress and environmental protection had worked with Democrats in pursuit of those goals. Were such Republicans to reemerge as a force in their party, they might push open the doors to allow a return to bipartisan efforts.[xxvi] Until that happens, however, Democrats will have to rely on themselves to push forward with legislative action (related to federal tax and appropriations policy), using the budget reconciliation process that requires majority support of the House and a simple majority (as opposed to 60 votes) in the Senate.[141,xxvii]

12.4.2 The Courts' Skepticism of the Administrative State

Perhaps the longest legacy of the Trump administration will be its pattern and profusion of judicial appointments, which will mean that any future legislation and regulations to usher in an energy transition will need to maneuver around the Supreme Court's current conservative majority, as well as around judges in the appellate and district courts who take a skeptical view of the administrative state (Chapter 10).

Trump appointees are likely to continue to shape the jurisprudence in the courts, leaning toward deregulation and thus favoring polluters, such as the oil and gas industry, over the renewable energy industry. Conservative judges' skepticism of the administrative state, combined with their tendency to focus on the text of a statute and to ignore its purpose, are likely to keep the space for environmental protection narrow (Chapter 10).

[xxv] Political scientists Norm Ornstein and Thomas Mann concur with Boehner's assessment. In their book, *It's Even Worse Than It Looks*, they write that by the 2012 presidential election, the majority of Republicans in Congress had embraced partisanship, rejected compromise and denied scientific evidence.

[xxvi] During the Trump era, there were a few cases of bipartisan legislation to support clean energy and environmental protection. These include legislation in 2019 to protect public lands (cosponsored by Republican Senator Lisa Murkowski and Democrat Senator Maria Cantwell) and legislation in 2020 to provide funds for clean energy investments (cosponsored by Murkowski and Democrat Senator Joe Manchin).

[xxvii] Democrats used the budget reconciliation process to enact the March 2021 American Rescue Plan, which funded the vaccination and healthcare responses to COVID-19 and provided assistance for households, communities and small businesses. Congressional Republicans subsequently touted the delivery of these benefits to their constituents as their accomplishments. Republicans, it may be noted, used the budget reconciliation process to enact Trump's 2017 tax cuts [137].

Justice Neil Gorsuch, nominated to the Supreme Court by Trump in January 2017, offered a narrow interpretation of the Clean Water Act's authority in *County of Maui v. Hawaii Wildlife Fund*.[142] He interpreted the act's requirement for permits when pollutants are discharged directly into navigable waters to mean that permits are not required when pollutants travel even a short distance through groundwater before being released into navigable waters. As Justice Stephen Breyer, a Clinton appointee, explained in his majority decision in the case,[xxviii] such a literal reading of the statute, without regard to the purpose of the act to protect water resources, creates a loophole that defeats the protective purposes of the statute.

Before Brett Kavanaugh was nominated, in July 2018, by Trump to the Supreme Court, he served on the Court of Appeals for the DC Circuit. In that capacity, in 2017 he opined a narrow interpretation of the EPA's powers under the Clean Air Act. At issue in *Mexichem Fluor v. EPA* was EPA's powers under a provision of the act to require companies to replace ozone-depleting substances with safe substitutes that "reduce overall risks to human health and the environment."[143] Judge Kavanaugh insisted that the word "replace" means a one-time substitution, and thus the EPA cannot require a company to switch twice, even when the initial substitute was found to be unsafe. Judge Robert Wilkins, an Obama appointee, disagreed, reasoning that the statute, read in its entirety, requires an *ongoing* process of assessing safe substitutes and thus permits the EPA to require companies to make that switch.

Judge Neomi Rao's dissent in *Natural Resources Defense Council v. Wheeler*,[144] also in the Court of Appeals for the DC Circuit, illustrates how judicial decisions can open the door to federal agencies' unchecked deregulatory actions. In 2018, the Trump-era EPA issued a rule that eliminated an Obama-era ban on the use of hydrofluorocarbons, a potent greenhouse gas. Rao – a Trump appointee – backed the EPA's dubious claim that it was only interpreting the 2017 *Mexichem* decision in issuing that rule. Chief Judge Sri Srinivasan, an Obama appointee, countered that the EPA was engaged in rulemaking in issuing a rule that went beyond the *Mexichem* decision and that the EPA must abide by the notice and comment procedure in rulemaking, (i.e., it must request public comments and respond to those comments).

Among the most important cases that will come before the courts, including the Supreme Court, during Biden's administration are climate-related cases that will affect the financial health of the oil and gas sector.[78] As of January 2021, 19 lawsuits had been filed by local and state governments across the United States

---

[xxviii] Justices Roberts and Kavanaugh sided with Justices Breyer, Ginsburg, Kagan and Sotomayor in that 6–3 decision.

against oil and gas companies, arguing that the companies' products caused adverse climate impacts and inflicted damages on local and state governments through flooding, intense storms and a rise in sea level.[145] If these cases reach the discovery phase, oil and gas companies will be forced to hand over documents to plaintiffs that may reveal the extent to which they knew about the harmful effects of their products and whether or not they misinformed the public. Such revelations might erode investors' willingness to fund oil and gas companies.

Justice Amy Coney Barrett's statements on climate change during her confirmation hearing have raised concerns about her approach to adjudicating climate-related cases. Barrett's statement that "I don't think that my views on global warming or climate change are relevant to the job I would do as a judge" is alarming given that judicial decisions rest on adjudicating both the facts and the law.[62,146] Barrett did not respond affirmatively when asked if she believed "humans cause global warming" and instead characterized global warming as "a very contentious matter of public debate" and "a matter of public policy ... that is politically controversial."

12.5 Reflections from Southwestern Pennsylvania

As I reflect on the long-term challenge of the US energy transition, the tug of war between two competing visions of energy and economic development in my home region of southwestern Pennsylvania echoes the contest in the United States as a whole. On the one side, many in the region continue to support the shale gas industry, which they regard as the economic backbone of the region, despite its financial woes (Chapter 2).[147,148] Pennsylvania's state legislature doubled down in supporting petrochemicals in the region with $25 million in tax credits for Shell's petrochemical complex that employs 600 people permanently and with the allocation of $670 million more in tax credits to entice other plants to locate in the state.[149] Reporting by the *Pittsburgh Post-Gazette* detailed how government and university officials and business groups worked together to woo ExxonMobil to set up a petrochemical plant in the region.[147,148,150]

On the other side, other voices in the region warn of the economic folly of continued reliance on natural gas extraction and petrochemicals, noting the limited jobs (both past and future) and poor prospects of these industries.[151,152] A coalition of grassroots organizations put forward their "Reimagine Appalachia" vision of diversifying the regional economy and shifting to renewable energy and incentivizing renewables-related manufacturing.[153]

The knowledge economy, anchored by higher-learning institutions in Pittsburgh,[154] can shore up demand for sustainable agriculture and the recreation and amenities sector, the latter benefitting from the remediation of the region's ecosystems.[153] Republicans from rural Pennsylvania, recognizing the economic opportunities that renewable energy presents, have sponsored bills that would authorize the operation of small-scale community solar projects that sell electricity to subscribers within close proximity, thus enhancing the opportunities for farmers to lease their land for these projects.[155,156]

With the shale bust in Pennsylvania in 2014, jobs and revenue to mineral owners and local government dried up, leaving lands pockmarked by an estimated 200,000 abandoned gas wells (from the shale bust and an earlier extraction era) that can contaminate water and land.[157] A study of the counties in the Marcellus Shale between 2008 and 2019 found little job growth in counties that were the major producers of natural gas.[151] Evidence continues to mount on the public health impacts of the shale industry, such as a state grand jury investigation in mid-2020 that faulted the state's Department of Environmental Protection for being too deferential to the shale industry.[158] The grand jury report criticized the conflicts of interests that arise when government officials join the industry as soon as they leave public office.[158] In terms of the costs versus the benefits for the average Pennsylvanian, drilling deeper into shale gas seems to have been a losing bet.

America's energy gamble to deepen its reliance on oil and gas has endangered the American people, the American economy and the planet as a whole. Pursuing the energy transition, despite the many obstacles confronting it, is the prudent way forward.

References

1. K. Dozier and V. Bergengruen. "Incited by the President, Trump Supporters Violently Storm the Capitol Time." *Time*, January 6, 2021 https://time.com/5926883/trump-supporters-storm-capitol.
2. D. McManus. "Opinion Column: Ted Cruz and Josh Hawley Are Torching the Constitution to Further Their Own Presidential Ambitions." *Los Angeles Times*, January 6, 2020. www.latimes.com/politics/story/2021-01-06/column-ted-cruz-josh-hawley-electoral-vote-challenge-constitution.
3. M. DeBonis and P. Kane. "House Hands Trump a Second Impeachment, This Time with GOP Support." *Washington Post*, January 14, 2021.
4. T. Snyder. "The American Abyss." *New York Times*, January 9, 2021.
5. M. Scherer. "Riot Elevates Long-Festering Republican Power Struggle." *Washington Post*, January 11, 2021.

6. Lardner, R. and M.R. Smith. "Records: Trump Allies behind Rally That Ignited Capitol Riot." *Associated Press News*, January 17, 2021. https://apnews.com/article/election-2020-donald-trump-capitol-siege-campaigns-elections-d14c78d53b3a212658223252fec87e99.
7. J. S. Hacker and P. Pierson. *American Amnesia: How the War on Government Led Us to Forget What Made America Prosper*. New York: Simon and Schuster, 2016.
8. F. Krupp. "George H. W. Bush, Environmental Hero: He Exemplified the Real Art of the Deal." *New York Daily News*, December 3, 2018.
9. G. Bush, 41st President of the United States. *Remarks to the Intergovernmental Panel on Climate Change*. Submitted to Intergovernmental Panel on Climate Change. February 5, 1990.
10. Governors' Wind & Solar Energy Coalition. *Letter Regarding Solar and Wind Energy in the US*. Submitted to President D. Trump. February 13, 2017.
11. *Fair Access to Financial Services: Final Rule*. 12 Code of Federal Regulations 55. Office of Comptroller of the Currency, Department of the Treasury. RIN 1557-AF05 (April 1, 2021).
12. J. A. Dlouhy. "'Stealth Bailout' Shovels Millions of Dollars to Oil Companies." *Bloomberg*, May 15, 2020.
13. S. M. Rosenthal and A. Boddupalli. "Heads I Win, Tails I Win Too: Winners from the Tax Relief for Losses in the CARES Act." *TaxVox*, April 20, 2020. www.taxpolicycenter.org/taxvox/heads-i-win-tails-i-win-too-winners-tax-relief-losses-cares-act.
14. P. Whoriskey, D. MacMillan and J. O'Connell. "'Doomed to Fail': Why a $4 Trillion Bailout Couldn't Revive the American Economy." *Washington Post*, October 5, 2020.
15. Bailout Watch, Friends of the Earth and Public Citizen. *Big Oil's $100 Billion Bender: How the US Government Provided a Safety Net for the Flagging Fossil Fuel Industry*. Report by L. Ross, A. Zibel, D. Wagner and C. Kuveke (September 2020). https://bailout.cdn.prismic.io/bailout/1b1e1458-bbff-49bc-a636-f6cbd47a88af_Big+Oils+Billion+Dollar+Bender.pdf.
16. Board of Governors of the Federal Reserve System. *Secondary Market Corporate Credit Facility Transaction Specific Disclosure Report* (June 11, 2020). www.federalreserve.gov/monetarypolicy/smccf.htm.
17. S. B. Raskin. "Opinion: Why Is the Fed Spending So Much Money on a Dying Industry?" *New York Times*, June 10, 2020.
18. G. Ludwig and S. B. Raskin. "Opinion: How the Fed's Rescue Program Is Worsening Inequality." *Politico*, May 28, 2020. www.politico.com/news/agenda/2020/05/28/how-the-feds-rescue-program-is-worsening-inequality-287379.
19. N. Groom. "States Ask Trump Administration to Pay Laid Off Oil Workers to Plug Abandoned Wells." *Commodities News*, May 6, 2020. www.reuters.com/article/us-global-oil-usa-wells-idUSKBN22I2KA.
20. J. W. Peters. "Heritage Foundation Says Trump Has Embraced Two-Thirds of Its Agenda." *New York Times*, January 22, 2018.
21. H. J. Feldman, senior director of Regulatory and Scientific Affairs. *Re: EPA-HQ-OA-2017-0190 (82 FR 17793) Comments in Response to the EPA's Solicitation of Input from the Public to Inform Its Regulatory Reform Task Force's Evaluation of Existing Regulations*. Submitted to S. K. Dravis, regulatory reform officer and

associate administrator in the Office of Policy at the Environmental Protection Agency. May 15, 2017.
22. H. J. Feldman, senior director of Regulatory and Scientific Affairs of the American Petroleum Institute. *Re: EPA-HQ-OA-2017-0190 (82 FR 17793) Comments in Response to the EPA's Solicitation of Input from the Public to Inform Its Regulatory Reform Task Force's Evaluation of Existing Regulations Attachment 1: API Comments on Specific Regulations*. Submitted to S. K. Davis, regulatory reform officer and associate administrator in the Office of Policy at the Environmental Protection Agency. May 15, 2017.
23. J. Mervis and W. Cornwall. "Lamar Smith, the Departing Head of the House Science Panel, Will Leave a Controversial and Complicated Legacy." *Science*, November 5, 2017. www.sciencemag.org/news/2017/11/lamar-smith-departing-head-house-science-panel-will-leave-controversial-and-complicated+&cd=5&hl=en&ct=clnk&gl=ca.
24. G. Metcalf. "The Impact of Removing Tax Preferences for US Oil and Gas Production." *Council on Foreign Relations*, August 2, 2016. www.cfr.org/energy-policy/impact-removing-taxpreferences-us-oil-gas-production/p38150.
25. J. Marn. "US Oil Production Growth Heading for a 'Major Slowdown,' As Capital Discipline and Weak Prices Play Out, IHS Markit Says." *IHS Markit*, November 6, 2019. https://news.ihsmarkit.com/prviewer/release_only/slug/energy-us-oil-production-growth-heading-major-slowdown-capital-discipline-and-weak-pri.
26. J. Stein. "Tax Change in Coronavirus Package Overwhelmingly Benefits Millionaires, Congressional Body Finds." *Washington Post*, April 14, 2020. www.washingtonpost.com/business/2020/04/14/coronavirus-law-congress-tax-change.
27. M. S. Schwartz. "As Election Nears, Trump Expands Moratorium on Exploratory Drilling in Atlantic." *National Public Radio*, September 26, 2020. www.npr.org/2020/09/26/917309717/as-election-nears-trump-expands-moratorium-on-exploratory-drilling-in-atlantic.
28. L. Lessig. *Republic, Lost: How Money Corrupts Congress – and a Plan to Stop It*. New York: Twelve, 2015.
29. R. L. Hasen. *Plutocrats United: Campaign Money, the Supreme Court, and the Distortion of American Elections*. New Haven, CT: Yale University Press, 2016.
30. R. E. Mutch. *Campaign Finance: What Everyone Needs to Know*. New York: Oxford University Press, 2016.
31. M. H. Goldberg et al. "Oil and Gas Companies Invest in Legislators That Vote against the Environment." *Proceedings of the National Academy of Sciences* 117, no. 10 (2020): 5111–5112.
32. T. Cruz. *In Support of the Federal Reserve and Treasury Taking Urgent Action to Ensure Access to Capital for America's Oil and Gas Industry*. Submitted to Secretary S. T. Mnuchin and Chairman J. Powell. April 24, 2020.
33. S. Mohsin and A. Natter. "Energy Chief Says Fed Was Asked to Expand Lending for Oil Firms." *Bloomberg*, May 12, 2020.
34. D. Grandoni. "The Energy 202: Oil and Gas Companies Stand to Gain from Fed Loosening Coronavirus Loan Rules." *Washington Post*, May 1, 2020.
35. C. Sunstein and A. Vermeule. *Law and Leviathan: Redeeming the Administrative State*. Cambridge, MA: Belknap Press of the Harvard University Press, 2020.

36. B. Emerson. *The Public's Law: Origins and Architecture of Progressive Democracy.* New York: Oxford University Press, 2019.
37. N. Oreskes and E. Conway. *Merchants of Doubt: How a Handful of Scientists Obscured the Truth on Issues from Tobacco Smoke to Climate Change.* New York: Bloomsbury Press, 2011.
38. T. Mann and N. Ornstein. *It's Even Worse Than It Looks: How the American Constitutional System Collided with the New Politics of Extremism.* New York: Basic Books, 2016.
39. E. Berman and J. Carter. "Policy Analysis: Scientific Integrity in Federal Policymaking Under Past and Present Administrations." *Journal of Science Policy & Governance* 13, no. 1 (September 2018). www.sciencepolicyjournal.org/uploads/5/4/3/4/5434385/berman_emily__carter_jacob.pdf.
40. L. Fredrickson et al. "History of US Presidential Assaults on Modern Environmental Health Protection." *American Public Health Association* 108, suppl. 2 (April 2018): S95–S103. www.ncbi.nlm.nih.gov/pmc/articles/PMC5922215/.
41. J. D. Leshy. "Public Land Policy after the Trump Administration: Is This a Turning Point?" *Colorado Natural Resource, Energy, & Environmental Law Review* 31, no. 3 (2020):471–507. www.colorado.edu/law/sites/default/files/attached-files/002_leshy_final_copy.pdf.
42. M. Weisskopf. "Bush Was Aloof in Warming Debate." *Washington Post*, October 31, 1992.
43. M. Hudson. "George Bush Sr. Could Have Got in on the Ground Floor of Climate Action – History Would Have Thanked Him." *The Conversation*, December 5, 2018. https://theconversation.com/george-bush-sr-could-have-got-in-on-the-ground-floor-of-climate-action-history-would-have-thanked-him-108050.
44. J. Chemnick and M. Quiñones. "Clean Air Act Debates Show How Much Politics Have Changed." *E&E News*, February 13, 2014.
45. J. E. Zelizer. *Burning Down the House: Newt Gingrich, the Fall of a Speaker, and the Rise of the New Republican Party 2020.* New York: Penguin Press, 2020.
46. A. C. Revkin. "Bush Aide Softened Greenhouse Gas Links to Global Warming." *New York Times*, June 8, 2005. www.nytimes.com/2005/06/08/politics/bush-aide-softened-greenhouse-gas-linksto-global-warming.html.
47. C. Mooney. *The Republican War on Science.* New York: Basic Books, 2005.
48. A. C. Revkin. "Climate Expert Says NASA Tried to Silence Him." *New York Times*, January 29, 2006. www.nytimes.com/2006/01/29/science/earth/climate-expert-says-nasa-tried-to-silence-him.html.
49. L. Fang and S. Horn. "Hillary Clinton's Energy Initiative Pressed Countries to Embrace Fracking, New Emails Reveal." *The Intercept*, May 23, 2016. https://theintercept.com/2016/05/23/hillary-clinton-fracking/+&cd=4&hl=en&ct=clnk&gl=ca.
50. S. Horn. "Obama Alums Are Pushing Fracked Gas Exports. That's Exactly What Trump Wants." *Desmog.com*, 2018. www.desmogblog.com/2018/02/02/obama-officials-trump-energy-dominance.
51. D. R. Caulton et al. "Toward a Better Understanding and Quantification of Methane Emissions from Shale Gas Development." *Proceedings of the National Academy of Sciences* 111, no. 17 (2014): 6237–6242.

52. A. Karion et al. "Aircraft-Based Estimate of Total Methane Emissions from the Barnett Shale Region." *Environmental Science & Technology* 49, no. 13 (2015): 8124–8131.
53. T. I. Yacovitch et al. "Mobile Laboratory Observations of Methane Emissions in the Barnett Shale Region." *Environmental Science & Technology* 49, no. 13: 7889–7895.
54. White House Office of the Press Secretary. *Fact Sheet: The Recovery Act Made the Largest Single Investment in Clean Energy in History, Driving the Deployment of Clean Energy, Promoting Energy Efficiency, and Supporting Manufacturing.* Obama Administration. February 25, 2016. https://obamawhitehouse.archives.gov/the-press-office/2016/02/25/fact-sheet-recovery-act-made-largest-single-investment-clean-energy.
55. J. E. Aldy. "Policy Monitor: A Preliminary Assessment of the American Recovery and Reinvestment Act's Clean Energy Package." *Review of Environmental Economics and Policy* 7, no. 1 (2013): 136–155.
56. T. Lim, T. S. Guzman, and W. M. Bowen. "Rhetoric and Reality: Jobs and the Energy Provisions of the American Recovery and Reinvestment Act." *Energy Policy* 137 (2020): 111182.
57. T. Lim, T. Tang and W. M. Bowen. "The Impact of Intergovernmental Grants on Innovation in Clean Energy and Energy Conservation: Evidence from the American Recovery and Reinvestment Act." *Energy Policy* 148 (2021): 111923.
58. D. Popp et al. "The Employment Impact of Green Fiscal Push: Evidence from the American Recovery Act." National Bureau of Economic Research working paper 27321 (2020). www.nber.org/papers/w27321.
59. R. J. Brulle. "Institutionalizing Delay: Foundation Funding and the Creation of US Climate Change Counter-Movement Organizations." *Climatic Change* 122 (2014): 681–694.
60. G. Supran and N. Oreskes. "Assessing ExxonMobil's Climate Change Communications (1977–2014)." *Environmental Research Letters* 12, no. 8 (2017): 084019.
61. J. Farell. "Corporate Funding and Ideological Polarization about Climate Change." *Proceedings of the National Academy of Sciences* 113, no. 1 (2015): 92–97. www.pnas.org/content/pnas/113/1/92.full.pdf.
62. J. Schwartz and H. Tabuchi. "By Calling Climate Change 'Controversial,' Barrett Created Controversy." *New York Times*, October 15, 2020.
63. *Oversight: Oil and Gas Development: Impacts of Business-as-Usual on the Climate and Public Health.* Subcommittee on Energy and Mineral Resources, Natural Resources Committee. 116th Congress, 1st Sess. July 16, 2019.
64. E2. *Clean Energy & COVID-19 Crisis: December 2020 Unemployment Analysis.* (January 13, 2021). https://e2.org/reports/clean-jobs-covid-economic-crisis-december-2020.
65. E. Penrod. "US Energy Sector Takes Beating from COVID-19, But Demand for Renewable Energy Surges." *Utility Dive*, August 14, 2020.
66. S&P Global: Market Intelligence. *The 2021 US Renewable Energy Outlook.* Report by L. Federico et al. (January 2021).
67. Energy Information Administration. *Annual Energy Outlook* (February 3, 2021). www.eia.gov/outlooks/aeo.

68. J. Guzman. "Renewable Energy Predicted to Surpass Natural Gas in US by 2045." *The Hill*, January 30, 2020.
69. S. Mufson. "General Motors to Eliminate Gasoline and Diesel Light-Duty Cars and SUVs by 2035." *Washington Post*, January 28, 2021.
70. K. Laing. "Automakers Vow to Spend $250 Billion on US Electric Vehicles." *Bloomberg*, March 30, 2021.
71. Energy Information Administration. *Financial Review of the Global Oil and Natural Gas Industry* (Washington, DC: 2020). www.eia.gov/finance/review/pdf/2020%20Financial%20Review.pdf.
72. Energy Information Administration. *Early 2020 Drop in Crude Oil Prices Led to Write-Downs of US Oil Producers' Assets*. Report by J. Barron (July 27, 2020). www.eia.gov/todayinenergy/detail.php?id=44516.
73. T. Randall and H. Warren. "Peak Oil Is Suddenly upon US." *Bloomberg*, December 1, 2020.
74. G. Smith. "Global Oil Demand to Hit a Plateau around 2030, IEA Predicts." *Bloomberg*, November 13, 2019.
75. G. Smith. "IEA Sees Oil Demand Suffering Long-Lasting Blow from Coronavirus." *Bloomberg*, October 13, 2020. www.bloomberg.com/news/articles/2020-10-13/iea-sees-oil-demand-suffering-long-lasting-blow-from-coronavirus.
76. A. Raval, L. Hook and D. Sheppard. "Why the IEA Is 'Calling Time' on the Fossil Fuel Industry." *Financial Times*, May 19, 2021.
77. International Energy Agency. *Net Zero by 2050: A Roadmap for the Global Energy Sector*. Report by S. Bouckaert et al. (2021). https://iea.blob.core.windows.net/assets/ad0d4830-bd7e-47b6-838c-40d115733c13/NetZeroby2050-ARoadmapfortheGlobalEnergySector.pdf.
78. M. B. Gerrard. "Climate Change Litigation in the United States: High Volume of Cases, Mostly about Statutes." In *Climate Change Litigation: Global Perspectives*, edited by I. Alogna, C. Bakker and J.-P. Gauci. 33–46. Leiden: Brill Nijhoff, 2021.
79. Grantham Research Institute for Climate Change and Environment, London School of Economics. *Global Trends in Climate Change Litigation: 2020 Snapshot*. Report by J. Setzer and R. Byrnes (July 3, 2020). www.lse.ac.uk/granthaminstitute/publication/global-trends-in-climate-change-litigation-2020-snapshot.
80. S. Sengupta. "Rebels Open New Fronts for Climate, and Win." *New York Times*, May 29, 2021.
81. M. Barry and K. Silverman-Roati, *June 2021 Updates to the Climate Case Charts, in Climate Law Blog*, Sabin Center for Climate Change Law, 2021. http://blogs.law.columbia.edu/climatechange/2021/06/08/june-2021-updates-to-the-climate-case-charts.
82. C. W. Backes and G. A. v. d. Veen. "Urgenda: The Final Judgment of the Dutch Supreme Court." *Journal for European Environmental & Planning Law* 17, no. 3 (2020): 307–321.
83. B. McKibben. "Citing Climate Change, BlackRock Will Start Moving Away from Fossil Fuels." *New Yorker*, January 16, 2020.
84. A. Barnard. "New York's $226 Billion Pension Fund Is Dropping Fossil Fuel." *New York Times*, December 9, 2020.

85. J. Kollewe. "Lloyd's Market to Quit Fossil Fuel Insurance by 2030." *Guardian*, December 16, 2020. www.theguardian.com/business/2020/dec/17/lloyds-market-to-quit-fossil-fuel-insurance-by-2030.
86. "The Biden Plan for a Clean Energy Revolution and Environmental Justice." Biden–Harris Campaign, 2020, https://joebiden.com/climate-plan.
87. J. K. Maldonado, B. Colombi and R. Pandya. *Climate Change and Indigenous Peoples in the United States: Impacts, Experiences, and Actions*. London: Springer, 2014. https://link.springer.com/content/pdf/10.1007/978-3-319-05266-3.pdf.
88. C. W. Tessum et al. "PM2.5 Polluters Disproportionately and Systemically Affect People of Color in the United States." *Science Advances* 7, no. 18 (2021). doi: 10.1126/sciadv.abf4491.
89. D. Moore and A. Litvak. "Biden's Oil Comments Fuel Long-Burning Debate over Pa. Energy Jobs." *Pittsburgh Post-Gazette*, October 24, 2020. www.post-gazette.com/news/politics-nation/2020/10/24/Joe-Biden-comments-on-oil-stoke-political-energy-jobs-Donald-Trump-fracking/stories/202010240029.
90. E. Lehmann. "The 2008 GOP Platform Spotlighted Climate Change; Some See an Eclipse Coming under Romney." *E&E News*, August 7, 2012.
91. C. Hulse. "GOP Party Platform at Odds with Some of McCain's Stances." *New York Times*, September 4, 2008.
92. Republican Platform Committee. *Republican Platform* (2008). www.eenews.net/assets/2012/08/06/document_gw_03.pdf.
93. A. C. Lin. "A Sustainability Critique of the Obama 'All-of-the-Above' Energy Approach." *Journal of Energy and Environmental Law* (Winter 2014): 17–25. https://gwjeel.com/wp-content/uploads/2014/05/lin_article_jeel_0114.pdf.
94. C. McGlade and P. Ekins. "The Geographical Distribution of Fossil Fuels Unused When Limiting Global Warming to 2°C." *Nature* 517 (2015): 187–190. www.nature.com/articles/nature14016.
95. K. B. Tokarska and N. P. Gillett. "Cumulative Carbon Emissions Budgets Consistent with 1.5°C Global Warming." *Nature Climate Change* 8 (2018): 296–299.
96. J. Biden. *Remarks As Prepared for Delivery by President Biden, Address to a Joint Session of Congress*. Submitted to US Congress. April 28, 2021.
97. Climate-Related Market Risk Subcommittee. *Managing Climate Risk in the US Financial System*. US Commodity Futures Trading Commission, Market Risk Advisory Committee (Washington, DC: 2020). www.cftc.gov/sites/default/files/2020-09/9-9-20%20Report%20of%20the%20Subcommittee%20on%20Climate-Related%20Market%20Risk%20-%20Managing%20Climate%20Risk%20in%20the%20U.S.%20Financial%20System%20for%20posting.pdf.
98. A. H. Lee, acting chair. *Playing the Long Game: The Intersection of Climate Change Risk and Financial Regulation*. Submitted to PLI's 52nd Annual Institute on Securities Regulation. November 5, 2020.
99. E. L. Grayson and P. L. Boye-Williams. "SEC Disclosure Obligations: Increasing Scrutiny on Environmental Liabilities and Climate Change Impacts." In *Environmental Issues in Business Transactions*, edited by L. P. Schnapf. Chicago, IL: American Bar Association, 2011.
100. Chemnick, J. "Cost of Carbon Pollution Pegged at $51 a Ton." *E&E News*, March 1, 2021.

101. White House Office of the President. *Executive Order on Tackling the Climate Crisis at Home and Abroad*. President J. Biden, 2021. www.whitehouse.gov/briefing-room/presidential-actions/2021/01/27/executive-order-on-tackling-the-climate-crisis-at-home-and-abroad.
102. C. Barnosky et al. *Reversing Trump Environmental Rollbacks: A 100 Day Analysis on the Biden Administration's Reversals*. Report by UC Berkeley Law School and Goldman School of Public Policy (May 2021). www.law.berkeley.edu/wp-content/uploads/2021/05/Trump-Rollbacks-Report.pdf.
103. H. Peris. "The Downfall of the 'Secret Science' Rule, and What It Means for Biden's Environmental Agenda." *Harvard Environmental & Energy Law Program*, March 5, 2021. https://eelp.law.harvard.edu/2021/03/final-secret-science-rule.
104. Congressional Research Service. *Department of Defense Energy Management: Background and Issues for Congress*. Report by H. L. Greenley (Washington, DC: July 25, 2019). https://fas.org/sgp/crs/natsec/R45832.pdf.
105. T. R. McDonald et al., "Biden Administration Executive Clean Energy Actions: Fleet Procurement." *Holland & Knight Insights*, February 5, 2021. www.hklaw.com/en/insights/publications/2021/02/biden-administration-executive-clean-energy-actions.
106. J. Eilperin and B. Dennis. "Biden Administration Launches Major Push to Expand Offshore Wind Power." *Washington Post*, March 29, 2021
107. White House Office of the Press Secretary. *Fact Sheet: Biden Administration Advances Expansion & Modernization of the Electric Grid*. Biden Administration (April 27, 2021). www.whitehouse.gov/briefing-room/statements-releases/2021/04/27/fact-sheet-biden-administration-advances-expansion-modernization-of-the-electric-grid.
108. J. Biden. *Remarks by President Biden on the American Jobs Plan in Pittsburgh*. March 31, 2021. www.whitehouse.gov/briefing-room/speeches-remarks/2021/03/31/remarks-by-president-biden-on-the-american-jobs-plan.
109. J. Biden. *Remarks by President Biden on the American Jobs Plan*. April 7, 2021. www.whitehouse.gov/briefing-room/speeches-remarks/2021/04/07/remarks-by-president-biden-on-the-american-jobs-plan-2.
110. E. Nilsen. "Joe Biden's $2 Trillion Infrastructure and Jobs Plan, Explained." *Vox*, March 31, 2021. www.vox.com/2021/3/31/22357179/biden-two-trillion-infrastructure-jobs-plan-explained.
111. Committee for a Responsible Federal Budget. *What's in President Biden's American Jobs Plan?* (2021). www.crfb.org/blogs/whats-president-bidens-american-jobs-plan.
112. Institute on Taxation and Economic Policy. *The 35 Percent Corporate Tax Myth: Corporate Tax Avoidance by Fortune 500 Companies, 2008 to 2015*. Report by M. Gardner et al. (March 2017). https://itep.sfo2.digitaloceanspaces.com/35percentfullreport.pdf.
113. Institute on Taxation and Economic Policy. *Corporate Tax Avoidance in the First Year of the Trump Tax Law*. Report by M. Gardner et al. (December 2019). https://itep.sfo2.digitaloceanspaces.com/121619-ITEP-Corporate-Tax-Avoidance-in-the-First-Year-of-the-Trump-Tax-Law.pdf.

114. Institute on Taxation and Economic Policy. *55 Corporations Paid $0 in Federal Taxes on 2020 Profits*. Report by M. Gardner and S. Wamhoff (April 2021). https://itep.sfo2.digitaloceanspaces.com/040221-55-Profitable-Corporations-Zero-Corporate-Taxes.pdf.
115. Z. Chen et al. "Green Stimulus in a Post-pandemic Recovery: The Role of Skills for a Resilient Recovery." *Environmental and Resource Economics* 76, no. 4 (2020): 901–911.
116. R. Wheeler. *Can Biden "Rebalance" the Judiciary?* Brookings Institute (2021). www.brookings.edu/blog/fixgov/2021/03/18/can-biden-rebalance-the-judiciary.
117. A. E. Marimow and P. Kane. "Senate Committee to Take Up Biden Judicial Nominees in Preview of Potential Supreme Court Fight." *Washington Post*, April 23, 2021.
118. Brookings Institute. *Build a Better Future for Coal Workers and their Communities*. Report by A. C. Morris (April 25, 2016). www.brookings.edu/wp-content/uploads/2016/07/Build-a-Better-Future-for-Coal-Workers-and-their-Communities-Morris.pdf.
119. Brookings Institution. *The Risk of Fiscal Collapse in Coal-Reliant Communities*. Report by A. Morris et al. (July 15, 2019). www.brookings.edu/research/the-risk-of-fiscal-collapse-in-coal-reliant-communities.
120. K. F. Roemer and J. H. Haggerty. "Coal Communities and the US Energy Transition: A Policy Corridors Assessment." *Energy Policy* 151 (2021): 112112.
121. Macey, J. and J. Salovaara. "Bankruptcy As Bailout: Coal Company Insolvency and the Erosion of Federal Law." *Stanford Law Review* 71 (2019): 879–962.
122. Columbia Center on Global Energy Policy. *Can Coal Make a Comeback?* Report by T. Houser et al. (April 2017). www.ourenergypolicy.org/wp-content/uploads/2017/05/Center_on_Global_Energy_Policy_Can_Coal_Make_Comeback_April_2017.pdf.
123. Congressional Research Service. *Health and Pension Benefits for United Mine Workers of America Retirees: Recent Legislation*. Report by J. J. Topoleski. IF11366 (November 20, 2019). https://crsreports.congress.gov/product/pdf/IF/IF11366.
124. Interagency Working Group on Coal and Power Plant Communities and Economic Revitalization. *Initial Report to the President on Empowering Workers through Revitalizing Energy Communities* (April 2021). https://netl.doe.gov/sites/default/files/2021-04/Initial%20Report%20on%20Energy%20Communities_Apr2021.pdf.
125. C. Mackie and L. Besco. "Rethinking the Function of Financial Assurance for End-of-Life Obligations." *Environmental Law Reporter* 50 (2020): 10573–10603.
126. K. B. Hall. "Decommissioning of Offshore Oil and Gas Facilities in the United States." *Charleston Law Review* 14, no. 3 (2020): 437–464.
127. M. Kang et al. "Reducing Methane Emissions from Abandoned Oil and Gas Wells: Strategies and Costs." *Energy Policy, Elsevier* 132, no. C (2019): 594–601.
128. Columbia Center on Global Energy Policy. *Green Stimulus for Oil and Gas Workers: Considering a Major Federal Effort to Plug Orphaned and Abandoned Wells*. Report by J. Bordoff et al. (July 20, 2020). www.energypolicy.columbia.edu/research/report/green-stimulus-oil-and-gas-workers-considering-major-federal-effort-plug-orphaned-and-abandoned.

129. J. Walsh. "14 GOP States Sue Biden Administration over Oil Restrictions." *Forbes*, March 24, 2021. www.forbes.com/sites/joewalsh/2021/03/24/14-gop-states-sue-biden-administration-over-oil-restrictions/?sh=3aa1f8bf25be.
130. J. Eilperin and D. Grandoni. "Biden Vowed to Ban New Drilling on Public Lands. It Won't Be Easy." *New York Times*, November 19, 2020.
131. Aspen Institute Economic Strategy Group. *Promoting Recovery after COVID-19*. Report by J. Furman et al. (June 16, 2020). www.economicstrategygroup.org/wp-content/uploads/2020/11/Promoting-Economic-Recovery-After-COVID-0615-FINAL.pdf.
132. Oxford Smith School of Enterprise and the Environment. *Will COVID-19 Fiscal Recovery Packages Accelerate or Retard Progress on Climate Change?* Report by C. Hepburn et al. Working paper 20-02 (May 4, 2020). www.smithschool.ox.ac.uk/publications/wpapers/workingpaper20-02.pdf.
133. Moody's Analytics. *The Macroeconomic Consequences: Trump vs. Biden*. Report by M. Zandi and B. Yaros (September 23, 2020). www.moodysanalytics.com/-/media/article/2020/the-macroeconomic-consequences-trump-vs-biden.pdf.
134. Penn Wharton Budget Model. *PWBM Analysis of the Biden Platform*. Report by A. Arnon et al. (September 14, 2020).
135. Moody's Analytics. *The Macroeconomic Consequences of the American Jobs Plan*. Report by M. Zandi and B. Yaros (April 2021). www.economy.com/getlocal?q=C228A0FF-2701-47B2-ADE0-D158B5866251&app=download.
136. Congressional Budget Office. *The Budget and Economic Outlook: 2018 to 2028*. Report by K. Hall et al. (April 2018). www.cbo.gov/system/files/2019-04/53651-outlook-2.pdf.
137. W. G. Gale et al. "Effects of the Tax Cuts and Jobs Act: A Preliminary Analysis." *National Tax Journal* 71, no. 4 (2019): 589–612.
138. C. Williams. "More Than 3 in 5 Voters Support Corporate Tax Hike to Fund Biden's Infrastructure Plan." *Morning Consult*, April 7, 2021.
139. K. Shepherd. "Biden's Climate Plan Doesn't Ban Meat. But Baseless Claims Left Republicans Fuming: 'Stay Out of My Kitchen.'" *Washington Post*, April 26, 2021.
140. J. Boehner. *On the House: A Washington Memoir*. New York: St. Martin's Press, 2021.
141. A. Rappeport and J. Tankersley. "Atop the Powerful Budget Committee at Last, Bernie Sanders Wants to Go Big." *New York Times*, January 12, 2021.
142. *County of Maui, Hawaii v. Hawaii Wildlife Fund*, 140 S. Ct. 1462 (Supreme Court 2020).
143. *Mexichem Fluor, Inc. v. Environmental Protection Agency, et al.*, 866 F.3d 451 (DC Cir. 2017).
144. *Natural Resources Defense Council v. Wheeler*, 955 F.3d 68 (DC Cir. 2020).
145. R. Frank. "Climate Change, Big Energy & the US Supreme Court – What Could Possibly Go Wrong?" *Legal Planet*, January 17, 2021. https://legal-planet.org/2021/01/17/climate-change-big-energy-the-u-s-supreme-court-what-could-possibly-go-wrong.
146. J. Nobel and A. Juhasz. "Op-ed: More Than 70 Science and Climate Journalists Challenge Supreme Court Nomination of Amy Coney Barrett." *Rolling Stone*, October 25, 2020.

147. A. Ladd. "Priming the Well: 'Frackademia' and the Corporate Pipeline of Oil and Gas Funding into Higher Education." *Humanity & Society* 44, no. 2 (October 2019): 151–177.
148. A. Litvak. "Wooing Petrochemical Plants in the Age of COVID-19." *Pittsburgh Post-Gazette*, July 20, 2020.
149. L. Legere. "Pa. Legislature Adopts $670 Million Tax Credit Bill for Petrochemical Plants." *Pittsburgh Post-Gazette*, July 14, 2020. www.post-gazette.com/business/powersource/2020/07/14/Pennsylvania-petrochemical-tax-credit-fertilizer-dry-natural-gas-Marcellus-Utica-fracking/stories/202007140122.
150. A. Litvak and L. Legere. "The Wooing of a Would-Be Petrochemical Plant." *Pittsburgh Post-Gazette*, September 21, 2020. www.post-gazette.com/business/powersource/2020/09/21/wooing-would-be-petrochemical-plant-ExxonMobil-Shell-ethane-cracker-Department-of-Community-and-Economic-Development/stories/202009200048.
151. Ohio River Valley Institute. *Appalachia's Natural Gas Counties: Contributing More to the US Economy and Getting Less in Return.* Report by S. O'Leary. (February 12, 2021). https://ohiorivervalleyinstitute.org/wp-content/uploads/2021/02/Frackalachia-Report-update-2_12_01.pdf.
152. Ohio River Valley Institute. *Risks for New Natural Gas Developments in Appalachia.* Report by P. Erickson and P. Achakulwisut (March 2021). https://ohiorivervalleyinstitute.org/wp-content/uploads/2021/03/Risks-of-new-natural-gas-developments-in-Appalachia_March-2021_Final_3.9.21.pdf.
153. Reimagine Appalachia. *A New Deal that Works for Us* (March 2021). https://reimagineappalachia.org/wp-content/uploads/2021/03/ReImagineAppalachia_Blueprint_042021.pdf.
154. Brookings Institute. *Capturing the Next Economy: Pittsburgh's Rise As a Global Innovation City.* Report by S. Andes et al. (September 13, 2017). www.brookings.edu/research/capturing-the-next-economy-pittsburghs-rise-as-a-global-innovation-city.
155. D. Youker and L. A. Elder. "Opinion: Community Solar Energy Will Create Jobs and Help Farmers in Pennsylvania." *Pennlive.com*, November 16, 2020. www.pennlive.com/opinion/2020/11/community-solar-energy-will-create-jobs-and-help-farmers-in-pennsylvania-opinion.html.
156. C. Smith. "Senate Bill Proposes Expansion of Pennsylvania's Renewable Energy Targets." *The Center Square*, March 29, 2021. www.thecentersquare.com/pennsylvania/senate-bill-proposes-expansion-of-pennsylvania-s-renewable-energy-targets/article_14bce05a-90d5-11eb-a13d-83676ce11122.html.
157. L. Legere and A. Litvak. "PA Faces New Wave of Abandoned Conventional Oil & Gas Wells." *Pittsburgh Post-Gazette*, April 5, 2020. http://liber.post-gazette.com/business/powersource/2020/04/03/allegheny-national-forest-drilling-pennsylvania-oil-wells-pipelines-abandoned/stories/202004030081.
158. Office of the Attorney General, Commonwealth of Pennsylvania. *Report 1 of the Forty-Third Statewide Investigating Grand Jury.* www.attorneygeneral.gov/wp-content/uploads/2020/06/FINAL-fracking-report-w.responses-with-page-number-V2.pdf.
159. J. M. Turner and A. Isenberg. *The Republican Reversal: Conservatives and the Environment from Nixon to Trump.* Cambridge, MA: Harvard University Press, 2018.

Index

Abandoned Mine Land Reclamation Economic Development project, 87
Abandoned Mine Land Reclamation Fund, 439
abandoned wells, 44
Abbott, Greg, 89
aboriginal lands, 165
"Accelerating Decarbonization of the US Energy System," 74
Administrative Procedure Act (APA), 167, 175, 177, 178, 272, 307, 349, 351, 352, 353, 354, 369, 372, 391, 398
administrative state, 379–381, 390, 392, 395, 466
   congressional delegation of powers to, 386, 388
   conservative ideology on, 389
   legal doctrine related to, 379, 381, 385
Affordable Care Act, 382
Affordable Clean Energy Rule, 70, 291, 297, 338, 356, 480
air pollution, 327
air quality, 41
Air Quality Standards Coalition, 287
Alabama, 40, 233
Alaska, 91, 112, 113, 156, 225, 235–237, 260
   crude oil production in, 219
   governmental support of drilling in ANWR, 224
   production of natural gas in, 219
   protection of lands of, 138
   threat to fisheries in, 220
   tourism and, 220
Alaska National Interest Lands Conservation Act, 118
Alaskan Permanent Fund, 236
Alberta, Canada, 175

Alito, Samuel, 163, 377
All Pueblo Council of Governors, 186, 195
Allegheny Defense Project v. Federal Energy Regulatory Commission, 127
Allotment and Assimilation Era, 158, 162
Alphabet, 67
America First Energy Plan, 2, 4, 19, 27, 28, 44, 464
   announcement of, 18
   land management and, 110
America First Offshore Energy Plan, 221
America's Natural Treasures of Immeasurable Quality Unite, Inspire, and Together Improve the Economies of States (ANTIQUITIES) Act, 196
American Bar Association, 7
American Clean Energy and Security Act, 416
American Electric Power, 89
American Energy Innovation Act, 442
American Innovation and Manufacturing Act, 444
American Jobs Plan, 481
American Legislative Exchange Council, 89
American Lung Association v. EPA, 70, 480
American Opportunity Carbon Fee Act, 428
American Petroleum Institute, 128, 273, 287, 288, 293, 297, 327, 334, 336, 341, 429, 468, 472
American Power Act, 289
American Recovery and Reinvestment Act (ARRA), 81, 416, 440, 473, 482
American Rescue Plan, 481
Americans for Carbon Dividends, 430
Americans for Prosperity, 89, 288
ancestral lands, 109, 141, 157, 164, 165, 181, 182

Angelle, Scott, 254
Annenberg Center, 396
Annual Energy Outlook, 475
Antiquities Act, 182, 185, 196, 373
Appalachia, 52, 86, 87, 103, 441, 459
Appalachian Regional Commission, 87, 441
Appalachian Trail, 382
Appalachian Voices, 84
Apple, 67
Arctic National Wildlife Refuge (ANWR), 230, 236, 238, 242, 468, 470
　Alaska governmental support of drilling in, 224
　Congressional approval of drilling in, 218
　economic value of, 229
　end of moratorium on drilling in, 3
　Gwich'in opposition to drilling in, 237
　lease sales in, 111, 241
　oil and gas drilling in, 117
　Trans-Alaska Pipeline System and, 236
　Trump administration support of drilling in, 224, 425
Arctic Ocean, 5, 218, 221, 236, 241
Arctic Slope Regional Corporation (ASRC), 236, 237
Arctic summer sea ice, 343
Arizona, 68, 187, 188, 195
Arizona Corporation Commission, 68
Army Corps of Engineers, 116
　408 permits, 171
　analysis of the Clean Water Rule of, 134
　approval of infrastructure projects of, 127
　approval process for Keystone XL pipeline and, 179
　denial of permits for export terminals, 180
　mandates of under Clean Water for All Act, 139
　powers under the 401 Certification Rule, 374
　review of Dakota Access pipeline of, 171
Asia, 180, 181
Assiniboine, 161
Atlantic Coast Pipeline, 168
Atlantic Ocean, 5, 218, 221, 225
　impact of America First Energy Plan on, 18
　seismic airgun testing in, 272
auto industry, 424
automation, 36

Bad River Band, 166
Baker, James, 350
Bakken Shale, 135, 166, 169

Balash, Joe, 227
Baltimore Gas & Electric Co. v. Natural Resources Defense Council, 355
Bank of America, 425, 435
Bannocks, 112
Barr, William, 396
Barragán, Nanette Diaz, 270
Barrett, Amy Coney, 491
Barrasso, John, 303
Barrow Whaling Area, 237
Beacon Center, 389
Bears Ears Expansion and Respect for Sovereignty Act, 195
Bears Ears National Monument, 5, 118, 121, 157, 158, 167, 181–186, 195, 197
Beaufort Sea, 221, 227, 258
Bell, Robert, 173
Bennet, Michael, 349, 432
benzene, 189, 286, 287
Bering Sea Elders Group, 237
Bering Strait, 237
Bernhardt, David, 119, 186, 233, 254, 264, 291
BH Group, 394
BHP Billiton, 29
Biden administration, 74, 78, 267, 356, 487
　clean energy transition of, 479
　Congressional opposition to, 487
　hope for climate action of, 446
　Plan for Tribal Nations of, 196
　promotion of renewable energy of, 479, 481
　promotion of wind power of, 481
　protection of local communities of, 483
　regulation of greenhouse gas emissions of, 480
　regulation of methane of, 431
　support of Native American clean energy projects, 197
　valuation of social cost of carbon, 341
　withdrawal of Science Transparency Rule, 307
Biden, Joe, 331
　announcement of American Jobs Plan, 481
　clean electricity standard proposed by, 482
　clean energy transition of, 477
　Donald Trump's denial of victory of, 463
　emissions reduction target proposed by, 444
　executive orders by, 197, 229, 331, 341, 376, 421, 426
　pledge to transition away from oil by, 478
　promises to tribes by, 196
　support for coal communities by, 485

support for environmental justice by, 477
support for rejoining the Paris Climate
   Agreement, 421
Biden–Harris Plan for Tribal Nations,
   197
biomass, 22, 58, 61
Birckhead v. FERC, 126
Bishop, Robert, 118, 182, 224
Bismarck, North Dakota, 173
Black Elk Energy, 257
Blackfeet Reservation, 188
Blackfeet tribe, 112
BlackRock, 34, 477
Blanket 4(d) Rule, 132
blowout preventers, 254, 259, 262, 263, 266,
   267, 272, 273
BlueGreen Alliance, 86
Blumenthal, Richard, 264
Boehlert, Sherwood, 416, 472
Boehner, John, 489
border carbon adjustments, 429
Boylan, James, 294, 296
BP, 227, 255, 430, 477
BP oil spill, 5, 234, 242, 253, 258–260, *See also*
   oil spills, Deepwater Horizon
   Obama administration and, 261, 473
   regulations and, 466
   safety reforms and, 266
   Trump administration and, 268
   waivers and, 273
Bradley Foundation, 394
*Brendale v. Confederated Tribes and Bands of
   Yakima*, 160
Breyer, Stephen, 377
Brighton, Colorado, 66
Brouillette, Dan, 35, 72
brownfield redevelopment, 440
Buchanan, Vern, 270
Bureau of Indian Affairs (BIA), 186, 187, 188,
   190, 193, 209
Bureau of Labor Statistics, 438
Bureau of Land Management (BLM), 71, 109,
   110, 112, 113, 117, 158, 181
   assessment of drilling in ANWR, 229
   changes under Biden administration of, 480
   court striking down 2018 Suspension Rule
      of, 352
   estimates on the social cost of carbon, 341
   FLPMA requirements of, 121
   leasing in ANWR, 238
   management of ancestral lands, 165

management of Bears Ears National
   Monument, 182
management of Chaco Canyon, 186
management of national monuments, 184
Native American lands and, 163
requirements under MLA of, 374
rescinding of Planning Rule 2.0, 118
Resource Management Plans for national
   monuments, 186
waste prevention on public lands, 347
Bureau of Ocean Energy Management
   (BOEM), 71, 241, 258, 261, 480
Bureau of Safety and Environmental
   Enforcement (BSEE), 254, 261, 262, 263,
   264, 266, 267, 268, 269, 272
   regulatory approach of, 262
Burr, Richard, 303
Bush administration
   anti-science attitude of, 472
   Bureau of Land Management and, 120
   cost–benefit analysis and, 336
   EPA and, 338
   non-threshold models and, 298
   particulate matter and, 334
   Trump administration and, 287
Bush, George H. W., 185, 225, 286, 345,
   415, 464
   science and, 302
Bush, George W., 2, 17, 66, 89, 112, 436,
   472
   National Energy Policy Development Group
      and, 21
   offshore drilling and, 225
   scientific integrity and, 291
   tax preferences and, 417

Calabrese, Steven, 391
California, 68, 180, 184, 194, 222, 225, 232,
   331, 355, 374, 423, 467
California Air Resources Board, 424
California v. Bernhardt, 376
Calvert, Ken, 302
Campo Kumeyaay Nation, 192
Canada, 20, 180, 237
Cannonball River, 174
Canter, Virginia, 119
Cantwell, Maria, 239
Canyonlands National Park, 187
Capuano, Linda, 475
Caputo, Shelley Moore, 303
carbon, 59, 341, 350, 429

carbon dioxide, 31, 70, 113, 344, 345, 417, 421, 429, 445
  carbon taxes and, 428
  oil recovery and, 439
  social cost of, 340, 341, 432, 480
carbon dioxide emissions, 31, 70, 113, 421
  carbon taxes and, 430
Carbon Pollution Transparency Act, 349, 432
carbon capture and sequestration (CCS), 420, 427, 443, 461
  RD&D and, 445
carbon taxes, 59, 428, 429, 431, 480, 486
  American Opportunity Carbon Fee Act and, 428
  benefits of, 429
  clean energy transition and, 483
  diversification programs and, 441
  economic incentive and, 431
  EPA and, 430
  oil industry support of, 430
  as policy instrument, 428
  regulations and, 430
  Republican support of, 429
  social cost and, 429
Carleton, Tamma, 433
Carper, Tom, 444
Carter administration, 354
Carter, Jimmy, 2, 471
Cartwright, Matt, 438
Cascade Siskiyou, 184
Casten, Sean, 435
Cato Institute, 389
Center for American Progress, 71, 115, 123
Center for Biological Diversity, 241
Center for Biological Diversity v. National Highway Traffic and Safety Administration, 340, 480
Center for Responsibility and Ethics and the Brennan Center for Justice, 137
Center for Rural Affairs, 77
Center for Sustainable Economy v. Jewell, 229
Chaco Canyon National Historical Park (CCNHP), 181, 186–187, 195
Chafee, Lincoln, 415
Chaffetz, Jason, 118, 182
Charmley, William, 307
Chatterjee, Neil, 124
Chemical Safety Board, 272
Cheney, Dick, 22, 472
Cherokee Nation v. Georgia, 161
Chevron, 40, 227

Chevron deference, 352, 386, 387
Chevron USA, Inc. v. Natural Resources Defense Council, Inc., 352
Chevron v. NRDC, 387
Cheyenne River Sioux Indian Reservation, 169, 173, 175
China, 345, 444
chlorofluorocarbons (CFCs), 444
Chmielewski, Kevin, 306
Christie, Mark, 124
Chu, Steven, 260, 423
Chukchi Sea, 222, 226, 227, 237, 241, 258
Chutkan, Tanya, 185
Cicilline, David, 424
Citibank, 435
Citigroup, 425
Citizens for a Sound Economy, 287
Citizens for Responsibility and Ethics in Washington, 119
civil society, 238, 275, 286, 308, 356
Clean Air Act, 69, 287, 292, 293, 399
  California's fuel efficiency standards and, 374
  carbon taxes and, 431
  Commerce Clause and, 388
  cost–benefit analysis and, 336
  Cost–Benefit Rule and, 331
  EPA and, 349, 375, 386, 387, 431
  EPA violations of, 352
  global social cost of carbon and, 345
  interpretation of, 490
  legislative history of, 336
  NAAQS and, 331
  particulate matter and, 336
  regulations and, 327
  Richard Nixon and, 471
  Trump administration and, 386
  zero emissions vehicle programs and, 423
Clean Air Act Amendments, 464
Clean Air Council v. Pruitt, 352
Clean Air Science Advisory Committee (CASAC), 293, 294
  Anthony Cox and, 293
  EPA and, 294, 295
  particulate matter standards and, 294
  review of ozone standards of, 296
clean electricity standard, 482
Clean Energy Jobs and Innovation Act, 439, 442
Clean Energy Transition and Environmental Justice Plan, 477

Clean Power Plan, 69, 291, 297, 307, 332, 334, 349
Clean Water Act, 111, 116, 127, 128, 130, 135, 139, 179, 189, 240, 356, 370, 376, 388, 399, 471
    401 Certification Rule and, 374
    aims of, 132
    Clean Water for All Act and, 139
    Environmental Protection Agency (EPA) and, 133, 134, 377
    infrastructure projects and, 373
    interpretation of, 490
    Native American lands and, 180
    permits and, 132
    pipelines and, 133
    protected waters and, 133
    Trump administration narrowing of, 134
Clean Water for All Act, 139
Clean Water Rule, 134, 135
Clements, Allison, 124
Clements, Joel, 306
Cliffs Natural Resources, 40
climate action, 6, 288, 289, 345, 415, 416, 417, 418, 420, 437, 446
    costs of, 418
climate change, 120, 178, 341, 399, 415, 416, 419, 420, 432, 473, 478, 491
    cases in the courts involving, 491
    casting doubt on scientific evidence for, 288
    climate risks, 428, 434, 435, 477, 480
    costs for future generations of, 327
    denial of, 474
    economic costs of, 350
    feasible action on, 417
    global costs of, 340, 341, 345
    impacts of blocking climate action on, 6
    impacts of disruption from, 418
    impacts of Keystone XL pipeline on, 175
    impacts on ocean food chain of, 419
    impacts on oil and gas infrastructure of, 435
    increased oil spills due to, 257
    Joel Clements' demotion after warnings about, 306
    lack of legislation on, 468
    legislative protections for public lands and, 138
    Lincoln Chafee's hearings on, 472
    multilateral approach to, 344
    need for action on, 396
    OCC and, 426
    oil companies' immunity from lawsuits, 430
    political pressure against, 288
    possibility of bipartisan legislation on, 446
    Republican party platform and, 416, 478
    responsibility for incorporation of costs of, 470
    Rod Schoonover's warnings about, 306
    social cost of, 341, 345, 432
    stance of George H. W. Bush on, 286
    State Department disregard of, 178
Climate Change Financial Risk Act, 435
climate denial, 471
climate disruptions, 415, 420, 421, 428, 430, 433, 435
climate legislation, 6, 289, 388, 416, 417, 418, 438, 442, 444, 446
Climate Risk Disclosure Act, 435
Climate Risk Scenario Committee, 435
climate risks, 435, 436
Clinton, Bill, 112, 184, 185, 225
coal, 188
    Appalachian communities and, 86, 441
    carbon taxes and, 428
    Clean Water Act and, 132
    Colorado and, 85
    consumption rates of, 20
    cost-competitiveness of, 63
    gas and, 69
    jobs in, 86, 114, 485
    Native American lands and, 181
    natural gas and, 274
    renewable energy and, 88
    Trump administration and, 70
coal development, 163
coal extraction, 7, 10
coal plants, 11, 41, 74, 76, 192, 332, 338, 485
    acid rain and, 329
    communities dependent upon, 86
    retirement of, 60, 85, 86
Coastal Zone Management Act, 232, 235
Cobell v. Salazar, 157
Coharie tribe, 168
Collins, Susan, 272, 302
Colorado, 42, 66, 82, 85, 139, 187, 188, 191, 350
Colorado General Assembly, 85
Commerce Clause, 110, 388
Commission on Civil Rights, 195
Commodity Futures Trading Commission, 479
community-owned solar energy, 84
community-owned wind farms, 80
Competitive Enterprise Institute, 287, 289

compliance costs, 2, 267, 268, 326, 328, 329, 337, 338, 339, 342, 347, 348
Comprehensive Master Plan for a Sustainable Coast, 234
Confederated Tribes and Bands of the Yakama Nation, 181
Confederated Tribes of Coos, 180
Confederated Tribes of the Umatilla Indian Reservation, 181
Congressional Budget Office, 8, 138
Congressional Review Act, 118, 468
Connector pipeline, 179
conservation, 113
Conservative Case for Carbon Dividends, 350
Consolidated Appropriations Act, 442, 444
Consortia-Led Energy and Advanced Manufacturing Networks Act, 438
Cook Inlet, 221, 227
Cooney, Philip, 288, 472
cooperative federalism, 132, 133, 233, 373
Coronavirus Aid, Relief, and Economic Security (CARES) Act, 32, 33, 34, 35, 50, 51, 242, 467, 470
Corrosion Proof Fittings v. EPA, 328
cost–benefit analysis, 326, 328, 330, 334, 337, 346, 350
  Trump administration and, 347
Cost–Benefit Rule, 331
Cotton Petroleum Corp. v. New Mexico, 193
Council for Environmental Quality, 373
Council of Economic Advisors, 433
Council on Environmental Quality, 130, 288
Council on Foreign Relations, 29
County of Maui v. Hawaii Wildlife Fund, 376, 490
Court of Appeals, 6, 389, 466, 484
COVID-19, 186, 240, 370, 373, 469, 474
Cow Creek Band of Umpqua Tribe, 180
Cox, Anthony, 293
Creek Reservation, 164
Crist, Charlie, 270
Crows, 112
crude oil, 1, 27, 66, 109, 114, 135, 136, 166, 167, 172, 287, *See also* oil
  Gulf of Mexico and, 219
  production in Alaska of, 219
Cunningham, Joe, 239

Daines, Steve, 397
Dakota Access pipeline, 110, 157, 158, 169–175

Dallas-Fort Worth, Texas, 67
Danly, James, 124
Dawes Act, 158
De Santis, Mark, 123
decarbonization, 74, 97
Deepwater Horizon, 258, 270, *See also* BP oil spill
DeFazio, Peter, 139
Delaware, 232
Delaware Tribal Business Committee v. Weeks, 162
democratic values, 9, 379, 382, 394, 396, 397, 463, 488
Department of Commerce, 86, 132, 221, 232, 441
Department of Defense, 193, 436
Department of Energy, 71, 187, 193, 440, 442, 443
Department of Environmental Protection Pennsylvania, 492
Department of the Interior, 131, 182, 187, 188, 192, 197, 227, 254
  Arctic National Wildlife Refuge (ANWR) and, 224, 242
  Florida and, 238
  Gulf of Mexico and, 255
  National OCS Plan and, 222
  offshore drilling and, 219, 239
  Outer Continental Shelf and, 221
  SEMS and, 261
  Trump administration and, 264, 272
  Waiver Rule and, 273
  waste and, 347
  the Well Control Rule and, 272
Department of Justice, 190, 376, 424, 467
Department of Justice v. Reporters Committee for Freedom of Press, 120
Department of Labor, 86, 102, 309, 426, 440, 453
Department of the Treasury, 40, 427, 433
deregulation, 2, 3, 4, 6, 8, 9, 242, 275, 278, 285, 289, 292, 308, 326, 327, 351, 374, 375, 378, 391, 431, 489
  auto industry and, 424
  judicial review and, 377
Devon Energy, 40
Diné Citizens against Ruining Our Environment, 188
Diné Citizens et al. v. Bernhardt, 187
discount rates, 297, 332, 340, 341, 342, 343, 351, 432, 433

distributed energy sources, 83
diversification, 4, 7, 86, 87, 88, 236, 418, 436, 437, 439, 441
Doctrine of Discovery, 161
Dodd-Frank Wall Street Reform and Consumer Protection Act, 436
Donors Capital Fund, 389
Donors Trust, 389
Dourson, Michael, 303
drilling, 218
drinking water, 26, 41, 44, 114, 139, 170, 173, 189, 190, 303
droughts, 419
due process, 381
Due Process Clause, 381
DuPont, 287

Earthjustice, 100, 136, 151, 153, 241
easements, 117, 127, 140, 166, 397
Ebell, Myron, 289
economic analysis, 6, 135, 297, 327, 328, 331, 332, 339, 340, 341, 349, 351, 353, 356
Economic Development Administration, 86, 87, 441
economic growth, 2, 42, 343, 379, 472
   economic development strategies, 37, 40
   oil and gas industry and, 475
ecosystems, 114
Edwards, John Bel, 235
Einhorn, David, 29
electricity cooperatives, 81, 89
Electricity Freedom Act, 89
electricity markets, 17, 58
   gas and, 63
   interstate electricity markets, 59, 69
   intrastate electricity markets, 59
   levelized cost of energy and, 61
   renewable energy and, 60
   wind energy and, 58, 61, 67
Electricity Reliability Council of Texas (ERCOT), 67
Elias, John W., 424
eminent domain, 117, 124, 127, 140
Employee Retirement Income Security Act (ERISA), 426
Enbridge Line 5 pipeline, 166
End Oil and Gas Tax Subsidies Act, 434
Endangered Species Act, 116, 128, 130, 131, 179, 232, 388, 389
Energy Act of 2020, 137
Energy Fuels Resources Inc., 184

Energy Innovation and Carbon Dividend Act, 430
Energy Policy Act, 193, 436, 472
energy poverty, 192
Energy Transition Act, 86
Eni facility, 221
Environment and Public Works Subcommittee on Environmental Pollution, 415
Environmental Defense Fund, 307
Environmental Defense Fund v. EPA, 307, 481
Environmental Economics Advisory Committee, 330
Environmental Integrity Project, 7
environmental justice, 171, 173, 472, 478, 479, 481
Environmental Law Institute, 7
environmental protection, 18
Environmental Protection Agency (EPA), 70, 190, 400
   abandonment of threshold models, 298
   ability to consider scientific evidence, 298
   accusations of HIPAA violations of, 301
   Affordable Clean Energy Rule and, 70
   aiding the oil industry, 307
   air quality standards and, 296
   American Petroleum Institute and, 336
   analysis of Clean Water Rule of, 135
   analysis of particulate matter reductions, 297
   Anne Gorsuch as administrator of, 471
   approach to promulgation of rules, 398
   ARRA grants for, 440
   assessment of Dakota Access pipeline of, 173
   assessment of hydraulic fracturing, 26
   attitude toward benefits from regulations, 338
   auto industry and, 423
   back to basics agenda of, 290
   basis for regulation of air quality standards, 349
   carbon dioxide sequestration projects and, 427
   Christy Whitman as head of, 291
   comments of Union of Concerned Scientists to, 306
   conclusions on cost of Mercury Rule, 339
   congressional oversight of, 302, 303
   criticism of Navigable Waters Protection Rule, 134
   defense of Strengthening Transparency in Regulatory Science rule, 299

Environmental Protection Agency (cont.)
  deregulatory actions under the Trump administration, 128, 302
  economic analysis of benefits of regulations, 135
  elements of 401 Certification Rule of, 132, 374
  endangerment finding on greenhouse gases, 378
  Environmental Economics Advisory Committee and, 330
  establishment of Energy Star, 27
  estimates on the social cost of carbon, 341
  Federal Housekeeping Act and, 301
  focus on monetized benefits of, 339
  guidance on compliance costs, 338
  issues of scientific integrity and, 292
  lack of transparency of, 299
  mandates of under Clean Water for All Act, 139
  members of conservative think tanks joining, 327
  Myron Ebell as head of transition team, 289
  NAAQS and, 331
  opposition of Anthony Cox to, 293
  opposition to 401 Certification Rule, 133
  opposition to oil and gas regulations of, 291
  opposition to Science Transparency Rule, 307
  opposition to Scott Pruitt as head of, 302
  ozone standard of, 293
  particulate matter review panel of, 294
  particulate matter standards of, 295, 334
  position on pollution permits, 377
  powers under the Consolidated Appropriations Act, 444
  process for setting NAAQS, 334
  prohibition of academic scientists from advising, 292
  promulgation of Affordable Clean Energy Rule, 338
  promulgation of Clean Water Rule, 134
  reanalysis of MATS, 297
  regulation of air quality, 375
  regulation of greenhouse gases, 70, 378
  regulation of mercury emissions, 338
  regulation of methane emissions, 352
  regulation of oil and gas extraction, 480
  regulatory powers to cut emissions of, 431
  repeal of regulations of, 297
  rescission of Clean Power Plan, 334
  rescission of Cost–Benefit Rule, 331
  review of ambient air quality standards of, 292
  review of ozone standards of, 296
  revision of cost–benefit analyses of, 331
  revocation of California's emssions waiver, 424
  ruling in *American Lung Association v. EPA*, 70, 481
  ruling in *Massachusetts v. EPA*, 387
  ruling in *Michigan v. EPA*, 328
  ruling in *United States Sugar Corp. v. EPA*, 336
  ruling in *Whitman v. American Trucking*, 386
  Science Transparency Rule of, 120, 468
  scientific review process of, 300
  selection of CASAC members, 293
  selection of emissions reduction systems, 292
  standards setting and health risks, 334
  standards setting for pollution, 336
  suspension of regulatory authority of, 430
  Trump administration and, 490
  violations of Clean Air Act of, 352
  zero emissions vehicle programs and, 423
Environmental Protection in the Trump Era, 7
Environmental Research, Development and Demonstration Authorization Act, 305
Environmental Rights Amendment, 43
EOG Resources, 29, 40
EPA Office of Chemical Safety and Pollution Prevention, 303
epidemiology, 120, 294
EQT, 29
equity, 34, 58, 59, 83, 342, 343
Estes, Nick, 174
ethane cracker plants, 31
Ethyl Corporation, 287
ethylene, 31
Europe, 31
European Parliament, 422
excise taxes, 90
Executive Order 13771, 331
Executive Order 13783, 341
Executive Order 13867, 177
Executive Order 13990, 331, 341
executive power, 10
  expansion of, 372
  preference to oil and gas industry with the use of, 372
exploratory drilling, 241, 254, 272, 273

Exxon Valdez oil spill, 268, 269
ExxonMobil, 29, 227, 260, 274, 430, 445, 491

Facebook, 67
Fair Access to Financial Services Rule, 426
Fair Returns for Public Lands Act, 138
Federal Communications Commission v. Fox Television Stations, Inc., 353
Federal Energy Regulatory Commission (FERC), 59, 69, 70, 72, 116, 140, 168, 180
  greenhouse gas emissions and, 125
  land management and, 117
  oil and gas infrastructure and, 124
  pipelines and, 124
  powers under the 401 Certification Rule, 374
  renewable energy and, 73
federal government, 8
  assistance of coal communities of, 86
  community-owned projects and, 81
  competing interests and, 217
  energy consumption of, 436
  greenhouse gas emissions and, 340, 480
  land management and, 109, 112
  Native American lands and, 161, 165
  OCSLA and, 221
  off-reservation rights and, 168
  offshore drilling and, 239
  renewable energy and, 59, 481
  treaty rights and, 164
  westward expansion and, 112
Federal Housekeeping Act, 301
federal judiciary, 371
Federal Land Policy and Management Act (FLPMA), 110, 113, 121
  national monuments and, 185
federal lands, 18, 71, 110, 112, 113, 163, 195, 236, 470, 486
  Congress and, 185, 269
  Keystone XL pipeline and, 178
  treaty rights and, 164
  Utah and, 182
Federal Railroad Administration, 136, 140
Federal Regulatory Certainty for Water Act, 139
Federal Reserve Bank, 3
federal tax credits, 91
Federal Trade Commission Act, 380
federalism, 390
Federalist Society, 365, 371, 372, 390, 391, 392–394, 395, 396, 403, 409
Feinstein, Dianne, 396

Feulner, Ed, 468
Fifth Amendment Takings and Just Compensation Clause, 126, 166
FirstEnergy, 89
Fish and Wildlife Service, 128, 131, 132, 165
fishing, 225, 226, 227, 230, 233, 236
fishing rights, 180, 181
Fitzpatrick, Brian, 442
Flood Control Act, 173
flooding, 419
Florida, 223, 225, 226, 230, 233, 239
  offshore drilling and, 270
  state waters of, 219
Food and Drug Administration, 380
For the People Act, 137
Foreign Commerce Clause, 178, 179
Fort Belknap Indian Reservation, 161, 178
Fort Berthold Indian Reservation, 188, 191
Fort Berthold Protectors of Water and Earth Rights, 188
Fort Laramie Treaties, 174, 178
Fort Madison, Iowa, 66
Fort Peck Reservation, 188
fossil fuel industries, 36
fossil fuel pipelines, 110, 116
Fourteenth Amendment, 381
fracking. *See* hydraulic fracturing
Framework Convention on Climate Change, 345
Frampton, Mark, 294, 296
France, 112
freedom, 390
Freedom of Information Act (FOIA), 120
FreedomWorks, 288
Freeport McMoRan, 235
Frelinghuysen, Rodney, 302
Frey, Christopher, 293
fuel taxes, 423

Gallego, Ruben, 195
Gardner, Cory, 397
Garland, Merrick, 392, 484
gasoline, 287, 307, 420
  California's fuel efficiency standards and, 374
  carbon taxes and, 430
  compliance costs and, 330
  fuel taxes and, 423
  gasoline demand, 425
  gasoline prices, 423
  social cost and, 340

Gateway Pacific Coal Terminal, 180
General Electric, 66
General Motors, 423, 476
General Service Administration, 193
generation and transmission cooperatives, 81, 82, 88
Geographical Information Systems, 118
George Mason Law School, 391
Georgia, 223, 230, 232
geothermal, 58, 61, 137, 443
Giant Sequoia National Monument, 185
Gibbstown, New Jersey, 136
Gingrich, Newt, 472
Ginsburg, Ruth Bader, 390
Gleason, Sharon, 376
Glick, Richard, 73, 125
Global Change Research Group, 288
global social cost, 344
global social cost of carbon, 332, 344, 345
Gold Butte, 184
Goldman Sachs, 425
Gorsuch, Anne, 471
Gorsuch, Neil, 377, 384, 387, 395, 398, 490
Gosar, Paul, 224
Government Accountability Office, 8, 121, 167, 229, 341, 346, 375, 446
Government Accountability Project, 306
Graham, Lindsey, 239, 289
Grand Canyon, 187
Grand Canyon Centennial Protection Act, 195
Grand Canyon National Monument, 185
Grand Forks, North Dakota, 66
Grand Junction, Colorado, 120
Grand Staircase Escalante National Monument, 112, 118, 121, 158, 167, 184, 185, 197, 470
Grassley, Chuck, 137, 416, 464
Great American Outdoors Act, 138
Great Britain, 112
Great Recession, 37, 289, 423
Great River Energy, 81
Great Sioux Nation, 174
Great Sioux Reservation, 174
greenhouse gas emissions, 11, 26, 31, 44, 69, 70, 73, 74, 129, 197, 350, 399, 415, 416, 418, 428, 430, 431, 476
   Affordable Clean Energy Rule and, 291
   auto industry and, 423
   Biden administration and, 480
   cap and trade proposal of 2009 and, 417
   carbon taxes and, 428, 430

   CCS and, 427
   Clean Power Plan and, 297
   climate risks and, 435
   Congress and, 446
   Consolidated Appropriations Act and, 442
   cost–benefit analysis and, 340
   costs of, 432
   EPA and, 378
   federal government and, 340, 480
   FERC and, 126
   George H. W. Bush and, 472
   Interagency Working Group and, 350
   Kyoto Protocol and, 417, 472
   lawsuits and, 476
   Obama administration and, 344, 473
   oil and gas extraction and, 288
   Paris Climate Agreement and, 421
   renewable energy and, 59
   Sabal Trail pipeline and, 126
   social cost of, 340, 431, 432
greenhouse gases
   regulations and, 327
Greenland ice sheet, 343
Greenstone, Michael, 271, 342, 432
Grijalva, Raúl, 196, 239
Grisham, Michelle Lujan, 195, 488
Gros Ventre, 161
gross domestic product (GDP), 26, 220, 429
Gulf of Alaska, 237
Gulf of Mexico, 5, 217, 219, 220, 221, 228, 229, 234, 248, 258, 267, 483
   blowout preventers and, 272
   BP oil spill and, 253, 259
   crude oil production in, 219
   exploratory drilling in, 273
   Florida and, 238
   George W. Bush and, 225
   National OCS plan and, 221, 222
   Obama administration and, 225
   oil spills and, 255
   risks of drilling in, 253
   threat to fisheries in, 220
   threat to tourism in, 220
   Trump administration and, 227, 268
Gulf of Mexico Energy Security Act, 234, 238
Gundy v. United States, 387, 400
Gwich'in tribe, 237

Haaland, Debra, 119, 138, 195, 196
Haliwa-Saponi tribe, 168
Halliburton, 22, 40, 259

Index 513

Hansen, James, 473
Hatch, Orrin, 182, 184
Hawaii, 138
Haynesville Shale, 234
Hazardous Materials Transportation Act (HMTA), 135, 136
Health Insurance Portability and Accountability Act (HIPAA), 301
Healthy Gulf, 241, 273
Heart River, 174
Heartland Institute, 89, 287
Heinrich, Martin, 119, 195, 488
Helping Expedite and Advance Responsible Tribal Home Ownership (HEARTH) Act, 160, 193
Heritage Foundation, 288, 289, 306, 327, 371, 468
Herrera v. Wyoming, 164
high-assay low-enriched uranium, 439
Hilcorp, 425
Hispanic Federation, 84
Hollis-Brusky, Amanda, 393
Hopi tribe, 187
Horner, Chris, 301
House Appropriations Committee, 302
House Committee on Energy and Commerce, 303
House Committee on Oversight and Reform, 140
House Committee on Science, Space, and Technology, 72, 301, 303, 304
House Democrat's Climate Crisis Report, 431
House Intelligence Committee, 306
House Judiciary Committee, 424
House Natural Resources Committee, 138, 224
House Natural Resources Committee Chair, 118
Hovenweep National Monument, 187
Hualapai tribe, 187
Huffman, Jared, 270
hunting rights, 164
Huntsman, Jon, 289, 474
Hurricane Ivan, 257
Hurricane Katrina, 257
Hurricane Sandy, 71
hurricanes, 257, 419
Hutchinson, Kansas, 66
hydraulic fracturing, 1, 17, 23, 26, 41, 110, 188, 303, 355, 377, 472, 478
 Biden administration and, 478
 Bureau of Land Management and, 123

 deregulation of, 355
 fracking fluids, 41
 horizontal drilling, 17
hydrocarbons, 235, 258
hydroflurocarbons (HFCs), 391, 444, 490
hydropower, 17, 20, 22, 58, 60, 68, 74, 443

idle iron, 257
Illinois, 67, 77, 169
income taxes, 40, 234
incumbent energy, 60, 88, 89
Independent Particulate Matter Review Panel, 294
Indian Mineral Development Act, 188
Indian Right-of-Way Act, 166
Indian Tribal Energy Development and Self Determination Act, 188
Indian Trust Asset Reform Act, 162
Indigenous Environmental Network, 178
Indigenous Environmental Network v Trump, 178
Indigenous Environmental Network v. Department of State, 177, 353
industrialization, 379
Inglis, Bob, 288
Integrated Scientific Assessment, 295
Interagency Working Group, 341, 344, 350, 432
Intergovernmental Panel on Climate Change, 421
International Energy Agency, 30, 49, 476, 497
International Organization for Standardization (ISO), 73, 427, 428
interstate commerce, 369
Intertribal Coalition, 182
Investment Tax Credit, 193, 443
Iowa, 58, 66, 67, 169
Iraq, 21

Jenkins, Evan, 350
Jicarilla Reservation, 188
John D. Dingell Jr., Conservation, Management and Recreation Act, 139
Johnson v. McIntosh, 161
Johnson, Bernice, 72, 301, 303
Jordan Cove LNG terminal, 110, 158, 169, 179–181
JP Morgan Chase, 425, 435
judicial branch, the, 10
judicial review, 127, 131, 162, 349, 351, 353, 354, 355, 375, 377, 379

judicial selection, 391
junk bonds, 33, 34
Just Transition, 86

Kagan, Elena, 382, 385, 387
Kaiparowits Plateau, 184
Kaktovik Iñupiat Corporation, 237
Kaktovik Whaling Area, 237
Kansas, 58, 66, 67, 68, 77, 90
Katahdin Woods and Waters, 184
Kavanaugh, Brett, 384, 490
Kennedy, Senator John N., 444
Kerry, John, 289
Keystone Pipeline Approval Act, 177
Keystone XL pipeline, 110, 116, 117, 128, 129, 157, 158, 167, 169
  Biden administration and, 197
  national monuments and, 5
  Native American lands and, 169, 175
  permit approval for, 353
Kigali Amendment to the Montreal Protocol, 444
Kinder Morgan, 40
King v. Burwell, 382
Kisor v Wilkie, 382
Koch Foundation, 389
Koch Industries, 89, 188, 288
Koch Seminar Network, 395
Kodiak, Alaska, 260
Kristol, Bill, 396
Kuwait, 422
Kyoto Protocol, 417, 472

Lac-Mégantic, Quebec, 135
LaFleur, Cheryl, 124, 125
Lake Oahe, 169, 171
Lake Superior Tribe of Chippewa Indians, 166
Lamb, Conor, 442, 488
Lame Bull Treaty, 178
Land and Water Conservation Fund, 139, 397
landowner rights, 140
landowners, 4, 53, 58, 63, 65, 67, 77, 79, 94, 99, 117, 124, 126, 148
  benefits of Thirty by Thirty goal for, 138
  fairness toward, 140
  FERC and, 127
Landry, Jeff, 235
Lankford, James, 350
Lazard, 63
leaded gasoline, 287
League of Conservation Voters v. Trump, 376

leases, 123
Legacy Fund, 39
legislative history, 125, 185, 336
Leo, Leonard, 371, 394, 396
Leshy, John, 117
liberty, 381, 384, 390
Lieberman, Joseph, 289, 416
Light v. the United States, 110
liquefied natural gas (LNG), 31, 33, 110, 116, 124, 130, 132, 136, 147, 205, 373, *See also* natural gas
  export of, 180
  Native American lands and, 157, 158, 180
  transportation of, 136, 140
liquid petroleum pipelines, 116
LNG-by-Rail Rule, 136, 140
local governments, 5, 38, 39, 42, 43, 58, 63, 67, 76, 77, 86, 242, 243
  American Rescue Plan and, 481
  clean energy transition and, 485
  coal and, 485
  job transitions and, 438
  Native American lands and, 193
  tax exemptions and, 79
Lochner v. New York, 381
Lone Wolf v. Hitchcock, 162
Los Angeles, 287
Louisiana, 40, 188, 233–235, 258
Louisiana Purchase, 112
Lower Umpqua Tribe, 180
Lucas, Frank, 304
Ludwig, Eugene, 32
Lugar, Richard, 288
Lumbee tribe, 168
Lummi Nation, 180
Lyng v. Northwest Indian Cemetery Protective Association, 166

Maine, 184
Mainstream Lending Program, 35
Malcolm, John, 371
Manchin, Joseph, 91, 119, 237, 303, 442
manipulative causality designs, 296
Marathon Oil, 307
Marathon Petroleum, 33, 34, 425
Marcellus Shale, 1, 7, 37, 44, 492
March 2020 price war, 19
Marine Mammal Protection Act, 232
Marshall, Thurgood, 161
Maryland, 110, 232
Mashpee Wampanoag tribe, 158

Index   515

Massachusetts, 68, 71, 158, 232
Massachusetts v. EPA, 70, 297, 378, 387, 399
Maui, 376
McCain, John, 416, 478
McClendon, Aubrey, 124
McConnell, Mitch, 3, 138, 371, 390, 484
McEachi, Donald, 432
McGahn, Don, 371, 390
McGirt v. Oklahoma, 162, 164
McNamee, Bernard, 124
MDU Resources, 40
media, the, 10, 26, 35, 72, 184, 268, 273, 301, 306, 307, 394, 395, 425, 470
Menendez, Robert, 271
Menominee Tribe v. United States, 164
mercury, 339
Mercury and Air Toxics Standards (MATS), 297, 298, 339
methane, 75, 113, 291, 327
   abandoned wells and, 44
   compliance costs and, 347
   fracking and, 41
   Obama administration regulation of, 431
   social cost of, 340, 341, 350
methane leaks, 3, 26, 41, 292, 346
Methane Rule, 123, 146, 273, 291, 332, 349
Methane Waste Prevention rule, 346
Metzger, Gillian, 384, 398
Mexichem Fluor v. EPA, 490
Mexico, 20, 112
MHA Nation, 191
Michigan v. EPA, 297, 328, 334, 338, 387
Microsoft, 67
Mikva, Abner, 389
Millennium Bulk Coal Export Terminal, 133, 181
Mineral Leasing Act (MLA), 123, 374, 375, 376
Mineral Management Service (MMS), 259, 261, 266
mineral owners, 5, 18, 38, 42, 76, 475, 492
mineral rights, 1, 38, 42, 109, 113, 157, 158, 191, 195
Mineral Rights Act, 123
mining, 42
Minnesota, 80, 81, 331
Minnesota v. Mille Lacs Band of Chippewa Indians, 164
Mississippi, 233
Mississippi Delta, 234
Mississippi River, 234

Missouri, 77
Missouri River, 169
Mitchel v. United States, 165
Mnuchin, Steve, 32
Moapa Southern Paiute Solar Park, 192
Montana, 112, 139, 175, 177, 181, 188, 194, 353
Montreal Protocol, 365, 431, 444
Moody's, 67, 94
Morgan Stanley, 425
Morris, Brian, 177, 178, 186, 307
Morton v. Mancari, 160
Motor Vehicle Manufacturers Association v. State Farm Auto Mutual Insurance Co., 354
Mountain States Legal Foundation, 186
Murkowski, Lisa, 91, 224, 237, 271, 425, 442
Murr v. Wisconsin, 381, 389
Murray Energy, 70, 291
Murray, Bob, 70
Muscogee (Creek) Nation, 164

Nadler, Jerry, 424
naphtha, 31
National Academy of Sciences, 236, 302
National Academies of Sciences, Engineering and Medicine, 74, 268, 287, 341, 349, 417, 422
National Ambient Air Quality Standards (NAAQS), 293, 296, 310, 313, 321, 331, 332, 334, 357, 386
National Association for the Advancement of Colored People, 84
National Association of State Fire Marshals, 136
National Energy Policy Development Group, 21
National Environmental Policy Act (NEPA), 111, 116, 125, 126, 129, 136, 139, 147, 179, 180, 232
   aims of, 128
   BP oil spill and, 259
   Dakota Access pipeline and, 170, 171
   effectiveness of, 130
   FERC and, 126
   interpretations of, 131
   judicial review and, 131
   Keystone XL pipeline and, 177, 178
   Native American lands and, 167
   oil and gas infrastructure and, 370
   Richard Nixon and, 471
   Trump administration and, 132
   Trump administration violations of, 272
   Trump and, 373

National Highway Traffic Safety
    Administration (NHTSA), 324, 423, 424
National Historic Preservation Act (NHPA),
    167, 187
National Labor Relations Act, 380
National Marine Fisheries Service, 132
National Ocean Industries Association
    (NOIA), 242, 264, 273
National Outer Continental Shelf (OCS) plan,
    220, 225, 227, 229
National Park Service, 139, 165, 382
national parks, 112, 115, 166, 181, 182, 397
    protection of, 139
National Petroleum Reserve, 230, 231,
    426, 481
National Renewable Energy Lab (NREL), 72
National Research Council, 258
National Transportation Safety Board, 136
National Wilderness Preservation System, 117
Native American lands
    Trump administration and, 156
Native Americans, 5, 119, 140, 158, 191, 199
    in Alaska, 156, 236, 237
    Allotment Era, 158, 162
    ancestral lands and, 164, 165, 181
    Article 6 of US Constitution and, 160
    assertion and retainment of sovereignty by,
        156, 159, 160, 161, 162, 168, 196
    Bears Ears National Monument and, 181
    Biden administration and, 197
    Chaco Canyon National Historical Park, 186
    Congress and, 193, 195, 470
    Congress's adverse actions toward, 162, 164,
        168, 174, 191
    contaminations from extractive activities on
        lands of, 189, 190
    Dakota Access pipeline and, 169
    Dawes Act and, 158
    Debra Haaland and, 195, 196
    economic diversification by, 188, 192
    federal agencies' duty to consult with, 167,
        170, 171, 173, 186
    impacts of oil and gas industry on, 157, 167,
        169, 175, 184, 186, 191
    land-buyback program, 157
    legal strategies pursued by, 163, 165, 166,
        167, 168, 170, 171, 173, 174, 175, 177,
        180, 185, 187
    national parks and, 112, 182
    poverty and, 191
    Termination Era and, 158, 162
    treaty rights of, 127, 157–180
natural gas, 1, 11, 23, 26, 71, 75, 76, 89, 90,
    114, 116, 130, 332, 441, 445, 492, *See also*
    liquefied natural gas
    air quality due to increased extraction of, 41
    American prosperity and support of, 19
    as a bridge fuel, 473
    consumption rates of, 20
    costs of renewable energy and, 63
    economic assessment of prices of, 18
    energy subsidies in Texas, 90
    expansion of under America First Energy
        Plan, 18
    export of, 1, 126
    forecast growth of, 475, 476
    glut of production of, 31
    impairment of economic development, 41
    import of, 17, 22
    job growth in Marcellus Shale region, 37
    jobs in, 87, 114, 438, 492
    marketing of, 467
    opposition to restraints on, 488
    opposition to shift away from, 75
    prices of, 18
    production growth of, 29
    production in Alaska of, 219
    reliance on, 491
    renewable energy and demand for, 31
    renewable energy in Texas and, 89
    rescission of the Methane Rule and, 274
    reserves on Native American lands, 157
    resiliency of, 71
    retirement of coal plants and, 60
    role of under American Jobs Plan, 482
    shift from coal to, 69
    social cost of methane and, 340
    state governments' protection of local
        communities, 43
    support of Department of the Interior
        for, 188
    tax policies on, 39
    taxes on, 428
    transportation of, 135
    Trump administration approach to
        production of, 217
    US reliance on, 17
    waste of, 121, 347, 375
Natural Gas Act, 124, 126
natural gas dependency, 415
natural gas development, 38, 110, 132, 186,
    192, 232, 426

natural gas drilling, 117, 157, 187, 376, 468
natural gas expansion, 19, 23, 27, 36, 42, 44, 233, 395, 417
natural gas extraction, 18, 19, 39, 40, 109, 110, 157, 188, 372, 468, 474, 478
  air quality and, 41
  boom-bust economy of, 191
  communities' overreliance on, 38
  contaminants and, 41
  decline of, 113
  greenhouse gas emissions and, 288
  Louisiana and, 233
  mineral rights and, 42
  Native American lands and, 157, 181
  oceans and, 217
natural gas extraction workers, 37
natural gas generation, 75
natural gas industry, 31
  CARES Act and, 36
  Congress and, 468
  cost–benefit analysis and, 329
  Federal Reserve and, 33
  fisheries and, 220
  Louisiana and, 235
  PFAS and, 303
  regulations and, 285, 287, 326
  tax preferences of, 433
  tourism and, 220
natural gas infrastructure, 69, 111, 139, 168, 240, 370, 435
natural gas leasing, 111, 113, 114, 116, 119, 121, 139, 158, 186, 195, 197, 425
natural gas peaker plants, 63
natural gas pipelines, 72, 73, 127, 167
natural gas plants, 74, 76, 85, 88
natural gas production, 31, 109, 434
natural gas reserves, 110, 184
natural gas transmission lines, 116
Natural Resources Defense Council v. Wheeler, 490
Nature, 301
Navajo Generating Station (NGS), 194
Navajo Nation, 186, 187, 192, 193, 194, 195
Navajo Nation Oil and Gas Company, 194
Navajo Reservation, 188
Navajo Transitional Energy Company, 194
Navajo Tribal Utility Authority, 192
Navigable Waters Protection Rule, 134, 135, 139, 356
Necefer, Len, 194
Nelson, Bill, 239

Nevada, 112, 184, 187, 192
New Deal, the, 81
New Hampshire, 232
New Jersey, 232
New Jersey v. EPA, 387
New Mexico, 40, 42, 82, 85, 86, 158, 186, 188, 194, 195
New York, 68, 110, 477
New York Department of Environmental Conservation, 433
NEXUS pipeline, 126
Nigeria, 20
nitrous oxide, 350
Nixon, Richard, 2, 471
No Climate Tax pledge, 288
non-threshold models, 298, 300
North Carolina, 68, 223, 230, 239
North Coast Rivers Alliance, 178
North Dakota, 36, 39, 58, 66, 68, 88, 135, 166, 169, 188, 438
  MHA Nation and, 191
Northeast Canyons and Seamounts Marine National Monument, 184
Northern Cheyenne reservation, 163
Northern Cheyenne Tribe v. Hodel, 163
Northern Plains Resource Council v. US Army Corps of Engineers, 128
Northern Ute Reservation, 188
Northstar facility, 221
Northwest Sea Farms v. US Army Corps of Engineers, 164, 180
Norway, 261, 262
nuclear energy, 20, 63, 70
nuclear plants, 60, 76, 85, 86
NWP-12 permit, 128, 179

O'Connor, Sandra Day, 166
Obama administration, 19
  auto industry and, 423
  Bureau of Land Management and, 120
  Clean Power Plan and, 334
  energy strategy of, 473, 478
  exploratory drilling and, 273
  greenhouse gas emissions and, 344
  Interagency Working Group and, 341
  Keystone XL pipeline and, 175, 177
  Native American lands and, 157, 182
  offshore drilling and, 222, 225
  oil spills and, 261
  POWER program and, 86
  regulation of methane of, 431

Obama administration (cont.)
    social cost of carbon and, 433
    Trump administration and, 20, 196, 217, 254
    well inspections and, 268
Obama, Barack, 2, 177, 417, 478
Oberlin v. Federal Energy Regulatory Commission, 126
Occidental Petroleum, 40
occupancy title, 161
ocean economy, 220, 226
Oceana, 239
Office of Information and Regulatory Affairs, 292, 391
Office of Just Transition, 85
Office of Management and Budget (OMB), 329, 336, 337
Office of Technology Assessment, 472
Office of the Comptroller of the Currency (OCC), 426
Office of the Inspector General of the Department of the Interior, 375
off-reservation rights, 159, 160, 161, 168
offshore drilling, 233, 261
    Arctic National Wildlife Refuge and, 224
    financial management of, 218
    Florida and, 270
    OCSLA and, 220
    risks of, 253
    royalty rates and, 229
    United States Congress and, 224
offshore extraction, 217
offshore oil and gas sector, 218
Oglalla Lakota tribe, 175
Ohio, 1, 7
oil, 11, *See also* crude oil, shale oil
    American prosperity and support of, 19
    Biden's vow to transition away from, 478
    consumption rates of, 20
    costs of reliance on, 20
    expansion of under America First Energy Plan, 18
    impairment of economic development, 41
    import of, 17, 22
    jobs in, 36, 87, 114, 438
    production in the Gulf of Mexico of, 219
    reserves on Native American lands, 157
    social cost of methane and, 340
    state governments' protection of local communities, 43
    support of Department of the Interior for, 188
    tax policies on, 39
    transportation of, 135
    Trump administration approach to production of, 217
    US reliance on, 17
    waste of, 347
oil consumption, 31
oil demand, 31
oil dependency, 415
oil development, 38, 110, 186, 232, 426
oil drilling, 117, 157, 227, 376
    ANWR and, 468
    Chaco Canyon and, 187
oil expansion, 36, 44, 233, 395, 417
oil extraction, 18, 19, 36, 39, 40, 109, 110, 157, 188, 372, 468, 474, 478
    air quality and, 41
    boom–bust economy of, 191
    communities' overreliance on, 38
    contaminants and, 41
    decline of, 113
    greenhouse gas emissions and, 288
    Louisiana and, 233
    mineral rights and, 42
    Native American lands and, 157, 181, 188
    oceans and, 217
oil extraction workers, 37
oil industry, 2, 31
    ambient air quality standards and, 292
    CARES Act and, 36
    compliance costs and, 330
    Congress and, 468
    cost–benefit analysis and, 329
    EPA and, 307
    Federal Reserve and, 33
    fisheries and, 220
    greenhouse gas emissions and, 297
    Interagency Working Group and, 341
    Louisiana and, 235
    PFAS and, 303
    regulations and, 285, 287, 326
    tax preferences of, 433
    tourism and, 220
    transition away from, 478
oil infrastructure, 111, 139, 168, 240, 370, 435
oil leasing, 113, 116, 121, 139, 158, 195, 197, 425
oil pipelines, 127, 167
Oil Pollution Act, 261
oil production, 40, 434
oil reserves, 110, 184

oil spills, 129, 135, 171, 190, 234, 241, 255, 257, 273, *See also* BP oil spill
  Congress and, 261
  costs of, 232, 271
  Rex Tillerson on, 260
  risk management and, 258–260
  Shell and, 260
Oklahoma, 36, 40, 42, 58, 66, 77, 87, 164, 188, 438
Omnibus Parks and Public Lands Management Act, 112
Oregon, 179, 180, 181, 184, 225, 331
Organ Mountains-Desert Peaks and Rio Grande, 184
Organization of the Petroleum Exporting Countries (OPEC), 18, 29
Osage Reservation, 188
Ostler, Jeffrey, 174
Outer Continental Shelf Lands Act (OCSLA), 219, 220, 221, 240, 264, 272, 373, 376
ozone, 301, 336

Pacific Ocean, 5, 18, 112, 138, 218, 221, 226
Pacific Remote Island Marine National Monument, 184
Paiute Tribe, 192
Paiutes Indian Reservation, 192
Pallone, Frank, 303
Panama Refining Co. v. Ryan, 400
Papa, Mark, 29
Papahānaumokuākea National Monument, 185
Paris Climate Agreement, 3, 74, 344, 345, 417, 418, 421
particulate matter, 301, 334, 336, 339
particulate matter standards, 294
partisanship, 8
Partnerships for Opportunity and Workforce and Economic Revitalization (POWER) program[152], 86
Patagonia, 185
Pembina, 180
Pendley, William Perry, 119, 158, 186
Pennsylvania, 1, 7, 39, 491
People for the Ethical Treatment of Property Owners (PETPO) v. US Fish and Wildlife Service, 402
Permian Midland Basin, 30
Perry, Rick, 66, 70, 89
petrochemicals, 17, 31, 35, 491
PHH Corp. v. Consumer Financial Protection Bureau, 384

Philadelphia, 136
Physicians for Social Responsibility v. Pruitt, 292
Pienta, Allison, 395
Pine Ridge Reservation, 175
Pioneer Natural Resources, 30
Pipeline and Hazardous Materials Safety Administration (PHMSA), 136, 140, 172
Pittsburgh, 1
Plains and Eastern Clean Line, 77
Plan for Clean Energy Transition and Environmental Justice, 7
plenary powers, 162, 164, 168
$PM_{2.5}$, 294, 295, 306, 335
Pollutant-Specific Significant Contribution Finding For Greenhouse Gas Emissions Rule, 431
poly-fluoroalkyl substances (PFAS), 303
Porcupine caribou, 237
Posner, Richard, 393
poverty, 81, 191, 236, 249
Powder River Basin, 181
Price, David, 270
private land management, 111
private lands, 5, 8, 109, 110, 111, 113, 115, 478
  Congress and, 470
  Dakota Access pipeline and, 171
  management of, 117
  Native American lands and, 195
  oil and gas extraction and, 474
  protection of, 136, 140
private property, 140
procedural rights, 170, 171
Production Safety Systems Rule, 262, 266, 270
Production Tax Credit, 193, 443
Project On Government Oversight, 264
property, 168
Property Clause, 110, 178, 185, 195
property rights, 229, 388, 390
property taxes, 68, 79, 83, 236
Protecting and Securing Florida's Coastline Act, 238
Prudhoe Bay oilfields, 230
Pruitt, Scott, 291
public health and the environment
  economic analysis of, 331
  oil industry impacts on, 40
  state government protection of, 43
  statutory interpretation of laws on, 385, 386
  Trump administration attacks on, 294
public land management, 111

public lands, 111, 369
  Congress and, 470
  congressional oversight of, 110
  conservation and, 115
  federal government and, 109
  management of, 117
  Native American lands and, 181, 195
  protection of, 136, 138, 140
  Thirty by Thirty goal and, 138
  Trump administration and, 140
Public Lands Initiative, 182
Public Utility Commission of Colorado, 350
Public Utility Commission of Texas, 67
PUD no.1 v. Washington Department of Ecology, 374
Pueblo tribe, 187
Pueblo, Colorado, 66
pulmonology, 294
Pure Food and Drug Act, 380

Qatar, 31

radioactivity, 41
Rancho Viejo LLC v. Norton, 401
Rao, Neomi, 391
Rapanos v. United States, 134, 399
Raskin, Sarah Bloom, 32
ratepayers, 67, 71, 82, 84, 85, 86, 125
Reagan administration, 345
  automatic seatbelts and, 354
  Chevron deference and, 387
Reagan, Ronald, 112, 326, 468, 471
RE-AMP, 82
Reason Foundation, 389
recreation, 109, 113, 118, 217, 225, 492
regulations
  assessment of, 327
  BP oil spill and, 466
  Congress and, 285
  cost benefit analysis and, 326, 338
  economic analysis of, 327–330
  history of, 379
  legal doctrine on, 379, 385, 386, 388
  polluters and, 465
  profits and, 285
  social cost and, 340
  Trump administration and, 290
Regulations from the Executive in Need of Scrutiny (REINS) Act, 397
Regulatory Reform Task Force, 128
Regulatory Right-to-Know Act, 329

regulatory takings doctrine, 388
Reichert, David, 224
Reimagine Appalachia, 441, 491
Reliance, 29
renewable energy, 4, 10, 25, 69, 75
  adoption of, 63
  America First Energy Plan and, 18
  Appalachia and, 491
  assistance for deployment of, 442
  benefits of, 58
  Biden administration promotion of, 479
  Bradley Foundation and, 394
  carbon taxes and, 59
  clean electricity standard, 482
  Clean Power Plan and, 332
  Congress and, 436, 469
  consumption rates of, 20
  corporations and, 68
  cost-competitiveness of, 60
  courts and, 489
  decarbonization and, 74, 442, 477
  Department of Defense procurement of, 436
  distributed energy sources and, 83
  development of, 59
  diversification and, 4
  DOE and, 71
  DOE grants for, 440
  Donald Trump and, 464
  electricity and, 58
  electricity cooperatives and, 81
  energy transition and, 464
  equity and, 76
  expansion of, 59, 60
  feasible expansion of, 74
  federal government and, 436, 481
  FERC and, 69, 73
  forecast growth of, 475
  fossil fuels in combination with, 439
  gas prioritization as compared to, 31
  G. W. Bush administration and, 22
  hydraulic fracturing and, 23
  incumbent energy and, 88, 89
  investment in, 439, 477, 482
  job transitions and, 437, 438, 482
  jobs in, 58, 65, 67, 114, 416, 438, 440, 473
  Kansas and, 90
  lack of support for, 22
  land management and, 137
  levelized cost and, 63
  in low-income households, 418
  Minnesota and, 80

Native American lands and, 187, 191, 193, 194
natural gas and, 26, 76
Obama administration and, 27, 473, 478
oil and gas extraction and, 113
omnibus appropriations bill and, 442
pipelines and, 125
political opposition to, 88
politicization of, 488
price competitiveness of, 474, 475
projections of generation of, 30
RD&D and, 442, 443, 445
research on, 72
resiliency of, 71
in rural America, 79
Rural Energy for America Program and, 84
Rural Utility Service and, 82
states' pursuing of, 68
storage as complement to, 438, 443, 474
success of investments in, 443
Texas and, 66, 89, 90
transition to, 475
Trump administration and, 59, 73, 465
Trump and, 69
voter support for, 91
wind energy and, 60
renewable energy development, 193
renewable energy expansion, 85
renewable energy generation, 17, 79
renewable energy goals, 85, 86
renewable energy projects, 69, 76, 79, 137, 187, 192, 193, 194, 197, 475, 482
Renewable Portfolio Standards, 59, 66
Republicans for the American Outdoors Act, 397
research, development, and deployment (RD&D), 442, 443, 444, 445, 446
reservation lands, 168
reservation rights, 159, 160, 161, 168
residential investment tax credit, 84
resilience, 71
Resources Conservation and Recovery Act, 286
Restoring Community Input and Public Protections in Oil and Gas Leasing Act, 138
Revitalizing the Economy of Coal Communities by Leveraging Local Activities and Investing More (RECLAIM) Act, 439
Richards, Ann, 66
right of occupancy, 161, 165

Rivers and Harbors Act, 127, 180
Roberts Court, 381–385
Roberts, John, 383
Roberts, Owen, 381
Rocky Mountain Institute, 76
Rogers, Yvonne Gonzalez, 376
Rooney, Francis, 238, 270
Roosevelt, Franklin D., 380
Roosevelt, Theodore, 112, 185
Rose Atoll Marine National Monument, 184
Rosebud Indian Reservation, 175
Rosebud Sioux Tribe, 178
Rosebud Tribe v. Trump, 178
Ross, Wilbur, 222
royalty payments, 121, 236
royalty rates, 121, 138, 218, 227, 229, 239
Rubio, Marco, 238, 270
Ruby Mountains, 187
Rural Energy for America Program, 84
Rural Utility Service, 82
Russia, 19, 112, 422

S.D. Warren Co. v. Maine Board of Environmental Protection, 374
Sabal Trail pipeline, 126
sacred sites, 158, 184
Safe Coasts, Oceans, and Seaside Towns (COAST) Act, 270
Safe Drinking Water Act, 472
Safer Affordable Fuel-Efficient (SAFE) Vehicles Rule, 323, 324, 424
safety and environmental management systems (SEMS), 261, 262, 266
Sagebrush rebels, 112
Salerno, Vice Admiral Brian, 262
sales taxes, 236
San Antonio, 67
San Diego, 192
San Juan coal plant, 86
Saudi Arabia, 19, 20, 30, 422
Scalia, Antonin, 134, 386
Schatz, Brian, 435
Schlotterbeck, Steve, 29
Schultz, Debbie Wasserman, 270
Science Advisory Board (SAB), 292, 299, 300, 305, 330
Science Integrity Act, 304
Science Transparency Rule, 120, 307, 468
scientific evidence, 298
scientific integrity, 72, 286, 290, 291, 292, 302, 304, 306, 307, 308, 470, 484

Scientific Review Committee, 432
Scott, Rick, 238, 239, 270
Seattle, Washington, 260
section 401 permits, 130, 133, 181
Securities and Exchange Commission (SEC), 124, 435, 436, 480
Seila Law v. Consumer Financial Protection Bureau, 382
seismic airgun blasting, 232, 272
Senate Committee on Energy and Natural Resources, 271
Senate Judiciary Committee, 394
Seneca Nation, 159
separation of powers, 371, 382, 384, 387, 390, 392, 393, 398, 401
severance taxes, 39, 40, 90
Sex Offender Registration and Notification Act (SORNA), 400
shale boom, 1, 4, 9, 17, 19, 29, 37, 41, 44, 115, 191
shale bust, 19, 191, 478, 492
shale development, 42
shale extraction, 9, 18, 23, 26, 43
shale gas, 478
shale industry, 1, 3, 18, 23, 26, 29, 31, 39, 41, 485, 492
   America First Energy Plan and, 19
   financial foundations of, 19
shale oil, 19, 26, 27, 29, 36, 110, 219, 464, 478, *See also* oil
Shell, 29, 227, 254, 255, 430, 476, 477, 491
   the Arctic Rule and, 265
   oil spills and, 260
Shell Arctic expedition, 5
Shoshones, 112
Siemens-Gamesa, 66
Sierra Club, 88, 241, 272
Sierra Club v. Babbitt, 131
Sierra Club v. the Federal Energy Regulatory Commission (FERC), 126, 340
Sinema, Kyrsten, 119
Sioux Nation, 174
Siuslaw Tribe, 180
Small Business Administration, 86
Smith, Lamar, 288, 301, 304, 468
Smoky Hill River, 174
social cost of carbon, 340, 350, 429, 433, 480
social pressure, 242
solar energy, 58, 464, 475
   distributed energy sources and, 83
   expansion of, 68

fossil fuels and, 69
gas and, 63
Investment Tax Credit and, 193
Native American lands and, 194
natural gas and, 76
new construction of, 63
solar farms, 68, 79, 80, 90
solar panels, 88
Sotomayor, Sonia, 163
South Carolina, 138, 223, 230, 232, 239
South Dakota, 169, 175
Southern Environmental Law Center, 239
sovereignty, 156, 157, 158, 160, 162
Spain, 112
Spill Response and Prevention Surety Act, 271
Spy Island, 221
Srinivasan, Sri, 490
Standard Oil, 287
Standing Rock Sioux Reservation, 157, 169, 171, 173
Standing Rock Sioux Tribe, 173
stare decisis, 378
State Department, the, 116
state governments, 23, 39, 42, 58, 59, 79, 80, 130, 242
   American Rescue Plan and, 481
   attacks on regulatory authority of, 132, 232, 373, 465
   clean energy transition and, 485, 486
   community-owned projects and, 81
   distributed energy sources and, 83
   federal government and, 217
   land management and, 110
   Native American lands and, 187, 193
   OCSLA and, 221
   pipelines and, 116
   renewable energy and, 91
   renewable energy goals and, 85
   royalty payments and, 121
   Trump administration and, 218
state lands, 236
State Management of Federal Lands and Waters Act, 224
state sovereignty, 373
Statoil, 29
statutory interpretation
   conservative judicial appointments on, 389
   presumption against stringent regulations, 385
   presumption limiting agency powers, 387

Supreme Court views on the administrative
    state and, 381, 385
Steele City, Nebraska, 175
Stevens Treaties, 163
Strengthening Transparency in Regulatory
    Science rule, 299
Subcommittee on Energy and Mineral
    Resources, 240
Sullivan, Dan, 271
Sunstein, Cass, 378
Sununu, John, 415
Surface Mining Control and Reclamation Act
    of 1977, 87
surface owners, 5, 38, 42, 475
surface rights, 38, 42, 237
*Swinomish Indian Tribal Community v BNSF
    Railway Company*, 166

tax abatement, 79
tax credits, 59, 81, 193, 416, 420, 423, 427,
    443, 482, 491
Tax Cuts and Jobs Act, 40, 117, 224
tax exemptions, 79
tax preferences, 417
tax rates, 429
tax revenues, 39, 40, 63
taxes, 40, 249
    border carbon adjustments and, 429
    fear-mongering and, 488
    Louisiana and, 235
    national parks and, 115
    Native American lands and, 193
    tax policy and, 39
Taylor Energy, 257
TC Energy, 177, 179
Tea Party, 289
Tee-Hit-Ton Indians v United States, 165
Te-Moak Western Shoshone tribes, 187
Tennessee, 77
Termination Era, 158, 162
Tesoro Savage Oil terminal, 181
Texas, 22, 30, 41, 42, 58, 87
    blackout, 89
    crude oil and, 219
    fossil fuel industries and, 36
    job transitions and, 438
    Louisiana and, 233
    renewable energy and, 67, 79
    royalty payments and, 121
    solar energy and, 66, 68
    state waters of, 219

wind energy and, 66, 67, 89
winter freezes of, 71
Texas Senate, 90
textualism, 393
Thomas, Clarence, 377, 384
Thornberry, Mac, 139
threshold models, 298, 299
Tillerson, Rex, 29, 260
Tillis, Tom, 239, 303
Tohono O'odham Nation, 158
Tomblin, Earl Ray, 89
Total, 29
tourism, 220, 222, 225, 227, 230, 233, 236, 237,
    270, 473
Trans-Alaska Pipeline System, 236
TransCanada, 175
transmission line projects, 79
Transocean, 259, 263, 266, 267, 272
transparency, 299
transportation, 58
treaty lands, 141, 172, 175
Treaty of Canandaigua, 159
treaty rights, 116, 127, 157, 163, 164, 167, 170,
    171, 173, 174, 175, 177, 179, 180, 195
Tribal Energy Program, 187
Tribal Energy Resource Agreements
    (TERAs), 188
tribal sovereignty, 196
Tri-State, 82
Trump administration, 5, 7, 8, 9, 71, 197
    adverse impacts to cooperative federalism
        of, 373
    America First Energy Plan and, 18, 463
    Arctic National Wildlife Refuge and, 229
    the Arctic Rule and, 267
    Army Corps of Engineers and, 127
    attacks against scientific integrity by, 6,
        289–294, 304
    blocking climate action by, 420–428
    blocking of nonfossil-fuel private sector of,
        425, 465
    blocking of scientific evidence by, 298
    Bush administration and, 20
    Chaco Canyon and, 186, 187
    Clean Water Act and, 376
    climate denial and, 417
    climate disruptions and, 420
    comments of Union of Concerned Scientists
        against, 306
    Congressional Review Act and, 468
    cost–benefit analysis and, 329, 347

524 Index

Trump administration (cont.)
  curbing states' rights, 2, 132, 232
  decision not to support oil and gas workers, 36
  defunding of Chemical Safety Board by, 272
  deregulation and, 6, 391
  discount rates and, 342
  distortion of norms of economic analysis of, 331
  economic policy of, 469
  Ed Feulner opinions on, 468
  environmental law and, 139
  EPA and, 291, 302, 305, 377, 398
  FERC and, 124
  Gulf of Mexico offshore drilling leases, 227
  harm to trust in federal agencies by, 370
  Heritage Foundation and, 289
  ideology of, 285, 390, 466
  impacts on environmental law of, 128–137
  judicial selection and, 389, 489
  Keystone XL pipeline and, 178
  land management and, 111, 121, 123
  legal interpretation and, 372, 373, 376, 386
  legal interpretations of regulatory statutes, 131, 132, 134, 136, 374
  long-term impacts of, 463
  Methane Waste Prevention Rule and, 346
  mismanagement of government agencies of, 10, 119, 158, 254, 268, 285, 288, 291, 292, 306, 468
  narrowing of Clean Water Act by, 134, 375
  narrowing of Endangered Species Act by, 131
  narrowing of social cost of carbon by, 344
  National OCS Plan and, 222
  National Petroleum Reserve Alaska and, 230
  Native American lands and, 156, 157, 168
  natural resource management and, 347
  Obama administration and, 20, 196, 222, 227
  obstruction of climate action of, 423
  offshore drilling and, 218, 220–225, 232, 233, 239, 240, 270
  oil and gas leasing on public lands by, 116, 119–124
  opposition to Clean Power Plan by, 334
  opposition to renewable energy by, 59, 69, 187
  presidential permit for Keystone XL pipeline and, 177
  previous administrations and, 19
  proposed cuts to coal communities assistance, 87
  public land mismanagement by, 110–111, 119–124
  regulatory rollbacks of, 42
  removal of economists from advisory boards by, 330
  reversal of safety reforms by, 264–268
  risks of deregulatory actions of, 140
  safety regulations and, 254
  scientific integrity and, 290, 306
  skewing economic analysis by, 327, 330–345, 351
  subversion of CARES Act funds by, 36, 466
  support for carbon sequestration and storage by, 427
  support for coal industry by, 11, 70
  support for crude rail transportation by, 135
  support for Dakota Access pipeline by, 171
  support for nuclear industry of, 70
  support for oil and gas infrastructure by, 124–128
  support for privatization of reservations by, 158
  United States congressional oversight of, 356, 467
  unorthodox statutory interpretation by, 369–375
  waste and, 347
  Waste Prevention Rule and, 375
  well inspections reductions by, 268
"Trump on Earth," 7
Trump, Donald, 2, 19
  appointment of federal judges of, 6, 389, 466
  campaign promises of, 417
  CARES Act and, 35
  claims of emergency powers under NEPA of, 373
  distortion of delegated powers from Congress of, 372
  executive orders by, 331, 370, 373
  Interagency Working Group on the Social Cost of Carbon and, 341
  Keystone XL pipeline and, 177
  opposition to renewable energy of, 69, 464
  removal of protections for Native Americans, 184
  reshaping of federal judiciary of, 371, 389, 390, 391, 396, 466, 489
  shrunk national monuments, 158, 221
  support for drilling in ANWR of, 425
  tumultuous presidency of, 463
  weakening of regulations, 290

Index

Tucson Herpetological Society v. Salazar, 354
Tunica Biloxi Reservation, 188

Udall, Tom, 137, 138, 158, 195
Uintah and Ouray Reservation, 188
UN Framework Convention for Climate Change, 415
Unalaska, Alaska, 260
Uncompahgre Reservation, 112, 182
Union of Concerned Scientists, 306
United Kingdom, 261, 262
United Nations Environment Program, 75
United States, 111, 180
   America First Energy Plan and, 19
   carbon dioxide emissions of, 344
   energy independence of, 18
   natural gas and, 1
   participation in global efforts of, 421
   responsibilities to Native Americans of, 162
   Thirty by Thirty goal and, 138
United States Coast Guard, 258
United States Commodities Trading and Financial Commission, 436
United States Congress, 8, 112, 116
   accounting for costs of infrastructure buildout, 117
   Bears Ears National Monument and, 182
   BP oil spill and, 261
   clean energy transition and, 479
   climate disruptions and, 428
   the Constitution and, 369
   environmental justice and, 479
   federal lands and, 269
   FLPMA and, 185
   global social cost of carbon and, 345
   national monuments and, 185
   Native American lands and, 162, 193, 195
   OCSLA and, 221
   offshore drilling and, 224, 239
   pandemic recovery and, 19
   regulations and, 285
   renewable energy and, 193, 436, 469
   science and, 302
   separation of powers and, 387
   support for Clean Water for All Act of, 139
   support of drilling in Arctic National Wildlife Refuge, 224
   support of Thirty by Thirty goal of, 138
   the Supreme Court and, 370
   transition from oil and gas dependency and, 446
   Trump administration and, 350, 356, 465, 467
   United States Constitution and, 385
   views of administrative state by, 379
United States Constitution, 110, 160, 178, 369, 379, 385
United States Department of Agriculture, 82
United States District Court of Alaska, 241
United States Energy Information Administration, 75
United States ex rel Huaipai Indians v. Santa Fe Pacific Railroad, 165
United States Federal Reserve Bank, 32, 33
United States Fish and Wildlife Service, 230
United States Forest Service, 165, 181
United States Gulf Coast, 31
United States State Department, 175, 178
United States Sugar Corp. v. EPA, 336
United States Supreme Court
   Chevron deference and, 352
   Clean Air Act and, 297, 328
   Clean Power Plan rule, stay of, 69
   Clean Water Act and, 370, 374, 376
   Congress and, 370, 382, 383, 390
   Dakota Access pipeline and, 175
   decisions pertaining to Native Americans' legal rights, 159–164
   doctrinal developments by, 382, 385
   Exxon Valdez decision, 274
   Keystone XL pipeline and, 129, 175, 179
   Mercury and Air Toxics Rule ruling, 338
   Native American lands and, 165
   pipelines and, 128
   regulations and, 328
   separation of powers and, 387
   standards for judicial review, 353, 354, 355
   stare decisis and, 378
   treaty rights and, 164
   Trump appointment of justices to, 6, 389
   views on the administrative state by, 379, 381
United States v. Kagama, 162
United States v. the Sioux Nation, 174
United States v. Lopez, 401
United States v. the Sioux Nation, 162
United States v. Winans, 161
uranium, 157, 182, 184, 188, 445, 482
uranium mining, 182, 184, 187
US Forestry Service v. Cowpasture River Preservation Association, 382
US v. Jicarilla Apache Nation, 163

US v. Montana, 167
US-China Climate Accord, 344
Utah, 71, 112, 118, 182, 187, 188
Ute Mountain Ute Reservation, 188
Ute tribe, 187, 188
Utility Air Regulatory Group v. EPA, 399
Utqiagvik, 237

Valley of the Gods, 184
Venezuela, 20
Vermont, 84
vertical drilling, 17
Vestas, 66
Virginia, 68

Waiver Rule, 273
Walker, Bill, 237
Walker, Justin, 70
Warren, Elizabeth, 264, 435
Washington, 68, 139, 180, 225, 350, 381
Washington v. United States, 163
waste, 347
Waste Prevention Rule, 347, 375
wastewater, 23, 41, 115, 123, 189, 190, 234, 286, 376
wastewater spills, 41
water loss, 419
water rights, 156, 160, 161, 170, 173
Watt, James, 471
Waxman, Henry, 416
wealth, 8
Well Control and Blowout Preventer Rule, 262, 263, 266, 267, 269, 270, 272, 273
Wells Fargo, 425, 435
West Antarctic ice sheet, 343
West Coast Hotel Co. v. Parrish, 381
West Virginia, 1, 7, 36, 42, 88, 89, 91, 438
wetlands, 115
    Louisiana and, 234

Wheeler, Andrew, 184, 291
Wheeler, Russell, 484
White House Council of Economic Advisors, 350
White, Rear Admiral Jonathan, 258
Whitehouse, Sheldon, 394, 396, 428
Whitman v. American Trucking Association, 328, 386
Whitman, Christy, 291
wildfires, 419
Williston, 39
wind developers, 67
wind energy, 58, 59, 60, 81, 464
    excise taxes and, 90
    fossil fuels and, 69
    incumbent energy and, 89
    Midwest and Plain states and, 67
    natural gas and, 63, 76
    new construction of, 63
    Production Tax Credit and, 193
    state legislators' support of, 66
    Texas and, 89
wind farms, 66, 67, 79, 80, 90
wind power, 475, 481
Wind River Reservation, 188
wind turbines, 88
Winters v. United States, 161
Wisconsin, 81, 166
Worcester v. Georgia, 160, 161
Wyalusing, Pennsylvania, 136
Wyoming, 36, 88, 112, 164, 181, 188, 194, 438

Yakima Nation, 160
Yellowstone National Park, 112

Zero Zone v. Dept. of Energy, 345
Zinke, Ryan, 119, 184, 232, 291
Zukunft, Admiral Paul, 258
Zuni tribe, 187

CPSIA information can be obtained
at www.ICGtesting.com
Printed in the USA
BVHW041748110122
626011BV00002B/41